The Prussian Army - to 1815

THE PRUSSIAN ARMY
TO 1815

Digby Smith

Schiffer Military History
Atglen, PA

Book Design by Ian Robertson.

Library of Congress Control Number: 2003116855

Printed in China.
ISBN: 0-7643-1990-6

Published by Schiffer Publishing Ltd.
4880 Lower Valley Road
Atglen, PA 19310
Phone: (610) 593-1777
FAX: (610) 593-2002
E-mail: Info@schifferbooks.com.
Visit our web site at: www.schifferbooks.com
Please write for a free catalog.
This book may be purchased from the publisher.
Please include $3.95 postage.
Try your bookstore first.

In Europe, Schiffer books are distributed by:
Bushwood Books
6 Marksbury Avenue
Kew Gardens
Surrey TW9 4JF
England
Phone: 44 (0) 20 8392-8585
FAX: 44 (0) 20 8392-9876
E-mail: Info@bushwoodbooks.co.uk.
Free postage in the UK. Europe: air mail at cost.
Try your bookstore first.

Contents

Foreword

This work covers a considerable span of extremely active European military history, including the Silesian Wars and the Seven Years War, during which Frederick the Great was carving out his realm at the expense of his neighbours on all fronts. Of necessity, the details of the many wars prior to 1792 have had to be reduced to mere outlines, to allow us to concentrate on the Revolutionary and Napoleonic era. The book then describes the collapse of the once-invincible Prussian army in the white heat of the 1806 campaign, with Napoleon's convincing victory at Jena on 14 October and the truly astonishing feat of Marshal Davout's isolated III Corps at Auerstaedt that same day, where he crushed the main body of the Prussian army single-handedly. Very few of the old Prussian regiments survived this maelstrom of destruction and the proud state of Prussia was humbled in the dust at Napoleon's feet. From 1807 to 1812, the remnants of the Prussian army (and the state) were examined, probed, purged of outdated commanders, tactics, policies and practices and rebuilt to a dramatically modernised format. When the reader examines the regimental histories' section of this book, he may well be surprised at the number of the Colonels-in-Chief who 'died in post'. This was the logical outcome of a state in which most of the landed aristocracy (from whom alone officers of the army could be drawn) were without adequate alternative income to military service and were forbidden to sell their land or to practice a trade. This produced a true 'caste system', almost as rigid as that of the Hindu religion. This was aggravated by the practice of allowing officers 'owning' units down to company level to run them as their private businesses, paring expenses on things such as clothing, equipment, and training to maximise their own personal profits. This was a common feature of many military systems of the time and went on until its evils became too obvious to overlook. Continuing in military service until a ripe old age was the only way for many of these veteran officers to survive. This led to the disasters of 1806, when Napoleon and his proven team of young senior commanders wrecked the Prussian army, led largely as it was by physically – sometimes also mentally – infirm geriatrics. The average age of the senior Prussian commanders in October 1806 was 58,8 years; that of Napoleon's team was 40,5 years. By 1813 this gap of 18,3 years had been reduced to just 4 years. It demonstrates how Napoleon and his trusted marshals, still in harness in 1813 as in 1806, were now facing a team of much more youthful, energetic opponents, who had learned many hard lessons in the intervening years and were able and willing to put them into practice on the field.

Promotion in the Prussian army prior to the catastrophe of 1806 was almost literally a matter of stepping into dead men's boots. It was only when pensions were introduced that older officers could afford to retire and the average age of the officer corps was significantly reduced.

Perhaps the greatest exception which proved the rule of ensuring that older officers should retire was Field Marshal Bluecher. Although aged 73 in 1815, the old warhorse stood up well to the rigours of campaigning in the fields of Ligny and Waterloo and was inspired with a deep hatred of Napoleon and all that he stood for. He also had an amazingly modern ability to inspire the soldiers and officers that he led to achieve great things.

The fortunes of each surviving old regiment and of the many new ones, which were spawned in the battles of the Wars of Liberation from 1813 to 1815 are carefully recounted here. Links have also been provided to enable the reader to follow these regiments through to the Imperial German army of 1914. Naturally, space for this project is not unlimited; a conscious decision has thus been made to lend greatest emphasis on the period 1792 – 1815. There is, however, a complete listing in Annex A of all battles, clashes and sieges in which the Prussian army was involved up to 1815. Inevitably, the coverage of the individual regiments is not uniform in scope. Some of the better known are much more comprehensively documented than others. Sufficient details of the early history of the dress, regimental genealogy, commanders, colours, standards and uniforms to satisfy all but the most dedicated specialist have however been included. All information has been taken from the best available German language sources; much of it appears in English for the first time.

In many cases, the original German names of the characters mentioned in the book have been retained to preserve the authenticity of the material.

There were numerous contradictions in details of the dates of the raisings of various regiments and in the names and ranks of their colonels in chief. It should be noted that prior to 1806, the seniority of a regiment, and its place in the line of battle, was decided by the seniority of its colonel in chief. In all such cases, for the early period of history, I have followed the data given in the *Stammliste* (Official Army List) of 1806. It is hoped that this book will introduce many enthusiasts to this fascinating army and its turbulent history.

Finally, I wish to thank my friend Rudolf Hoelzel, in Willich, Germany, for his assistance in verifying some of the more complex details of this book.

Digby Smith,
Thetford 2003

Abbreviations Used in this Book

B battle
Bn battalion
C clash
Chef the Colonel-in-Chief of a regiment
Col colonel
D defence
FAB Foot Artillery Battery
FM field-marshal
GdI General der Infanterie
GdK General der Kavallerie
Gen general
GFM general field-marshal
GFZM General-Feldzeugmeister
GL general-leutnant
GM general major
Gren grenadier
Grsn garrison
HAB Horse Artillery Battery
IR Infantry Regiment
Lt Col lieutenant-colonel
LW Landwehr
LWIR Landwehr-Infanterie-Regiment
LWKR Landwehr-Kavallerie-Regiment
Neumaerk Neumaerkischisches (New Mark or Brandenburg)
NKR National-Kavallerie-Regiment
Ostpr Ostpreussisches (East Prussian)
Regt regiment
S siege
Schles Schesisches (Siesian)
Sk skirmish
Sqn squadron
Westpr Westpreussisches (West Prussian)

Introduction: Development of the Uniforms

From 1660 the province of Prussia was freed from Polish control and made the seat of the Hohenzollern family. It was a very modest beginning for the power that would cause the world to combine to defeat Hitler's Germany in the Second World War. The Prussia of 1660 was a poor state, ravaged by the Thirty Years War and with very restricted natural resources, no commercial manufacturing base or traffic in goods to speak of. It was an absolute monarchy in which the peasants – as in most other states of this period - had minimal rights. It was joined politically to the Electorate of Brandenburg.

An essential character in the development of the Prussian state and army to the position they occupied on the European stage by 1806 was King Frederick William I , 'The Soldier King' , born in 1688, who ruled Prussia from 1713 - 1740. He was the son of the first 'King in Prussia', Frederick I (the elector Frederick III of Brandenburg). In the mid-XVII Century Prussia consisted of the provinces of East Prussia, Brandenburg (the Old and the New Mark, on the Elbe and Oder), the Duchy of Cleve-Juelich on the lower Rhine and the Counties of Mark and Ravensburg in Westphalia. East Pommerania was added shortly after this point.

In 1701 the Emperor of the Holy Roman Empire agreed to the province of Prussia becoming a kingdom, ruled by the Elector of Brandenburg. In 1713 King Frederick William I launched a scheme to found a commercial entity in his state based on the production and export of wool. If commerce was in its difficult infancy at this point, the army was already a mature and relatively efficient organisation and performed well in the

Wars of the Spanish Succession from 1701 – 1713, fighting largely as mercenaries for other powers and generating much-needed income for the impoverished state.

By the time of his death in 1740 Frederick William I had acquired the province of Geldern in the west, amassed a fortune of 9 million Thaler, increased his army to 83,000 men and forged it into a formidable weapon.

When not on campaign or on ceremonial or security duties, the soldier and his family were put to work, spinning and weaving wool for export. The quality and price of Prussian woollen cloth gradually came to be internationally recognised, so much so that in 1725 the state won a contract to provide green woollen cloth to clothe the entire Russian army.

At this time each Prussian soldier was issued with a new coat every year; this provided a market for the wool and the over 50,000 used, but still serviceable garments, which came onto the market annually, were then snapped up by the civilian populace. It soon came to be the national costume: dark blue in areas where infantry and artillery units were quartered, white in cavalry garrison towns in the northeastern corner of East Prussia.

Until 1740 Brandenburg-Prussia was a scattered collection of isolated territories; East Prussia was separated from Pommerania, and the commercially most profitable provinces of Westphalia, including the towns of Cleve, Geldern, Minden, Ravensberg and Wesel, lay far off to the west. The state was vulnerable to the many and changing whims of European politics and indefensible with the means at its disposal.

Towards the end of the Great Northern War (1720) Prussia began to develop the first 'nation in arms' of Europe. In 1705, Frederick Wilhelm I had experimented with the raising of a militia for home defence. It was to be a voluntary organisation, and in order to attract recruits, all volunteers were to be exempt service liability in the regular army. The scheme was abandoned when this failing became apparent.

During the War of the Spanish Succession (1701 – 1714), attempts to increase or even to maintain the strength of the 15,000 soldiers which existed in 1701 very soon degenerated into regular press-gang tactics, when few volunteered to enlist only to die of malaria in northern Italy, which was where many of the Prussian regiments served. To counteract the effects of the losses occurred by campaign attrition in the regiments, it soon became established custom to recruit five men per company above the required establishment at the start of a campaign.

This gave the regiment a good chance of being at full establishment even by the time it had marched several hundreds of miles from its base to the theatre of the conflict.

It was in 1713 that Frederick William I's General of Cavalry von Derfflinger (son of a farmer in Oberoesterreich) made the suggestion that each regiment should be allocated a fixed, permanent area within which it would be garrisoned in peacetime and from which it would draw its recruits. This scheme was not adopted at this point. It was also at this time that the Prussian king developed his mania for filling the ranks of his guards with tall men, some of them actually kidnapped into service, from inside and outside his own kingdom.

Following the end of the Great Northern War in 1716, the Prussian king decided to maintain the strength of his army, as economically as possible, by the introduction of the *Kruemper* system. This term comes from the wool production industry and implies the compacting of the fibres in the cloth (shrinking) to produce a better quality fabric. Under this plan, men were called up for basic training to their regiments and some of them would be released back into civilian life (without army pay) when that training was completed. This practice was contained in the army regulations of 1726 for the first time, and it was this system which allowed Prussia to evade the 42,000 limit on the size on their army by Napoleon in 1808.

The *Kruemper* system was applied only to the infantry; the cavalry and artillery were all full-time soldiers.

Regiments also still competed among themselves for the available manpower; regular manhunts were mounted by the various recruiting teams. To put an end to this evil, King Frederick William I introduced the *Kantonsystem*, as originally suggested by Derfflinger twenty years ago, in 1733. The basic workings of this scheme are explained below.

The word comes from the French *Canton* and means: a stretch of land or a district. Following a stastical survey of the kingdom, each infantry regiment stationed in East Prussia was given an area enclosing 5,000 'hearths' or households from which to draw its recruits. In Brandenburg and Westphalia it was 8,000 hearths per regiment. Cavalry regiments having five squadrons received 3,500 – 4,000 hearths. The regimental *Kanton* was subdivided into company slices so that the men of each village would go into a sub-unit at time of mobilisation together with their friends and acquaitences.

Parish registers were used as the administrative basis for the scheme. Young lads were registered by the company commander and given a red neck cloth and the regimental hat pompon to wear. After confirmation, the youths were sworn in; according to strategic need, more or fewer *Kantonists* would be called up for training. Having completed two years' service, the man would then be sent back to his home town as a '*Beurlaubter'* or a member of the regiment on extended vacation.

He remained liable for military service as required; theoretically for the rest of his life. But his status had now changed; he was now a trained soldier and no longer liable to the vagiaries of feudal liabilities of the simple peasant in society.

There were certain exemptions to military service under normal circumstances, and these were dictated by financial status and the commercial priorities of the state. Exemptions included all those who owned their own homes and farms, '*Manufakturiers'* or those engaged in managing industrial processes, all workers in the woollen and cloth industry, state civil servants, the clergy and capitalists with at least 10,000 Thalers (this was later reduced to 6,000 Thalers). In order to encourage the development of East Prussia, which was largely underpopulated at this point, those who volunteered to move there and farm, or work as carpenters were also exempt from military service, as were the bargees working the waterways of all parts of the country. There were certain other local exemptions from time to time.

Of course, in times of national emergency, such as in 1758 during the Seven Years War, when the Russians invaded, it was a case of all hands to the pumps and only the clergy, bargees and cloth workers remained exempt.

Nothing however, could save those unfortunate enough to be taller than five feet nine inches from military service; the king's grenadiers was where they landed!

There was a saying in Prussia at this time, which mothers drummed into their sons: 'Don't dare to grow or the recruiters will get you!'

Normally, *Kantonisten* could form up to one third of the total of the men in a regiment; only the reliable among them were liable to be *'beurlaubt'* or sent on long leave for up to ten months at a time. Unreliable *Kantonisten* were treated like the foreigners who served in the Prussian army and were kept under close watch in barracks to prevent desertion. Here, they formed a pool of manpower for training new recruits, guard, police and fire-fighting duties and hygiene monitoring patrols.

The manpower requirements of the army had to be considered in relation to those of domestic industry; the army could not be allowed to strip the country of all able-bodied young men; the shortfall – about 50% - had to be covered by recruiting foreigners.

To differentiate between *Kantonisten* and others in the regimental rolls, the former were entered in black ink, the others in red. Although the red group included the foreigners, it also included Prussian nationals who had volunteered for military service even though exempt, and those sons of members of the regiment, who were fed and clothed at regimental expense and usually graduated through the status of drummer boys to being adult soldiers. Some of these had achieved commissions and at least two (Groeling and Rohdich) became generals.

Due to this over-simplification in the regimental rolls of the black and red entries, the actual numbers of real non-Prussians in the Prussian army has often been over-estimated.

This also brings to our notice that Prussian regiments, like all regiments of all armies of this era, were communities that dragged a large tail of dependents along with them; wives, children, sutlers, whores and washerwomen. Not only were these people part of the 'regimental baggage' in the garrisons, some of them would accompany the regiment on campaign, as has been recorded by eye-witness accounts from Spain and even from Russia in 1812.

The result of this *Krumper* system was to form a reservoir of trained military manpower, which could be called upon at short notice to swell the ranks of the field army in emergencies. Even when at home, the *Beurlaubten* men had their uniforms and sabres with them, which they wore to church every Sunday and holiday.

In a minority of cases, the commander (owner) of a company was also the feudal lord of the *Kantonisten*. This occurance was highest in East Prussia, with about half of the company commanders falling into this category; but this practice was frowned upon by the king, who normally demanded a separation of these functions as was the case in the upper echelons of the civil service.

A feature of the *Kanton* system was that the army became an instrument of social reform in Prussia. By virtue of the fact that the army had first call on the services of all the life-long soldiers, whether they were on leave or not, and by means of the fact that civilians had no powers to discipline soldiers, the military intervened directly between the peasant and his old feudal lord.

No longer was a landowner able to punish 'his' peasants at will; as soldiers they remained under military discipline and wore at least part of their regimental uniforms at all times, even if on long leave. The landowner was forced to approach the regiment and lodge a complaint against the man if a dispute arose between them. The case was then investigated and dealt with by the regiment.

It should also be mentioned that classical serfdom, where the peasant was the outright property of his feudal lord, to be sold or disposed of at will, no longer existed in Prussia in 1713, although the custom was alive and well in some neighbouring states such as Holstein, Swedish Pommerania and Mecklenburg.

Two groups in Prussia at this time still owed services to landlords: those who owned property but had legal obligations to their landlords and could only move away under forfeiture of that property, and those peasants who owned no property at all. The second group were often liable to work for their landlords without pay. The numbers liable to this imposition was highest in the east of the kingdom and reduced the farther west one came. As from 1713, the kings of Prussia tried to minimise and to eradicate serfdom. When Prussia took over Silesia in 1743 at the end of the First Silesian War, the existing feudal rights of landowners were abolished, but an attempt to do the same in Pommerania in 1763 (at the end of the Seven Years War) failed in the face of the united resistance of the local nobility, who managed to convince the king that the entire economy of the province would collapse if his reforms were introduced.

The last vestiges of such feudal conditions were swept away in Prussia by the *Landrecht* Act of 1794, which reformed the ownership of the lands.

The harsh military discipline of the Prussian army of this period seems to have been overstated; Frederick the Great instructed his officers, that their men should regard them as their fathers, not their executioners.

King Frederick William I was a man of utter honesty and a painstakingly exact administrator of his realm. He was also an autocrat with a violent temper, subjecting his family, and others, to indescribable treatment and beatings. In a modern society he would have been thrown into prison or a mental institution. His complex personality allowed a love of painting and music to exist quite happily alongside his violence.

As a young man he had served in the War of the Spanish Succession and had been present at the bloodbath of Malplaquet in 1709, where the Duke of Marlborough and Prince Eugene of Savoy had won their terrible pyrrhic victory over the French. This was one of the events which convinced him that his kingdom must have a first class army in order to survive.

In 1740 he left his son an army of some 90,000 men, 4 % of the population; an extremely high percentage for states in Europe at that time.

The man who took up the reins of Prussia upon the death of his father in 1740 was his son, Frederick II, later to be dubbed 'The Great'. Frederick was born on 21 January 1712; he was an artistic, musically talented (homosexual?), cordially hating and being hated – and regularly beaten - by his father, who recognized and detested his son's sexual inclination.

Frederick was the eldest surviving son of this severe father, who soon realised that his cherished vision of a cloned version of himself was never to be. Nevertheless, he tried with all his might to bend the boy to his form, closeting him in the remote castle of Wuesterhausen, some miles south of Berlin, with two trusted senior army officers (the 65-year-old General Finck von Finckenstein and a somewhat younger Swedish colonel von Kalkstein) as his tutors and mentors.

Despite being taught German – and being forbidden to learn Latin – Frederick learnt, and loved, French, taught himself Latin and seemed to revel thoroughly in doing the exact opposite, in every respect and at every opportunity, of that which his father desired.

Not surprisingly, his violent, bigoted and intolerant father paid him back, at every possible occasion, in spades. There were repeated public humiliations and beatings, which did nothing to alter the attitude or behaviour of either party.

Regardless of this extraordinary situation, the young prince continued to be groomed for the throne and in 1725 he was commissioned into the Potsdam Giant Guards. Six years later, he was made Colonel-in-Chief, or *Chef* of the infantry regiment Nr 15.

By 1730 his marriage, to Princess Amalia, a daughter of the future King George II of England, had seemed to have been arranged, as part of a double package, whereby his sister, Wilhelmina, would also be betrothed to the Prince of Wales. The English / Hanoverian court welcomed the idea of both matches.

Frederick William however, rightly sensed that the Imperial Austrian court would take a dim view of this new Anglo-Prussian alliance and this was in fact the case.

When, in April 1730, the British envoy, Sir Charles Hotham, arrived in Berlin to press the case for the package deal, he was told that the young Frederick was still too young.

Hotham persisted, so did Fredrick William; finally, Hotham left for England on 12 July to report his failure to close the twin marriages.

By this point, young Frederick was desperate to escape from his virtual imprisonment in the Prussian court.

A possible opportunity arose in the summer of that year when he was to accompany his father on a tour of Prussia's western provinces. Believing that the time had come, Frederick planned to escape to France over the River Rhine, aided by a close confidant, a page of the court, Peter von Keith and a young officer, Hans-Hermann von Katte, son of the Prussian Field Marshal Hans-Heinrich von Katte, who was to accompany the party.

On 10 August the party reached Bonn on the Rhine; the attempt was set to be made the next day.

Word however leaked out; Keith was interrogated and confessed; he then escaped to England and was given asylum until returning to Prussia as a lieutenant-colonel in December 1740, following Frederick's accession to the throne.

Frederick and von Katte were not so lucky. Frederick was confronted by his enraged father and subjected to another public beating, placed in arrest, deprived of his sword, taken back to Berlin and later interrogated by his father. His sister Wilhelmina was also beaten.

A court-martial was convened under the presidency of Lieutenant-General von Schulenburg; they commended the unfortunate prince to his father's tender mercies.

Frederick was deprived of his commission and condemned to virtual solitary confinement in Cuestrin fortress, allowed a strictly limited diet of religious and non-fictional books.

His fate might have been much worse but for the intercession of the courts of England, Russia, Sweden and of many of the senior ranks of the army.

Katte junior was also tried and condemned to life imprisonment by the court. This was not good enough by half for the king; he upped the sentence to death and no amount of pleading could shift him. Katte was executed by being beheaded by a sword in Cuestrin fortress on 6 Nov 1730. Frederick was forced, on his terrible father's exact instructions, to watch his execution.

Following this event, the prince was imprisoned there from 1730 – 1732 and only released when he was able to convince his father that he was truly repentant. Frederick was then made *Chef* of the 15th Infantry Regiment in Neuruppin and immersed himself in learning the trade of a soldier.

In 1733 young Frederick was married off to Elisabeth Christina von Braunschweig-Bevern and lived in Rheinsberg until assuming the throne of Prussia in 1740. The marriage was without issue.

In 1733 Prussia was involved in the **War of the Polish Succession**; supporting the Emperor Charles VI and the candidacy of the Elector of Saxony against that of France and Louis XV's father-in-law, Stanislas Leszczynski.

Prince Eugene of Savoy commanded the imperial forces on the Rhein and young Frederick was taken under his wing during the rather brief campaign around Philippsburg fortress.

The future king of Prussia had a most strongly developed sense of duty and regarded himself to be the first servant of his people. His top priority in this respect was national security; for this, he needed a strong army and this presupposed a population larger than that which Prussia then had and more economic resources than she then commanded.

There was a Prussian claim to the Silesian duchies of Brieg, Jaegerndorf, Liegnitz and Wohlau and these territories were populous and rich in minerals; they were vital, in Frederick's mind, to the stronger Prussia that he wished to create.

At the end of May 1740 Frederick William I died and Frederick became king.

He inherited an army of 83,000 men. He was determined to increase this force numerically and in its efficiency at the expense of the 'buttons and bows' that his father had so loved. The 6th Foot (the Giant Grenadiers) were reduced from three battalions to one.

Frederick was soon to need his army.

On 20 October 1740 Emperor Charles VI of the Holy Roman Empire died and his daughter, Maria Theresia became head of the house of Hapsburg. Frederick judged that the moment had come to enforce his claim to the Siesian provinces. This led to the **First Silesian War** which lasted from 1740 – 1742. Peace negotiations between Prussia and Austria followed; the First Silesian War was over and peace was signed at Breslau in June 1742. Austria retained the areas of Pardubitz and Koeniggraetz.

By good military luck and cold, diplomatic self-interest, Frederick had achieved almost all of his limited, well-defined political aims.

Peace was brief; Maria Theresia could not accept the loss of her rich provinces and this led to the **Second Silesian War** from 1744 – 1745. Again it ended in defeat for Austria. Frederick's ownership of Silesia and Glatz was confirmed; he recognised the Pragmatic Sanction and the claim of Franz von Lothringen to the throne of the Holy Roman Empire. There was now peace until 1756.

The conflict which broke out in 1756 was to last until 1763 and was to involve all major European states and was to extend to all areas under their control; in Europe, the Americas, and the East and West Indies. With much justification, the **Seven Years War** can be called the first world war. It was to see the defeat of France on all fronts and the further reinforcement of Prussia's position as a first rate power.

Frederick's iron will, and his amazingly professional army had finally worn down all their enemies. The Peace of Hubertusberg was signed by the warring parties on 15 February 1763.

Prussia's next conflict was the **War of the Bavarian Succession 1778-1779.**

On 3 January 1778 the Austrian diplomat Graf von Kaunitz, on insistance from his master, the new Holy Roman Emperor Joseph – not entirely with the full backing of his mother, the Empress Maria-Theresia, signed a treaty with the Elector Palatine, Karl Theodor of Bavaria, recognising Karl's rights as Elector of Bavaria in return for the entire province of Niederbayern, about one third of Bavaria's total area. It later transpired that Karl Theodor had been pressured into putting his signature to the document. The rest of the Imperial German statelets quivered; who would be swallowed up next? They looked to the newly emergent power in northern Germany – Prussia – to put an end to Austria's expansionist policy.

Prussia and Saxony allied against this action. Most of the actions took place in the border areas of Silesia, Saxony, Bohemia and Moravia. The Saxons would operate under Prussian command, on their right flank; Frederick commanded the army of Upper Silesia himself. War was declared on 3 July 1778, but consisted of a great deal of tactical pussyfooting with minimal casualties to either side.

Due to the low level of hostilities, this is also known as the Potato War, from a skirmish in which the opposing sides actually threw potatoes at each other.

On 15 January 1799 Austrian troops invaded Bavaria, but the action was half-hearted, largely due to the fact that Austria's war chest was empty and the Empress was sick of the war.

On 11 March 1779 peace negotiations began in Teschen, a small town on the borders of Silesia, south of Breslau. Peace followed in the May; Bavaria was restored to its former size.

In 1787 Prussia sent forces to stifle the nascent republican movement in Holland; the successful expedition yielded little martial glory.

The War Against Revolutionary France 1792 – 1795

When the revolutionary government of France pledged to export its new political system throughout Europe, thus directly threatening every crowned head in the continent, those threatened heads banded together in the First Coalition to defend themselves. They also sought to free the imprisoned Louis XVI. As in all coalitions, each partner jealously guarded their own interests as a matter of priority, so allied military cooperation was scarcely as effective as it might have been.

Prussia joined the common cause against France but was also distracted by the prospect of gaining more territory in the east where the moribund Polish state was about to be torn to pieces by its neighbours Russia, Austria and Prussia. Initially Prussian forces were active on the middle Rhine, their first action of any size being on 11 August 1792 where the Duke of Brunswick's advanced guard surprised and overthrew a French force at Sierck-les-Baines on the upper Moselle, taking 31 men, a gun and a colour. Their next victorious clash was at Fontoy eight days later and on 23 August the town of Longwy, southeast of Luxemburg capitulated to an Austro-Prussian force. They then moved on to take Verdun on 2 September. The invaders pushed slowly on towards Paris through Champagne in appalling weather, scattering French force at Montcheutin on 15 September, but the advance came to a grinding halt at Valmy five days later at the famous 'cannonade', when the Duke of Brunswick, whose heart was not in the invasion, threw in the towel and headed back to the Rhine after losing some 200 men on the field. The French, under Generals Dumouriez and Kellermann, lost 300 men.

Valmy was the first sign that the Prussian command (and the army) was no longer the magnificent fighting machine that it had been at the end of the Seven Years War.

There was a minor clash at Limburg on the River Lahn on 10 November as the elated French followed up the retreating invaders; they had also taken the Electorate of Mainz on 21 October, together with its mighty citadel and Frankfurt on Main the next day. The von Katte Dragoon Regiment Nr 4 and a battery of artillery were involved in the recapture of Frankfurt on 2 December 1792. The armies went into winter quarters.

The 1793 campaign opened on 6 January for the Prussians when general the Erbprinz von Hohenlohe-Ingelfingen re-took the small town of Hochheim (part way between Frankfurt and Mainz) from General Sedillon. To the horror of most of Europe, the French revolutionary government executed King Louis XVI on 21 Jan 1793. Allied resolve to crush the savages was greatlt strengthened, but party politics continued to bedevil their efforts.

The allied spring offensive on the middle Rhine began with a Prusso-Hessian victory at Waldalgasheim near Bingen on 27 March, where they took six guns for very minor losses. Three days later they won another victory at Alsheim, north of Worms, where Prussian von Eben ambushed a French column under General Munnier, killing 300, capturing amost 1,000 and taking four guns.

The Austrians and Prussians on the middle Rhine were joined by contingents from Bavaria, Hessen-Darmstadt, Hessen-Kassel and Saxony and moved to surround and besiege the French-held city of Mainz. The siege lasted from 10 April until 23 July, when the 8,000 survivors of the garrison were allowed to march out back to France having sworn not to fight against the allies for one year. They were promptly sent to fight the royalist rebels in the Vendée.

The allies now advanced carefully westwards; the next action was at Kettricher Hof on 20 August, where they bundled the French out of a key post which cut communications between the armies of the Rhine and of the Moselle. That same day, to the southwestwards, the Austrian general Graf Wurmser stormed the Weissenburg lines (held by part of the *Armée des Vosges* under General Illier) with a mixed force of Austrians, Hessians (Kassel) and French *émigrés*. The struggle for this line of fortifications went on until 13 October, when Wurmser finally broke through.

Meanwhile, the Duke of Brunswick, with 8,000 men and 54 guns of the main Prussian field army, defeated General Moreau's 34,000 men and 36 guns at Pirmasens on 14 September, inflicting losses of over 2,000 killed, wounded and captured, 19 guns and 29 ammunition wagons for losses of only 167 killed and wounded themselves. This stunning victory was not followed up; the Prussian king had issued a directive that operations were not to be pressed too hard in view of the possible need for more troops in Poland.

Wurmser's success on 13 October caused the defeated French to pull back south to a new defensive line at Hagenau. The defeated commander, General Carlenc, was arrested and imprisoned; General Pichegru replaced him as commander of the *Armée du Rhin* and General Hoche's *Armée de la Moselle* was brought south to cooperate with him to restore the situation on the upper Rhine and to relieve the fortress of Landau.

On 17 November the Prussians attempted to rush the impregnable fortress of Bitche, 30 km south of Zweibruecken with a force of volunteers from several regiments; they were bloodily repulsed.

Meanwhile, Hoche, with the divisions of Huet and Taponnier, some 20,000 men, was defeated at Biesingen, 20 km west of Zweibruecken, on 17 November by General von Kalckreuth with 13,000 Prussians, Saxons and troops from the Electorate of Trier. The French lost some 800 men, the allies just over one hundred. There followed the battle of Kaiserslautern from 28-30 November, where Hoche was again defeated by the Duke of Brunswick's allied army. The French lost over 3,000 men, two guns and a colour; the allies lost under 800. As usual, there was no pursuit, so the French were allowed to run away and fight on other days.

General Hoche, with 35,000 men, determined to throw Wurmser's Austro-Bavarian-Prussian force of about 8,000 out of the Weissenburg lines and to lift the blockade of Landau. From 6 to 26 December 1793 the French launched a series of hammer blows on the entrenchments, biting off section after section. Wurmser had no answer and his pleas to the Prussians for more aid fell on deaf ears.

The will to win of the French commanders opposing him was strengthened by the sure knowledge that prison and the guillotine awaited their failure as surely as night followed day. With his outnumbered men suffering from lack of supplies and ammunition, the Austrian was forced back over the Rhine

The Prussian forces under the Erbprinz von Hohenlohe-Ingelfingen blockading Landau suffered badly from sickness and lacked siege artillery; on 23 December, with a general French advance gathering momentum against them, they abandoned the blockade and withdrew.

During 1794 the campaign in the Austrian Netherlands (now Belgium) turned against the allies and Austria lost the will to hold on to that province at all costs. The United Provinces (now Holland) had an ineffective army and a disaffected populace; nothing could save that state from enemy occupation.

On the upper Rhine, the campaign began on 23 May with two actions; firstly there was a minor French victory at Schifferstadt, 8 km north of Speyer on the Rhine, where General Michaud with 20,000 men defeated Hohenlohe-Ingelfingen's 30,000 allies. But this defeat was more than compensated for by General von Moellendorf's victory with 46,000 men over General Ambert's 5,000 at Kaiserslautern. The French lost 1,000 men, 17 guns and two colours; the allies only about 150 men.

The numerically superior French forces on all fronts were now exerting immense pressure with their waves of conscripts against the strictly limited professional armies of their foes; French losses were quickly replaced; the allied armies dwindled and wasted away.

On 12 and 13 July the Prussian post of 6,000 men on the Schaenzel hills, 22 km southwest of Kaiserslautern was overrun by General Gouvion St Cyr with 9,000. The allies lost nine guns and fell back eastwards over the Rhine.

In September the allies made another thrust westwards; from 17-20 September they fought and won the battle of Kaiserslautern against General Schaal's 16,000 men of the *Armée du Rhin*. Hohenlohe-Ingelfingen, with 30,000 men, inflicted a stinging defeat, causing over 4,000 casualties and taking four guns and three colours for the loss of some 400 men.

Action now stagnated in this sector until 25 December, when the French, under General Vachot, took the Austrian-held bridgehead of Mannheim on the Rhine.

In the east meanwhile, Prussia, Russia and Austria fell upon what remained of Poland in a feeding frenzy in which each tore off the areas closest to their borders and the ancient state vanished.

Action began with a Polish victory over the Russians at Raclawice on 4 April, where General Kosciuszko, with 6,000 men inflicted a heavy defeat on General Tormassow's 5,000, causing 2,000 casualties and taking 12 guns. The Russians then failed to take Warsaw in an action from 17 – 19 April. Prussian involvement was limited mainly to minor, scrappy actions to impose their new authority. The main clash was on 6 June at Rawka, where Prussian General von Favrat commanded a combined Russo-Prussian force of 16,000 men and defeated Kosciuszko with his 26,000 men, mainly peasants. On 15 June General von Elsner took Krakau for Prussia, and on 18 July General von Guenther again defeated

Kosciuszko's 6,000 raw volunteers at Demnicki. In late September there was another minor clash at Labiszyn and on 2 October General Dombrowski caught and defeated Prussian Oberst Szekely at Bromberg. The strategic balance was restored on 1 November when General von Favrat captured General Grabowsky and his entire force of 1,700 men. A combined Russo-Prussian force under the Russian General Fersen forced the capitulation of 12,000 Poles under General Wawrzecki at Radosczyce on 18 November and the Polish resistance flickered out five days later when Prince Poniatowski surrendered at Sochaczew.

In early 1795 King Friedrich Wilhelm II of Prussia concluded a separate peace with France and withdrew his army to the east to concentrate on digesting what had once been Poland.

The War of 1806-1807

Prussia had dithered on the sidelines throughout the war of 1805 in which Napoleon had crushed Austria at Ulm and then Austria and Russia at Austerlitz on 2 December at Austerlitz, The Battle of the Three Emperors. In early 1806 Napoleon took the minor states of the now-defunct Holy Roman Empire and formed them into the 'Confederation of the Rhine' with himself as its protector. At one stroke he had created a willing, pliant *cordon sanitaire* between France and her major competitors on mainland Europe: Austria, Prussia and Russia and formed a pool of military manpower which further increased France's military predominance on the continent.

With extraordinarily poor timing King Friedrich Wilhelm III then decided to take the advice of the 'War Party' in his government and to present Napoleon with an ultimatum, demanding that the Emperor should not stand in the way of a 'North German Confederation (to be formed under Prussia's leadership), that he withdraw the French army behind the Rhine and declaring war on France on 9 August. Saxony and Saxe-Weimar were persuaded to support Prussia and the ageing Duke Karl Wilhelm Ferdinand of Brunswick, now 71, was again placed in command of 150,000 men, spread out over a front of 250 km facing southwards in an arc from Eisenach to Schleitz in the Thueringian forests. The allies had little clear idea of exactly what their aim was and less idea of where the enemy was and what his intentions were.

Napoleon had concentrated 160,000 men in the area of Bamberg and Bayreuth in Franconia in eight corps and a heavy cavalry reserve by 5

October; he struck northeastwards behind a screen of light cavalry, not yet certain of the Prussian army's aims or location. To be prepared for any eventuality, he marched his army in the famous *bataillon carre* formation, a mobile square, capable of reacting to a threat from any quarter. As the situation became clearer, he intended to envelope the allied left flank, but even on the fateful 14 October he still did not know the true enemy dispositions.

As the allies began to comprehend where and what the threat was, they tried to concentrate to their rear, north of Jena. On 9 October Bernadotte's corps cavalry ambushed Tauentzien's rearguard at Schleitz and caused them over 550 casualties; Tauentzien fell back north on Jena.

Next day, Lannes' V Corps clashed with Prinz Louis Ferdinand's Avantgarde at Saalfeld and defeated them, killing the young prince and causing some 1,800 casualties and taking 33 guns from the 8,000-strong allied force for a loss of only 200 of their own men. Five days later, in thick early morning mists, the famous twin battles of Jena and Auerstaedt were fought about 20 km apart.

Napoleon and his main army crushed the corps of Prinz Hohenlohe-Ingelfingen and General von Ruechel at Jena, causing unknown, but very heavy casualties to the 50,000-strong allied force and taking 66 guns for losses to themselves of about 6,000 men.

At Auerstaedt, Marshal Davout's III Corps (30,000 men) took on and defeated the main body of the allied army, some 53,000 men. Allied losses here were 10,000 killed and wounded and 3,000 captured, 115 guns and 35 colours and standards; Davout lost about 7,000 men. The Duke of Brunswick was mortally wounded early in the battle.

Bernadotte's I Corps had famously spent the entire, monumental day between both battles, participating in neither.

The critical allied failures in the campaign so far had been vast overconfidence and incompetence at the level of the high command, but on that fateful day in October 1806, the much-vaunted Prussian army of Frederick the Great vanished from the battlefield into the pages of history.

The remnants of the Saxon army and the contingent of Saxe-Weimar fell away from their ally and went home. The demoralised fragments of the Prussian army fled eastwards or surrendered meekly to the French in a string of shameful capitulations over the next few weeks. Only the fiery Bluecher held his command together and conducted a fighting withdrawal, but he was cornered at Ratkau, near Luebeck, on 7 November by the three corps of Bernadotte, Murat and Soult. Without food or ammunition he was forced to capitulate.

In East Prussia was a small surviving corps of the Prussian army under General l'Estocq, and these carried on the fight against the French with the tardy support of a Russian army under General Bennigsen into 1807.

The French pursued the Prussians into East Prussia, a poor, sparsely populated, underdeveloped area offering little for the support of either army. The Prussian-held fortresses in Silesia were reduced one by one by Jerome's IX Corps while Napoleon tried to destroy the Russo-Prussians in the mud and snows of the icy deserts in which the opponents grappled.

Prussian forces were involved in the following actions. Waltersdorf on 5 February, where General von Kluechzner's brigade was badly mauled by Marchand's 1st Division of Ney's VI Corps; Preussisch-Eylau on 7 and 8 February, where both sides claimed a victory; Braunsberg on 26 February, where part of l'Estocq's corps, with three Russian battalions and a battery of Russian artillery were defeated by Dupont's 1st Division of Bernadotte's I Corps; Kanth on 14 May, where a small Prussian force under Major von Losthin ambushed and defeated a Bavarian-Saxon force under Lefebvre-Desnouettes; Spanden on 4/5 June, where an allied assault on a bridgehead was beaten off, Heilsberg on 10 June, where the allies beat the French and Koenigsberg on 14 June, where the Prussians held off a French assault on the same day that Napoleon defeated the Russians decisively at Friedland.

In the ensuing peace negotiations, the Prussian monarch was a very minor partner, on the sidelines whilst Napoleon and Alexander forged an alliance that was to last until the irresistible pressures of Britain's blockade of world trade caused it to collapse in 1811, leading to Napoleon's fateful gamble in invading Russia the following year.

Prussia was stripped of her western provinces, which went to make up the kingdom of Westphalia for Jerome Bonaparte, the Emperor's brother. She was occupied, humiliated, subjected to massive fines, looted and her army reduced to that of a third-rate power.

1

The Prussian Military Reforms After 1806-1807

The *Kanton* and *Kruemper* system had remained in being, unaltered, until 1763, when increased exemptions were introduced and some trained soldiers were declared subject again to feudal duties and obligations. Despite this, the system survived the storms of the Napoleonic era and was of critical importance for the great struggles of the Wars of Liberation of 1813 - 1815.

Following the treaties of Tilsit (9 Jul 1807) and Paris (8 Sep 1808), Prussia was humbled, crushed, her shattered army limited to 42,000 men, the state crippled by the imposition of massive fines, her cities occupied by French troops, whose costs had to be borne by Prussia. Napoleon ensured that all activities of the Prussian government were closely monitored by the army of spies with which he infested the country. He interfered with the appointments that the king made and ensured that officials hostile to France were replaced by more neutral figures.

The Electorate of Saxony had been raised to the status of a kingdom by Napoleon and granted nominal rule over the Grand Duchy of Warsaw, which Napoleon had created as a sop to the aspirations of those many Poles who had fought so hard and so well for him. As these two states were separated by Prussian territory, the Saxons were granted the right to use a military road across it to communicate with Warsaw.

Although not allowed to attend the negotiations leading up to the Treaty of Tilsit, Frederick Wilhelm III was bound by their outcome. Prussia lost about half its pre-war population and area. The only fortresses remaining in Prussian hands were Colberg, Cosel, Glatz, Graudenz, Pillau and Silberberg.

The king was outraged that his senior officers had failed him so miserably in so many cases. The commanders of those fortresses, formations and regiments, which had been tainted by 'shameful' capitulations were placed before courts-martial, several were sentenced to death, but the king usually relented and commuted the sentences to imprisonment.

The old system, whereby commissions were granted only to those of noble birth was scrapped and commoners were now eligible to become officers if they demonstrated the right qualities.

On 15 July 1807, the Prussian king invited Scharnhorst and Graf von Lottum to form a team to investigate the causes of the 1806 debacle and to present plans for the thorough reform of the army.

The team which was picked was entitled the *'Militair-Reorganisations-Kommission des Heeres'* or Army Reorganization Commission and included the following major players. Another committee was appointed to consider civil reforms.

Karl Leonhard Ludwig von Borstell. Born on 30 Dec 1773 in Tangermuende 34 years old in 1807. Entered Prussian military service 29 May 1790 as a Kornett in the Kuerassiers. He fought against France on the Rhine from 1792 – 1795 and won the *Pour le Merite* at Pirmasens. By 1800 he was in the Garde du Corps and was a major in 1806 and was distinguished in the battle of Auerstedt. He escaped to East Prussia to join the king, and on 11 May 1807 joined Bluecher's corps on the island of Ruegen. On 16 May 1807 he was appointed to be an ADC to the king and on 20 Sep of that year to be a member of the Military Reorganization Committee. Von Borstell was a reactionary and argued violently and frequently with Scharnhorst . The latter finally went to the king and forced him to choose between them ; Borstell left the Committee in the September.

August Wilhelm Anton Neidhardt von Gneisenau. Born on 27 Oct 1760, thus 47 years old in 1807. He had studied in the Jesuit college in Wuerzburg and spoke French , English , Polish and some Italian. He then went to Erfurt; here, there were running fights between the students and the cobblers and in one such brawl Gneisenau apparently killed a cobbler. He at once went to the Austrian recruiting office in the town and entered Austrian military service on 1 Dec 1778 in the Wurmser Hussars; the unit left for Bohemia the next day. One day , Gneisenau caught his squadron commander molesting a local girl ; in the ensueing scuffle both men drew their swords - Gneisenau had to flee a second time. In 1779 he transferred to the service of Anspach as cadet in the Jaegers; from 1782 – 1783 he served in America but the fighting there was over. Back in Bayreuth, Gneisenau found peacetime soldiering too boring so he resigned and applied for Prussian service; in 1786 Gneisenau was commissioned onto the king's staff. In June 1787 he transferred to Fuesilier Bn von Schurff Nr 15. In 1794 - 1795 he fought in Poland; at Rawka and Skala and at the blockade of Czenstochau. Gneisenau-like Scharnhorst-was a keen student of military history and of current military developments; he studied Napoleon's campaigns and the new tactics used by the French. On 11 Apr 1807 he received the king's order to take command in the besieged fortress of Colberg where the commandant, Lucadou, was seen as too passive. As soon as he inspected the place, Gneiseau recognised the weak spots and correctly predicted where the assaults would come. He at once started taking corrective action. The siege lasted from 20 Mar - 2 Jul 1807 before the French withdrew. On that last day, Marshal Mortier (commanding the siege) doffed his hat to Gneisenau on the ramparts in recognition of his spirited defence. During the siege, Gneisenau wrote a paper outlining the need for new regulations for the light infantry which he presented to the king on 27 Feb 1807, and which were approved by him for implementation on 5 March of the same year. On 25 Jul 1807 he was appointed a member of the Committee of Reorganization and began to work together with Scharnhorst and Stein. He and Scharnhorst were soon at daggers drawn with the more conservative committee members. On 14 Jan 1808 the disagreements on the Committee had now become so bitter that Gneisenau tendered his resignation; the king refused to accept it. When Grolmann and von Boyen joined the Committee, the reformers won the upper hand.

Friedrich Wilhelm Graf von Goetzen. Born on 20 Jan 1767 in Potsdam, he was 40 in 1807. Goetzen grew up together with with the future King Friedrich Wilhelm III and they became friends. In 1782 he entered Prussian military service as Estandartenjunker, Kuerassier-Regiment Nr 11 (Leib-Karabiniers); on 17 Apr 1792 he transferred to HusBn von Frankenberg Nr 11. During the siege of Mainz, King Friedrich Wilhelm II had von Goetzen join his suite to learn the arts of war; on 3 May 1794 he was ennobled as a Graf. During 1798 he suffered sever health problems and handed in his resignation but the king refused to accept it. In the war of 1806/7 he took active part in the defence of Silesia and on 7 Nov 1806 was sent by the king to report on the conduct of the defence of Colberg under Oberst von Lucadou. After the fall of Breslau on 6 Jan 1807, the king wrote to him urging him to spare no efforts to hold the remaining Silesian fortresses. All this work wrecked his health and as soon as the war was over, he went on long sick leave. On 16 Dec 1807 – appointed a member of the Military Reorganization Committee at Scharnhorst's instigation in place of Oberst von Borstell.His health caused him to take sick leave again, but he was a valuable member of Scharnhorst's team, even from his sickbed.

Eberhard Friedrich Fabian, Freiherr von Massenbach. Born on 3 May 1753, he was 54 in 1807. He entered Prussian military service as *Junker* in the 6th Dragoons in 1766; fought in the war in Poland in 1794/5 in the clashes at Gabnicki and Plonsk; repeatedly distinguished. In 1801 he was appointed commander of the newly raised 13th Dragoons. Von Massenbach was a talented cavalry commander. In the war of 1806/7 he took active part in the defence of Danzig. In June 1807 he was promoted to GM as a reward for his part in the defence of Danzig. He was appointed to the Army Reorganization Commission and was a member of the reactionaries, opposing most reforms.

Heironymus Franz Seraph Roedlich. Born on 30 Sep 1767 in Bruenn, Moravia; he was 40 in 1807. He entered Austrian military service in 1785 as Kadett in the 4th Dragoons and served in the war against the Turks in 1778/9. On 4 Nov 1791 he transferred to the UlR Meszaros Nr 1, and in 1796 he transferred again to the 2. Walachian Grenz-IR and then on to the general staff. He fought in Italy against the French at the battles of Arcole and Rivoli and the clashes at Fontaniva and Lanove. On 23 Sep 1798 he went with FML Mack to Neapolitan service; fought against the French at Civita Castellana and the capitulation of Capua. 1 Jan 1799 – transferred to the 2nd Ulans. In October 1800 Roedlich retired as a major until 6 Feb 1802 when he reentered Austrian service as *Platzmajor* (garrison commander) in Cattaro. In 1804 he transferred to the Liccaner Grenz-IR Nr 1until retiring again on 11 Sep 1806. In 1807 he reentered Austrian service. The Prussian reformer Graf von Goetzen met Roedlich in Vienna and learned that he had experience in planning and implementing the arming the populace and thought that he would be of great value in the organization of the *Landwehr* in Prussia. Roedlich was allowed to leave Austrian service and transfer into that of Prussia in August 1807. He was placed in the suite of the army as a lieutenant-colonel; he soon became a well respected member of the reform group of von Goetzen, von Scharnhorst, von Stein and von Gneisenau.

Johann David von Scharnhorst. Born on the 12 Nov 1775, he was only 32 in 1807, but already a well-known and respected military writer. On 1 May 1771 he entered the *Kriegsschule* (War School) of the Graf of Lippe and Bueckeburg on the island of Wilhelmstein fortress in lake Steinhudermeer. By 1775 he was a *Kondukteur* (senior warrant officer of artillery). He then was commissioned into the Hanoverian infantry and later instructed in the *Kriegsschule* in Hanover . In 1783 he toured Bavaria , Austria and Prussia and wrote reports on the states of their armies , he judged that of Bavaria to be very bad , Austria's to be very good. He criticised the Prussian army for having too many different types and calibres of cannon.

Several nations wanted his services, but Oberst von Lecoq led the Prussian attempt to win him over and offered him a commission as a major with a salary of Thlr 3,000 p.a. as opposed to his Hanoverian pay of Thlr 1,000. On 5 May 1801 Scharnhorst was accepted into Prussian service as Lieutant-Colonel in the 3rd Foot Artillery Regiment. Much of the Prussian officer corps was hostile to him, but the king, Prinz Louis Ferdinand, Bluecher, Ruechel and others recognized his talents and supported him. He ran the autumn officers' military academy and his pupils included Clausewitz,Ruehle and Tiedemann. Scharnhorst's literary output continued; he wrote a handbook on artillery tactics and several other manuals as well as continuing to produce his Military Journal. He suggested forming divisions in the army and other improvements in infantry and cavalry tactics. On 16 Sep 1802 Scharnhorst was ennobled, and on 26 Mar 1804 he was appointed Quartermaster General of the Prussian army. He was a great admirer of Napoleon's and held discussions of the new French tactics with his students. He suggested the introduction of general liability for military service in Prussia and the establishment of a national militia, but the king could not bring himself to follow his advice. He did however, agree to the increase of the establisment of the *Kriegsschule* and Scharnhorst was empowered to examine and approve all suggested improvements in artillery equipment and tactics. In the 1805 mobilisation of the Prussian army,von Scharnhorst was attached to the Duke of Brunswick and had to participate in the annexation of his homeland. In the 1806 campaign he served on the staff the Duke of Brunswick. He made suggestions for the conduct of the forthcoming campaign but Massenbach's opposition was too strong, thus he fought at Auerstaedt in a relatively isolated position, from which he could do little to affect the outcome of the battle. Scharnhorst then served on l'Estocq's staff in the 1807 campaign.

The Peace of Tilsit shattered von Scharnhorst and initially he considered leaving Prussian service. Czar Alexander offered him a post but he refused to accept it. He also considered going into British service but on 17 Jul 1807 he was promoted GM,and on 25 Jul 1808 appointed Chairman of the *Militair-Reorganisations-Kommission des Heeres*. Unfortunately for the army, von Scharnhorst (who was Bluecher's Chief of Staff in 1813) was mortally wounded in the battle of Gross-Goerschen on 2 May 1813 and died in Prague on 28 June of that year.

Other military members who served on the Commission were Lieutenant-Colonels von Boyen (who became Minister for War in 1814), Bronikowski, Graf Lottum and Major Grolmann.

There were also two civilian members in the Commission: Schoen and Freiherr von Stein. This latter figure was to become the mainspring of the civil reforms of the Prussian state in this era. Born in 1757, he had studied in Hanover and was very pro-British. He entered the Prussian civil service and in 1796 was president of the Westphalian chamber of commerce. In 1804 he became the Prussian Minister of Finance. Stein was openly hostile to France and Napoleon insisted upon his removal from office and the confiscation of his family estates near Nassau. In 1809 continued pressure from Napoleon forced von Stein to flee to Austria until 1812 when he accepted the post of advisor to Czar Alexander.

At the Congress of Vienna in 1815, King Frederick William III renaged on his promise to von Stein to introduce a more liberal representational constitution to the country and von Stein resigned from political life.

Despite this late setback, he and von Scharnhorst's team had radically modernised the Prussian army and the state by their efforts.

The king drew up the following list of measures and reforms that he wished to see introduced:

- Identification and punishment of all officers guilty of dishonourable conduct in the 1806 campaign.*
- Opening up careers as officers to all candidates who could show that they possessed the appropriate qualifications.
- Improved uniforms.
- Reduction of the baggage trains at all levels.
- Improved field supply systems.
- Revision in the cavalry establishments.
- Production of a new set of Articles of War.
- Production of improved tactical regulations for all arms of the army.
- Abolition of the practice of allowing soldiers to buy freedom from guard duty.

Scharnhorst had called on his old friends Gneisenau, Grolmann, Goetzen and Boyen to help him in the massive task of reconstructing the Prussian army, but Gneisenau and Boyen were so disheartened by the state of Prussia that they fell away in 1809. Before they left however, they supported von Scharnhorst in breaking the opposition of the old reactionary generals, several of whom were then removed from the commission. In the following years, von Scharnhorst's team reorganized the Prussian army and succeeded in inducing the king to bring in universal military service liability as he had finally managed to convince the king that Prussia's salvation lay in an army of native patriots and not mercenaries. Such a radical change meant a complete revision of the entire social, economical and political systems in Prussia.

The nobility—who in the past had enjoyed the monopoly of supplying officers to the army—were now allowed to sell their lands to anyone they chose and were free to adopt any profession that they wished. If the peasant lads were to be allowed to join the army, they would have to be freed from their massive responsibilities towards the landowners and this was accomplished by abolishing feudal privileges and obligations.

* In fact some 208 officers of the total of 6,915 serving in 1806 – and this includes the infantry, cavalry, artillery, pioneers, engineers, general staff and the *Kriegs-Kollegium* - were found to be guilty of dishonourable conduct in 1806-1807. These investigations were finally abandoned in early 1813. 3,900 officers of the 'old' army served in the Wars of Liberation from 1813 – 1815; a surprisingly high figure.

The foundations were also laid for the formation of the Landwehr. Flogging as a military punishment was abolished and a more liberal code of military justice was introduced. All able-bodied Prussian male citizens of certain age groups were theoretically liable for military service; exemptions were minimised and the recruitment of foreigners was forbidden. Birth and status were no longer alone the keys to entry into the officer class; however, as literacy was a prerequiste for a commission, and literacy was still limited, it did not dramatically widen the potential pool of new officer candidates, but anyone who thought they had the required qualities could apply for a commission. Officer candidates now had to serve for three months as privates before being commissioned. It was now possible for serving NCOs and men to apply to a regimental examination board to become officers. There could be up to fourteen such ensigns in each infantry regiment and eight cornets in each cavalry regiment. Candidates from among these junior officers for the desired commission then had to go to Berlin to pass another examination on military and general topics.

On 10 March 1809 the king asserted the right to appoint regimental commanders on the basis of merit as opposed to seniority alone. Dead mens' boots were now *passé*.

The old established custom of entering young nobles into the regimental records as officer aspirants when they were only ten or twelve years old, so that they would have advanced through time promotion to be senior lieutenants when they actually joined the regiment when they were sixteen or so was stopped.

The army was organized into brigades containing elements of all three arms. In fact, in 1813, these brigades were almost as large as the divisions in the French army. The brigade headquarters was now the authority responsible for administration of the units which constituted it. At army level there was a War Commissariat to which each brigade sent a representative so that requirements could be ordered in bulk at lower unit prices that hitherto.

In the old days, each captain bought his company's muskets when he took office as '*Inhaber*' or owner and had the right to sell his men items of equipment. The reforms put a stop to this; the muskets were sold to the state and all equipment sales were handled at regimental level. The pay of men on long leave no longer went into the company *Inhaber's* pocket but to the state and the practice of the men on leave working as artisans was restricted. The loss of earnings of such men was compensated for by issuing them increased rations.

The War Commissariat was also responsible for the supply of rations in the field. Permanent Train Depots were set up in Berlin, Breslau, Colberg and Koenigsberg and each brigade was to receive 32 bread wagons on mobilisation. Horses and drivers were mostly added at the time of mobilisation.

The number of vehicles accompanying the army was drastically curtailed; now each company had only one pack horse to carry the personal items of all the officers in the sub-unit. Each battalion was now allowed only three baggage wagons.

A cavalry regiment now had only two baggage wagons and each squadron had two pack horses.

The system of supplying bread and rations to the army in the field from magazines was however retained so the resulting army train was still considerable.

At the humiliating Peace of Tilsit, the Prussian army had been reduced to a maximum strength of 42,000 men by Napoleon. On 31 July 1807 von Scharnhorst presented to the king a revival of the old *Kruemper* system, in a plan to allow the rapid expansion of this small force in case of need. In the first three years of the plan, 20 able-bodied, trained men of each company were to be '*beurlaubt*' i.e. sent on long leave; in each subsequent year, ten men per company were to be sent on long leave. The missing men were to be replaced by new recruits.

This measure was initially adopted on 31 July 1807, but not until 3 April 1809 was in extended to all regiments of the army. An ever-increasing pool of trained men would thus be formed who could easily be remobilized if and when needed. On 25 Sep 1807 the Army Reorganization Commission accepted this plan and proposed to the king that the army should be organized into three, geographically-based corps: Brandenburg-Pommerania, Prussia and Silesia. Dictated by Napoleon's limitations of the strength of the army, this plan had to be revised and the following system of brigades was adopted: East Prussian, West Prussian, Pommeranian, Brandenburg, Lower Siesian and Upper Silesian.

On 6 August 1808 the king ordered that the system should make a modest start in that each company was to rotate from 3 – 5 men in this year. The new recruits were to be drilled for one month then released; they were subsequently drilled on Sundays and holidays by officers and NCOs sent out into the towns for the purpose.

This system was popular with the farmers because the conscripted men were only absent for a short period in peacetime. Army officers were less enthusiastic about the brevity of the training.

As of 1809, the units of each brigade were billeted in concentrated garrisons and this gave the opportunity for all-arms training to be carried out for the first time.

Gradually, these rotation quotas were increased. The reports of October 1812 show that 33,337 *Kruemper* were trained and fit for duty in the field and a further 3,087 were fit for garrison duty. Earlier estimates of available trained manpower were wildly exaggerated; some spoke of 150,000 being achieved.

The infantry regiment was now defined as two companies of grenadiers, two battalions of musketiers and a battalion of fusiliers (light infantry) each of four companies and a depot company. Minimum strength of an infantry company was to be 50 men. The Order of 10 October 1807 laid down that the infantry company would consist of five officers, 15 NCOs, three or four musicians and 170 men. In peacetime 50 men were to remain with the colours, 120 were to be sent on long leave.

The tactical line was to be formed in three ranks; the third rank was to be trained in skirmishing and twenty of the best shots in each company were to be equipped with rifles and trained as sharpshooters.

Contrary to much public opinion, the Prussian skirmishers (the Fusilier battalions) had performed well in 1806 (but there were then too few of them), so not too much reformation was needed in that arm of the service. In 1806 only 8,7% of the line infantry were light infantry; by 1812 it was 33%.

In November 1807 the infantry totalled 37,754 officers and men. The old battalion guns were withdrawn from service.

A cavalry regiment now had four squadrons (except the Ulans who had eight), each with a minimum strength of 62 men. The new cavalry squadron establishment was for six officers, 15 NCOs, three trumpeters and 132 troopers, of whom 72 were to be sent on long leave in time of peace. Each squadron had 125 horses.

In all there were 86 squadrons in November 1807 with 12,060 officers and men, which was about a third of the pre-1806 total.

The artillery was not really ready for the field again until 1809. By November 1807 there were 3,633 officers and men in the field and fortress artillery. In 1809 the field artillery was organized into three brigades each of twelve, six-pounder foot companies (batteries) and three horse companies, each consisting of six cannon and two 7-pound howitzers. Batteries of eight howitzers were also formed and some of the artillery pieces were of British origin, imported through Colberg. Drivers of the gun teams were now to be members of the artillery and no longer provided by the transport corps. This whole scheme improved the mobility of the artillery in the field.

Tactics were overhauled and a greater choice of combat formation, rather than the automatic selection of line was stressed. Greater inter-arm cooperation was also emphasised.

Thousands of prisoners of war of the 1806 – 1807 campaign were returned to Prussia in 1809; the fit among them were used to bring the weaker brigades up to strength.

This gave a total of 54,180 all ranks in the army; within a year this total had dropped to 50,000.

In addition to the standing army, Prussia adopted a militia or '*Landwehr*' on the Austrian model. Initially, entry to the *Landwehr* was limited to those who could place a deposit of 100 Thalers in the regimental account, and who could arm and clothe themselves; in the cavalry they had to provide their own horses. There was no artillery in the *Landwehr*.

These proposals were opposed by the two civilian members of the Commission, who feared that there would be too many exemptions from service liability in the regular army and that the danger would arise of the *Landwehr* becoming a rich man's club.

Apart from purely military reforms, Scharnhorst introduced and improved elementary schooling in the regiments to improve literacy and general education.

The new uniforms were the topic closest to the king's heart; those adopted showed strong Russian influence.

Napoleon was well informed of the activities of effective reformers such as Scharnhorst and Stein and on 6 June 1810 he forced the king to dismiss them both. Scharnhorst was replaced by Colonel von Hake as head of military administration. On 14 Jan 1811, von Scharnhorst was appointed chairman of a commission charged with producing new drill regulations for all arms of the army. On 23 Jul 1811, he was given leave to tour the fortresses in Pommerania and Silesia and to take the waters in the spas of Silesia and Bohemia. This whole scheme was a cover to allow him to go to St Petersburg and to win the czar over to the Prussian side in the increasingly threatening political climate. On 24 Sep von Scharnhorst met with Alexander and in November he returned to Prussia with a draft treaty. Friedrich Wilhelm III, characteristically, dithered , largely due to ignorance of the Austrian position. Von Scharnhorst was sent secretly to Vienna in December 1811 and held negotiations with Metternich, the Austrian Chancellor, but failed to convince Austria and returned to Berlin on 24 December.

This ended the fight to introduce universal conscription for a time. It was not until early 1813 that the old *Kantonpflicht* (which meant that the nobility and many of the peasants were exempt from military service liability) was abolished. The more general military service conscription liability was now introduced in the face of entrenched resistance from the king and his conservative advisers Gentz and Niebuhr, who feared a 'French Revolution scenario' if the more progressive military reformers got their way.

Not all areas of the country welcomed conscription; Silesia and West Prussia were particularly hostile to the system and desertion among those called up was high in these provinces.

Napoleon, still meddling in Prussia's affairs in 1811, unwittingly helped the rebuilding of their army along, when he demanded that Frederick William raise a coastal defence force to guard against possible British amphibious landings on the Baltic coast. The measure was also aimed at combating the smuggling of British colonial goods, which was rife and which undermined his Continental System.

Some 11,700 of the soldiers on long leave were called up to create the force, but instead of pure military training (which was what the army hoped they would do) they spent most of their time building defences.

As Napoleon's demands for a Prussian corps to support his invasion of Russia in 1812 grew, most of this manpower was recalled to the colours. In February 1812 France and Prussia entered into a military alliance and Prussia was committed to providing 20,842 men for the adventure.

This left a theoretical 22,000 soldiers in Prussia according to the terms of the Treaty of Paris. In reality, there were some 38,300 when the trained reservists were included.

On 5 March 1812 von Scharnhorst resigned from the Drill Commission in protest against the fact that the king had entered into a treaty with Napoleon against Russia.

On 15 May 1812 von Scharnhorst was permitted to tour the battlefields in the Austrian empire. As soon as he heard of Napoleon's defeat in Russia, he plunged into a whirwind of activity to increase weapons production and the readiness of the army.

An Artillery Testing Commission was established to overhaul the design and production of artillery pieces; new cannon foundries were set up and each of the three artillery brigades there was a factory to produce the newly-designed gun carriages, limbers and other vehicles.

In order to ensure that the new drill regulations would be adopted throughout the army as quickly as possible, *'Normal'*, or standard regiments of cavalry and infantry were set up in May 1811. After they had mastered the new evolutions, they were used to demonstrate them to other units.

One of Scharnhorst's measures was to insist that the New Prussian musket of 1809 should have a calibre of 18,57 mm to allow foreign and captured ball ammunition to be used in the weapon.

Muskets, weapons, clothing and equipment were supplied to Prussia in 1813 and 1814 from many sources; the various calibres of these weapons are shown below.

Austria – 17,58 mm, Russia – 18 mm (and captured French muskets from Russia – 17,52 mm), Sweden – 19,9 mm and Britain – 19,35 to 20, 05 mm.

Of course all items of equipment were in very short supply; many of the *Landwehr* marched and fought barefoot and in rags.

On 3 Feb 1813 the formation of '*Freiwillige-Jaeger-Detachements*' (Volunteer Rifle Detachment), attached to all regiments and designed to attract young men from the propertied classes, academics and the professions to the colours, was anounced. Volunteers could choose which regiment they wished to serve in. Some 7,000 volunteers came forward in 1813.

Von Scharnhorst also headed the organization of the *Landwehr* and the *Landsturm* in the kingdom.

On 9 February all exemptions from military service liability were suspended for the duration of the coming war and general conscription was introduced. At this point the strength of the standing army (excluding *Landwehr* and *Landsturm*) was 65,700 men with 1,659 artillery pieces (many of which were old or unserviceable). In the event, some 213 pieces were taken into the field in early 1813.

Conscription was generally a success and soon added over 37,000 men to the army.

On 11 March 1813 Scharnhorst was promoted to Lieutenant General and appointed Quartermaster General on Bluecher's staff ; on 17 Mar 1813 he published his 'V*erordnung ueber die Orginisation der Landwehr'* which called up all able-bodied men from 17 to 40 years of age and on the 21 April he published his'V*erordnung ueber die Orginisation der Landsturm'* in which he stated all youths who have grown a beard are liable for service in the *Landsturm'*. Theoretically, the *Landwehr* were liable to serve only within their home province; this stipulation was rapidly forgotten.

Several volunteer units were also raised; these included the Elb-Infantry Regiment, Hellwig's Freikorps, Luetzow's Royal Prussian *Freikorps*, the Battalion Reuss and Rieche's Foreign Legion. These units were later to be formed into new regiments of the standing army. The flame of Prussian nationalism had at last been lit.

Another volunteer organization formed in each province at this time were the 'National Cavalry Regiments'; members were expected to provide their own uniforms, weapons and mounts.

On 30 December 1812 General von Yorck had signed the Convention of Tauroggen with General von Diebitsch in Courland and took the Prussian contingent of Macdonald's X Corps out of the *Grande Armée.*

This forced the indecisive king's hand and he broke with France, but it was not until 4 March that Prince Eugene evacuated his French troops from Berlin.

On 27 February von Scharnhorst left Berlin, at Czar Alexander's invitation to visit Imperial Headquarters in Kalisch to discuss joint plans against Napoleon. The Czar later said: 'Never have I seen a man with such a clear sense of planning and comprehensive knowledge as Scharnhorst.'

The result of this conference was agreement that the old Kutusov was to be Commander-in-Chief of the allied armies; Prussia was to command one of the two armies to be formed; von Bluecher was to command the Army of Silesia, with von Scharnhorst as his Chief of Staff. The stage was set for an epic international struggle, which would end with Napoleon's first abdication.

The Order of the Iron Cross was established on 10 March 1813 and existed as a coveted decoration until 1945. As Napoleon had said: 'It is with baubles that men are led.'

On 17 March 1813 King Frederick William III issued his proclamation '*An Mein Volk*' (To my People) announcing his final break with Napoleon and calling for popular support for the impending war.

The new army was soon to be put to the test; it was not to be found wanting.

The 1812 Campaign in Russia

Prussia was forced to provide a division for Napoleon's adventure; it was commanded by General von Yorck and allocated to Marshal Macdonald's X Corps to operate in Courland, before the Baltic port of Riga. In order to spread the experience of current military operations as far as possible, most units which took part in this campaign were composite bodies, formed of battalions or squadrons from two regiments. Only the Leib-Infanterie-Regiment was mobilised in its entirety as the '4th Combined Infantry Regiment' in the 2nd Brigade. Action was low-key and combat losses were relatively light, the Russians having instructions not to press the Prussians too hard. The first action involving the Prussians was at Eckau on 19 July. As the *Grande Armée's* remnants withdrew across the Beresina, X Corps fell back to the southwest.

The Wars of Liberation 1813 – 1815

The 1813 Campaign

Yorck's single handed action had plunged King Frederick William III into a serious crisis. His country was still full of French troops, their spies and their allies. Finances were in a parlous state, weapons, ammunition, equipment and uniforms all in very short supply. Luckily for Prussia, the wreck of the *Grande Armée* could do nothing more than fall back before the Cossack scouts and most Prussian territory was clear of the enemy before too long.

As mobilisation rushed ahead, von Yorck's corps was soon in action; the first of any note being on 5 April east of Magdeburg fortress, where a Russo-Prussian force beat elements of the V and XI Corps.

On 21 April Spandau fortress surrendered to the Russo-Prussians, but the first serious action took place at Luetzen (Gross-Goerschen, 20 km southwest of Leipzig) on 2 May. Here, Napoleon defeated the allies, who had failed to exploit their great advantage in cavalry over their opponents. The allies withdrew to the east in good order.

On 20/21 May the next serious battle was fought at Bautzen and Wurschen on the River Spree, 52 km northeast of Dresden; again it was a French victory, but their lack of cavalry meant that there was no effective pursuit, the usual Napoleonic cavalry screen was not available to hide his movements or to gather intelligence.

Both sides needed a pause in operations to repair damage, bring up new formations and to conduct diplomatic negotiations in the hopes of gaining new allies. An armistice was agreed on 4 June and lasted until 10 August.

By that time, Austria had abandoned Napoleon and joined the allies, Sweden had also declared against France and Prussia was now fielding four corps.

The allies formed three armies, each being carefully built to contain contingents from all three major powers. The main army was that of Bohemia, commanded by Prinz Karl zu Schwarzenberg, the second being the Army of Silesia under Bluecher's command and the third being the Army of the North, commanded by Bernadotte in his new capacity of Crown Prince of Sweden. The three allied rulers, Czar Alexander, Kaiser Franz of Austria and the king of Prussia travelled with Schwarzenberg.

There followed the mighty battles of the dramatic autumn campaign: the Katzbach on 26 August, where Macdonald's Army of the Bober was badly defeated by Bluecher; Dresden on 26/27 August, in which Schwarzenberg's main Army of Bohemia was beaten by Napoleon, Kulm on 30 August, where Vandamme's I, XIV Corps and I Cavalry Corps were wrecked by the Army of Bohemia, and Dennewitz on 6 September in which Ney was defeated by von Buelow's III (Prussian) Corps, with some help from von Tauentzien's IV (Prussian) Corps.

The allied strategic aim was to wear Napoleon's forces down by employing the principle that whichever of the three armies were confronted by him would withdraw, while the other two closed in on his flanks, like terriers baiting a bull. The plan worked excellently and gradually led to the great confrontation at Leipzig from 16 – 19 October. Here, Napoleon stood with his back to the difficult defile of the swamps formed by the Elster and Pleisse rivers and was decisively defeated, even if the Swedish army scarcely took part, thanks to Bernadotte's almost treasonable activities.

Napoleon fell back westwards to the Rhine, Bavaria abandoning him just after the great battle and opposing him at Hanau on 30/31 October.

On New Year's Day 1814 Bluecher led his army across the Rhine and into France.

1814 This spring campaign was to see Napoleon at his most brilliant as his enemies closed in on Paris and Bluecher at his most careless and impetuous as he recklessly threw himself after his hated foe.

1815 Napoleon interrupted the victors at their deliberations at the Congress of Vienna when he left his place of exile on Elba and returned to France to oust Louis XVIII and to place himself on the throne of France again.

The allies were united in opposing him and gathered their forces to topple him. The armies first ready and closest to Paris were those of Wellington and Bluecher in the Netherlands; Napoleon decided to strike at them first, aiming to separate them, then destroy each in detail.

Due to his excellent security and the speed of his advance, he almost succeeded.

His plan was to throw Ney at Wellington's army while he struck at Bluecher, about 30 km away from Wellington to the east. He aimed then to drive the Prussians off to the east and to unite with Ney to crush Wellington.

On 15 June his cavalry surprised von Zieten's I (Prussian) Corps, which was in extended encampments northeast of Charleroi on the River Sambre. Surprise was complete and the Prussians were thrown back to Ligny, on the main body of their army. Next day, Napoleon won the victory of Ligny, during which Bluecher was caught under his fallen horse and knocked out. His Chief of Staff, von Gneisenau, gave the fateful order that night, that the Prussian army would not fall back along its lines of communication to the east, but would withdraw northwards to join up with Wellington at Waterloo as had been just previously agreed.

That same day, 16 June, Ney had bungled his job of crushing Wellington. Thanks to the Dutch General Perponcher, the allies had held onto the strategically vital crossroads at Quatre-Bras all day and given Wellington the time to reinforce them there, then to withdraw north to his chosen position at Mont St Jean, or Waterloo.

The 17 June was spent by Napoleon closing up to Waterloo, whilst Marshall Grouchy, with the III and IV Corps and the II Cavalry Corps were sent off to clear the beaten Prussians away to the east.

In fact, it was only von Thielmann's III (Prussian) Corps that they were chasing, the rest of Bluecher's battered army was marching to make their dramatic entrance onto the field of Waterloo through the villages of Frischermont, La Haie and Papelotte, deep in Napoleon's right flank at the dramatic climax of that fateful battle.

On the day that Napoleon was finally defeated, von Thielmann's corps was caught at Wavre by Grouchy and badly handled, but the game was almost over.

Napoleon's army dissolved and fled the field of Waterloo, but fighting went on until into September of 1815 as the many individual fortresses along France's borders, held by committed Bonapartists, were subdued one by one.

On 20 June, von Pirch's II (Prussian) Corps tried to storm Namur, (a fortified city) and received a bloody nose for their troubles

Tactics

Infantry Tactics

As Frederick the Great ascended his throne at the end of 1740, the system of linear tactics on the battlefield had been well established for about fifty years.

Miltary tactics are dictated by the capabilities of the weapons available; not much was to change as far as the infantryman's muzzle-loading, smoothbore musket was concerned up to the time of the American Civil War.

Rifled small arms were already readily available in the Seven Years War, but they were expensive to produce and their greater accuracy compared to the musket was heavily outweighed by the much slower rate of fire (about a shot every two minutes) and the need to clean the weapon thoroughly after about twenty shots. These were still specialist weapons, to be used only by the best marksmen in this era; indeed, Napoleon never did make much use at all of rifled firearms.

The aim of all military tactics was – and remains – to make optimum use of the weapons available.

Sabres, swords, pikes and pistols were used for close quarter fighting and were of little interest in deciding battles. Junior officers of the era carried spontoons or pikes as did senior NCOs

The men were armed with smoothbore, muzzle-loading muskets, with an effective combat range of about two hundred feet, capable of being fired at a steady rate of about two shots per minute on the battlefield.

Frederick William I was obsessed with increasing the rate of fire of his infantry and drilled them mercilessly until they could regularly achieve three shots per minute with their smoothbore muskets, which he frequently timed on his pocket watch. This rate was never matched by other European armies of the era; most of them attaining just two shots in the same time.

It should be noted that these rates of fire were achieved under parade-ground conditions; on the battlefield, with the attendant noise, imminent danger and possible adverse weather conditions, the rate of fire would rapidly drop, the neat vollies become more ragged, the aim worse. As firing continued, debris would build up in the musket barrels, reloading would become more difficult and time consuming, flints would wear down, ramrods would be dropped or fired off in mistake, the priming powder would be spilt instead of all going into the pan, the touchhole would block up.

In the early period (up to about 1730 in the Prussian army) the wooden ramrods, used to push the ball and charge down the barrel of the muzzle-loader, would frequently break, thus rendering the weapon useless. Frederick William I realised the defect and had them replaced by iron ramrods. These were to demonstrate their advantages over the Austrians in the battle of Mollwitz.

Another factor leading to the inevitable decay of the rate of fire of an infantry unit in battle at this time was the flint. Even the best flint would lose its efficiency and fail to produce the vital spark after 30 – 50 shots. Replacing the flint would take minutes, even under parade-ground conditions.

Two factors were of great importance in a campaign: speed of movement and the ability to bring the maximum amount of firepower to bear on the enemy.

Rapidity of infantry movement was limited by the physical abilities of the men and the need to have them keep in formation. Thus, for movement out of effective range of hostile weapons, the column, on a frontage of three or four men, was quickest and most efficient.

When going into action it was vital to be able to bring as many firearms into effective use as possible. Up to about 1750 this meant deploying into lines, parallel to those of the enemy's, of four ranks. Frederick improved on this by reducing the number of ranks to three. In the Napoleonic Wars the British were to extend the principle by reducing their lines to two ranks deep only.

The great problem with moving lines of troops over broken, rough ground was that the pace had to be slowed or the line would disintegrate into a series of incoherent blobs of men. The regimental colours, usually carried in the centre of the line, provided a vital marker for the flanking sub-units to use to maintain their alignment.

The speed of advance of such a line would be about seventy paces per minute, just as still now used by the British Guards at the annual Trooping of the Colour on Horseguards Parade.

Given the characteristics of the musket: smoothbore, muzzle-loading, black powder charge, flintlock ignition; the maximum rate of fire on the battlefield under optimum conditions was two to three shots per minute. This rate would decay as time passed for the reasons explained above.

The EFFECTIVE range of a smoothbore musket, that is to say, the range at which a hit on an opposing infantry line, likely to kill or wound the enemy were likely, was about 300 paces. The MAXIMUM range was much further, but chances of getting a hit were negligible.

The effective range of the smoothbore pistol was not much above 20 paces.

The rifled weapons caused their projectiles to spin as they were fired; this spinning ball had much more stability than the smoothbore projectile, thus the EFFECTIVE range of the rifle was much higher (up to 600 paces) but the maximum range of a rifle would be the same as that of a smoothbore using the same charge of powder of like quality.

Cavalry Tactics

Following the end of the First Silesian War Frederick worked hard on his military machine; in particular, he was very dissatisfied with the performance of his cavalry. They were amateurs; the officers were ignorant of their duties, the standard of horsemanship were low; the ability of regiments to manoeuvre on the field poor; the cost of maintaining the extremely large mounts of the Kuerassier regiments was very high. Frederick felt that he was not getting much of a bang for his bucks where his cavalry was concerned.

He instituted a commission, under his leadership, to bring about the much-needed reforms.

The height of future cavalry remounts was reduced to 15 hands* for Kuerassiers; Prussian studs or those in Holstein were the preferred sources. For the dragoon and hussar regiments, horses as small as 14 hands* were to be bought; many of these were traditional Tartar horses from old Poland.

The height limits for men in the Kuerassier regiments were also reduced.

Mounted and dismounted drill sessions for cavalry regiments were increased, the standards of care of the horses much improved. None of this was revolutionary, but the fact that the king was forcing the programme through gripped the imagination of all good cavalry officers, NCOs and men; they set to work with a will.

Frederick also reviewed the cavalry tactical manual and weaponry with an aim of improving the shock action of the units on the battlefield.

Hitherto, most cavalry trotted up close to the enemy line and discharged their carbines or pistols into them and then trotted back. The sword was rarely used; 'shock action' - entire squadrons or regiments crashing as one into the enemy – was unknown.

The pace of the charge was now to be increased from a trot, through a canter and into a gallop at 200 paces from the enemy line. Cavalry would no longer use their firearms as the weapon of choice and the curved sabres of the Kuerassiers and dragoons were replaced by straight-bladed, double-edged swords, designed to inflict serious injury.

Artillery Tactics

The smoothbore, muzzle-loading field artillery went through the same process as the musket. The field pieces became slightly lighter, more standardised and more mobile, but in principle, the improvements in guns and their projectiles developed were marginal for many years.

At the time of the First Silesian War the artillery of the Austrian army was generally considered to be the best in Europe. It had been brought to this level of efficiency under the leadership of General of Cavalry Prince Josef Wenzel Liechtenstein. By the time of the Seven Years War (1756-1763) its reputation was so admired that the French sent General Gribeauval to Vienna from 1757-1762 to study it.

* A hand = 4 inches

When he returned to France, Gribeauval set to work to design his integrated, modular system which was adopted by the French army in 1765 and was to become so famous under Napoleon forty years on.

Frederick also studied the Austrian artillery system and used some of its principles for the improvement of his own.

There were to be two innovations in artillery techniques in the Napoleonic era, both British. The first was the introduction of Colonel Shrapnel's new explosive shell, filled with musket balls; the other the use of Congreve's rockets.

Until 1807 Prussian infantry and cavalry regiments were known by the names of their colonels-in-chief (or *Chefs*) rather than by their numbers. These names are shown in the combat histories. Following the destruction of the Prussian army by Napoleon's *Grande Armée* in the campaigns of 1806 / 1807, the newly-formed army regiments became known by their numbers and titles, less than by the names of the Chefs, although these worthies continued to be appointed into the First World War. Indeed, the British army maintains the custom today. The regiments are listed below by the numbers in use up to 1807. Regiments formed after 1807 are shown after the old regiments.

The combat history of each regiment shows the campaigns for which they were mobilised followed by the dates and names of the individual actions in which they took part. Clothing and appointments were decorated with the Prussian eagle or the crowned royal cipher. This was 'FWR' for Friedrich Wilhelm I, the 'Soldier King' (1713-1740); Frederick the Great (1740-1768) used 'FR'; his successor, Friedrich Wilhelm II (1786-1797) used 'FWR'as did his son, Friedrich Wilhelm III (1797-1840).

In time of peace, the regiments were organized into geographical inspectorates, subject to critical periodic scrutiny by inspector generals of infantry, cavalry and artillery, to ensure that they were being administered effectively. These inspectorates and garrisons changed with the passage of time; they are shown, as are the towns in which the units were garrisoned, as valid in 1806. In the case of the garrison regiments, which were disbanded in 1788 to form the fusilier battalions, they are valid for that year. Particularly in modern-day Poland, Russia and other eastern areas once under Prussian rule, the names of these garrison towns have been changed out of all recognition.

Uniforms. The first signs of a generally applied uniform policy in the Prussian army, initially mentioned in the reign of the Elector Georg Wilhelm of Brandenburg in 1632, was probably in 1670. The infantry coats were initially red or dark blue; in 1691 dark blue was decreed for all units. Regiments were distinguished by the colour of their 'facings', that is to say, the cuffs, collars, and the shoulder strap on the left shoulder and by their buttons, which were of brass or tin. Further distinction was often achieved by the colour of the waistcoats and breeches (known as the 'small clothes'); white, buff and yellow being the colours mostly selected. The coat turnbacks were often the same colour for all regiments, as were the belts. The latter were usually white for regiments of the line and black for light infantry, fusiliers and Jaegers.

Indeed, since beginning to investigate this topic in earnest, it is surprising, that as late as 1808, there was a very high degree of variation in uniform detail, often within a regiment itself. This frequently extended to what badges were worn to distinguish NCOs. Some regiments had them wear lace edgings to the cuffs, others omitting these and giving them gold or silver buttonholes instead. All regiments retained the gold or silver hat edging and the black and white pompons, fist straps and plumes. Some regiments of the line were further distinguished by various patterns and colours of lace loops to the buttonholes. For officers these were silver or gold according to the button colour. The patterns of these laces varied quite often. Until 1808 there were varying numbers of buttons on the lapels according to regiment; some wore six on each lapel, some eight, some ten. The number of buttons on the cuff also varied according to the style. Swedish cuffs had two buttons, Brandenburg and Prussian cuffs usually three. The 'Prussian' cuff was a cross between the Brandenburg and the Swedish style, it had a cuff flap, only the top two buttons of which were visible; the third button was inside the turned up cuff. There were two buttons under the right lapel, two in the small of the back, usually three under each hip pocket flap and often one at the junction of the turnbacks. The lace loops described in the regimental sections relate only to that worn by privates. Drummers, musicians, NCOs and officers of each regiment wore their own patterns.

Within each foot regiment (1 through 28) there were two types of soldiers: the musketeers (the main part of the regiment) and the grenadiers, who formed the elite and were usually the tallest men. Originally, the grenadiers carried grenades, often of glass, which they threw at the enemy. This very close range, dangerous work, and the uncertain quality of the fuses, helps to explain the elite status. As it was impossible to sling their muskets and throw grenades effectively without dislodging the large cocked hats, grenadiers adopted the nightcap, or fatigue cap, as their headwear and over the years these developed into imposing works of art, with metal front plates in the regimental button colour, of the Napoleonic era. For Catholic states, the front plate was fur-trimmed; Protestant states tended not to have fur on their grenadier caps. By the end of the War of the Spanish Succession (1713), the use of hand grenades in the field had largely fallen into disuse, but grenadiers retained the badges of their trade as marks of honour. These included the mitre cap, the metal fuse case on the bandolier, grenade badges on the cartridge pouch and on the hat and the carrying of a sabre.

The Prussian regiments 29 through 49 were raised as Fuesilier regiments and usually recruited smaller men than the regiments of foot; they also used shorter flintlocks. Instead of the musketier tricorns, the normal soldiers wore metal-fronted mitre caps up to 1787, or until they were converted to musketeers as was sometimes the case. These were similar to the grenadier caps, but the cloth headpiece did not run up to the top of the plate and there was no coloured tassle on the top. Instead, the headpiece was low, rounded and topped off with a metal spike in the shape of the upper half of a flaming grenade. The grenadiers in these fusilier regiments wore their usual grenadier caps as for other regiments; the metal plates were identical within a regiment. These sartorial differences were the only things differentiating the fusiliers of this period from the musketiers of the foot regiments. They were not trained for skirmishing

The fusilier battalions raised in 1787 however, were trained for skirmishing and wore black belts as opposed to the white of the foot regiments. From 1782 special muskets had been produced for sharpshooters and these weapons were 11 centimeters shorter than those of the musketiers. The butt was also set at a sharper angle than that of the 'Kuhfuss' (cow's foot, as the musket was known) but experts considered that the angle should have been sharper still, to permit accurately aimed fire to be given.

The bicorns and mitre caps were then replaced by the Kaskett in 1787. The front of the Kaskett was decorated with metal badges in the button colour; a flaming grenade for grenadiers, the cipher 'FWR' for musketiers. The twenty-four fusilier battalions raised at the end of the XVIII Century had the Prussian eagle badge on their Kasketts. In 1799 the Kaskett was superseded by the old-style bicorn for musketiers and fusiliers, and a new model cap for grenadiers. This new grenadier cap had a black leather front plate and peak. The peak was edged in the brass, the front plate bore a brass bottom band with the Prussian eagle in black and over this was a brass flaming grenade. The edges of the front plate were trimmed with a worsted roll, a white plume was worn on the left hand side. The head band of the cap was covered in cloth of the facing colour, edged in the button colour. The back of the leather front plate was painted in the facing colour; light blue for those regiments with white facings. Grenadier officers wore no items to distinguish them from musketier officers.

The national cockade, worn on the bicorn, was black within white.

The fusilier battalions adopted the shako in the years 1802 – 1803. This was of black felt, with a black leather peak. Around the top was a white band with a regimental pompon; beneath this was the metal Prussian eagle badge in the button colour.

Until about 1806 it was usual for the knee-breeches to be worn with gaiters, white in summer, black in winter. After this point, white or grey, buttoned overall trousers came into popularity.

In the 1790s white waistcoats and breeches were introduced for all regiments to replace the buff ones previously worn by some regiments. The 3rd, or depot, battalions of the regiments, formed in 1787 by disbanding the garrison regiments, wore the uniform of their regiment but without a plate on the cartridge pouch. They also carried no sabres.

Drummers and musicians of the Prussian army in the late XVIII Century were distinguished by a special braid decoration to sleeves, cuffs, lapels and buttonholes. This lace varied from regiment to regiment in design and was also worn on the 'swallows nests' at the top of each sleeve. Drummers and musicians carried no firearms.

By Order of 3 March 1787 each musketier company was to have the ten best shots classified as sharpshooters and armed with rifled muskets. They were trained in patrolling and skirmishing.

It was the custom for NCOs and officers to carry canes as a badge of office and spontoons or halberds instead of muskets. Their status was also shown by special black and white/silver straps and tassels worn on their sword hilts, and by silver or gold braid edging to collars and cuffs. NCOs wore white plumes with black tips, officers' plumes had black bases.

Officers' coats were longer than those of the enlisted men, and their skirts reached to the knee.

As further badges of office officers wore gorgets at their throats and silver and black silk waist sashes when on duty. Until 1808 there were no particular badges of rank for Prussian officers.

The facings and button colours shown in the following table were worn from 1756 onwards.

The only infantry regiments to survive the catastrophe of 1806 intact were: von Ruechel Nr 2, von Ruets Nr 8, von Schoening Nr 11, von Besser Nr 14, von Diericke Nr 16, von Ploetz Nr 42 (minus its 3rd Musketier Battalion), von Hamberger Nr 52, and de Courbiere Nr 58.

The 3rd Musketier Battalions of the following regiments also survived: von Kalkreuth Nr 4, von Owstein Nr 7, von Treskow Nr 17, vacant von Borcke Nr 30, von Kropff Nr 31 (the grenadiers also survived), von Alvensleben Nr 33, von Thile Nr 46 (the grenadiers also survived), von Grawert Nr 47, von Sanitz Nr 50, von Kauffberg Nr 51, Jung-Larisch Nr 53, von Natzmer Nr 54, von Manstein Nr 55 and the Feld-Bataillon von Chlebowsky Nr 60.

Due to the pursuit of economy in the war ministry, it often occurred that newly-raised regiments would be presented with colours, standards or guidons from stocks of items handed in in the past by other regiments. Thus it happened that the colour of these items did not match the facings of the new unit. It also meant that old pattern items were carried along with current items.

Of the Fusilier battalions the following (and their depots) survived largely intact: Wakenitz Nr 3, Rembow Nr 6, Bergen Nr 11, Stutterheim Nr 21, Schachtmeyer Nr 23 and Buelow Nr 24.

The battalions Pelet Nr 14 and Ruehle Nr 15, as well as the depots of battalions Keyserlingk Nr 1, Wedel Nr 5, Sobbe Nr 18, Ernest Nr 19 and Invernois Nr 20 also remained. Remnants of battalions Borel du Vernay Nr 9, Knorr Nr 12 and Hinrichs Nr 17, plus their depots were available as were nine companies of Jaegers.

During the war the following new formations, reserve battalions, had been raised from escaped POWs, recruits and volunteers: six in East Prussia, six in West Prussia, three each in Pommerania and the New Mark, one in Silesia. Several *Freikorps* ('private armies', outside the state's military establishment) were also raised from volunteers in besieged cities such as Danzig.

Of the cavalry, the following had survived: Kuerassier regiments Nr 4 and 13, and remnants of the others. In the dragoons, the regiments Nrs

3, 5, 6, 8, with remnants and the depots of several others. Of the hussars, two regiments and their depots (Nrs 4 and 5) were left intact, and the Bluecher hussars Nr 8, although having capitulated at Ratkau, re-formed itself when all ranks escaped and made their way to East Prussia. By January 1807, one Kuerassier, two dragoon and three hussar brigades had been formed.

Each infantry regiment now had a depot company, two grenadier companies, two musketier battalions and a fusilier (or light) battalion each of four companies. The fusilier battalions were now combined into the new regiments as their light battalions.

In 1808 new uniform regulations, as shown below, were issued for the newly formed regiments.

Musketier Privates. Black felt shako with white top band and black-within-white pompon; black leather chinstrap. The front was decorated with the uncrowned royal cipher: 'FWR' in brass. For parades, white cords and thin, black plumes were worn. The double-breasted dark blue coat had two rows each of eight brass buttons, dark blue cuff flaps with three buttons and two buttons in the small of the back. The bottom cuff flap button was always worn open. The turnbacks were red and plain. Collars and cuffs were in the colour of the province in which the regiment was raised. These were: East Prussia – brick red, West Prussia – crimson, Pommerania – white, Brandenburg – poppy red, Silesia - white. The Colberg Regiment had white facings. The shoulder straps of the first regiment in each province were white, red for the second, yellow for the third and light blue for the fourth. The buttons on the shoulder straps bore the company number in Arabic numerals. Breeches were grey, the short, black gaiters had black leather buttons. Belts were white with brass buckles, the cartridge pouch had a round brass plate with the eagle on it. The brass-hilted sabre was borne in a brown leather sheath on the waist belt. On campaign the grey greatcoat – if not worn – was carried rolled and strapped to the top of the brown calfskin pack.

The sabre strap and tassel were white for all companies; the 1st Battalion also had white crowns and bodies to the tassel. Company identity was shown by the colour of the wreath: 1st - white; 2nd - yellow; 3rd – light blue; 4th - red.

In the 2nd Battalion the crown and wreath were in the company colours as shown below: 1st - green; 2nd - yellow; 3rd – light blue; 4th - red.

The fusilier battalion had the same colours as the 2nd Battalion, but the body of the tassel was also on in the company colour.

Musketier NCOs. As for privates but with gold top band to shako, gold braid to the top of the cuffs and to the front and bottom of the collar, white gloves and a black and white sabre strap and tassel. Their black plumes had a white tip.

Musketier Drummers and Musicians. Swallows nests in the provincial facing colour with white lace. It is thought that Silesian and Pommeranian regiments wore poppy red swallows nests. Drums were brass, with red and white hoops and black sticks. On the body of the drum was the crowned royal cipher within trophies of arms.

Musketier Officers. Company officers wore the shako in finer construction and materials. At each side was a small Prussian eagle in gilt metal. Gold chains ran from these side eagles to the frontal pompon and to the top of the back of the shako. Their cords were silver and black as were their waist sashes.

The pompon was black and silver, and all officers wore the black and silver cockade with gold loop and button. Their coat tails reached to the back of the knee, and there were horizontal, false pocket flaps on the hips, each with two buttons. They carried swords in brown sheaths on white slings with gilt hilts and silver and black sword strap and tassel. They wore boots. In October 1808, rank badges were introduced for officers and were worn on the shoulder straps, which were in the facing colour and edged in red. A lieutenant had a single black and silver lace along the centre of the strap; a captain one at either side, and a major's strap was edged all the way around with the same lace.

Fusiliers. All details as for the musketiers except as noted below. Instead of the 'FWR' badge they wore a black-within-white cockade, with regimental button and loop. For NCOs this loop was of brass.

Belts were black and worn crossed. Those armed with rifles had sword-bayonets.

Grenadiers. All details as for the musketiers except as noted below. The shako had a brass Prussian eagle plate; the shoulder strap buttons bore the company number in Roman numerals. They wore black and white sabre strap tassels.

Colours and Standards

In the early days of the Brandenburg-Prussian army each company and squadron carried a colour or standard respectively. Those of the Brandenburg regiments bore the red eagle of that province or the sceptre. Prussian regiments bore the black eagle of Hohenzollern. In the reign of king Friedrich I the red eagle was phased out, and the black eagle was now accompanied by the mottos NEC SOLI CEDIT until Frederich the Great replaced this with PRO GLORIA ET PATRIA. There was a wide variety of colour schemes among the regiments of the army. There seems to have been no relationship between the facings worn by a regiment and the colours they carried. In each corner was the royal cipher (now 'FWR') within golden laurel and palm wreaths. The king's colour was white with a round central disc, containg the eagle (usually black) under a scroll, all under a crown and enclosed in laurel branches. The gilt finials were in the form of spearpoints, pierced to show the royal cipher. The cravats and tassels were black and silver. If there was some uniformity in the infantry, the cavalry standards were a riot of colour.

In 1787 King Friedrich Wilhelm II reduced the number of colours carried by each infantry regiment to one king's and three regimental. The 1st Battalion bore the king's colour and one regimental, the 2nd Battalion carried two regimental colours. The colour borne in the front rank of each battalion was known as the 'Avancierfahne', that in the third rank was the 'Retirierfahne'.

Infantry colours were painted on silk and required frequent renewal.

The colours presented as of 1813 were standardised for all regiments; the king's colour was white with black centre, gold eagle, white scroll edged gold, gold embroidery. Regimental colours black, white centre, black eagle, light blue scroll edged gold, gold embroidery.

Details of the colours carried by the regiments raised after the disaster of 1806 will be found in the sections relating to those units. Basically, they were issued with some new items, but mainly they received colours randomly chosen from stocks of old items still available in the various fortresses in Prussia.

2

The Infantry Regimental Histories

1. Regiment zu Fuss.
Raised in 1619. Title in 1806 – **Graf Kunheim.**
Berlin Inspectorate. Garrisons: Berlin, Straussberg.

Uniforms. Poppy red facings, Prussian cuffs, lapels and turnbacks, white buttons; on the lapels were six buttons in pairs. Red stock. In 1806 officers had embroidered silver lace loops, two under each lapel, two on each pocket flap and four in the small of the back. Officers' lapels, cuffs and cuff flaps were edged in silver lace as were their hats. NCOs had plain silver lace edging to lapels, cuffs and cuff flaps. Privates wore plain white lace loops, six on and two under each lapel, two on each cuff flap and two in the small of the back. White waistcoat and breeches. Grenadier cap plates up to 1787 were white metal; the cloth backing was dark blue, the pompon and lace white.

Colours. On the King's colour the cloth was white, the central disc was orange, the eagle black, silver decoration. Regimental colours: white centre, black eagle, orange field, silver decoration. The four colours were lost in the capitulation of Ratkau on 7 Nov 1806. White pike staffs.

Campaign History. Raised in 1619 by the Elector (Kuerfuerst) Georg Wilhelm of Brandenburg and taken into Prussia service with him when he came to Prussia in 1627. Commander in 1619 was Captain von Calchum; next year Captain von Burgsdorf took over. In 1652 Captain

Pierre de la Cave became the commander. In 1655 there were five companies with 2,000 men, based on the palace guards, the 'Maerkischen' (from Brandenburg) and the 'Preussischen' (from Prussia). **1656** In Poland; B Warsaw. In 1657 Oberst von Poelnitz was commander. **1663** the capture of Magdeburg. In 1676 it gave men to raise IR Nr 7. **1677** S and capture of Stettin. 1679 Commander Oberst von Wrangel. By 1684 it absorbed IR Nr 5 and had 30 companies. 1684 Commander was GL von Schoening. In 1685 it gave men to help raise IR Nr 12. **1686** In Hungary against the Turks; S Ofen. **1689/1690** Against France on the Rhine. **1692** two battalions in the Netherlands, one on the Rhein and one in Hungary. 1691 Commander was GFM von Flemming. By 1692 it consisted of four battalions of Brandenburgers and two of Prussians, and fought in B Steenkerken. In 1693 the commander was GFM von Barfuss; the regiment was in action at Neerwinden. In 1698 one battalion was detached to form IR Nr 18. **1704** Against France in the War of the Spanish Succession on the Danube. In 1705 there was one battalion of 'Grenadier-Leibgarde' and one of 'Fuesilier-Leibgarde'. In 1708 one battalion was in the Netherlands at the capture of Lille. Next year it was with Prince Eugen's army with Marlborough; capture of Dornick, B Malplaquet, capture of Mons. **1710** S Bethune, Aire. **1713** Kuerfuerst Friedrich Wilhelm took companies to form IRs Nrs 23 and 24 and the regiment lost its guard status; first Chef was FM Alexander Hermann Graf von Wartensleben. **1715** Against Sweden; S Stralsund. **1723** Chef GFM Kaspar Otto von Glasenapp. **1740-1742** The First Silesian War against Austria. **1740** Grenadiers in Gren Bn Kleist (1/25). 29 Dec-9 Mar 1741 S and storm of Glogau. **1741** 7 Apr C Lassoth. 10 Apr B Mollwitz. 10 Aug capture of Breslau. 23 Aug Sk Schoenheide. 8 Sep C Nimptsch. **1742:** Col Johann Christoph Friedrich Graf von Hakke. Grenadiers in Gren Bn Kleist (1/22). 10-14 Mar C Goeding-Ungarisch Brod. Apr – May in Upper Silesia. **1744 – 1745** Against Austria. **1744** Grenadiers in Gren Bn Tauentzien (1/23). 2-18 Sep S Prague. 6 Sep C Beraun. 19 Sep-4 Oct the advance over Moldauthein. 9-19 Oct withdrawal to Beneschau. 26 Oct-9 Nov withdrawal to Neu-Kolin. 20 Nov-8 Dec withdrawal to Silesia. **1745** 4 Jun B Hohefriedberg; 17 Jun capture of Neustadt, Bohemia. 4-12 Sep D Neustadt. 23 Nov C Katholisch-Hennersdorf. **1756 – 1763** The Seven Years War; against Austria and other states. **1756** Chef GL Hans Karl von Winterfeldt (mortally wounded at Moys on 7 Sep 1757) the regiment was at Pirna; no action. Grenadiers in Gren Bn Baendemer (1/23). **1757** 6 May B Prague (distinguished), thence blockade of the city until 20 Jun; 27 Jul C Aussig; 24 Aug C Kottau; 29 Oct C Markranstaedt; 5 Nov B Rossbach; 5 Dec B Leuthen; 6 Dec cannonade at Klein-Mochbern; 22 Dec C Landshut. **1758** GM Johann Sigismund von Lattorf. 3 May-2 Jul at Olmuetz; 25 Aug B Zorndorf; 14 Oct B Hochkirch (distinguished). **1759** Sep-Nov on campaign in Saxony. Grenadiers in Gren Bn Rathenow (1/23). **1760** Chef GM Karl Christoph von Zeuner. 13-22 Jul failed S Dresden; 15 Aug B Liegnitz ; 17 Sep C Hochgiersdorf; 2 Nov Sk Schoena; 3 Nov B Torgau. **1762** Grenadiers in Gren Bn Poseck (1/23). 9 Mar raid on Gretha; together with IR Nr 23 they were attacked by over 4,000 Austrians, but beat them off. 12 May C Doebeln; 13 May C Arnsdorf; 15 May C Frauenstein; 14/15 Oct C Brand; 29 Oct B Freiberg. 1768: Chef Oberst Erich Julius von Koschenbahr. 1776: Chef Oberst Christoph Friedrich von Bandemer. 1778: Chef GM Ludwig Gottlieb von Kalckreuth. 1778: Chef GM Hans Erich von Bornstedt. The regiment was known as 'Alt-Bornstedt' to distinguish it from IR Nr 20, 'Jung-Bornstedt'. **1778-1779** Against Austria in the War of the Bavarian Succession; Grenadiers in Gren Bn Graf Schlieben (1/13). Employed in raids into Bohemia; no major actions of note. 1792: Chef Oberst Johann Ernst Graf von Kunheim; he was created Graf in 1798 and promoted GL in 1806. **1793-1794:** Against France. **1792:** 23 Mar – 22 Jul S Mainz; 28 – 30 Nov B Kaiserslautern. **1794:** 13 Jul B Trippstadt. Also present at Mainspitz, Buergerinsel, Gustavsberg, Kettricherhof, Pirmasens, St Ingbert, Landau, Bitsch, Frankenthal, Hochspeyer, Mosalbertal. **1806**: 14 Oct B Auerstaedt. Parts of the regiment capitulated at Erfurt, 16 Oct, Prenzlau 28 Oct and Magdeburg. The remnants of the regiment were captured on 6 Nov, as part of Bluecher's corps at Schwartau, near Luebeck and not re-raised.

2. Regiment zu Fuss.

Raised on 20 Dec 1655. Title in 1806 – **von Ruechel.**
East Prussian Inspectorate. Garrisons: Koenigsberg in Prussia, Pillau.

Uniforms. Light brick red facings, yellow buttons, red stock, Prussian cuffs; in 1806 they wore 6 crimson lace loops with white tassels; two under each lapel and two in the small of the back. NCOs wore gold laces with tassels instead of crimson. Officers' lapels and cuffs were edged in gold lace as were their hats. Buff small clothes. Brass grenadier front plate and ornaments, up to 1787, light brick red headband, yellow cloth backing, white tape, red within black pompon.

Colours. King's colour white with black centre, gold eagle, white scroll edged gold, gold decoration. Regimental colours black, white centre, black eagle, light blue scroll edged gold, gold decoration. White pike staffs.

Campaigns. As with IR Nr 1, the original unit was raised by Kurfuerst Georg Wilhelm of Brandenburg and brought by him into Prussian service. 1655 General Wachtmeister Bogislav von Schwerin, raised at 8 companies from new recruits in Colberg; 1656 Chef Fuerst von Radziwill; fought in B Warsaw in Poland. 1659 reduced to 6 companies; 1660 reduced to 3 companies, but took in 3 companies of the disbanded IR von Hundbeck; 1663 reduced to 4 companies. In 1669 it was split into two battalions; Gen-Lt Graf Friedrich von Doehnhoff was Chef of one of them, Oberst von Flemming Chef of the other. Flemming went into Saxon service when the two battalions were reunited in 1696.

1674 against France in Alsace 8 Aug storm of Wasselnheim; 31 Dec Tuerckheim. **1675-1679 against Sweden in Pommerania and Prussia** Under Kurfuerst Friedrich Wilhelm the Great. 15 Jun 1675 Rathenow; 27 Oct-10 Nov S of Wolgast; Jun-Aug **1676** S Anklam and capture of Loecknitz; **1677** 27 Jun-26 Dec S Stettin; **1678** 24 Nov-1 Dec Def Memel; **1679** 19 Jan pursuit over the Kurische Haff (2 coys); 1679 reduced to 4 companies. **1680** 80 men acting as marines captured the Spanish frigate *Carl II* in Ostend; **1683** 2 coys in the war against the Turks; 10 Nov storm of Szeczeny. **1684 against the Turks 5 Oct** Kamenjec-Podolsk, thence withdrawal until 11 Oct; 30 Dec 1685 split into two battalions. **1686 against the Turks in Hungary** (2nd Bn) 24 Jun-2 Sep S Ofen. **1689-1697 against France. 1689** On the Rhine. 12 Mar Uerdingen (1st Bn); 24-27 Jun S of Kaiserswerth (2nd Bn); 25 Aug-12 Oct S Bonn; 11 Oct storm of Bonn; then to Hungary against the Turks. **1692** Oct (1st Bn) relief of Charleroi; 16-28 Sep (1st Bn) S Huy; 24 Sep storm of Huy; **1695** Jul-1 Sep S Namur; 30 Aug (1st Bn) storm of Namur. **1691-1692 against the Turks in Hungary.** (2nd Bn) 19 Aug Szalankemen; 11 Oct-17 Nov S Gross-Wardein; 17 Oct storm of Gross-Wardein. **1693-1697 against the Turks in Hungary.** (2nd Bn) **1693** 9 Aug-12 Sep S Belgrade; **1694** 11 Sep-2 Oct D camp of Peterwardein; **1696** Aug many lost as POW at Titul and Sabila; **1697** Chef Oberst Graf Otto Magnus von Doehnhoff. 11 Sep B Zenta; **1698** 14 Oct-10 Nov S Elbing (2nd Bn). **1702-1714 The War of the Spanish Succession against France. 1702** On the Rhine. All actions 2nd Bn. S Kaiserswerth (the 2 companies which were given to 2nd Bn, IR von Sydow in 1703). **1705** 28 Sep-5Oct S Hagenau; 7

May 1707 grenadiers given to IR Markgraf Philipp; **1708** In the Netherlands. 10 Jul Oudenaarde; 14 Aug-23 Oct S Lille; 28 Aug Wynendael; 18 Dec-2 **1709** S Ghent; 27 Jun-3 Sep S Tournay; 11 Sep Malplaquet; **1710** 3 May-26 Jun S Douay; 10 Jun Liege; **1711** 21 Aug-13 Sep S Bouchain. **1705** The regiment was in Italy. 19 May Soseto; 1 Jun on the Naviglio; 12 Jul Soncino; 16 Aug Cassano; **1706** 19 Mar Calcinato; 7 Sep Turin; Sep capture of Novara and Milan; 6 Oct Ghiera. **1708** In the Netherlands; S Ryssel. Jan 1713 the regiment was now known as Alt-Doehnhoff. **1715 against Sweden.** 14 Jul-22 Sep S Stralsund; 16 Nov Stresow. 1718 Chef GM Erhardt Ernst von Roeder**, 1733-1735 against France.** On the Rhine. 1 May 1735 2 grenadier companies raised; Grenadiers – 10 Oct 1735 Speyer. **1742 against Austria.** The grenadiers were detached to the Gren Bn Pfuhl (2/4). Regiment - 17 May Chotusitz. 1743 GL Samuel von Schlichting. **1744-1745 against Austria.** The grenadiers were detached in the Gren Bns Gaudy, then Tresckow, then St Surin, then Kleist von Wuerttemberg. **1744** The main body of the regiment in Bohemia; B Chotusitz. Grenadiers in Gren Bn Pfuhl (2/43). 29 Aug-17 Sep S Prague. **1745** 14 Feb C Habelschwerdt; suffered heavy losses. 4 Jun B Hohenfriedberg; 30 Sep Soor. 1750 Chef GM Hans Wilhelm von Kanitz. **1756-1763 Seven Years War against Austria and others.** The grenadiers were detached to the Gren Bns Manstein, then Nesse, then Natalis. The main body of the regiment fought against the Russians: **1757** Grenadiers in Gren Bn Manstein (2/Grsn R 2). The main body of the regiment: 26 Aug Plibischken; 30 Aug Gross-Jaegersdorf; 30 Sep Tilsit; 29-31 Dec B Demmin; **1758** 15 May-18 Jun B Stralsund; 25 Aug Zorndorf; 27 Oct (2nd Bn) Greiffenberg. **1759** 11-18 Jan S Demmin; 1-14 Apr (2nd Bn) S Peenemuende; 4 Jul Neustadt; 23 Jul B Kay; 12 Aug B Kunersdorf; after the battle, the remaining grenadiers were combined into the Gren Bn Nesse (2/Grsn R 2 and Grsn Rs 1/11). 4 Dec C Maxen; one battalion captured.**1760** 7/8 Oct Mariendorf and Schoeneberg; 3 Nov Torgau; 4 Jun-15 Nov D Colberg camp; 25 Nov Petershagen; 8 Dec Treptow. **1762** Against the Austrians. 21 May (grenadiers) Gehringswalde; 2 Aug (grenadiers) Brix; 16 Aug Main body B Reichenbach; 1 Sep-29 Oct S Schweidnetz; 29 Oct (grenadiers) Freiberg. 1769 Chef GL Joachim Friedrich von Stutterheim. **1778-1779 against Austra.** The grenadiers were in the Gren Bns Collrepp then Rauter (2/16). 11 Sep 1778 (2 coys) Leopold. 1783 Chef GL Heinrich Wilhelm von Anhalt; 1786 Chef GM Victor Amadeus Graf Henckell von Donnersmark; 1 Jun 1787 reorganized into 1 grenadier and 2 musketier battalions each of 4 companies;

Oct 1787 depot battalion raised from men of the Garrison R von Bose (*1713); 1793 Chef GL Wilhelm Magnus von Bruenneck; 1796 3rd Bn raised; 1 Jun 1799 reorganized as 2 Gren Coys, 3 Musketier Bns, each of 4 coys; 1805 Chef GL Ernst Friedrich Wilhelm von Ruechel. **1806 against France.** In l'Estocq's Reserve Corps. The grenadiers were in Gren Bn Schlieffen. 23 Dec (parts of the regiment) Biezun; 25 Dec Soldau. **1807** 8 Feb Preussisch-Eylau; 18 Feb (a detachment) Bartenstein; 13 Jun (1st Bn) Gollau; 14 Jun (2nd Bn and Grens) Koenigsberg. 20 Nov 1807 reorganized as 2 Gren Coys, 2 Musk Bns each of 4 coys; 1Jan 1808 received the Fus Bn von Bergen as Lt Bn; 1 Dec 1809 renamed Fus Bn; 7 Sep 1808 **Inf R Nr 1, 1. Ostpreuss IR**, 19 Nov 1809 **1. Ostpreuss IR GM von Stutterheim**, Jan 1810 name of the Chef dropped from the title. **12 Jan 1812 2nd and Fus Bns sent on campaign in Mobile IR Nr 1 in von Yorck's corps against Russia.** 5 Aug (9th Coy) Schlock, and St Annen (12th Coy); 7 Aug (12th Coy) Wollgund; 22 Aug (Fus Bn) Schlock; 27 Sep (2nd Bn) Dahlenkirchen; 29/30 Sep Graventhal and Eckau; 3 Dec (Fus Bn) Dahlbingen. 6 Mar 1813 regiment re-united; in 1813 the drill depot (* 25 Jun 1811) becames 3rd Mus Bn; 27 Dec 1812 captured; reraised as Res Bn Nr 1; 2 Mar 1813 3rd Mus Bn; 1 Jul 1813 1st Bn, 1st Res IR (see IR Nr 13); 7 Jun 1813 received the 3rd Res Bn, 3rd Ostpreuss IR as 1st Res Bn (see there and IR Nr 24); 7 Jun 1813 received the 4th Res Bn, 1st Schles IR as 2nd Res Bn (see there and IR Nr 18); 14 Oct 1814 Gren coys given to form Kaiser Alexander Gren R; 5 Nov 1816 **1. IR (1. Ostpreuss.)**.

1813 against France. St Dizier 30 Jan; took a gun. Dannigkow (Moeckern) 5 Apr; Merseburg 29 Apr; Gross-Goerschen. Following this action, the 1st and 2nd Bns were combined into the Fuesilier-Bn of the 2. Ostpreussisches IR. Koenigswartha 19 May; Bautzen 20/21 May; Loewenberg 21 Aug; Katzbach 26 Aug; Moeckern 16 Oct ; losses were so heavy in this battle, that no officers were left who were fit for duty; FW Martins took command of the regiment.

1814 against France. Chateau-Thierry 12 Feb; Laon 9 Mar; the regiment took 5 guns and 25 ammunition wagons. Paris 30 Mar; The regiment took 2 guns.

On 14 Aug 1814 the two grenadier companies went into the '**Kaiser Alexander Garde-Grenadier-Regiment' Nr 1.**

1815 against France. The regiment was mobilised but saw no action.

3. Regiment zu Fuss.
Raised 1665. Title in 1806 – **von Renouard.**
Magdeburg Inspectorate. Garrisons: Halle, Quedlinburg (grenadiers).

Uniforms. No lapels; poppy red facings, yellow buttons, red stock, Prussian cuffs; prior to 1787 no lace decoration, thence 6 black and white lace loops with tassels (two under where each lapel would have been and two in the small of the back); white small clothes. Officers had 12 buttons where each lapel would have been and a narrow gold lace edging to the hat. Soldiers had only seven buttons on each lapel. Tin grenadier front plate up to 1787, with brass cartouche and crown and black eagle. Brass side ornaments, tin grenade at the rear. White headband and cloth backing, yellow tape, mixed black and white pompon.

Colours. King's colour white with yellow centre, black eagle, white scroll edged silvel, silver decoration. Regimental colours yellow, white centre, black eagle, yellow scroll edged silver, silver decoration. Brown pike staffs.

All four colours were lost at the capitulation of Magdeburg on 11 Nov 1806.

Campaign history. Raised in 1665 by Oberst Johann von Faergel at three battalions from new recruits. **1672** On the Rhine, against France. **1674** In Alsace C Tuerkheim. **1675** In Pommerania against Sweden; took part in the capture of Wolgast, Usedom, Garz and Triebsee. **1676** Again in Pommerania; at the capture of Anklam, Demmin and Wollin. **1677** S and capture of Stettin. **1678** capture of Ruegen Island and S Stralsund. **1679** Chef GFM Fuerst Johann Georg von Anhalt-Dessau. **1686** In Hungary against the Turks; S and capture of Ofen. **1688** One battalion went into Dutch service; the others were on the Rhine at the battle of Uerdingen and the capture of Neuss, Rheinbergen, Kaiserswerth and Bonn from the French. **1690** The battalion in Dutch service was at the clashes at Fleury, Brussels, Loewen and Namur. **1692** The 'Dutch' battalion was in action on the left flank in the clash at Steenkerken and suffered heavy loss. **1693** Chef Oberst Fuerst Leopold von Anhalt-Dessau, son of Johann Georg 'Der Alte Dessauer'. The regiment was in the B Neerwinden 19 Jul. **1694** the battalion in Dutch service was in action at the defence of Liege and Mastricht and at the capture of Huy. **1695** The 'Dutch' battalion was in action at S and storm of Namur. In the war of the Spanish Succession (1701 1714) it supported the capture of Campen, Kaiserswerth, Liege, Roermond and Venlo and the siege of Rheinbergen. **1703** The regiment was at the blockade of Geldern, the storm of Bonn; then went to the Danube under command of the Imperial General Styrum at C Hoechstaedt. **1704** under Prinz Eugen in B Hoechsteadt; capture of Landau. **1705** The regiment was in Italy B Cassano. **1706** C Calcinato; capture of Reggio; relief of Turin; capture of Milan, Novarra, the Ghiera pass and Pizzighettone. **1707** At Toulon and Susa. **1708** Part of the regiment was in the Papal states, the others in the Delphinate, at the capture of Crilles and Fenestrelles. **1715** The regiment marched back to Pommerania and S Stralsund. **1719** 3rd Bn raised. **1740/1741.** 29 Dec-9 Mar (Grenadiers in Gren Bn Bolstern (3/27)) S Glogau and storm on 9 Mar; 10 Apr B Mollwitz; 10 Aug capture of Breslau; 23 Aug Schoenheide. **1742.** Mar/Apr in Hungary; Hradisch. 10 Apr C Kremsier; Apr/May (Grenadiers only, in Gren Bn Bolstern (3/21)) on frontier duty at Jaegerndorf. **1744.** 2-18 Sep (Grenadiers in Gren Bn Buddenbrock (3/6)) S Prague; 18 Sep-4 Oct advance through Moldauthein; 8-17 Oct retreat to Beneschau; 24 Oct advance to Marschowitz; 19 Nov C Teltschitz; 27 Nov-1 Dec rearguard actions from Staudenz to Trautenau. **1745.** Jan-Feb frontier duty at Peterswaldau; 13 Feb C Alt-Waltersdorf; 14 Feb C Habelschwerdt; 4 Jun B Hohenfriedberg; 30 Sep B Soor. After the battle, the grenadiers were combined into Gren Bn Kleist (3/6 and 15/18). 25 Nov capture of Goerlitz. 15 Dec B Kesselsdorf. **1747** Chef FM Fuerst Leopold (II) Maximilian von Anhalt-Dessau, 'Alt-Anhalt' to distinguish it from IRs Nrs 10, 22 and 27. **1751** Chef GM Erbprinz Leopold Friedrich Franz von Anhalt-Dessau. **1756-1763** Against Austria and others in the Seven Years War. **1756** 10-13 Sep (grenadiers) at Pirna; took in many Saxons. 19 Sep capture of Aussig; 1 Oct B Lobositz. **1757** 6 May B Prague; 6 May-20 Jun S Prague; 21 May 1st C on the Angelka; 24 May sortie in the Mansfeld gardens; 2/3 Jun 2nd C on the Angelka; 18 Jun B Kolin; 20 Jun rearguard action at Weissen Berg; 3 Jul C Wellemin; 17 Aug bombardment of Hirschfelde; 7 Sep C Moys Grenadiers in Gren Bn Hacke; 26 Sep bombardment of Barschdorf; 22 Nov B Breslau; 5 Dec B Leuthen; 8-20 Dec S Breslau. **1758** Chef GM Hennig Alexander von Kahlden, mortally wounded at Zorndorf. Grenadiers in Gren Bn Plotho (3/6). 2 May-2 Jul at Olmuetz; 11/12 Jul rearguard action at Neu-Holitz; 11 Oct C Jenkwitz; 14 Oct B Hochkirch. **1759** Chef GM Prinz Franz Adolph von Anhalt-

Bernburg. Jul-Aug (Grenadiers in Gren Bn Hacke (3/6)) in the camp at Schmottseiffen; 23 Jul B Kay; 25 Sep C Hoyerswerda. **1760** 13-22 Jul S Dresden; did not perform well here. 9 Aug advanced guard C at Adelsdorf; 15 Aug B Liegnitz (distinguished); 2 Nov advanced guard C Schoena; 3 Nov B Torgau. After the battle, the grenadiers were combined into the Gren Bn Hacke (3/6 and 15/18) **1761** no actions. **1762** Adelsbach; Leuthmannsdorf. 12 Aug-25 Sep covering force for Schweidnitz. **1778/1779** Against Austria. Grenadiers in Gren Bn Blomberg (3/6). 31 Jul C Tollenstein. **1784** Oberst Heinrich Ernst von Leipziger. **1788** Chef GM Leopold von Thadden. **1792** Valmy; surrender of Longwy; capture of Verdun; storm of Frankfurt / Main. **1793** C Stromberg; blockade of Mainz; blockade of Landau. **1794** C Kaiserslautern; C Trippstadt; C Lautern; C Mainz. **1800.** Chef GM Johann Jeremias von Renouard. **1806.** In Wartensleben's Centre Corps of the main Prussian army. Auerstaedt 16 Oct. The regiment surrendered in Magdeburg and was not re-raised.

4. Regiment zu Fuss.
Raised in 1671. Title in 1806 – **von Kalckreuth.**
West Prussian Inspectorate. Garrisons: Elbing, Marienburg (grenadiers and three battalions).

Uniforms. Red facings up to 1787, then light blue to 1797 then orange, yellow buttons, red stock, Prussian cuffs. No lapels. Officers wore 18 embroidered gold lace loops with open tassels; two under where each lapel would have been, two in the small of the back, three on each cuff flap and two on each pocket flap. NCOs wore two gold braid loops under each lapel, two on each cuff flap and two in the small of the back; they had no gold lace edging to cuffs or cuf flaps. Soldiers had white tape loops with six narrow light blue lines and tassels, two under each lapel, two above each cuff and two in the small of the back. Buff small clothes. Brass grenadier front plate and ornaments, up to 1787, red headband, yellow cloth backing, red tape edged royal blue, royal blue pompon with a red ring in it.

Colours. King's colour white with upright, wavy yellow cross, red centre, black eagle, white scroll edged gold, gold decoration. Regimental colours red with upright wavy yellow cross, white centre, black eagle, red scroll edged gold, gold decoration. Brown pike staffs.

All four colours were lost at the capitulation of Magdeburg 11 Oct 1806.

Campaign History. Raised from men of the garrison of Kuestrin in Berlin in 1671; eight companies strong. Chef GFZM Christian Albrecht Graf zu Dohna; died in office. **1677** Chef Oberst Johann Albrecht von Barfuss; in 1698, as GFM, he received IR Nr 1. **1686** In Hungary against the Turks; S and capture of Ofen. **1689** Against France, on the Rhine; capture of Bonn. **1691** In Hungary against the Turks; C Salankemen. **1695** In the Netherlands; S Namur. **1698** Chef GM Christoph Graf zu Dohna; retired as GdI. In 1713 the regiment was dubbed GdI 'Jung-Dohna', to distinguish it from IR Nr 16. **War of the Spanish Succession.** Against France. **1708** B Oudenaarde; capture of Lille (Ryssel). **1709** B Malplaquet; capture of Mons. **1715** Against the Swedes in Pommerania. Capture of the Peenemuende redoubts, heavy losses. 1716 Chef Oberst Jakob von Bechefer (1724 GL); he later received IR Nr 5. 1721Chef Oberst Georg Rudolf von Glaubitz (1740 GL); he received a garrison battalion. **1740** Chef Oberst Konrad Heinrich von der Groeben; retired as GM. **1742** Grenadiers in Gren Bn Pfuhl (2/4). 16 May C Neu-Kolin. B Chotusitz. **1744** Chef GM Samuel von Polentz; he received IR Nr 13. Grenadiers in Gren Bn Kleist (4/29). 2-18 Sep S Prague; 18 Sep-4 Oct advance through Moldauthein; Grenadiers in Gren Bn Sydow (4/22); 8-18 Oct retreat to Beneschau; 26 Oct-9 Nov retreat to the Elbe; 20 Nov-6 Dec retreat to Silesia. **1745** 1745 Chef GM Christoph Graf zu Dohna; he later received IR Nr 23. At this point, IR Nr 38 was Jung-Dohna and IR Nr 44 was Alt-Dohna. Later in 1745 Chef GM Karl Eduard von Kalnein (1756 GL); died in office as GL. Jan-Feb border security duty at Silberberg; 13 Feb C Alt-Waltersdorf; 14 Feb C Habelschwerdt; 4 Jun B Hohenfriedberg; after the battle, the grenadiers in Gren Bn Finck (4/24). 30 Sep B Soor; 23 Nov C Katholisch-Hennersdorf. **1756-1763** Against Austria and others in the Seven Years War. **1757** Chef GM Karl Friedrich von Rautern; he retired. Grenadiers in Gren Bn Polentz (4/16). 30 Aug B Gross-Jaegersdorf. **1758** Chef GM Georg Friedrich von Kleist; he retired as GL. Grenadiers in Gren Bn Willemey (4/16). 5 Jan capture of the redoubt at the Anklam ferry; 10 Jan-18 Jun S Stralsund; 25 Aug B Zorndorf; the regiment lost very heavily here. 22 Sep-3 Oct defence of the bridgehead at Schwedt on the River Oder; 15 Nov C Eilenburg. **1759** 1 Jan capture of Dammgarten; 9 Jan-15 May S Stralsund; 28 Aug capture of Wittenberg; 31 Aug capture of Torgau; 5 Sep C Trachau near Dresden; 8 Sep C Torgau (grenadiers only); 13 Sep capture of Leipzig; 16 Sep cannonade of Alt-Tanneberg; 21 Sep C Korbitz; 25 Oct C Vogelsang; 29 Oct C Pretzsch on the River Elbe; 13 Nov bombardment of Nossen; 16 Nov C Oberhesslich; 20/21 Nov C and capitulation of Maxen (grenadiers only). **1760** Grenadiers in Gren Bn Beneckendorff (4/16 and 41/44). 14 May C Baerenwalde; 17 May C Belgard; 29 May defence of Koeslin; 3 Jun C Belgard; 19 Sep C between Brieg and Glogau. **1761** Jun-Nov actions around Kolberg; 19 Sep storm and defence of the green redoubt; 18-29 Nov advance on Belgard, combats between Regenwalde and Petershagen; 30 Nov rearguard action at Naugard; 11/12 Dec combats at Drenow and Spie; 13-17 Dec rearguard action at Treptow-Stargard. **1762** Grenadiers in Gren Bn Thielau (4/16). 3 Jan C Neu-Kalen (Malchin); 24 Jun-8 Aug actions in Upper Silesia; 12 Aug-10 Oct defence of Schweidnitz. C Leutmannsdorf. **1764** Chef GM von Thadden; received IR Nr 33; later GL and Governor of Glatz. **1774** Chef GM Friedrich von Pelkowsky; in 1781 he was promoted GL and made Commandant of Colberg. **1778** Against Austria. Grenadiers in Gren Bn Hausen (4/53) in the King's army. 20 Nov C Kunzendorf. **1782** Chef GM Baron Albrecht Dietrich Gottfired von und zu Egloffstein (Graf in 1786); retired as GL. **1789** Chef GM Benjamin d'Amaudruez; died in office as GL. **1794** In Poland; the advance on Warsaw and siege. C Wola. The regiment was split up into several detachments. **1797** Chef Oberst Wilhelm Heinrich Adolph von Kalkreuth (1806 GL; 1807 FM). **1806** Against France. 17 Oct Halle; 11 Oct capitulation of Magdeburg; part of the regiment surrendered here. Other members of the regiment had fought on with Bluecher until the capitulation of Ratkau on 7 Nov. The regiment was dissolved and not re-raised.

5. Regiment zu Fuss.
Raised in 1672. Title in 1806 – **von Kleist.**
Magdeburg Inspectorate. Garrison: Magdeburg.

Uniforms. Pale buff facings and small clothes, Prussian cuffs, yellow buttons, red stock. Officers wore two gold lace loops with loose tassels under each lapel, two on each cuff flap and two in the small of the back. NCOs had two plain gold braid loops under each lapel, two on each cuff and two in the small of the back. Their cuffs and cuff flaps were also edged ib gold. Privates had two orange lace loops with white tassels under each lapel and two in the small of the back. Brass grenadier front plate and ornaments up to 1787, dark blue headband, buff cloth backing, white tape edged red, red within white within yellow pompon.

Colours. King's colour white with upright wavy red cross, yellow centre, black eagle, white scroll edged gold, gold decoration. Regimental colours yellow with upright wavy red cross, white centre, black eagle, yellow scroll edged gold, gold decoration. Brown pike staffs.

All four colours were lost at the capitulation of Magdeburg on 11 Nov 1806.

Campaigns. Raised in 1672 by Oberst Hans Adam von Schoening; for Kurprinz Karl Emil of Brandenburg (1674 GM; died of the plague at Strassburg, fighting the French). **1674** von Schoening became Chef, (transferred in 1691 to Saxon service as GFZM). In 1684 the regiment gave one company to help raise IR Nr 12 and was then absorbed into the Liebgarde; later IR Nr 1. 1707 regained its independence as the 'Leibregiment'; Chef in 1707 GL Georg Abraham von Arnim (1728 FM) retired. In 1713 it lost the title 'Leibregiment' and became 'von Arnim'. **1672** Against France in Alsace. **1674** Against France in Alsace. **1677** Against the Swedes in Pommerania; S Stettin. **1686** In Hungary against the Turks. S Ofen. **1689** Against France on the lower Rhine. War of the Spanish Succession; against France. **1705** Marched to Italy to serve under Fuerst Leopold von Anhalt-Dessau; B Cassano; B Calcinato. **1707** In the Dauphinate. **1708** S Exilles; S Port Louis; storm of Fenestrelles. **1715** In Pommerania against the Swedes; S Stralsund. **1731** Chef GL Jakob von Bechefer; died in office. **1731** GL Christoph Heinrich von der Goltz; died in office. **1734/1735** On the Rhine against France. **1739** Chef Oberst Johann Heinrich von Wedell (1740 GM; fatally wounded at Chotusitz). **1740/1741.** Grenadiers in Gren Bn Goetze-Winterfeldt (5/21). 28 Dec-9 Mar S and storm of Glogau; 10 Apr B Mollwitz. **1742** Chef Oberst Anselm Christoph von Bonin; died in office as GL. 10-15 Mar combats at Goeding and Ungarisch-Brod. Grenadiers in Gren Bn Geist (5/20) were at B Chotusitz on 17 May. **1744.** 2-18 Sep S Prague; 9 Oct rearguard C Moldauthein; 19-16 Oct in the camp at Beneschau; 2 Dec rearguard C Starkstadt. Grenadiers in Gren Bn Jeetze (5/20). **1745.** Jan-Apr border security duty at Friedland and Geiersdorf; 4 Jun B Hohenfriedberg; 30 Sep B Soor; 16 Oct rearguard C Schatzlar. **1755** Chef GL Herzog Ferdinand von Braunschweig (1758 FM); resigned his office in 1766. The regiment was known as 'Alt- Braunschweig' to differentiate it from IR Nr 39). **1766** Chef GL Friedrich Christoph von Saldern; died as Governor of Magdeburg. **1756-1763 Against Austria and others in the Seven Years War. 1756** Grenadiers in Gren Bn Jung-Billerbeck (5/20). 10-13 Sep at Pirna; 22/23 Sep capture of Tetschen; 1 Oct B Lobositz; 16 Oct rearguard action at Hellendorf. **1757** 6 May-20 Jun S Prague; 21 May 1st C on the Angelka River; 24 May sortie in the Mansfeld gardens; 2/3 Jun 2nd C on the Angelka River; 24 Aug C Kotta; 17 Oct relief of Berlin; 31 Oct C Merseburg; 5 Nov B Rossbach. **1758** 22 May-22 Jun-raid into Franconia; 31 May raid on Bamberg; 31 Jul C St Sebastiansberg. **1759** 11 May advanced guard combat at Himmelkron; 16 May capture of Bamberg; 31 May C Muenchberg; 12 Aug B Kunersdorf; 16 Sep cannonade at Alt Tanneberg; Grenadiers in Gren Bn Ripp (5/20 and 7/30). 21 Sep C Korbitz; 7 Nov cannonade at Zoethain. **1760** Grenadiers in Gren Bn Jung-Billerbeck (5/20). 22 Jun C Gross-Dobritz; 13-22 Jul S Dresden; 15 Aug B Liegnitz; 2 Nov advanced guard combat at Schoena; 3 Nov B Torgau; heavy losses. After the battle, the grenadiers were combined in Gren Bn Rathenow (1/23 and 5/20). **1761** 4-13 Apr raid into the Voigtland. **1762** 12 May C Doebeln; 13 May C Arnsdorf; 1 Jun C Reichstaedt; 29 Oct B Freiberg. **1766** Chef GL Friedrich Christoph von Saldern. **1778** Grenadiers in Gren Bn Kamecke (5/20). 30 Jul-26 Sep raid into northern Bohemia; 4-6 Dec advance to Brix. **1785** Chef GL Christian August von Langfeld; died in office. **1789** Chef GL Ludwig Karl von Kalckstein; died as GFM and Governorof Magdeburg. **1800** Chef GL Franz Casimir von Kleist; GdI and Governor of Magdeburg in 1806. **1806.** Auerstaedt 16 Oct. The regiment was captured in Magdeburg on 11 Nov 1806 and not re-raised.

6. Grenadier-Garde-Bataillon.
Raised in 1673. Title in 1806: **Grenadier-Garde-Bataillon.**
Potsdam Inspectorate. Garrison: Potsdam.

Uniforms. Parade uniform: scarlet collar and cuffs; on each lapel six gold laces, pointed to the rear; four on each flap of the Brandenburg cuff, two in the small of the back. Pale buff small clothes. Officers had eight embroidered gold lace loops on each lapel, four on the cuffs, two on the pocket flaps and two in the small of the back. Their hats had a narrow gold edging; they wore no collars until 1801. The NCOs wore white stocks, the men red. The regiment received this uniform from Friedrich Wilhelm I; it was worn only on 1st Review days, Sundays and holidays. Tall, brass grenadier front plate (taller than the other regiments), headband and ornaments, red cloth backing, gold tape, red within white pompon. This cap was worn until 1799.

From 1800 the regiment had an undress uniform for daily wear. Scarlet collar, cuffs and lapels. The officers wore eight golden loops on, and two under each lapel, two on each cuff and two in the small of the back. NCOs had the same distribution of gold laces and their loops also had tassels. Privates had the same decoration, less the two under each lapel, and their laces were pointed to the rear and without tassels. Now the grenadier companies wore the new-style caps, the centre companies wore tricorns with gold edging and loop.

Colours. King's colour: ground and central medallion white, black eagle, gold palm and laurel leaves and corner ciphers, light blue scroll, gold decoration. Battalion colour: ground and scroll white, medallion cornflower blue, gold decoration. Yellow pike staffs.

Both colours were lost at the capitulation of Erfurt on 16 Oct 1806. The regiment was not re-raised.

Campaign history. Raised in 1673 at two battalions for the protection of Berlin, in answer to the Swedish invasion; first Chef Kurprinz Friedrich von Brandenburg. **1677** Against the Swedes in Pommerania; S Stettin. In 1685 it provided men for the new IR Nr 12. **1686** In Hungary against the Turks; S and capture of Ofen. **1689** Against France on the Rhine; the storm of Kaiserswerth and Bonn. One battalion now went into Dutch service; it fought at Leuse in 1691 (covered the Dutch retreat) and was so badly mauled at B Steenkerken that it had to be withdrawn form the order of battle. In 1695 it was at S Namur; 1696 S Ghent. 1697 B Oudenaarde; 1702 S Kaiserswerth; 1703 S Rheinbergen; 1704 S Geldern, B Hoechstaedt; 1706 S Menin; 1708 C Oudenaarde; B Ryssel (Lille) and Wynendael; the Scheldt crossing and the capture of Ghent. 1709 B Malplaquet; capture of Dornick and Mons. 1710 S Betune; 1710 S Bethune; 1711 S Bouchain, Douai and Aire. 1712 S Landrecy; capture of Moers. 1715 Against the Swedes in Pommerania. **1745** The battalion fought at B Hohenfriedberg and B Soor. We now rejoin the 'Prussian' battalions. 1701 Regiment Kronprinz; 1702 men were provided from the regiment for IRs Nrs 8 and 19. In 1704 the 3rd Bn was raised and the 4th followed in 1710. The 3rd Bn later became IR Nr 22 in 1713. 1711 The Chef was now Kronprinz Friedrich Wilhelm I. 1713 The Kronprinz raised the regiment to guards status as his 'Leibregiment' (Lifeguards); 1717 combined with the so-called 'Great Grenadier Battalion' (*1710) into the Potsdam 'Riesengarde' (Giant Guards); 1733/1734 wing grenadier companies were raised for the three battalions of the regiment. This custom was extended to all other regiments next year. In 1740 Friedrich II reduced the regiment to one battalion and re-titled it 'Grenadier-Garde' (Nr 6). It contained the tallest men of the old regiment. From the older men of the old regiment a garrison regiment was raised. Men were also given to IR Nr 35. 1740 2nd Chef GM Gottfried Emanuel von Einsiedel; died in office as GL. **1741.** Apr-Oct in camp at Goettin; **1742** at Potsdam. 1745 2nd Chef Oberst Wolf Friedrich von Retzow; died as GL on 5 Nov 1758. **1744** 2-18 Sep S Prague; 19 Sep-4 Oct advance to Moldauthein; 9-19 Oct retreat to Beneschau; 24 Oct-9 Nov retreat to Neu-Kolin; 20 Nov-6 Dec retreat to Silesia.

1756-1763 Against Austria and others in the Seven Years War. **1756** 11 Sep-16 Oct at Pirna. B Lowositz (grenadiers only). **1757** 6 May-20 Jun S Prague; 24 May sortie in the Mansfeld gardens; 19/20 Jun rearguard combat at Prague; 5 Nov B Rossbach; 5 Dec B Leuthen; stormed the graveyard. 7-20 Dec S Breslau (grenadiers only). **1758** 6 May-1 Jul at Olmuetz; 14 Oct B Hochkirch; heavy losses. **1759** Jul-Aug in the camp at Schmottseiffen; Sep-Nov on campaign in Silesia and Saxony. 2nd Chef GM Friedrich Christoph von Saldern; received IR Nr 5 in 1766 as GL. **1760** 13-22 Jul S Dresden; 15 Aug B Liegnitz; 3 Nov B Torgau; led von Zieten's thrust into the Austrian flank between the two ponds. **1761** 20 Aug-25 Sep in the camp at Bunzelwitz. **1762** 20 Jul storm of Ohmsdorf castle; 21 Jul B Burkersdorf; 4 Aug-10 Oct protection of Schweidnitz; 16 Aug C Reichenbach. 1766 2nd Chef Oberst Hans Sigismund von Lestwitz; retired as GM. **1778/1779.** In the reserve corps of the army of the King; no actions. 1779 2nd Chef GM Friedrich Wilhelm von Rohdich; by 1796 he was a GL; the son of a sergeant of the Potsdam Riesengarde, he was raised to the nobility by the King. He became 1st War Minister and President of the Upper War College and died as GdI. **1792** On the Rhine against France. **1793** C Bingen; C Kettricher Hof; C Pirmasens and Fischbach. B Moorlautern; the raid on Bitsch fortress (112 all ranks). **1794** B Kaiserslautern. 1796 2nd Chef GM Friedrich Adam Dietrich von Roeder; retired as GL. 1798 Chef Oberst Friedrich von Ingersleben; died in office as GM. 1801 the title became the Grenadiergardebataillon, without the name of its new 2nd Chef, Oberst Karl Ludwig Edler von Le Coq, who was GM in 1806. **1806** 14 Oct at Auerstaedt but not engaged. Most of the men withdrew into Erfurt, where they surrendered on 16 Oct, the rest surrendered at Prenzlau on 28 Oct. Contrary to his stated practice of not re-raising regiments that surrendered, King Friedrich Wilhelm did raise a guards regiment in 1808 (Nr 8 in the seniority list); in June 1813 it became the 1st Garde-Regiment zu Fuss.

7. Regiment zu Fuss.
Raised in 1677. Title in 1806 – **von Owstein.**
Pommeranian Inspectorate. Garrison: Colberg (three battalions), Stettin.

Uniforms. Rose red facings, white buttons, eight, evenly-spaced, on each lapel; no lace loops, Swedish cuffs, buff small clothes. Tin grenadier front plate and ornaments up to 1787, white headband and tape, deep pink cloth backing and pompon. Officers' hats had a narrow silver lace edging. NCOs had silver lace to the cuffs.

Colours. King's colour white with narrow, deep pink corner piles, light blue centre, black eagle, white scroll edged silver, silver decoration. Regimental colours light blue with narrow pink corner piles, white centre, black eagle, light blue scroll edged silver, silver embroidery decoration. Brown pike staffs.

All four colours were thrown into the moat at Luebeck to prevent their capture on 6 Nov 1806, but the French succeeded in recovering them.

Campaign history. 1677 raised as the Leibregiment of the Kurfuerstin (Electoress) Dorothea von Brandenburg from some men of the Leibgarde (IR Nr 1 in 1806) and new recruits. 1688 the Kufuerstin gave the regiment to her son, Markgraf Carl Philipp von Brandenburg (died in 1695 as GL). From 1677 to 1693 this regiment accompanied IR Nr 6 on all campaigns. **1694** In Italy. **1695** Chef Markgraf Christian Ludwig von Brandenburg (1697 GL; died in 1734). The regiment was at S Casal. **1697** On the Upper Rhine; S Ebernburg. **1698** Occupied Elbing to take possession for Kurfuerst Friedrich III von Brandenburg. The War of the Spanish Succession against Austria and France. **1702** storm of Kaiserswerth. **1703** S Geldern. **1704** B Hoechstaedt, blockade of Landau. **1705** In Italy; B Cassano. **1706** B Calcinato; capture of Reggio; relief of Turin. **1708** S Lille. **1710** S Aire. **1715** In Pommerania against the Swedes; S Stralsund. 1734 Chef Oberst Friedrich Casimir von Botzheim (1737 GM): died in office. 1737 Chef Oberst Karl Wilhelm von Bredow (1740 GM); he later received a garrison regiment. **1740.** Grenadiers in Gren Bn Duering (7/30); 22-28 Dec at Glogau. **1741.** Chef Oberst Prinz August von Braunschweig-Bevern (1763 GdI; died in office as Governor of Stettin. From 1756 the regiment was known as 'Alt-Bevern' to differentiate it from the ex-Saxon IR Jung-Bevern). 1-3 Jan at Breslau; 9 Jan capture of Ohlau; 10 Jan-7 Apr S Brieg; 13 Jun rearguard combat at Alt-Grottkau; 23 Aug Sk Schoenheide. The regiment was at B Mollwitz. **1742** Grenadiers in Gren Bn Uchlander (7/19). 17 May B Chotusitz. **1744** Grenadiers in Gren Bn Lepel (7/30). 2-18 Sep S Prague; 19 Aug-4 Oct advance to Moldauthein; 8-19 Oct retreat to Beneschau; 26 Oct-9 Nov retreat to the Elbe; 9-19 Nov defence of the river line at Weletau; Grenadiers in Gren Bn Grumbkow (7/21). 22/23 Nov rearguard combat at Neu-Biczow; 2 Dec rearguard action at Starkstadt. **1745** 4 Jun B Hohenfriedberg; 30 Sep B Soor (grenadier only); 2-9 Dec with Winterfeldt's detachment. **1756-1763 Against Austria and others in the Seven Years War. 1756** Grenadiers in Gren Bn Kanitz (7/30). 8/9 Oct C Poetzscha-Wehlen. The regiment was at B Lobositz. **1757** 6 May B Prague (grenadiers only); 7 May-20 Jun S Prague (grenadiers); 21 Jun-3 Jul guarding the magazines in Jung-Bunzlau. 14 Jun C Reichstadt; 17-21 Jul rearguard actions north of Boehmisch-Kamnitz; 13 Sep capture of Erfurt; 26 Oct C Leipzig; 18 Jun B Kolin; heavy losses. 29 Oct C Markranstaedt; 31 Oct ambush at Weissenfels; 5 Nov B Rossbach (grenadiers only); 26 Nov raid on Leitmeritz. **1758** 31 May capture of Bamberg; 31 Jul C St Sebastiansberg. 25 Aug B Zorndorf (not grenadiers) heavy losses. **1759** 15 Apr C St Sebastiansberg; 16 May destruction of the magazines at Saaz; 20 Apr rearguard action at Brummersdorf; 11 May advanced guard action at Himmelkron; 16 May capture of Bamberg; 31 May C Muenchberg; 23 Jul B Kay (not grenadiers) heavy losses; 12 Aug B Kunersdorf; after the battle, the remaining grenadiers were combined into Gren Bn Ripp (5/20 and 7/30). 16 Sep cannonade of Alt-Tanneberg; 21 Sep C Korbitz; 7 Nov C Zoethain. **1760** 13-22 Jul S Dresden; 20 Aug C Strehla; 2 Oct C Eupen; 7 Oct C Zehlendorf; 3 Nov B Torgau. **1761** Grenadiers in Gren Bn Lubath (7/30). 21 and 25 Aug raids south of Nossen. **1762** 12 May C Nossen; 29 Oct B Freiberg; took five guns and a colour. 29 Nov capture of Nuremberg. **1778** In the army of the king. Grenadiers in Gren Bn Owstein (7/8). 30 Jul-28 Sep raids into Bohemia; 31 Jul advanced guard combat at Tollenstein. 1781 Chef GM Karl Ludwig von Winterfeld; died in office. 1784 Chef GM Wilhelm Heinrich Freiherr von der Goltz; died in office as GL. 1789 Oberst Karl Eduard von Tiedemann; received IR Nr 55. 1790 Chef GM Karl Philipp von Owstein (1806 GL). **1806 Against the French.** The musketiers surrendered with Bluecher's corps at Ratkau on 7 Nov; the grenadiers capitulated in Magdeburg on 11 Nov. The regiment was not re-raised. **1807** – remnants of the regiment in Colberg (3rd Bn and some others) formed the 1st Bn of the new Colberg IR. Titles. 1 Jul 1813 - 'Colbergisches IR Nr 10'; the number changed to 9 that same day. See Serial Nr 10 in the new regiments section below for subsequent history.

8. Regiment zu Fuss.
Raised in 1677. Title in 1806 – **von Ruits.**
Warsaw Inspectorate. Garrisons: Warsaw, Lowicz (three battalions).

Uniforms. Scarlet facings, yellow buttons, red stock. On the lapels were 11 narrow white lace loops with two dark blue lines, two more under each lapel and two in the small of the back. Officers wore golden lace loops, shaped like a horizontal figure eight. 16 on the two lapels, four under each lapel, two in the small of the back and three on each sleeve. Their hats had a narrow gold edging. Prussian cuffs, white small clothes. Brass grenadier front plate and ornaments up to 1787, red headband, dark blue cloth backing, tape with red central stripe, flanked on both sides by white then yellow stripes, red within white pompon, with yellow peripheral spots. NCOs had no blue and white laces; instead they wore two golden laces under each lapel and two in the small of the back; no gold lace edging to the cuffs.

Colours. On 25 Jun 1772 ten new colours were presented; on 27 Feb 1787 six of these were handed in to the arsenal in Stettin. King's colour white upright wavy black cross, black centre, gold eagle, white scroll edged gold, gold decoration. Regimental colours black with upright wavy white cross, white centre, black eagle, light blue scroll edged gold, gold decoration. Brown pike staffs.

One colour of the 2nd Bn was lost in combat at Braunsberg on 26 Feb 1807. On 1 Oct 1809 the three remaining colours, now reduced to tatters, were handed in to the arsenal in Colberg and replaced by four of the old regiment von Ruits. On 14 Oct 1814, the regimental colour of the 2nd Battalion was given to the Pommeranian Gren Bn and went on to the Gren R Kaiser Franz. The other three remained with the regiment until 1918.

Campaigns. Raised on 20 Feb 1677 from detachments from the garrisons in the Mark Brandenburg as a march battalion of reinforcements for use in East Prussia against the Swedes. 1679 increased to a regiment under Oberst Johann von Zieten; when he died in 1688 the regiment was divided into two. GM Fuerst (Anton) Guenther von Anhalt-Zerbst received the 1st Bn, Oberst von Horn the 2nd Bn. In 1697 the 2nd Bn was reduced to one company. 1702 the 1st Bn was organized into twelve weak companies (taking two from IR Nr 6) and transferred into Dutch service until 1713. On its return, it gave two companies to help raise IR Nr 21. **1688-1697 Against France. 1689** 3 Mar (1st Bn) Uerdingen; 4/5 Mar (1st Bn) Linn; S Kaiserswerth; 6 Sep-12 Oct (1st Bn) S Bonn. **1690** at von Zieten's death, the commander, Oberst Fuerst Anton-Guenther von Anhalt-Zerbst, took over as Chef (1705 GL). 1 Jul B Fleurus. **1691** 19 Sep Leuze. **1692** 25 May-30 Jun D Namur; 3 Aug C Steenkerken. **1693** 19 Jul B Neerwinden. **1694** 18-27 Sep S Huy. **1695** 2 Jul-3 Aug S Namur. **1697** 15 May-5 Jul D Ath. **1701-1713 Against France and others in Dutch service in Germany and the Netherlands. 1702** 15 Apr-15 Jun S Kaiserswerth; 29 Aug-25 Sep S Venlo; 5 Oct C Stephanswerth; 6 Oct C Roermond. **1703** 6 Feb C Rheinbergen; 24 Apr-15 May S Bonn; S Maastricht; 17-25 Aug S Huy. **1704** 13 Aug C Hoechstaedt; S Trier, Saarburg. **1705** relief of Liege; 11 Jul C Huy. **1706** 23 May B Ramillies; 23 Jul-22 Aug S Menin; 5 Sep C Dendermonde. **1708** 11 Jul B Oudenarde; 14 Aug-23 Oct S Lille. **1709** 27 Jun-28 Jul S Tournay; 11 Sep B Malplaquet; 25 Sep-20 Oct S Mons. **1710** 22 Apr-27 Jun S Douay; 15 Jul-28 Aug S Bethune; 6-29 Sep S St Venant; 6 Sep-9 Nov S Aire; 7 Aug-10 Sep S Bouchain. **1712** 8 Jun-30 Jul S Le Quesnoy; 24 Jul C Denain. 1714 Chef Oberst Prinz Christian August von Anhalt-Zerbst (1742 FM); he died in office as GFM.

1715 Against Sweden 17 Jul-24 Sep S Stralsund. **1740-1742 Against Austria** The grenadiers were in Gren Bn Saldern (8/36). **1742** Grenadiers in Gren Bn Itzenplitz (8/24) and fought in 1 Mar C Wissewitz; 17 May B Chotusitz; C Skalitz. **1744-1745 Against Austria** The grenadiers in Gren Bn Grumbkow until Oct 1744 thence Gren Bn Schoening. **1744** 10-16 Sep S Prague; 26 Oct C Kammerburg; Grenadiers in Gren Bn Grumbkow (8/21). 9 Nov B Pardubitz; 22 Nov (grenadiers only) C Neu-Bidschow; 23 Nov (grenadiers only) C Nechanitz; 27 Nov C Pless. **1745** 14 Feb (one bn only) C Habelschwerdt; 4 Jun B Hohenfriedberg; 28 Jul (a detachment) C Ziegenhals; 20 Aug-28 Sep S Cosel; 28 Sep C Jaegerndorf; 30 Sep (grenadiers only) B Soor; 14 Oct C Troppau; 20 Oct

C Hultschin; 15 Dec (grenadiers only) B Kesselsdorf. 1747 GM Johann Otto von Treskow; he retired in 1754. The regiment was known as 'Alt Treskow' to differentiate it from IR Nr 32. **1756-1763 Against Austria and others in the Seven Years War. 1756** Chef GM Georg Friedrich von Amstell, killed at Prag in 1757. The grenadiers were in Gren Bn Alt-Billerbeck (8/46); **1757** In Silesia. Chef GM Karl Friedrich Baron von Hagen, mortally wounded at Hochkirch on 14 Oct 1758, where the regiment was distinguished. 12 Mar C Busch-Allersdorf; 21 Apr Reichenberg; 24 Apr (a detachment) C Liebenau; 6 May B Prague; distinguished, with heavy losses. 6 May (grenadiers only) C Arnsdorf; 9 May-20 Jun S Prague; 14/15 Jul (grenadiers only) C Gabel, Barschdorf and Nimburg; 9 Sep C Naumburg; 22 Nov C Breslau; 5 Dec B Leuthen; 8-12 Dec S Breslau. **1758** 27 May-2 Jul blockade of Olmuetz; 28 Jun (grenadiers only, in Gren Bn Alt-Billerbeck (8/46)) C Gundersdorf; 30 Jun (grenadiers only) C Domstaedtl; 30 Jun (grenadiers only) Cs Gibau, Heiligenberg, Pohorz; 7 Jul (grenadiers only) C Krenau; 11/12 Jul (grenadiers only) C Holitz; 25 Aug (grenadiers only) B Zorndorf. 14 Oct B Hochkirch; at the end of the battle, the regiment, together with 2nd Bn, IR Nr 19 defended the village and secured the retreat of the Prussian army at cost of heavy losses. Grenadiers in Gren Bn Alt-Billerbeck and Rohr (8/46 and Grsn R 6/Grsn R 8). **1759** GM Julius Dietrich von Queiss; he died in office as GL. 27 Jul C Conradswalde; 12 Aug B Kunersdorf; 28 Aug (grenadiers only) C Sprottau; Sep C Berensdorf; 25 Sep (grenadiers only) C Hoyerswerda. **1760** 21 Jul (grenadiers only) C Korbitz; 19 Jul (grenadiers only) S Dresden; 21 Jul (grenadiers only) C Priesnitz; 15 Aug (grenadiers only) C Liegnitz; 3 Sep (a detachment) C Jauernick; 18 Sep C Neu-Reussendorf; 2 Nov (grenadiers only) C Schoena; 27 Oct (grenadiers only) C Radis; 3 Nov B Torgau. Grenadiers in combined Gren Bn Nimschoefsky (8/46 and 33/42). **1761** Until the clashes at Hartha on 27 Jun and 29 Sep 1762 and the battle of Freiberg on 29 Oct 1762 only the grenadiers were engaged. 5 Apr C Plauen. **1762** 12 May C Rosswein; 13 May C Arensdorf; 27 Jun C Hartha; 17 Jul Johnsdorf; 2 Aug Toeplitz; 29 Sep C Hartha; 1 Oct C Katzenhaeuser; 22 Oct C Gross-Schirma; 29 Oct B Freiberg. 1769 Chef Oberst Levin Friedrich von Hacke; died in office as GL and Governor of Stettin. 1778-1779 Against Austria. The grenadiers were in Gren Bn Owstein (7/8). **1778** The regiment was in the army of Prinz Heinrich. 28 Sep C Nickelsberg. **1779** 5 Feb C Brix. 1785 Chef GL Johann Georg Wilhelm Freiherr von Keller; died in office as Governor of Stettin. 1786 Chef GM Johann Andreas Anton von Scholten; died in office. 1791 Chef GM Franz

Otto von Pirch; he transferred to IR Nr 22. 1795 Chef GM Friedrich Leopold von Ruits (1806 GL).

1806 Against France. The regiment was in l'Estocq's army reserve corps and survived the war. The grenadiers were in Gren Bn Massow. No actions in 1806. **1807.** 5 Feb (grenadiers only) C Waltersdorf; 8 Feb C Wackern and Frisching; 26 Feb C Braunsberg ; losses not known, but all the Polish recruits deserted, the regiment was scattered and the 2nd Bn lost a colour. 10 May C Steegen.

1808 – vacant, then - 'Neumaerkisches Reserve-Bataillon. It was combined with the 'Grenadier-Bataillon von Braun' and men of the disbanded regiments 'von Treskau' Nr 17, 'Prinz von Oranien' Nr 19, 'von Pirch' Nr 22 and 'von Puttkammer' Nr 36. **7 Sep 1808 – '1. Pommersches IR Nr 2'.** For details of the histories of the old regiments, see the individual entries above. **Grenadier-Regiment Koenig Friedrich Wilhelm IV (1. Pommersches) Nr 2. Page 286, Vol I.** Together with the grenadiers of the 'Colberg' IR, the regimental grenadiers formed the Grenadier-Bataillon 'von Wangenheim'. On 14 Oct 1814 the two grenadier companies went to join the later 'Kaiser Franz Garde-Grenadier-Regiment' Nr 2.

1812. The 2nd and Fusilier Bns were in Russia with Macdonald's X Corps as part of the 3rd Combined IR. Wolgund 7 Aug; C Dahlenkirchen 22 Aug; Sk Kiopen 29 Sep. The Combined IR was disbanded on 10 Mar 1813.

1813. In von Buelow's III Corps, Army of the North. The Fues-Bn was with Wittgenstein's Avantgarde. 24 Mar (Fusiliers only) C Werben; 2 Apr (Fusiliers only) Lueneburg . They took 6 guns and two colours of the Saxon IR Prinz Maximilien. 5 Apr C Vehlitz (1st and 2nd Bns); 8 Apr-11 May blockade of Magdeburg; 12 May (Fusiliers) B Goehrde; 28 May Hoyerswerda (1st and 2nd Bns); 30 May (Fusiliers) Cs Ochsenwerder and Nettelnburger Schleuse; 4 Jun C Luckau; 17 Aug Cs at Baruth, Mueckendorf and Zesch; 21 Aug C Mellin; 23 Aug B Gross-Beeren (Dennewitz), they took 5 guns and 15 ammunition wagons. 26 Aug C Schmielkendorf; 27 Aug C Hohenschlennen; 3 Sep C Theissen; 6 Sep B Dennewitz 6 Sep (1st, 2nd Bns and Grenadiers); the regiment took 12 guns and a colour of the Bavarian 9th LIR 'Ysenburg'. The grenadiers took two other colours. 24 Sep B Wartenburg; 15 Sep-11 Oct blockade of Wittenberg; 2 Oct C Elster; 16-19 Oct B Leipzig. The whole regiment was in action here. They took 11 guns at the Ranstaedter Tor; in all, they captured 63 guns, 66 ammunition wagons, 12 generals, 18 colonels, 300 officers and 8,000 R&F. Fr-Jg Gauer took Gen Reynier prisoner. 26 Nov-

26 Dec S Wesel; 12 Dec C Rees. **1814.** Hoogstraten; Sweweghem 2 Mar; Sweweghem 7 Mar. **1815.** Ligny 16 Jun; Waterloo 18 Jun.

9. Regiment zu Fuss.
Raised in 1677. Title in 1806 – **von Schenck.**
Westphalian Inspectorate. Garrisons: Hamm; Soest (grenadiers), Unna (three battalions).
Titles. 7 Jun 1808 – IR 'von Borcke' Nr 30; 1 Jul 1813 - 'Colbergisches IR Nr 10; the number changed to 9 that same day. See Serial Nr 10 in the new regiments section below for subsequent history.

Uniforms. Scarlet facings, yellow buttons, red stock; officers' lapels and cuffs were edged in gold embroidery and their hats had a narrow gold edging. Prussian cuffs, white small clothes. In 1806 other ranks wore six, straight white loops, in pairs, on each lapel, two under each lapel; two above the white-edged cuffs and two in the small of the back. NCOs wore five gold lace loops on each lapel in the form of a horizontal '8' and two above each cuff. No gold edging to the cuffs. Brass grenadier front plate and ornaments up to 1787, red headband and cloth backing, red within yellow tape, red within orange pompon.

Colours. King's colour white with upright wavy red cross, dark green centre, black eagle, white scroll edged gold, gold decoration. Regimental colours dark green with upright wavy red cross, white centre, black eagle, dark green scroll edged gold, gold decoration. Brown pike staffs.

All four colours were lost at the capitulation of Erfurt on 16 Oct 1806.

Campaign history. Raised in 1677 as the garrison of Minden from men of other Prussian regiments. First Chef was GM Ernst von Poellnitz; he died in office. 1684 Chef GM Heinrich Baron de Briquemault; he died in office. 1692 Chef Oberst von Boys; killed at Neerwinden. 1693 Chef GM von Schlabrendorff; later GL and Governor of Kuestrin. In 1695 a battalion of IR Alt-Hanstein increased the unit to a regiment until 1697 when it was reduced to a battalion again. In the War of the Spanish Succession it went into Dutch service until 1713. **1702** S Kaiserswerth. **1703** Chef Oberst von Wulffen; killed at the Schellenberg. The regiment fought at Rheinbergen and B Hoechstaedt. 1704 Chef Oberst von Lattorf; died in office. **1706** S Menin. **1708** B Oudenarde. 1708 Chef GM Etienne du

Trossel; died in office. 1714 Chef GL Johan Sigismund Freiherr von Heiden; resigned his office. 1719 Chef Oberst Jonas Casimir von Auer; died in office as GM. 1721 Chef Oberst Johann du Buisson; died in office. 1726 Chef Oberst Christoph Rudolf von Schliewitz; GM in 1730, died in office. 1732 Chef Oberst von Waldow; died in office. 1735 Chef Oberst Otto Friedrich von Leps; died as GdI. **1745** Grenadiers in Gren Bn Hagen (9/10). 29 Nov invasion of Saxony Grenadiers in Gren Bn Sydow (9/27); 5 Dec capture of Torgau; Grenadiers in Gren Bn Kleist (9/27). 15 Dec B Kesselsdorf. 1747 Chef GM Johann Christ Rulemann Freiherr Quaad von Wickeradt (mortally wounded at Lowositz).

1756-1763 Against Austria and others in the Seven Years War. **1756** Chef GM Friedrich Ludwig von Kleist (killed at Breslau), the regiment was called 'Jung-Kleist' to differentiate it from IR Nr 27. Grenadiers in Gren Bn Moellendorf (9/10). 20 Apr capture of Grafenstein castle; 21 Apr C Reichenberg; 6 May B Prague; 18 Jun B Kolin; 14/15 Jul D Gabel; 1 Oct B Lobositz (1st Bn); heavy losses. **1757** 21 Apr C Reichenberg (1st Bn and grenadiers); 6 May B Prague; S Prague (musketiers only); 18 Jun B Kollin (grenadiers); 5 Nov B Rossbach (musketiers) **1758** Chef GM Juergen Friedrich von Oldenburg; died in office. That same year GM Nikolaus Lorenz von Puttkamer succeeded him; he later resigned his office. The musketiers were in Prinz Heinrich's corps. 28 Jun C Gundersdorf; Grenadiers in Gren Bn Baehr (9/10). 30 Jun C Domstaedl; 14 Oct B Hochkirch; 30 Oct rearguard action at Lauban. **1759** Chef GM Friedrich August von Schenckendorff; resigned his office. 23 Jul B Kay; 12 Aug B Kunersdorf; 11 Sep occupation of Goerlitz; 25 Oct C Vogelsang; 29 Oct advanced guard clash at Pretzsch. 20 Nov C Maxen; most of the regiment was captured here. **1760** Re-raised with one battalion. As of 1760 the regiment was called 'Jung-Schenckendorff' to differentiate it from IR Nr 22. 11 Jun-31 Jul with Prinz Heinrich's army on the Rivers Oder and Warthe. 1-6 Aug relief of Breslau; 17 Sep C Hoch Giersdorf; 2 Nov C Schoena; 3 Nov B Torgau (grenadiers). After the battle, the grenadiers in Gren Bn Baehr (9/10, 29/31). **1761** In Saxony; no actions. **1762** In Neisse garrison. 12 May C Doebeln; 29 Sep C Klingenberg; 15 Oct C Konradsdorf; 17 Oct C Klein-Waltersdorf; 29 Oct B Freiberg. 1763 Chef GM Karl Friedrich von Wolfersdorff; died as GL in 1781. **1778** 9-11 Aug in Platen's corps on the advance to Leitmeritz; Grenadiers in Gren Bn Bandemer (9/10). 27-31 Aug advance to Melnik; 24 Sep retreat to Saxony. 1782 Chef GM Alexander von Budberg; retired. **1787** In Holland at C Oudekerken and Duevendrecht. **1792** Chef Oberst Wilhelm Dietrich von

Mannstein; retired as GM. The regiment was on the Rhine and in Champagne; Valmy; Hochheim; Frankfurt / Main. **1793** 6 Jan C Hochheim; took twelve cannon. 19 Jun – 23 Jul S Mainz; C Zahlbach redoubt. Blockade of Landau. **1794** Sk Obrigheim (grenadiers); 3 Jan Frankenthal (musketiers); 23 May C Diedesheim; Sk Fenningen; 23 May B Kaiserslautern. 1796 Chef Oberst Christian Ludwig von Winning; he received IR Nr 23. 1796 Chef Oberst Johann Friedrich von Brehmer; died as GM. 1802 Chef Oberst Johann Friedrich von Schenck (GM that same year). **1806.** Jena 16 Oct. The regiment surrendered in Erfurt and was not re-raised.

10. Regiment zu Fuss.
Raised in 1683. Title in 1806 – **von Wedell.**
Westphalian Inspectorate. Garrisons: Bielefeld, Herford (grenadiers), Minden (three battalions).

Uniforms. Until 1798 lemon yellow facings were worn; they then changed to light blue. Originally no lapels. In 1801 they changed again to lemon yellow. White buttons (eight evenly-spaced on each lapel) and small clothes. Officers wore two embroidered gold lace loops with loose tassels under each lapel, three on the cuff, three on the pocket flap and two in the small of the back. They had a thin silver edging to the bicorn. NCOs wore two plain, silver loops under each lapel and two in the small of the back; privates wore the same configuration as the NCOs, but the tapes were white with three red stripes and white tassels. Tin grenadier front plate and ornaments up to 1787, with lemon yellow headband, cloth backing and pompon, silver tape.

Colours. King's colour white with light green centre, black eagle, white scroll edged gold, gold decoration. Regimental colours light green, white centre, black eagle, light green scroll edged gold, gold decoration. Brown pike staffs.

All four colours were lost when the regiment surrendered at the capitulation of Erfurt on 16 Oct 1806.

Campaign history. For the grenadiers see IR Nr 9. 1683 raised at battalion strength in Ostpreussen by Oberst Prinz Alexander von Kurland, who was killed at the siege of Ofen. 1685 increased to a regiment of two battalions for the war against the Turks. Chef Prinz Ferdinand von Kurland, after the death of his brother at Ofen in 1686, where the regiment was distinguished. He left the service of Brandenburg. 1690 The regiment was split into two; one battalion became IR Nr 16 and some companies of IR von Spaen were absorbed instead. Chef of IR Nr 10 was GM Friedrich Freiherr von Heiden (or Heyden); retired as GdI. **1695** In the Netherlands; S Namur 1702 the regiment gave men to reinforce IR Nr 9. For the duration of the War of the Spanish Succession, the regiment was in Dutch service until 1713. 1703 Chef GL Erbprinz Friedrich von Hessen-Cassel (1704 GdK); he died in 1751 as king of Sweden. **1706** S Menin. **1709** capture of Tournay; B Malplaquet; capture of Mons. **1710** S Douai; S Aire. **1711** S Bouchain 1713 the regiment gave two companies to raise IR Nr 21 and adopted the Prussian regimental organisation of ten companies. 1714 Chef GM Prinz Georg von Hessen-Cassel, after his brother had become King of Sweden. (1723 GL). **1715** In Pommerania; S Stralsund. Chef GM Georg, von Hessen-Kassel, brother of the previous incumbent. He retired as GdI. In 1729 it gave men to help raise IR Nr 31. 1730 Chef GM Prinz Dietrich von Anhalt-Dessau (FM in 1747). **1741** B Mollwitz. **1742** B Chotusitz. **1745** B Kesselsdorf; distinguished, heavy losses. 1750 Chef Oberst Dietrich Erhard von Knobloch (1756 GM); died in office. **1756-1763 Against Austria and others. 1757** Chef GM Gottlob Ernst von Pannwitz; retired. B Prague (grenadiers); S Prague; 18 Jun B Kolin (grenadiers); C Moys; 22 Nov B Breslau; 5 Dec B Leuthen; S Breslau. **1758** S Olmuetz; B Hochkirch. 1759 Chef GM Friedrich Wilhelm von der Mosel; died in office. **1760** 2nd Bn was ambushed at Topliwoda near Neisse, but beat the enemy off. **1761** 20 Jun C Landshut, the musketiers fought bravely but forced to surrender. C Domstaedtl (grenadiers), B Torgau (grenadiers). **1762** The regiment was exchanged. 21 Jul C Leutmannsdorf; B Friedberg (grenadiers). 1768 Chef Oberst Eggert Christian von Petersdorff; later GM and commandant of Colberg. **1779** In the army of Prinz Heinrich; C Brix in Bohemia. 1781 Chef Oberst Sylvius Ferdinand von Stwolinsky; died in office. **1787** Chef GM Friedrich Wilhelm von der Marwitz; died in office. In Holland; capture of Gorkum fortress (grenadiers); C Amstelveen; C Ouderkerke; C Duevendrecht; C Dymerbrueck. 1788 Chef GM Gisbert Wilhelm Freiherr von Romberg (1806 GL and Governor of Stettin). **1792** on the Rhine and in Champagne; Valmy. **1793** C Altstadt; C Eschweiler; C Herzogenhand; C Zweibruecken. Raid on Bitsch fortress (300 men). **1794** C Meckenheim; C Altdorf; C Fenningen; C Edinghofen; Sk Roth; B Kaiserslautern. 1800 Chef GM Wilhelm Albrecht von Burghagen; died in office. 1800 GM

Gottlieb Ehrenreich von Besser; he received IR Nr 14. 1803 Oberst Karl Alexander von Wedell (1806 GM). **1806.** Jena 14 Oct. Erfurt 16 Oct. The regiment was not re-raised.

11. Regiment zu Fuss.
Raised on 18 Aug 1685. Title in 1806 – **Prinz Heinrich von Preussen.** East Prussian Inspectorate. Garrison: Koenigsberg in Prussia.

Uniforms. Crimson facings, white buttons red stock, Prussian cuffs. In 1756 it seems that the regiment had yellow buttons; by 1788 they had changed to white. Officers had in 1806 eighteen embroidered silver lace loops with loose tassels, distributed as for IR Nr 10, and a narrow silver edging to the hat. NCOs had two plain silver loops under each lapel and two in the small of the back; no silver edging to the cuff. Other ranks had white lace loops with two narrow blue stripes outside two narrow crimson ones, with white tassels to the rear: two under each lapel, two over each cuff and two in the small of the back. White small clothes. Tin grenadier front plate and headband up to 1787, white cloth backing, red tape and pompon.

Colours. First recorded presentation in 1714; 15 Jun 1773-10 new colours received; of these, 6 were returned in May 1787. 16 May 1810 four new colours presented, the old ones were handed in. By AKO of 28 Mar 1812 they handed over the Retirierfahne of the 1st Bn to the 1st Ostpreussisches Gren Bn (see Gren R Nr 1). On 12 Jan 1813 they handed the Retirierfahne of the 2nd Bn to the arsenal in Graudenz. On 2 Feb 1817, by AKO of 15 Jun 1815, the fusilier battalion was given (wrongly) the Retirierfahne of the 1st Bn 1st Ostpreussische IR, originally presented on 19 Mar 1769. This had been handed in to the Colberg arsenal on 12 Mar 1813, and bore the 1813 campaign medal. Brown pike staffs.

King's colour white with upright wavy red cross red centre, black eagle, white scroll edged gold, gold decoration. Regimental colours red with upright wavy white cross, white centre, black eagle, red scroll edged gold, gold decoration.

One colour of the 1st Bn was lost in combat at Soldau on 23 Dec 1806.

Campaigns. Raised in 1685 from 2 companies of the IR Alt-Holstein (Herzog August von Holstein-Beck), 2 companies of the IR GL von Spaen and 1 newly-raised by Herzog Friedrich Ludwig von Holstein-Beck (Jung-Holstein); he handed the regiment on to his son. In 1688 it gave men to help raise IR Nr 20. 20 Sep 1688-into Dutch service with two battalions until 1697. **1689-1692 Against France.** 1st Bn only. **1689** On the Rhine. 12 Mar Uerdingen; 22-26 Jun S Kaiserswerth. **1691** 1-6 Jun defence of Liege. **1691-1697 Against France.** In the Netherlands; 2nd Bn only. **1691** in Charleroi; **1692** 17-22 Oct D Charleroi. **1693**? Jul Beaumont; 11 Sep-11 Oct D Charleroi. **1695** 26 Jul the 2nd Bn captured in Dixmuiden.

1693-1698 Against the Turks in Hungary. 1st Bn only. **1693** 12 Aug-12 Sep S Belgrade. **1694** 19 Sep-2 Oct defence of the camp at Peterwardein. **1697** 11 Sep Zenta. 1697 reduced to 2 companies. 1702 the regiment gave men to reinforce IR Nr 17. 1703 raised to 10 companies. **1705-1706 Against France.** 100 detached grenadiers only. **1705** 19 May Soseto; 10/11 Jun Soncino; 16 Aug Cassano. **1706** 19 Apr Calcinato; 6 Sep Pianezza; 7 Sep Turin; 5-21 Oct S Pizzighettoni. **1705 Against France on the Rhine.** A combined bn only. 24 Sep Drusenheim; 29 Sep-5 Oct S Hagenau. **1708-1713 Against France.** 1st Bn only. **1709** 11 Sep Cesanne. **1715 Against Sweden in Pommerania.** From mid Jul - 28 Nov S Stralsund; 15-17 Nov capture of Ruegen island; 15 Nov Stresow. 1721 Chef Prinz Friedrich Wilhelm von Holstein-Beck; died as GFM. 1 May 1735 2 grenadier companies raised. **1741-1742 Against Austria. 1741** Grenadiers in Gren Bn Lattorf (11/33). **1742** Grenadiers in Gren Bn Kanitz (11/14). 17 May Chotusitz. **1744-1745 Against Austria. 1744** Grenadiers in Gren Bn Huelsen, later von der Trenck (11/14). **1745** 14 Feb (grenadiers in Bn Trenck) Habelschwerdt; 22 May (musk bns) Bratsch; 4 Jun (the whole regiment) Hohenfriedberg; 12 Sep (a detachment) Neustadt; 30 Sep (grenadiers only) Soor. 1749 Chef GM Lorenz Ludwig von Below; as GL he was removed from his post for reputedly failing at the battle of Zorndorf on 25 Aug 1758. **1756-1763 Against Austria and others.** The grenadiers were in the Gren Bn Gohr, from 24 Sep 1757 Petersdorf) **1757** Grenadiers in Gren Bn Gohr, later Petersdorff (11/14). 29 Aug (grenadiers) Puschdorfer forest; 30 Aug Gross-Jaegersdorf; heavy losses. 29-31 Dec (2nd Bn and grenadiers) block of Demmin; **1758** Chef GM Johann Karl Freiherr von Rebentisch appointed on 12 Sep, after von

Below had been sacked. 7-18 Mar S Peenemuende redoubts; 4/5 Apr (a detachment) defence of the same; 15 Aug (grenadiers) Kuestrin; 25 Aug B Zorndorf; heavy losses. 3 Oct (grenadiers) Passkrug; 14 Oct (1st Bn) Krischa and Nechern; 15 Nov (grenadiers) Eilenburg; **1759** 21 May Grussau; (grenadiers in Gren Bn Beyer, 1760 Oppen (11/14)) 23 Jul Kay; 12 Aug (the same) Kunersdorf; 21 Sep (the same) Meissen; 20 Oct Pretsch; 20 Nov Maxen, the regiment was captured. **1760** One battalion was re-raised. 13-29 Jul (grenadiers) S Dresden; 31 Jul-4 (the re-raised battalion) Aug D Breslau; 20 Aug (grenadiers) Strehla; 7 Oct Grenadiers in Gren Bn Oppen (11/14)) Mariendorf and Schoeneberg; 3 Nov (grenadiers) Torgau. **1762** 2 Aug-9 Oct (the new battalion) S Schweidnitz. 1763 regiment completed; 1 Jun 1787 regiment set at 1 grenadier and 2 musketier battalions each of 4 companies, and a depot battalion of 3 companies of the Grsn R von Bose (*1713) 1763 Chef GM Hans von Tettenborn; retired as GL. 1774 Chef GM Jakob Ruediger von Zastrow; died in office. **1778-1779 Against Austria.** Grenadiers in Gren Bn Hertzberg (11/14). The regiment and the grenadiers in Gren Bn Herzberg. 1782 Chef GM Friedrich Alexander von Rothkirch; retired as GL. The regiment was known as 'Jung-Rothkirch' to differentiate it from IR Nr 32. 1785 Chef Oberst Karl von Voss; retired as GL. 26 Dec 1790 – Chef Friedrich Karl Ludwig Herzog von Holstein-Beck; went into Russian service in 1797. **1794 - 1795. Against Poland.** 30 Oct (2nd Bn only) Gawrichi. 1796 3rd Musk Bn raised, also from the Grsn R von Bose. 1 Jun 1799 reorganized at 2 grenadier companies and 2 musketier battalions each of 5 companies, the 3rd of 4 companies. They fought at Pionki, where Gen. Grabowsky and 4,000 men were captured. 1 Mar 1798 – Chef Ernst Sigismund von Schoening; GM in 1806. **1806. Against France.** The grenadiers of the regiment combined with those of IR 'von Brueneck' Nr 2 as Grenadier-Bataillon von Schlieffen.23 Oct 1806 – title now 'vacant von Schoening'. 23 Dec (a detachment) Biezun; 25 Dec (1st and 2nd Bns) Soldau. **1807.** The grenadiers of the regiment combined with those of IR von Brueneck Nr 2 as Grenadier-Bataillon von Kurowski. 7/8 Feb B Preussisch-Eylau; 10 Feb (grenadiers and a detachment) B Friedland; 13 Jun Gollau; 14 Jun C Koenigsberg; 8 Mar 1807 – IR **Prinz Heinrich**; The regiment was also present at Wackern, Allenau, Bartenstein, Heilsberg, Kahlberg, Gollau, Gross-Karschau, Mostbude, Schoenbusch and Pillau.

1 Jan 1808 reorganized at 2 grenadier companies, 2 musketier battalions each of 4 companies received the Fus Bn von Renouard as the light battalion (renamed 'Fus Bn' on 1 Dec 1809). 7 Sep 1808: **IR Nr 3 (2nd**

Ostpreussisches). 14 Sep 1808: **2. Ostpreussisches IR 'Prinz Heinrich von Preussen..**

1812 Against Russia. 6 Mar 1812 1st Bn mobilised as part of the 1st Mobile IR. 29 May 1812 Fus Bn mobilised as 7th Fus Bn. 22 Aug (2 fusliler coys) Olai; 26 Sep C Dahlenkirchen; 27 Sep-2 Oct Sk Eckau and along the River Aa; 15/16 Nov (fusiliers) Baldonen and Dahlenkirchen; 18 Nov (1st Bn) Friedrichstadt.

1813 Against France. 17 Mar 1813 both battalions returned to the regiment. In 1813 a 3rd Musk Bn was raised. The regiment was in the 2nd Brigade, in von Yorck's I Corps, Army of Silesia.

Halle 28 Apr (Fuesiliers only); Wittenberg 17 Apr; Gross-Goerschen 2 May; Wittenberg 6 May; Ottendorf 10 May; Bautzen 20/21 May; Loewenberg 21 Aug; Goldberg 23 Aug; Katzbach 26 Aug; Hochkirch 4 Sep (only the Grenadiers fought here); Goldbach 22 Sep (only the Fuesiliers fought here); Wartenburg 3 Oct; B Moeckern 16 Oct The regiment took 4 guns, 1 howitzer and an ammunition wagon. Due to heavy losses, the Fuesiliers were combined with the Fues-Bn of the 1st Ostpreussisches IR. C Gohlis 18 Oct; Freiburg. **1814.** Chalons 4 Feb; Chateau-Thierry 8 Feb; Montmirail 11/12 Feb (only the grenadiers fought here) C Mery-sur-Seine 22/23 Feb (only the Fuesiliers fought here); B Laon 9 Mar; C Clay-Souilly 28 Mar (only the Fuesiliers fought here); B Paris. The regiment saw no action in 1815.

14 Oct 1814 the grenadier companies went to form part of the **Kaiser Alexander Gren R.** 5 Nov 1816 title changed to **3rd IR (2. Ostpreussisches) Prinz Heinrich von Preussen.**

12. Regiment zu Fuss.

Raised in 1685. Title in 1806 – **Herzog Friedrich Wilhelm von Braunschweig-Oels.**

Mark-Brandenburg Inspectorate. Garrisons: Prenzlow, Templin (grenadiers), Angermuende (three battalions).

Uniforms. Light brick red facings, yellow buttons, red stock. Officers had eight richly ornamented gold lace loops on each lapel, two under each lapel, three over each cuff, two on each pocket flap and six in the back. NCOs wore plain gold laces, pointed to the rear and with loose golden tassels, six, in pairs, on and two under each lapel, two above each cuff and two in the small of the back. No gold lace edging to the cuff. Other ranks had six wide white lace loops with tassels, in pairs, on each lapel; two under each lapel, two over each cuff and two in the small of the back. Prussian cuffs, buff small clothes (white in 1806). Brass grenadier front plate and ornaments up to 1787; the front is decorated with a large black eagle. Light brick red headband, dark blue cloth backing, gold tape, dark blue within red pompon.

Colours. King's colour white with narrow light green corner piles and centre, black eagle, white scroll edged gold, gold decoration. Regimental colours light green with narrow white corner piles, white centre, black eagle, light green scroll edged gold, gold decoration. Brown pike staffs.

Campaign history. Raised in 1685 in Wesel from contingents from IRs Nrs 1, 3, 4, 5, 6 and 7 for the Markgraf Philipp von Brandenburg-Schwedt (1697 GFZM). **1689** Against the French on the Rhine; S Rheinbergen, S Kaiserswerth; S Bonn. **1690** B Fleurus (one battalion). **1693** C Neerwinden. **1694** S Huy. **1695** capture of Namur. **1702** Capture of Kaiserswerth, Venlo and Roermond. **1704** B Hochstaedt. **1705** Went to Italy against the French; opposed crossing of Lake Garda; B Cassano. **1706** Relief of Turin. **1707** In Provence. S Toulon. **1708** In the Dauphinate; capture of Exilles and Fenestrelles. **1711** Chef Markgraf Heinrich Friedrich, son of Philipp; 1740 GM, he received IR Nr 42. The regiment was in Pommerania; S Stettin. **1715** In Pommerania **1740/1741** Against Austria in Bohemia. Chef Oberst Wolf Balthasar von Selchow; retired as GL. Grenadiers in Gren Bn Puttkamer (12/24). 29 Dec-3 Jan at Breslau; 9 Jan capture of Ohlau; 23 Jan capture of Troppau; 25 Jan C Graetz; 10 Apr B Mollwitz; 23 Jun rearguard action at Alt-Grottkau; 30 Jul C Zobten; 3 Aug raid on Schoenheide. **1742** Grenadiers in Gren Bn Jeetze (12/17). 10-14 Apr Cs at Goeding and Ungarisch-Brod; 17 May B Chotusitz (grenadiers). 1743 Chef Oberst Erbprinz Ludwig von Hessen-Darmstadt; left Prussian service. Until 1747 the regiment was called 'Alt-Darmstadt' to differentiate it from IR Nr 47 (1756 GM). **1744** Grenadiers in Gren Bn Luck (12/21). 2-18 Sep S Prague; 6 Sep C Beraun; 1 Oct C Moldauthein; 9-20 Nov secured the River Elbe at Neu-Kolin; 2 Dec rearguard action at Starkstadt. **1744/1745** Dec-Apr border patrols at Friedland and Rudolfswaldau; 22 May C Landshut; 4 Jun B Hohenfriedberg; 22 Jul C

Schmirsitz; 31 Jul Sk Chwalkowitz; 31 Jul Sk Wildschuetz; 1 Dec C Loewenbeg. B Kesselsdorf. Grenadiers in Gren Bn Waldau (12/39). **1756-1763 Against Austria and others in the Seven Years War. 1756** no actions. **1757** Chef GM Friedrich August von Finck; GL in 1759; was removed from his position. 20 Apr C Kratzau (grenadiers); 21 Apr B Reichenberg; 6 May B Prague; led the final advance. 14 May C west of Kolin; Grenadiers combined with 13/26. 5 Jun C Gang; 13 Jun C Kuttenberg (grenadiers); 18 Jun B Kolin (grenadiers only); Grenadiers in Gren Bn Waldau (12/39). 16-19 Jul retreat to Zittau; the Grenadiers were combined with those of 33/42, 47/Grsn R 7, Grsn R3/Grsn R4. 24 Sep-15 Dec in Glatz fortress; Grenadiers in Gren Bn Nomschoefsky (12/39, 33/42). **1758** 6 Apr C Friedland; Grenadiers in Gren Bn Pieverlingk (12/39). 4 May-1Jul before Olmuetz; 14 Oct B Hochkirch (grenadiers). **1759** Aug-Nov the advance into the Saxon Lausitz. **1760** The Grenadiers in Gren Bn Stechow. 13-22 Jul S Dresden; 12 Aug B Kunersdorf; 15 Aug B Liegnitz; 20 Nov C Maxen – the musketiers of the regiment were captured. **1761** One battalion of musketiers re-raised. Grenadiers in Gren Bn Goerne-Oelsnitz. 14 Sep C Kobylin; 15 Sep C Gostyn; 30 Sep C Koerlin; 2 Oct C Spie; 22 Oct rearguard action at Gollnow; 15 Nov cannonade at Greifenberg; 18-29 Nov the thrust at Belgard, including the actions at Regenwalde, Petershagen, 30 Nov C Naugard; 11 Dec C Drenow; 12 Dec C Spie; 13-17 Dec Cs at Treptow-Stargard. **1762** Two musketier battalions strong. Grenadiers in Gren Bn Goerne-Tempsky. 21 Jul B Burkersdorf; 4 Aug-9 Oct S Schweidnitz. 29 May 1763: Chef GM Johann von Wunsch; ex-leader of a *Freikorps*; died in office as GdI. **1778** Chef GM Franz Casimir von Kleist; as GL and Governor of Magdeburg, he received IR Nr 5. In the army of Prinz Heinrich, on the raid to Brix. Grenadiers in Gren Bn Broesigke (12/34). 28 Sep C Nickelsberg. **1792** On the Rhine and in Champagne; C Crune; Valmy; C Homburg; C Koenigstein. **1793** C Gleisweiler; C Modebacher valley; bombardment of Landau; C Lembach; C Scheerhohle; C Weissenburg; C Klembach. **1794** C Kaiserslautern; C Johanniskreuz; C Ingershausen; C Heltersberg. 1800 Chef Oberst Prinz Friedrich Wilhelm von Braunschweig-Oels; GM in 1806. He raised the Black Corps in 1809 and was killed at Quatre-Bras in 1815. **1806 Against France.** The regiment capitulated on 7 November with Bluecher's corps and was not re-raised.

13. Regiment zu Fuss.
Raised in 1687. Title in 1806 – **von Arnim.**
Berlin Inspectorate. Garrisons: Berlin, Spandau (three battalions).

Uniforms. White facings and small clothes, white buttons, red stock. Officers had two richly embroidered silver loops under each lapel, three on the flap of each Prussian cuff, two on each pocket flap and two in the small of the back. NCOs wore embroidered silver lace loops, with tassels, two under each lapel, two over each cuff and two in the small of the back; no silver lace edging to the cuff. Under each lapel other ranks wore two white laces with tassels, two above each cuff and two in the small of the back. They wore silver shoulder cords and a thin silver edging to the bicorn. Tin grenadier front plate and ornaments up to 1787, buff headband, dark blue cloth backing, white tape, yellow pompon. In 1799 the new grenadier cap had a light blue headband edged white.

Colours. King's colour white with narrow pink corner piles, black centre, silver eagle, white scroll edged silver, silver decoration. Regimental colours black with narrow pink corner piles, white centre, black eagle, black scroll edged silver, silver decoration. Brown pike staffs.

Both colours of the 1st Bn were lost at the capitulation of Prenzlau on 28 Oct 1806. On colour of the 2nd Bn was lost in the capitulation of Erfurt, the other at that of Magdeburg. The regiment was not re-raised.

Campaign history. Raised 1687 at 1 battalion from French refugees under Oberst Jacques L'Aumonier Marquis de Varenne in the area of Bielefeld, Essen, Herford, Soest and Werden. In **1689** it gave men to the formation of IR Nr 15 and was at S Bonn on the Rhine. **1690** S Ath. **1691** C Leuse. **1692** S Namur. **1693** B Oudenarde; B Landen. In 1697 it was reduced to one company, but in 1702 it was made up to two battalions (12 companies) by extracts from all other regiments and went into Anglo-Dutch service as marines until 1713, when they returned to Berlin. **1702** S Kaiserswerth. **1708** S Lille. **1709** S Dornick; B Malplaquet. 1713 Chef GL Marquis de Varenne; **1715** Chef GL Anton von Pannewitz; retired. The regiment was in Pommerania S Stralsund. 1722 Chef GM Alexander Graf von Doenhoff; retired as GL. 1740 Chef Oberst Friedrich Sebastian Wunibald Graf von Truchses-Waldburg (killed as GL at Striegau). **1740/1741** Against Austria in Silesia. 27 Dec-9 Mar S and storm of Glogau (grenadiers in Gren Bn Reibitz (13/19)); 10 Apr B Mollwitz; 11 Apr-3

May S Brieg; 23 Aug raid on Schoenheide. **1742** Apr-May in Prinz Dietrich's corps in Upper Silesia; the grenadiers were in Gren Bn Lattorf-Hagen (13/28). The musketiers were distinguished at B Chotusitz and B Hohenfriedberg. **1744** Against Austria in Bohemia; the Grenadiers were in Gren Bn Geist (13/26). 2-18 Sep S Prag; 18 Sep-4 Oct advance to Moldauthein; 7-17 Oct retreat to Beneschau; 26 Oct-9 Nov retreat to Neu-Kolin; thence retreat to Silesia. Grenadiers in Gren Bn Geist (13/37). **1745** Chef GM Samuel von Polentz; von Truchses had been mortally wounded at Kesselsdorf, where the regiment had only light losses. Jan-20 Apr border patrols at Giersdorf; 4 Jun B Hohenfriedberg; 30 Sep B Soor (grenadiers); 15 Dec B Kesselsdorf (musketiers). Nov-Dec winter campaign in Silesia and Saxony. 1746 Chef GM Philipp Bogislav von Schwerin; died as GL. The regiment was called 'Bogislav-Schwerin' to differentiate it from IR Nr 24 (Alt-Schwerin) and IR Nr 32 (Jung-Schwerin). 1751 Chef GM August Friedrich von Itzenplitz (also spelt Itzenblitz); mortally wounded at Kunersdorf as GL. **1756-1763 Against Austria and others in the Seven Years War.** In Saxony and Bohemia. **1756** Grenadiers in Gren Bn Finck (13/26). 11 Sep-16 Oct before Pirna. **1757** 6 May B Prague; musketiers also at the siege. 14 May C Kolin; 5 Jun C Gang; 13 Jun C Kuttenberg; 18 Jun B Kolin (grenadiers only); 24 Aug C Kotta; 29 Oct C Markranstaedt; 31 Oct C Weissenfels; 5 Nov B Rossbach; 5 Dec B Leuthen. **1758** 22 May-22 Jun raid into Franconia in Prinz Heinrich's corps; 22 Aug-2 Sep in the fortified camp at Pirna; 14 Oct B Hochkirch; the regiment lost three-quarters of its strength. 18 Oct to the main army in Silesia. S Schweidnitz (musketiers). **1759** Chef GM Friedrich Wilhelm von Syburg; later received IR Nr 16. 7 May C Nordhalben in Franconia; 8 May C Asch; 9 Jul C Krzyszkowo; 23 Jul B Kay; 12 Aug B Kunersdorf; Grenadiers in Gren Bn Kreckwitz-Humboldt (13/19/25). 16 Sep cannonade at Alt-Tanneberg; 21 Sep C Korbitz; 25 Sep ambush at Hoyerswerda (musketries); 29 Oct C Pretzsch (musketiers); 3 Nov C Nossen (grenadiers); 20/21 Nov C Maxen, the grenadiers were captured here. **1760** Grenadiers in Gren Bn Schwerin (13/26, 19/25). 3 Apr C Buckow-Zanow; 14 May C Baerenwalde; 26 May D Belgard; 3 Jun C Belgard; 15 Jun C Greifenberg; 15 Aug B Liegnitz; 18 Sep relief of Kolberg; 3 Oct C Pasewalk. The musketiers were at the siege of Dresden and B Torgau on 3 Nov (distinguished). **1761** 14 Feb cannonade of Koeslin; 5 Jun C Koerlin; 10-19 Jun actions around Belgard; 24 Aug-14 Nov in the camp at Kolberg; 18-29 Nov thrust at Belgard, combats between Regenwalde and Petershagen; 30 Nov rearguard action at Naugard; 11

Dec C Drenow; 12 Dec C Spie; 13-18 Dec rearguard actions from Treptow to Stargard; 30/31 Dec before Malchin. **1762** Czar Peter III of Russia. The officers now wore silver aiguillettes on the right shoulder. Grenadiers in Gren Bn Kalckstein (13/26). 3 Jan C Neu-Kahlden; 21 Jul B Buerkersdorf. 2 Aug C Teplitz; 14/15 Oct C Brand, the grenadiers were captured here. The musketiers were at C Leutmannsdorf and S Schweidnitz. **1763** Chef GM Friedrich Wilhelm Graf von Wylich and Lottum. 1774 Chef Oberst Gottlob von Braun. **1778** Jul-Oct raid into Bohemia; Grenadiers in Gren Bn Schlieben (1/13). Oct 1778-Mar 1779 at Jaegerndorf. **1795** Chef GM Alexander Wilhelm von Arnim (1806 GL).The regiment was in Poland. **1806** The 1st Bn and the grenadiers capitulated at Prenzlau, the 2nd went with Bluecher to Luebeck and surrendered there; the regiment was not re-raised.

14. Regiment zu Fuss.
Raised on 1 May 1688. Title in 1806 **– von Besser.**
East Prussian Inspectorate. Garrisons: Bartenstein, Schippenbeil, Preussisch-Holland (grenadiers), Friedland (three battalions).

Uniforms. Light brick red facings, yellow buttons, six, in pairs, on each lapel, two on each Prussian cuff flap, two under each pocket flap. Officers wore two large, plain gold loops, with pointed rear ends and loose tassels under each lapel and two in the small of the back. NCOs wore gold lace edging to cuff and flap, but no other lace decoration. Other ranks have six white, horizontal figure-of-eight lace loops with two narrow red lines on their lapels and two over each cuff. Prussian cuffs, white small clothes. Brass grenadier front plate and ornaments up to 1787, white headband and cloth backing, orange tape and pompon. The grenadier cap from 1799-1806 had the usual band in the facing colour, edged in the button colour.

Colours. King's colour white with crimson centre, black eagle, white scroll edged gold, gold decoration. Regimental colours crimson, white centre, black eagle, crimson scroll edged gold, gold decoration. Brown pike staffs.

Both colours of the 1st Bn were lost in combat at Waltersdorf on 5 Feb 1807.

Campaign history. Raised on 1 May 1688 from new recruits and given to Oberst Hans Georg von Belling; killed at Bonn. **1689 Against France.** On the Rhine, Chef Oberst Wilhelm von Brandt; died in office as GL and Governor of Magdeburg. 24 Jul-5 Oct S Bonn incl storm on 9 Oct. 1689 2nd battalion as garrison in Pillau. **1690** Kurfuerst Friedrich II sent the regiment to Hungary to help the Austrians fight the Turks; C Salankemen, S Peterwardein. **1696** In Brabant until 1696 when it returned to Prussia and took possession of Elbing. **1701-1714 Against France.** 1702 Chef Oberst Christoph Albrecht von Canitz, he resigned in 1711. **1702** On the Rhine. 16 Apr-15 Jun S Kaiserswerth; 9 Jun storm of Kaiserswerth; 29 Aug-22 Sep S Venlo; 6-9 Oct B Roermond; **1703** Storm of Bonn; 20 Sep Hochstaedt. **1704** On the Danube. 13 Aug Hoechstaedt; 26 Sep-22 Nov S Landau. **1705** Fuerst Leopold von Dessau took the regiment to Italy. 1 Jun Bedizzole; 10-12 Jul B Soncino; 16 Jun B Cassano. **1706** 19 Apr C Calcinato; 2-6 Aug S Carpi; 5-19 Aug S Reggio; 5 Sep Pianezza; 7 Sep relief of Turin. **1707** 26 Jul-22 Aug S Toulon; 20 Sep-3 Oct S Susa. **1708** 2nd Chef GM Prinz Friedrich Ludwig von Oranien (the regiment retained the name 'von Oranien' until 1713) 15-31 Aug S Fort Fenestrelle. **1709** 28 Jul Conflans. **1711** 2nd Chef GFM Graf Finck von Finckenstein; he became Chef in 1713 and died in office as GFM. 16 Sep Exilles. Chaumont and Mont Vallon. **1715 Against Sweden in Pommerania.** 17 Jul-24 Dec S Stralsund; 1 Aug capture of Usedom Island; 21/22 Aug storm of the Peenemuende redoubts; 23 Oct sortie combat; 4/5 Nov storm; 10-21 Nov (grenadiers) on Ruegen Island; 16 Nov Stresow; 5-12 Dec storm of Stralsund's outworks. 18 Dec sortie combat. **1734-1735 Against France.** No actions. 1736 Chef Oberst Andreas Joachim von Kleist; died in office. 1738 GFM Johann von Lehwaldt; died in office as GFM. 1740 Grenadiers in Gren Bn Sydow (later Burghagen) (3/14). **1741-1742 Against Austria.** The grenadiers were in the Gren Bn von Kanitz (11/14). **1742** In Bohemia. 17 May Chotusitz. **1744-1745 Against Austria.** The grenadiers were in the Gren Bn von Huelsen, then Trenck. **1745** 14 Feb (Gren Bn) Habelschwerdt; 4 Jun Hohenfriedberg; 30 Sep Soor. 1756-1763 **Against Austria and others in the Seven Years War.** The grenadiers were in the Gren Bn Gohr then Petersdorff, Beyer, Oppen (11/14) for all other campaigns. **1757** (Gren Bn) Puschdorfer forest; 30 Aug Gross-Jaegersdorf; 30 Sep Tilsit. **1758** 15 Aug (Gren Bn) Cuestrin; 25 Aug B Zorndorf. **1759** 20 Jun (GrenBn) Zuellichau; 23 Jul Kay; 12 Aug

Kunersdorf; heavy losses. 21 Sep Meissen; 20 Nov (1st and 2nd Bns) Maxen. **1760** 13-30 Jul (Gren Bn) S Dresden; 18-20 Aug (Gren Bn) Strehlen; 5 – 9 Oct defence of Berlin. 3 Nov (Gren Bn only) Torgau. **1761** 4 Sep-14 Nov (1st and 2nd Bns) defence of Colberg. **1762** 12 May (1st and 2nd Bns) Doebeln; 21 May (1st and 2nd Bns) Chemnitz. Freiberg. 1768 Chef Oberst Reichsgraf Friedrich zu Anhalt; he resigned his office and went into Saxon, then Russian service. 1776 Chef GM Johann Christian Wilhelm von Steinwehr; retired. **1778-1789 Against Austria.** The regiment was in Prinz Heinrich's army; the grenadiers were in the Gren Bn Hertzberg. No actions. 1782 Chef Oberst Viktor Amadeus Graf Henckel von Donnersmark; later received IR Nr 2. 29 Dec 1786 – Chef GL Franz Ludolf Ferdinand von Wildau; died in office as GL. **1794 Against Poland.** Dembnicki 18 Jul (Gren Bns); took a howitzer, several cannon and many prisoners. Rezmy 18 Sep; Lachy-Sielce 8 Oct (Gren Bns); Magniszewo 25/26 Oct (one coy); and again on 26 Oct (5 coys). The regiment also fought at Kolno, Warsaw (17/18 Apr, Gren Bn); Biendunga, Ostrolenca, Nowy-Dwor and Gawrichi. At Gross-Magniczewo on 26 Oct they took Gen. Woitschinsky and 500 men, a howitzer and three guns. For its services in this campaign the regiment received several *Pour le Merites*, gold and silver medals and special medals from the East Prussian and Masurian nobility. 29 Dec 1794 – Chef Oberst Karl Leopold von Larisch; he received IR Nr 26. 6 Jul 1795 – Chef Oberst Georg Prinz zu Hohenlohe-Ingelfingen; died in office as GM. 20 Dec 1803 – Chef GM Gottlieb Ehrenreich von Besser. **1806. Against France.** The grenadiers were part of the Grenadier-Bn von Fabicky. The 2nd Musketier-Bn was in Graudenz, the 3rd in Danzig. From 10 Mar – 24 May 1807 these battalions were in Danzig. The 1st Bn and the grenadiers also fought at Soldau and Gross-Schimanen this year. 25 Nov 1806 – the regimant was vacant.**1807.** C Waltersdorf 5 Feb. The 1st Bn and the grenadiers were present here; they were taken prisoner, then released. These battalions also fought at Wackern, Koenigsberg, and Mostbude. Dirschau 23 Feb. On 20 Nov 1807 remnants of the regiment went to form a new regiment; 24 Dec 1807 – Chef GM August Ludwig von Stutterheim; 7 Sep 1808 – **3.Ostpreussisches IR Nr 4**; see the new regiments, Serial Nr 6 below. The grenadiers were together with those of IR Nr 16, also in the Reserve Corps. They went into the 2. Ostpreussisches Gen Bn in 1808 and in 1814 became the **3rd Bn, Kaiser Alexander Gren R Nr 1; see Ser Nr 2c in the new regiments section below.**

15. Garde-Regiment zu Fuss.

Raised in 1689. Title in 1806: **Garde-Regiment zu Fuss.**
Potsdam Inspectorate. Garrison: Potsdam.
This was a guards regiment, not part of the line. The colonel-in-chief was the king.

Uniforms. In the Seven Years' War: no lapels, poppy red facings and turnbacks, Swedish cuffs, plain white buttons, silver lace, yellow waistcoat and breeches. The 1st Battalion wore the following uniform. The lapels and collars of officers' coats were edged in silver lace. They also wore richly decorated silver lace loops, with loose tassels, to the buttonholes: eight on and two under each lapel, two on each cuff, two on each pocket flap and two in the small of the back. The silver buttons were slightly domed and all of one size. Officers' bicorns were edged with a wide, scalloped silver lace and was decorated with a white feather plume and a silver brooch. They wore silver shoulder cords. NCOs' coats had the same number of buttons, each decorated with a silver lace loop with a tassel. Their collars and lapels were edged in silver and their hats were also edged in scalloped silver lace, a silver lace loop. Privates had no tassels to their silver buttonhole loops and no silver edging to their collars and cuffs. Their hats had a plain, thin silver edging. Officers of the 2nd and 3rd Battalions had no tassels to their buttonhole loops; NCOs in these battalions had the same lace loop decoration as those in the 1st except that there were none on the pocket flaps and the lace was plain, with a silver tassel. Privates had no tassels, but wore the silver edging to the hat. In 1798 the waistcoat became white. Grenadier cap plates up to 1787 were white metal; the 1st Battalion had red cloth backing and silver lace, red within white pompon. From 1787 – 1798 the cuffs were Brandenburg, thence Swedish. The grenadier cap plate of all three battalions had silver front bands and peak edging and silver grenades. The top and bottom of the headpiece of the cap was also edged in silver. Their plumes were of cut feathers.

Colours. Prior to 1808, both the King's and regimental colours were white with narrow, vertical silver stripes. On the King's colour the central disc was silver, the eagle black, the scroll edged in gold. On the regimental colours they were light blue. The embroidery was mainly silver, but the royal ciphers were gold and the wreath around the central field was mixed gold and silver. Yellow pike staffs.

Both colours of the 1st Bn Leibgarde were lost at the capitulation of Prenzlau. Of the four colours of the Regiment Garde, one was lost on the retreat, probably on 17 Oct; the other three were taken in the capitulation of Magdeburg.

On 24 Apr 1808 four colours of the new pattern were presented. The fields were white the central disc orange (cloth of silver for the king's colour), in the disc the black eagle under a light blue scroll with the inscription: PRO GLORIA UND PATRIA. Silver decorations. Yellow pike staffs. The gold finials contained the royal cipher.

Campaign History. Raised in 1689 at 8 companies from men of the Regiment Varenne Nr 13 (*1687), for Oberst Karl Philipp Freiherr von Wylich und Lottum. (died as GFM in 1718). The regiment was called 'Alt-Lottum' from 1718 to differentiate it from IR Nr 25. **1690** One battalion sent to Hungary to fight the Turks; C Salankemen; C Peterwardein. In 1693 it gave the battalion of the regiment in Hungary to form IR Nr 17. In 1695 it took over a battalion from the Battalion von Doerfling. In 1702 it gave a further two companies to IR Nr 17. Nr **1701-1711: Against France in the War of the Spanish Succession, on the Rhine and in the Netherlands**. **1702** 18 Apr-15 Jun S Kaiserswerth. **1704** On the Danube. 13 Aug B Hochstaedt; 12 Sep-23 Nov S Landau. **1706** 23 Jul-22 Aug S Menin; 19 Sep-4 Oct S Ath. **1708** In the Netherlands. 11 Jul B Oudenaarde; the regiment was on the allied right wing and decided the battle in their favour. S Winnendaal; 11-30 Dec S Ghent. **1709** 11 Sep B Malplaquet; S Bruges. **1715: Against Sweden in Pommerania;** no actions of note. 1718 – Chef Oberst Friedrich Wilhelm Freiherr von Coenen; died in office. 1720 – Chef Oberst Christoph Heinrich von der Goltz; he later received IR Nr 5. 29 Feb 1731 – Chef Kronprinz Friedrich von Preussen; in 1740 he became King Friedrich II of Prussia and the regiment received the title of 'Guards'; he was succeeded in 1786 by King Friedrich Wilhelm II. In this same year, almost all the 2nd Bn was given to raise IR Nr 34. This was the start of the King's search to fill the ranks of the three-battalion regiment with the tallest men from all regiments of the army. **1741-1742:** Against Austria. **1741** In Bohemia. 10 Apr B Mollwitz (1st Bn) the king led the battalion forward himself. Half the officers of the battalion were killed or wounded and of the 800 men who started the action, only 180 were fit for duty afterwards. **1742** 17 May B Chotusitz (2nd, 3rd Bns). **1744 – 1745:** Against Austria; the regiment sent only the grenadiers and a detachment of 60 men of the 3rd Bn into the field. **1744** In Bohemia. 4 Jun B

Hohenfriedberg. 2-18 Sep (grenadiers, in Gren Bn Wedel (15/18) only) S Prague; 30 Sep B Soor. 19 Nov Teltschitz (Gren). **1745** 4 Jun Grenadiers in Gren Bn Buelow (15/18) B Hohenfriedberg; 30 Sep B Soor; 25 Nov (Gren) capture of Goerlitz. **1756 – 1763: Against Austria and other states in the Seven Years War**. **1756** 10 Sep-16 Oct Blockade of Pirna; 1 Oct B Lobositz. **1757** In Silesia and Bohemia. 6 May B Prague, thence blockade of the city until 20 Jun; 21 May B Angelka River (Gren); 24 May Mansfeld; 3 Jun Angelka (Gren); 18 Jun B Kollin (1st Bn) heavy losses; Grenadiers in Gren Bn Kleist; 20 Jun Weissen Berge; 3 Jul Wellemin (Gren); 8 Aug B Loebau; 7 Sep Moys (Gren); 5 Nov B Rossbach (musketiers); 22 Nov B Breslau (Gren); 5 Dec B Leuthen; 7-20 Dec siege of Breslau. **1758** 20 May-1 Jul S Olmuetz (3rd Bn only); 14 Oct B Hochkirch (1st, 2nd Bns and the grenadiers; the latter were captured); 30 Jul-4 Aug def of Breslau (1st Bn). **1759** Grenadiers in Gren Bn Anhalt (15/18). 25 Sep C Hoyerswerda (grenadiers only). **1760** In Saxony. 13-22 Jul (grenadiers and 3rd Bn only) S Dresden and C Adelsdorf on 9 Aug. 15 Aug B Liegnitz (2nd, 3rd Bns and grenadiers); 2 Nov (grenadiers only C Schoena; 3 Nov B Torgau (2nd, 3rd Bns and grenadiers). After the battle the grenadiers were combined into the Gren Bn Anhalt-Winning (3/6 and 15/18). **1761** 3rd Bn and Grenadiers (in Gren Bn Anhalt-Winning) only: 15 Aug C Wahlstatt; 20 Aug-25 Sep in the camp at Bunzelwitz. **1762** In Silesia. 19 Jul B Burkersdorf (2nd, 3rd Bns and grenadiers). Thence (grenadiers only) 4 Aug-10 Oct covering Schweidnitz; 16 Aug C Reichenbach (3rd Bn and grenadiers only). **1778-1779:** Against Austria in the army of the King; no actions of note. The grenadiers were in Gren Bn Apenburg (15/18). 1786 Chef now King Friedrich Wilhelm II. **1793-1794:** Against France on the Rhine; grenadiers only. **1792:** 23 Mar – 22 Jul S Mainz; 28 – 30 Nov B Kaiserslautern. **1794:** 13 Jul B Trippstadt. Also present at Mainspitz, Buergerinsel, Gustavsberg, Kettricherhof, Pirmasens, St Ingbert, Landau, Bitsch, Frankenthal, Hochspeyer, Mosalbertal. 1797 title changed to ' Leibgarde' (the 1st Bn was formed from the 1st Bn Kronprinz and the Regiment Garde); 2nd Bn formed of the 2nd Grenadier Coy and the old Koenigsregiment; 3rd Bn formed from drafts from other regiments. 1797 Chef now King Friedrich Wilhelm III. **1806:** Fluegel-Gren-Kie, Grenadiers of the Regiment Garde Nr 15 and the grenadier company of the Gren Garde-Bn Nr 6; known together as the 1st Gren Bn. 14 Oct B Auerstaedt. Parts of the regiment capitulated at Erfurt, 16 Oct, Prenzlau 28 Oct and Magdeburg. 4 Nov 1806 – title changed to 'Fussgarde'; the depot of the regiment was formed from a detachment of the 1st Bn

Leibgarde which had avoided capitulation, and other survivors of the Regiment Garde, Grenadier-Garde-Bn, Regiment des Koenigs and other regiments. On 24 Jan 1807 – strength two companies; 27 Jun 1807 – Bn Garde (4 coys). 9 Nov 1808 – **Regiment Garde zu Fuss Nr 8**; formed in two battalions. 1st and 2nd Coys from the old 1st Bn Leibgarde; 3rd and 4th from the old Gren. Garde-Bn; 5th, 6th, 7th, 8th from the old Regiment Garde. 17 Mar 1809 **– Garde-Fus-Bn** raised; **1 Jul 1813 – 1st Garde-Regt zu Fuss. 1813:** B Gross-Goerschen 2 May. B Bautzen 20/21 May. B Leipzig 16 – 18 Oct. **1814.** Paris, 30 Mar. **1815** No actions.

16. Regiment zu Fuss.

Raised in 1690. Title in 1806 – **von Diericke.**

East Prussian Inspectorate. Garrisons: Braunsberg, Preussisch Holland (grenadiers), Muehlhausen (three battalions).

Uniforms. Light red facings, yellow buttons, red stock. Officers had nine embroidered golden loops (in threes) on and two under each lapel, four on each Brandenburg cuff flap, two on each pocket flap and two in the small of the back. Their hats had a narrow gold edging. NCOs wore six buttons and gold lace loops (in pairs) on each lapel, but no loops elsewhere and no lace on the cuffs. Other ranks had white lace loops with narrow red outer edges and a narrow black central line, two under each lapel and two in the small of the back. Tassels (white, black and red) to the rear ends of the button holes. Prussian cuffs, white small clothes. Brass grenadier front plate and ornaments up to 1787, red headband, buff cloth backing, red within black and white mixed pompon. The tape has a black central stripe, flanked to each side by white then black stripes.

Colours. King's colour white with orange upright wavy cross and centre, black eagle, white scroll edged gold, gold decoration. Regimental colours orange with upright wavy white cross and centre, black eagle, light orange scroll edged gold, gold decoration. Brown pike staffs.

Campaigns. Raised in 1690 from a battalion of IR Nr 10 and the Pillau Garrison Battalion for Oberst Alexander Burggraf zu Dohna, Freiherr von Wartenburg (Alt-Dohna). In 1702 it gave a company to help raise IR Nr 17. **1701-1713 War of the Spanish Succession against France on the Rhine and in the Netherlands. 1705** 11 Jul storm of Huy. **1706** 23 Jul – 22 Aug S and surrender of Menin. **1708** 11 Jul B Oudenaarde. **1709** 28

Jun-28 Jul S Tournay; 11 Sep Malplaquet; 25 Sep- 20 Oct S and capture of Mons. **1710** 27 Jun storm of Douay; 6 Sep-10 Nov S Aire. **1711** 22 Aug-13 Sep S Bouchain. **1712** 17-29 Jul S Landrecy. 1728 Chef Oberst Kurt Friedrich von Flans; died in office as GFM. **1734-1735 Against France in the War of the Polish Succession; on the Upper Rhine. 1735** the raid on Speyer. **1741-1743 Against Austria. 1741** The grenadier company in Gren Bn Hauss-Ruitz (16/34). **1742** Apr – march from Berlin to Bohemia; Grenadiers in Gren Bn Kahlbutz (16/34). 14 May guarding the magazines in Nimburg; 17 May B Chotusitz. **1744-1745 Against Austria. 1744** The grenadier company was in Gren Bn Kleist von Jung-Schwerin (16/ 1st Grsn R). S and capture of Prague. 11 Sep C Wartha. **1745** 9 Feb Ratibor; 4 Jun (Grens) Hohenfriedberg; S Cosel (musketiers) 30 Sep (Grens) B Soor. 1748 Chef GM Burggraf Christoph zu Dohna; died in office as GL. **1756-1763 Against Austria and others.** (Grenadiers in Gren Bn von Polenz, then von Kleist, 1758 von Willemay, 1760 von Benkendorf, 1762 von Thielau). **1757** Grenadiers in Gren Bn Polentz (4/16). 30 Aug B Gross-Jaegersdorf. **1758** Grenadiers in Gren Bn Kleist (4/16). 25 Aug B Zorndorf; lost almost half its strength. **1759** 12-31 Jan S Anklam; 23 Jul B Kay; 12 Aug B Kunersdorf; 8 Sep (Grens) Torgau; 29 Oct (Grenadiers in Gren Bn Willemy). Pretsch; 20 Nov (Grens) Maxen. **1760** Grenadiers in Gren Bn Beneckendorf (4/16 and 41/44). The 2nd Bn was in skirmishes at Schindelmuehle and near Berlin. 3 Nov B Torgau (2nd Bn). **1761** In Pommerania. 25 Aug-15 Nov (1st and 2nd Bns) defence of the camp at Colberg; 11 Dec (whole regiment) 11 Dec Spie. **1762** Chef GM Friedrich Wilhelm von Syburg; died in office. Grenadiers in Gren Bn Thielau (4/16). 6 Jul C Adelsbach and Friedland; 21 Jul storming of the heights of Leutmannsdorf – took six cannon; 16 Aug (grenadiers) C Reichenbach; 4 Aug-11 Oct S Schweidnitz. 1770 Chef Oberst Christian Ernst Wilhelm von Bork; retired as GM. 1776 Chef GM Ludwig Freiherr von Buddenbrock; died in office. **1778-1779 Against Austria. 1778** Grenadiers in Gren Bn Collrep-Rautter (2/16). 21 Sep Forstberg. 1782 Chef GM Karl Adolph Schott von Schottenstein; retired.. 1785 Chef Oberst Johann Wilhelm Friedrich Moritz Freiherr von Romberg (Alt-Romberg); retired as GM and commandant of Wesel. 1790 – Chef Oberst Ernst Wilhelm von Gillern; retired. 1792 – Chef GM Friedrich Wilhelm Heinrich von Hausen; retired as GL. 1800 Chef Oberst Otto Friedrich von Diericke; GM in 1806. **1794** In Poland. **1806 Against France.** The grenadiers of the regiment were combined with those of the IR von Besser Nr 4 as the Grenadier-Bn von Fabecky. During 1806 and 1807 this battalion fought at Soldau, Gross-Schiemanen, Waltersdorf, Wackern, Preussisch-Eylau, Heilsberg, Braunsberg, Gollau and Koenigsberg as part of L'Estocq's corps. **1807.** 10 Mar – 25 May: the musketiers of the regiment were in the defence of Danzig. On 7 Sep 1808 remnants of the regiment went to form the **4. Westpreussisches IR von Dierické**. See Serial Nr 6 **Grenadier-Regiment Koenig Friedrich I (4. Ostpreussisches Nr 5).** in the new regiments for subsequent history. The grenadiers became the 2. Ostpreussisches Gren Bn in 1808; in 1814 they then became the **3rd Bn, Kaiser Alexander Gren R Nr 1**; see Ser Nr 2c in the new regiments' section below.

17. Regiment zu Fuss.
Raised in 1693. Title in 1806 – **von Treskow.**
West Prussian Inspectorate. Garrisons: Danzig, Schidlitz, Stolzenberg (three battalions).

Uniforms. White facings and small clothes, yellow buttons, red stock, Prussian cuffs. Officers wore six gold loops (in pairs) on and two under each lapel, two over each cuff, two on and one over each pocket flap and two in the small of the back. Their hats had a narrow gold edging. NCOs' lace loops were in the same layout as for officers, but no pocket flap. The gold laces were plain, pointed at the rear and with golden tassels. They wore no gold lace edging on the cuffs. Other ranks wore white lace loops with six narrow red stripes and white tassels. The layout was as for NCOs. Brass grenadier front plate and ornaments up to 1787, white headband and cloth backing, white tape with red pattern, white within red within green pompon. The grenadier cap, introduced in 1799 had a light blue headband edged yellow. From 1787 – 1798 a seventh buttonhole was added to each lapel.

Colours. King's colour white with narrow crimson corner piles and centre, black eagle, white scroll edged gold, gold decoration. Regimental colours crimson with narrow white corner piles and centre, black eagle, crimson scroll edged gold, gold decoration. Brown pike staffs.

All four colours were lost under honourable circumstances in a clash at the Doelauer Heide, near Halle, on 17 Oct 1806. The regiment was not re-raised.

Campaign history. Raised at 5 companies for the campaign in Hungary against the Turks, under Oberst Balthasar Friedrich von Sydow from a battalion of IR Nr 15. **1695** C Peterwardein. **1697** 11 Sep B Zenta 1702 the regiment went into Dutch service until 1713 and had to be reinforced with men from IRs Nrs 11, 15 and 16. 1703 von Sydow sold the regiment to Oberst Friedrich Wilhelm von Grumbkow; died in office as GFM. During the War of the Spanish Succession the regiment fought at: **1703** B Hochstaedt; S Bonn; S Huy. 1705 on the Moselle. **1706** 23 May B Ramillies. **1708** 11 Jul B Oudenaarde; the regiment was captured by the French but quickly exchanged. **1709** 11 Sep B Malplaquet; S Mons. **1710** 22 Apr 27 Jun S Douai. 1713 gave two companies to help raise IR Nr 21. **1715** In Pommerania; S Stralsund. 1739 Chef Oberst Ernst August de la Chevallerie Baron de la Motte; retired as GL. **1740/1741** Grenadiers in Gren Bn Wedel (17/18). **1741** 9 Jan capture of Ohlau; 5 Jan-2 Apr S Neisse; 23 Aug raid at Schoenheide; 28 Aug C Nimptsch; **1742** Grenadiers in Gren Bn Jeetze (12/17). 10-14 Mar C Goeding-Ungarisch Brod; 17 May B Chotusitz (one battalion only). **1744** 2-18 Sep S Prague, Grenadiers in Gren Bn Sydow (17/22). 1 Oct C Moldauthein, Grenadiers in Gren Bn Kleist von Wuerttemberg (12/17). 4-18 Oct rearguard actions to Beneschau, thence to Leitmeritz and Melnik; 23 Nov-4 Dec retreat to Silesia; 2 Dec C Marschendorf. **1744/1745** Dec-Apr border patrols at Liebau; 14 Feb C Habelschwerdt; 4 Jun B Hohenfriedberg; 11 Jul C Neustadt U/S; 26 Aug-5 Sep S Cosel; 28 Sep C Jaegerndorf; 30 Sep B Soor; the regiment suffered heavy losses. 20 Oct C Hultschin; 6 Dec C Schwarzwaldau. 1748 Chef Oberst Adam Friedrich von Jeetze; retired as GL. Until 1752 the regiment was known as 'Jung-Jeetze to differentiate it from IR Nr 30. **1756-1763 Against Austria and others in the Seven Years War. 1756** Chef GM Gerd Heinrich von Manteuffel; retired as GL. Grenadiers in Gren Bn Grumbkow (17/24). 13-17 Sep at Pirna; 30 Sep C Paschkopole, Grenadiers in Gren Bn Puttkamer (17/22); 1 Oct B Lobositz; heavy losses. **1757** Grenadiers in Gren Bn Wrede (17/22). 6 May B Prague; 7 May-20 Jun S Prague; 18 Jun B Kolin. Kremzow took over the Gren Bn. 19 Jul cannonade of Zittau; 17 Sep C Moys; 25 Oct C Leipzig; 26 Oct C Linday dam; 29 Oct C Markranstaedt; 5 Nov B Rossbach; 5 Dec B Leuthen; 7-20 Dec S Breslau heavy losses. **1758** 3 May-1 Jul at Olmuetz; 28 Aug B Zorndorf; 14 Oct B Hochkirch. **1759** 11-18 Jan S Demmin; 20 Jan-15 May S Stralsund, Grenadiers in Gren Bn von der Tann; 20 Jul C Radewitsch; 23 Jul B Kay (grenadiers); 12 Aug B Kunersdorf, Grenadiers in Gren Bn v. d. Busche (17/22 and Grsn Rs 6 and 8). **1760** Grenadiers in Gren Bn Wobersnow (17/22). 15 Mar C Neustadt / Upper Silesia; the regiment and one sqn of the 6th Dragoons were attacked by 5,000 enemy troops but beat them off causing them over 800 casualties for a loss of only 170. 17 Jun and 23 Jul Cs Landshut (grenadiers); 3 Nov B Torgau; lost almost 1,000 men and took 4 cannon. 1761 Grenadiers in Gren Bn Rothenburg (17/22 and 24/34). May-Jul at Glogau; 15 Sep C Gostyn; 30 Sep C Koerlin; 2 Oct C Spie; 22 Oct rearguard action at Gollnow; 15 Nov cannonade at Greifenberg; 18-29 Nov thrust at Belgard, including clashes at Regenwalde and Petershagen; 11 Dec C Drenow; 13-17 Dec rearguard action at Treptow-Stargard. **1762** Grenadiers in Gren Bn Rothenburg (17/22). 21 Jul B Burkersdorf; 4 Aug-10 Oct S Schweidnitz. The musketiers were at B Friedberg on 29 Oct. 1764 Chef GM Karl Gustav Freiherr von Rosen; died in office. 1772 Chef GM Konstantin von Billerbeck; died in office as GL. **1778** In the army of the king. Grenadiers in Gren Bn Below (17/22). 5-15 Jul raid into Bohemia; 23 Jul-Mar 1779 border security in Upper Silesia; 11 Aug C Glomnitz; 16 Aug C Heydepiltsch1. 1786 GM Heinrich Gottloeb von Kenitz; the regiment was known as 'Jung-Kenitz' to differentiate it from IR Nr 39. 4 Dec 1786 GM Wilhelm Magnus von Bruenneck; he received IR Nr 2. 1793 GL Karl Albrecht Friedrich von Raumer; retired. 1796 Oberst Karl Friedrich von Langen; died as GM1802 GM Karl Peter von Tresckow. **1806** The regiment was in Duke Eugen von Wuerttemberg's Reserve Corps, and was caught in isolation against the River Saale by Bernadotte's I Corps and destroyed. The regiment was not re-raised.

18. Regiment zu Fuss.

Raised in 1698. Title in 1806 **Regiment des Koenigs.**

Potsdam Inspectorate. Garrisons: Potsdam, Spandau (three battalions).

Uniforms. Rose red facings and turnbacks, white buttons and small clothes, Prussian cuffs. Officers wore six embroidered silver loops with tassels on and two under each lapel, two over each cuff, two on each pocket flap, four in the small of the back, and a narrow silver edging to the hat. NCOs' lace were of plain silver braid with tassels; six (in pairs) on and two under each lapel, two on each pocket flap and two in the small of the back. In 1806 their silver edgings to cuffs and cuff flaps had been removed. Their hats were edged in silver. Privates had six wide, silver lace loops with tassels on each lapel, two narrow ones over each cuff, two on each pocket flap and two in the small of the back. Tin grenadier front plate and ornaments up to 1787, rose red headband and cloth backing, white tape, green within red within white pompon.

Colours. King's colour white with narrow, red wavy corner piles, light blue centre, black eagle, white scroll edged silver, silver decoration. Regimental colours light blue with narrow, red wavy corner piles, white centre, black eagle, light blue scroll edged silver, silver decoration. Brown pike staffs.

All four colours were lost at the capitulation of Prenzlau on 28 Oct 1806.

Campaign history. Raised in 1698 as the Bataillon Grenadier-Garde from a battalion of IR Nr 1; Chef Oberst von Pannewitz; later Governor of Peiss. 1703 increased to a regiment under Chef Oberst Daniel von Tettau. 1709 Chef Oberst David Gottlieb von Gersdorff. **1709** B Malplaquet. 1713 reduced to a battalion; six companies were detached to form IR Nr 23. **1715** In Pommerania against the Swedes. 1716 converted to LIR Nr 18 and the ranks filled with 600 Swedish prisoners of war from the capitulation of Stralsund. The regiment lost the title of Guards. 1731 Oberst Georg Volrath von Kroecher; became Governor in Geldern. 1738 Oberst Christian Reinhold von Derschau; died in office as GM. **1740** Grenadiers in Gren Bn Wedel (17/18) until winter; thence with their parent regiment until after the battle of Chotusitz (17 May 1742). For a short time, the grenadiers were in Gren Bn Fouque (18/37/38). 1742 Chef Oberst Prinz August Wihelm von Preussen (since 1744 Prinz von Preussen; crown prince); died in office on 12 Jun 1758, in disgrace for his retreat to Bautzen. The regiment remained vacant from 1758 – 1764. Since 1744, the grenadiers were always together with the wing companies of the 2nd and 3rd Bns Garde. **1744** The regiment covered the retreat from Prague. C Solonitz; C Beraun. For combat history see that of IRs Nrs 15 and 17, **1745** B Hohenfriedberg; destroyed a Saxon infantry regiment. B Kesselsdorf. **1756-1763 The Seven Years War against Austria and others. 1756** 11 – 15 Oct Pirna **1757** 21 Apr B Reichenberg; 6 May B Prague; S Prague. 22 Nov B Breslau; 5 Dec B Leuthen; S Breslau. **1758** 25 Aug B Zorndorf; 14 Oct B Hochkirch, very heavy losses. **1760** S Dresden; 15 Aug B Liegnitz; C Hohengiersdorf; 3 Nov B Torgau. **1762** 26 Jul B Burkersdorf; B Reichenbach. 1764 Chef Prinz Friedrich Wilhelm von Preussen; succeeded to the throne in 1786 and died in 1797. In 1787 the regimental title became 'von Preussen'. **1779** C Neustadt in Upper Silesia. The regiment was isolated in the town and challenged to surrender by General Wallis with 12,000 men. It fought them off and escaped intact. 1786 Regiment von Preussen. 1790 Chef the Kronprinz von Preussen; 1797 'Regiment des Koenigs'; raised into guard status. From 1744 onwards, the grenadiers were always deployed together with those of the Wing Companies of the 2nd and 3rd Bns of the Guard; see IRs Nrs 15 and 17 for details. **1806.** Auerstaedt 16 Oct. The regiment surrendered at the capitulation of Prenzlau and was not re-raised.

19. Regiment zu Fuss.
Raised in 1702. Title in 1806 – **Prinz von Oranien.**
Berlin Inspectorate. Garrisons: Berlin, Cuestrin (three battalions).

Uniforms. Orange facings, yellow buttons, red turnbacks and stock. Officers wore two embroidered silver loops with loose tassels under each lapel, two in the small of the back and three on each Brandenburg cuff flap and pocket flap. They had a thin silver edging to the bicorn. NCOs' laces were of plain silver with tassels, in the same layout as for officers; silver hat edging; no lace edging to cuffs. By 1806 privates had lost the eight white lace loops with tassels from each lapel, but retained two under the lapels, two above the cuff and two in the small of the back. Prussian cuffs, white small clothes. Brass grenadier front plate and ornaments up to 1787, red headband, yellow cloth backing, white tape with orange pattern, orange within white pompon.

Colours. From 1740 to 1763: King's colour white with crimson Maltese cross and centre, black eagle, white scroll edged gold, gold decoration. Regimental colours crimson with white Maltese cross and centre, black eagle, crimson scroll edged gold, gold decoration. Brown pike staffs. The Colonel in Chief of the regiment, Margraf Karl von Brandenburg, was Master of the Order of St John.

Colours from 1763: King's colour white with very narrow upright cross. Regimental colours crimson with very narrow white 'union jack' cross. All other details as above.

All four colours were lost at the capitulation of Magdeburg on 11 Nov 1806.

Campaign history. Raised in **1702** from drafts from IRs Nrs 4, 6, 7 and 11 and the garrisons of Colberg and Kuestrin for Markgraf Albrecht von Brandenburg-Schwedt and placed in Dutch service until 1704. **1702** S Kaiserswerth. In 1703 it gave two companies to help raise IR Nr 10. **1706** 7 Sep B Turin. **1707** S Toulon (2nd Bn); the 1st Bn was in Brabant. **1708** B Oudenaarde (1st Bn). **1709** B Malplaquet (the whole regiment). **1710** S Mons (1st Bn); S Aire (2nd Bn). **1711** S Bouchain (1st Bn). **1715** In Pommerania; S Stralsund; storm of Peenemuende; heavy losses. 1731 Chef Oberst Markgraf Karl von Brandenburg-Schwedt, son of Markgraf Albrecht; 1747 GdI, died in 1763. **1740/1741** Grenadiers in Gren Bn Reibitz (13/19). 28 Dec-9 Mar before Glogau; 9 Mar storm of Glogau; 10 Apr B Mollwitz; heavy losses. 11 Apr-3 May S Brieg; 10 Aug capture of Breslau; 23 Aug raid on Schoenheide. **1742** B Chotusitz (grenadiers in Gren Bn Uchlander (7/19)). **1744** 2-18 Sep S Prague; 18 Sep-4 Oct advance to Moldauthein, Grenadiers in Gren Bn Finckenstein (19/25). 8 Oct-7 Dec retreat to Silesia; 5/6 Dec C Braunau; 7 Dec D Braunau. **1744/1745** Dec-Apr border security at Wuest-Giersdorf; 1 May C Hirschberg; 22 May C Landshut; 1-5 Jun in garrison in Striegau; 4 Jun B Hohenfriedberg; 17 Jun advanced guard action at Neustadt; 30 Sep B Soor; heavy losses. Nov-Dec winter campaign in Saxony and the Lausitz. **1756-1763 Against Austria and others in the Seven Years War. 1756** 11-18 Sep before Pirna, Grenadiers in Gren Bn Ramin (19/25). 29 Oct rearguard action at Hellendorf. **1757** 6 May-20 Jun B then S Prague; 10 Jun C in the Mansfeld gardens; 20 Jun rearguard action at Weissen Berg; 24 Aug advance to Kotta; 17 Oct Sk at Thyrow near Berlin; 31 Oct C Merseberg; 4 Nov cannonade at Schortau; 5 Nov B Rossbach; 5 Dec B Leuthen, very heavy losses. Grenadiers in Gren Bn Heyden; 6 Dec cannonade at Klein-Mochbern; 24 Dec capture of Troppau. **1758** 16 Feb advance on Graz; 18/19 Feb C Troppau; 31 Mar – 18 Apr S Schweidnitz; 3 May-1 Jul before Olmuetz; 22/23 Jun C Littau; 7 Jul rearguard action at Kroenau; 11/12 Jul rearguard action at Neu-Holitz; 14 Oct B Hochkirch; 30 Oct rearguard action at Lauban. **1759** 12 Aug B Kunersdorf; very heavy losses. 16 Sep cannonade at Alt-Tanneberg, Grenadiers in Gren Bn Kreckwitz (19/25, 13/26); 21 Sep C Korbitz; the grenadiers took 11 cannon and a colour. 29 Oct C Pretzsch, grenadier battalion commander now Homboldt; 13 Nov cannonade at Nossen; 20/21 Nov C and capitulation of Maxen. **1760** The musketiers were at S Dresden and B Torgau; the grenadiers in Gren Bn Schwerin (13/26, 19/25) at the following actions 3 Apr C Buchow-Zanow; 14 May C Baehrenwalde; 26 May D Belgard; 3 Jun C Belgard; 15 Jun C Greiffenberg; 18 Sep relief of Kolberg; 3 Oct C Pasewalk. The regiment was at Strehla on 17 Aug and at B Torgau in Ziethen's corps on 3 Nov; heavy losses. **1761** 14 Feb cannonade at Koeslin; 5 Jun C Koerli; 10-19 Jun Cs at Belgard; 24 Aug-14 Nov in the camp at Colberg; 18-29 Nov thrust at Belgard, actions at Regenwalde and Petershagen; 30 Nov C Naugard; 11 Dec C Drenow; 12 Dec C Spie; 13-18 Dec rearguard actions at Treptow-Stargard. **1762** 2 Jan C Neu-Kalden near Malchin; 12 May break through at Doebeln, Grenadiers in Gren BnWoldeck (19/25); 14/15 Oct C Brand; 29 Oct B Freiberg. 1763 GM von Tettenborn; received another regiment that same year. 1763 GL Herzog Friedrich August von Braunschweig; retired as GL. **1778** Jul-Oct campaign in Bohemia; Grenadiers in Gren Bn Loeben (19/25) and in the following actions: 15 and 26 Nov Cs Weisskirch; 28 Nov storm of the barricades at Jaegerdorf. 1794 GM Karl Ludwig Bogislav von Goetz; died in 1806 as GL. The regiment was in Poland. **1806** GL Fuerst Wilhelm Friedrich von Oranien-Fulda. 14 Oct B Auerstaedt. The regiment surrendered at Magdeburg and was not re-raised.

20. Regiment zu Fuss.
Raised in 1688. Title in 1806 – **Prinz Louis Ferdinand von Preussen.**
Magdeburg Inspectorate. Garrison: Magdeburg.

Uniforms. Scarlet facings with eight evenly-spaced yellow buttons on each lapel, red stock. White small clothes; Prussian cuffs. Officers wore embroidered golden lace buttonhole loops with loose tassels, eight on and two under each lapel, two each over the cuff, on the pocket flap and in the small of the back. Their hats had a narrow gold edging.

Brass grenadier front plate and ornaments up to 1787, scarlet headband, dark blue cloth backing. The tape has a red central stripe, flanked to both sides with dark green then white bands. The pompon colours, from the centre, are red, white, dark green, red. NCOs had no edging to their lapelsand no lace loops; they had gold edging to their hats, cuffs and cuff flaps. Privates wore no lace loops, but lapels , cuffs and cuff flaps were edged in white tape with three narrow blue stripes.

Colours. King's colour white with dark green centre, black eagle, white scroll edged gold, gold decoration. Regimental colours dark green, white centre, black eagle, dark green scroll edged gold, gold decoration. Brown pike staffs.

One colour of the 2nd Bn was lost at the capitulation of Erfurt on 16 Oct 1806, the other at the capitulation of Prenzlau. No certain news has survived as to the exact fate of the two colours of the 1st Bn.

Campaign history. Raised in 1688 from men on IRs Nrs 7 and 9 and from Oberst du Pless-Gouret's free company in the Spandau garrison. for GM von Schomberg. He was killed in the battle of Marseillein 1693 and the regiment was reduced to one company. In 1706 it was increased to a battalion; Chef GL von Borstel; died in office. 1711 Chef GM Ulrich Christoff von Stille, commandant of Magdeburg; died in office as GL. 1713 increased to a regiment of two battalions. **1715** In Pommerania; S Stralsund. 1728 Chef Oberst Wilhelm de Laujardiere; died in office. 1731 Oberst David Juergen von Grevenitz, GM in 1740; took over IR Nr 40. **1740/1741** Grenadiers in Gren Bn Buddenbrock (20/22). 29 Dec-3 Jan at Breslau; 9 Jan capture of Ohlau; 1741 Chef Oberst Johann August von Voigt; died in office as GM. 1-7 Apr S Brieg; 10 Apr B Mollwitz; 13 Jun rearguard action at Alt-Grottkau; 23 Aug raid on Schoenheide; 28 Aug C Nimptsch. **1742** Chef Oberst Johann Kaspar von Hertzberg; killed at Kesselsdorf on 15 Dec 1745. Grenadiers in Gren Bn Geist (5/20). 10-14 Mar Cs at Goeding and Ungarisch-Brod; 17 May B Chotusitz. **1744** 2-18 Sep S and capture of Prague, Grenadiers in Gren Bn Jeetze (5/20); 18 Sep –4 Oct advance to Moldauthein; 9 Oct rearguard action at Moldauthein; 20-30 Nov retreat to Silesia. 26 Oct reorganization: Grenadiers in Gren Bn Lepel (20/26). **1744/1745** Dec-Apr border security at Friedland and Falkenberg. 14 Feb C Habelschwerdt; 22 May C Landeshut; 4 Jun B Hohenfriedburg; 31 Aug Sk Chwalkowitz; 11 Sep C Neustadt; 25 Nov capture of Goerlitz. 15 Dec B Kesselsdorf. 1745 Chef GM Franz Andreas von Borck; retired as GL.Until 1747 the regiment was known as 'Jung-Borck' to differentiate it from IR Nr 29. **1756-1763 Against Austria and others in the Seven Years War. 1756** 24 Sep capture of Teschen castle. 1 Oct B Lobositz (1st Bn and grenadiers). Chef GM Berend Asmus von Zastrow; killed at Aussig on the advance into Bohemia on 25 Apr 1757. 1757 Chef GM August Gottlieb von Bornstedt; retired. 18 Jun B Kolin; lost almost 800 men. 5 Nov B Rossbach (grenadiers); 22 Nov B Breslau (musketiers); 5 Dec B Leuthen. **1758** S Olmuetz. 14 Oct B Hochkirch, lost over 500 men. **1759** GM Otto Ludwig von Stutterheim; retired as GL. The regiment was called 'Jung-Stutterheim' to differentiate it from

IR Nr 30. In the camp at Schmottseiffen. 12 Aug B Kunersdorf; 21 Sep C Gorbitz. **1760** 15 Aug B Liegnitz (grenadiers only); 3 Nov B Torgau in Ramin's brigade, lost almost 600 men. **1761** 2 Apr C Saalfeld. **1762** 29 Oct B Freiberg. 1784 Chef GM Matthias Wilhelm von Below; became Governor of Stettin. 1786 GM Dietrich Eugen Philipp von Bornstedt; died as GL. The regiment was called 'Jung-Bornstedt' to differentiate it from IR Nr1. **1779** 5 Feb C Brix (grenadiers). 1793 GM Prinz Ludwig zu Baden; 1795 GM Prinz Louis Ferdinand von Preussen; killed at Saalfeld on 10 Oct 1806. **1806.** Auerstaedt 16 Oct; the regiment surrendered in Magdeburg on 11 November and was not re-raised.

21. Regiment zu Fuss.

Raised in 1713. Title in 1806 – **Herzog Karl Wilhelm Ferdidand von Braunschweig.**

Magdeburg Inspectorate. Garrisons: Halberstadt, Quedlinburg (grenadiers), Magdeburg (three battalions).

Uniforms. Scarlet facings, white buttons, originally buff small clothes, changed to white by 1787, red stock. Officers had embroidered silver lace loops with loose tassels; two under each lapel, three on the cuff flap, three on each pocket flap and two in the small of the back. Their hats had a narrow silver edging and on their silver gorgets was the black, flying Prussian eagle. NCOs had wide silver hat edging and silver buttonhole lace with tassels; eight on and two under each lapel, three on each cuff flap and two in the small of the back. No lace edging to cuffs. Privates had the same distribution but their laces were white with two narrow scarlet lines; their tassels were white. Brass grenadier cap front plate and ornaments up to 1787, scarlet headband, black lace with white edging, buff cloth backing. The pompon colours, from the centre out, are black, white, black and red.

Colours. Up to 1802 - King's colour white with crimson centre, upright, wavy crimson cross, black eagle, white scroll edged gold, gold decoration. Regimental colours crimson, with black, upright wavy cross, white centre, black eagle, crimson scroll edged gold, gold decoration. Brown pike staffs.

In 1802 new colours were presented. The king's colour was white with narrow yellow corner piles, light blue centre, black eagle, white scroll edged silver, silver laurels and grenades, gold crowns and ciphers. Across the centre, to either side of the central disc, was: 'C.W.F.' and 'Hz.B.'(Carl Wilhelm Ferdinand Herzog von Braunschweig). The regimental colours were of the same design, but the centre was white, the field light blue, the corner piles yellow.

One colour of the 2nd Bn was lost at the capitulation of Erfurt on 16 Oct 1806, the other at the capitulation of Prenzlau. No certain news has survived as to the exact fate of the two colours of the 1st Bn.

Campaign history. Raised in 1713 from two companies each of the five regiments which had been in Dutch service(8, 9, 10, 13, 17); given to GM Ernst Graf von Doenhoff (1715 GL; died in office) the regiment was called 'Jung-Doenhoff' to differentiate it from IR Nr 2. **1715** In Pommerania.; S Stralsund. From 1722 it became 'Alt-Doenhoff' to differentiate it from IR Nr 13. 1724 Chef Oberst Heinrich Karl von der Marwitz (1742 GdI; died on 21 Dec 1744). **1740/1741** Grenadiers in Gren Bn Goetze-Winterfeldt (5/21). 29 Dec-9 Mar S Glogau; 8/9 Mar storm of Glogau; 10 Apr B Mollwitz. **1742** Grenadiers in Gren Bn Bolstern (3/21). Mar/Apr in Ungarisch-Hradisch; 10 Apr rearguard action at Kremsier; May- march to Bohemia. **1744** Chef GM Asmus Ehrentreich von Bredow (died in office as GL). Grenadiers in Gren Bn Finckenstein (21/25); Chef GM Asmus Ehrentreich von Bredow. 23-30 Aug advance to Lobositz; 2-18 Sep S Prague Grenadiers in Gren Bn Luck (12/21); 6 Sep C Beraun; 18 Sep-4 Oct advance to Moldauthein, Grenadiers in Gren BnGrumbkow (8/21); 8-19 Oct retreat to Beneschau; reorganization: Grenadiers in Gren Bn Grumbkow (7/21). 9-19 Oct defence of the River Elbe; 22/23 Nov rearguard action at Neu-Biczau; 2 Dec rearguard action at Starkstadt. **1744** Dec-Apr at Neisse fortress. **1745** 4 Jun B Hohenfriedberg; 30 Sep B Soor (grenadiers); 2-9 Dec with Winterfeldt's detachment. 15 Dec B Kesselsdorf (musketiers). **1756-1763 Against Austria and others in the Seven Years War. 1756** GM Johann Dietrich von Huelsen (1758 GL; died in office as Governor of Berlin). 14 Sep-20 Oct occupation of Dresden, Grenadiers in Gren Bn Lengefeld (21/27). 1 Oct B Lobositz. 21-26 Oct in Aussig; 29 Oct rearguard action at Hellendorf. **1757** 6 May-20 Jun S Prague, Grenadiers in Gren Bn Dieringshofen. 2/3 Jun C on the Angelka River; 18 Jun B Kollin (musketiers); the regiment was caught by Austro-Saxon cavalry and lost 727 killed and wounded and 250 captured. The regiment lost its colours. 20 Jun rearguard clash at Prague; 17 Aug cannonade at Hirschfelde; 7 Sep C Moys (grenadiers); after this action, Grenadiers in Gren Bn Dieringshofen (21/23/34). 5 Nov B Rossbach (1st Bn of musketiers only); 5 Dec B Leuthen (grenadiers); 7-20 Dec S Breslau. **1758** Grenadiers in Gren Bn Dieringshofen (21/27); 16 Apr storm of the Galgen fort at Schweidnitz (grenadiers), 3 May-1 Jul S Olmuetz; 11 Oct C Jenkwitz; 14 Oct B Hochkirch; 30 Oct rearguard clash at Lauban. **1759** 26 Mar ambush at Greiffenberg, grenadiers captured. C Basberg (musketiers); 12 Aug B Kunersdorf (musketiers, over 700 casualties). 20 Nov C Maxen; the rest of the regiment captured. **1760** The regiment was re-raised. In Silesia. 2 Jul thrust on Maerkisch-Friedland, Grenadiers in Gren Bn Kleist (21/27, 37/40); 6 Aug C Hundsfeld (grenadiers); 19 Sep C between Glogau and Brieg. 3 Nov B Torgau (musketiers). **1761** In Pommerania; 24 Aug-15 Nov in the camp at Kolberg; 19 Sep storm and defence of the Green Redoubt; 18-29 Nov advance against Belgard, actions between Regenwalde and Petershagen and (30 Nov) rearguard clash at Naugard; 11 Dec C Drenow; 12 Dec C Spie; 13-17 Dec rearguard clashes at Treptow-Stargard. **1762** 3 Jan C Neu-Kalden near Malchin; 23 Jun-8 Aug in Upper Silesia; Grenadiers in Gren Bn Budberg (21/27). 11 Aug-10 Oct covering Schweidnitz fortress. 1767Chef Oberst Karl Magnus von Schwerin (1760/1761 commander of the Grenadier Bn 13/26, 19/25; he became Chef of IR Nr 43). 1773 GdI Erbprinz Karl Wilhelm Ferdidand von Braunschweig; in 1780 he became the reigning Herzog Karl Wilhelm Ferdidand von Braunschweig and was mortally wounded at Auerstedt. **1778** Grenadiers in Gren Bn Holstein (21/27). 30 Jul-28 Sep on campaign in northern Bohemia; no actions. **1792** On the Rhine and in Champagne against France. Valmy. On the Moselle; in Koblenz, together with IR Nr 41 and Fus Bn Nr 1, held the vital post of Cartause to allow the army time to form up. **1793** C Edinghofen; Pirmasens; 17 and 20 Aug C Kettricher Hof; 14 Sep B Pirmasens, took several guns including three with teams. 13 Oct C Bodenthal (grenadiers); 16 Nov storm of Bitsch (323 volunteers, of whom 200 were lost); 28-30 Nov B Kaiserslautern. **1794** 23 May B Kaiserslautern. **1806.** Auerstaedt 14 Oct. The regiment capitulated and was not re-raised.

22. Regiment zu Fuss.
Raised in 1713. Title in 1806 – **von Pirch.**
Pommeranian Inspectorate. Garrisons: Stargard, Coeslin (grenadiers), Damm (three battalions).

Uniforms. From 1787 the facings were poppy red; yellow buttons, Prussian cuffs, red stock. Officers wore six golden lace loops on and two under each lapel, three over the cuff, two on each pocket flap and two in the small of the back. Their hats had a narrow gold edging. NCOs had wide hat edging, two gold laces under each lapel, four on the cuff and two above it and two in the small of the back. Privates had white lace loops with a diced red central pattern and flanking narrow red lines, pointed at the rear. They had two under each lapel, two in the small of the back, two above each cuff, and a lace edging to cuff and flap. White small clothes. Brass grenadier cap front plate and ornaments up to 1787, red headband, dark blue backcloth, dark blue lace edged white. White pompon, with mixed red and dark blue centre.

Colours. King's colour white with upright light red, wavy cross, dark blue centre, black eagle, white scroll edged gold, gold decoration. Regimental colours dark blue, upright wavy red cross, white centre, black eagle, dark blue scroll edged gold, gold decoration. Brown pike staffs.

One colour of the 1st Bn was lost in a clash at Nieder-Roessla on 15 Oct. The three colours of the regiment were lost at the capitulation of Pasewalk on 29 Oct 1806.

Campaign history. Raised in 1713, mainly from men of the 1st Bn, IR Nr 6 (then the King's Regiment), the Freikompagnie of GM von Dorthe and a company that had been raised in Oderberg for GM Adrian Berndt von Borck, the old commander of the Regiment Kronprinz (FM in 1737) the regiment was called'Alt-Borck' to differentiate it from IR Nr 29. **1715** In Pommerania against the Swedes. Ruegen Island; 17 Jul-24 Sep S Stralsund. **1741** Oberst Prinz Moritz von Anhalt-Dessau (later Fuerst), the youngest son of the 'Alten Dessauer'. He was promoted to FM on the field of Leuthen. He was badly wounded at the battle of Hochkirch and taken prisoner; he died in 1760 as GFM. **1741** 10 Apr B Mollwitz; 27-31 Oct S Neisse. Grenadiers in Gren Bn Buddenbrock (20/22). 1742 Grenadiers in Gren Bn Kleist (1/22). **1744** Grenadiers in Gren Bn Sydow (17/22, then 4/22); 6 Sep C Beraun; heavy losses. **1745** Grenadiers in Gren Bn Plotho

(10/22, from November). 4 Jun B Hohenfriedberg; 30 Sep B Soor, heavy losses. **1756-1763 Against Austria and others in the Seven Years War.** Grenadiers in Gren Bn Puttkamer, then Wrede, Kremzow, von der Tann, Wobersnow, Sobeck and Rothenburg. 11 - 15 Oct Pirna; 1 Oct B Lobositz (grenadiers only). **1757** 21 Apr B Reichenberg; S Prague; 18 Jun B Kolin (lost over 1,100 men); 5 Nov B Rossbach (grenadiers only); 5 Dec B Leuthen (grenadiers). **1758** Regiment re-raised; 25 Aug B Zorndorf. Oct Colberg. **1759** 23 Jul B Kay (grenadiers only). 12 Aug B Kunersdorf (grenadiers). **1760** 23 Jun B Landshut; very heavy losses, the remnants of the regiment were captured. Exchanged, reformed and used at S Dresden (10-22 Jul) and B Torgau on 3 Nov; took several connon and colours. As a reward for their actions here, all field officers received the *Pour le Merite* and a gift of 500 Thalers. Chef GM Balthasar Rudolf von Schenkendorff (retired as GL); the regiment was called 'Alt-Schenkendorff' to differentiate it from IR Nr 9. From 1760-1763 it was mostly referred to as 'Fuerst Moritz'. **1761** No actions of note. **1762** 21 Jul C Leutmannsdorf; C Burkersdorf 1768 Chef GM Karl Christoph von Ploetz (died in office). 1776 Chef GM Friedrich Karl von Schlieben (died in office as GL). **1778/1779** Grenadiers in Gren Bn Below (17/22). 22 Nov C Weisskirch (the whole regiment). 1791 Chef GM Karl Friedrich von Klinkwstroem; he then received IR Nr 28. 1795 Chef GL Franz Otto von Pirch. **1806** The musketiers of the regiment capitulated at Prenzlau; the grenadiers gathered themselves in Colberg and formed the 'Bataillon Waldenfels'. In 1808 it became part of the Leib-Infanterie-Regiment, which became IR Nr 9, and, in 1813 Nr 8. In 1814 the Leib-Grenadier-Bn became the **1st Bn, Kaiser Alexander Garde-Grenadier-Regiment Nr 1.** See IR Nr 2 above and serial Nr 2 in Regiments Raised After 1806.

23. Regiment zu Fuss.

Raised in 1713. Title in 1806 – **von Winning.**

Berlin Inspectorate. Garrisons: Berlin, Bernau (three battalions).

Uniforms. From 1787 the facings were scarlet, but the lapels were in the coat colour. In March 1798 the facings changed to rose red, which was also the colour of the officers' turnbacks. White buttons and small clothes. Officers wore silver lace loops with loose tassels as for IR Nr 10. Their hats had a narrow silver edging. NCOs wore plain silver braid loops, two under each lapel, two in the small of the back; wide silver hat edging, no lace to the cuffs. Privates wore two laces under each lapel, two over each cuff and two in the small of the back. Up to 1798 these white loops were round ended, thence they were white with three narrow, light blue lines and the ends were pointed and decorated with white tassels. Tin grenadier cap front plate and ornaments up to 1787, white headband, dark blue back cloth, red lace, pompon with mixed red and white filling.

Colours. Up to 1747: King's colour white with light blue centre and upright, wavy light blue cross, black eagle, white scroll edged gold, gold decoration. Regimental colours light blue with white centre and upright, wavy light blue cross; black eagle, light blue scroll edged gold, gold decoration.

From 1747: King's colour: white field and centre, light blue corner fields and scroll, black eagle, silver embroidery. Regimental colours: as for the king's colour but with a light blue centre.

All four colours were lost at the capitulation of Erfurt on 16 Oct 1806. Black pike staffs.

Campaign history. Raised in 1713 from 6 companies of IR Nr 13 and 4 from the Fusilier-Garde Nr 1. Chef GM Anton von Kamecke; retired. **1715** In Pommerania against Sweden. Ruegen Island; storm of the Peenemuende redoubt; S Stralsund. 1716 Oberst Johann von Forcade (1729 GL); died in office as Governor of Berlin. 1729 Oberst Egidius Ehrenreich von Sydow (1743 GdI); retired as Commandant of Berlin. **1741** 10 Apr B Mollwitz; 27-31 Oct S Neisse. 1743 GM Alexander Ernst von Blanckensee (killed in 1744 at the battle of Soor). **1744** 6 Sep C Beraun (grenadiers; heavy losses). **1745** 4 Jun B Hohenfriedberg; 30 Sep B Soor, heavy losses. Chef 1745 GM Christoph Graf von Dohna, who was Chef of IR Nr 4, and who passed on to IR Nr 16. 1748 GM Friedrich Wilhelm Quirin von Forcade (GL in 1760), son of GL Johann von Forcade; died as GL. **1756 – 1763 Seven Years War. 1756** Chef Oberst Christian Ernst von Puttkamer (received Grsn R von Renzel Nr 7). 10 – 15 Sep Pirna. **1757** 21 Apr B Reichenberg; 6 May B Prague; lost over 600 men. S Prague. 5 Nov B Rossbach; 5 Dec B Leuthen, heavy losses. **1758** 16 Apr storm of Schweidnitz (grenadiers only); 25 Aug B Zorndorf; lost 400 men. 14 Oct B Hochkirch, heavy losses. **1759** In the camp at Schmottseiffen. 28 Jul C Friedland; took 700 prisoners. **1760** Bombardment of Dresden. 15 Aug B Liegnitz; 17 Sep C Hohengiersdorf (grenadiers); 3 Nov B Torgau, lost nearly 600 casualties. **1762** 9 Mar C Gretha, together with IR Nr 1 they were attacked by over 4,000 Austrians but beat them off. 7 Aug-9 Oct S Schweidnitz. 29 Oct B Freiberg (grenadiers). 1765 GM Christoph Friedrich von Rentzel; died in office as GL. **1778** 22 Nov C Weisskirch (grenadiers); many of the officers received the *Pour le Merite*. Oberst August Wilhelm von Thuena; 1786 GM Stephan von Lichnowsky; died in office as GL. **1794** In Poland, acted in several detachments. 1796 Oberst Christian Ludwig von Winning (GL in 1806). **1806.** Jena 16 Oct. The regiment was scattered and fled into Erfurt fortress where they surrendered. The regiment was not re-raised.

24. Regiment zu Fuss.
Raised in 1715. Title in 1806 – **von Zenge.**
Mark Brandenburg Inspectorate. Garrisons: Frankfurt on the Oder, Soldin (grenadiers), Crossen (three battalions).

Uniforms. Poppy red facings, yellow buttons, a red and white braid edged the lapels, cuffs and cuff flaps of the mens' coats. In 1806 officers wore eight evenly-spaced buttons on each lapel; two embroidered gold lace loops under each lapel, two over the Prussian cuffs, two on each pocket flap and two in the small of the back. NCOs had six buttons, in pairs, on each lapel, each with an embroidered gold loop; two buttons on the cuff flap and two in the small of the back. Wide gold lace hat edging; no gold lace to the cuffs. Privates had six red and white lace loops (in pairs) on and two under each lapel, two above the cuff, and two in the small of the back. The lapels, cuffs and flaps were also edged in this lace, which was white and had a central red worm, flanked by a thin red line to each side. White small clothes. Brass grenadier cap front plate and ornaments up to 1787, red headband, dark blue backcloth, white lace with red zig-zag pattern. Pompon colours, from the centre, red, white, dark blue, yellow.

Colours. King's colour white with dark green centre and upright, wavy, dark green cross, black eagle, white scroll edged gold, gold decoration. Regimental colours dark green, white centre and wavy cross, black eagle, dark green scroll edged gold, gold decoration. Black pike staffs.

Three colours were lost at the capitulation of Pasewalk, the other at Erfurt on 16 Oct 1806.

Campaign history. In 1713 two independent battalions had been raised; one for GM Johann Sigismund Freiherr von Schwendy in Berlin, the other for GM Berend Christian von Schoenebeck in Colberg. In 1715 these were united as a regiment undr GM von Schwendy (1717 GL); he retired in 1723. **1715** Against the Swedes at Stralsund. 1723 GM Kurt Christoph von Schwerin (1740 FM and Graf; killed in 1757 at Prague); **1740/1741** Grenadiers in Gren Bn Puttkamer (12/24). 30 Dec-3 Jan S Breslau; 9 Jan capture of Ohlau; 23 Jan capture of Troppau; 25 Jan C Graetz; 10 Apr B Mollwitz; 28 Apr-3 May S Brieg (musketiers); 23 Jun C Alt-Grottkow; 30 Jul C Zobten; 2 Aug raid to Schoenheide. 27-31 Oct S Neisse (musketiers). **1742** Grenadiers in Gren Bn Itzenplitz (8/24) 17 May B Chotusitz. From 1743 the regiment was called 'Alt-Schwerin' to differentiate it from IRs Nr 31 and 32. **1744** Grenadiers in Gren Bn Kahlbutz (24/Grsn R 5). 27 Aug capture of Tetschen castle; 2-18 Sep S Prague, Grenadiers in Gren Bn Kahlbutz (24/27) 12 Sep storm of the Ziska; 18 Sep-4 Oct advance to Moldauthein; 15-19 Oct removal of the bakery of Tabor; 27 Nov C Jaromier. 3 / 4 Dec rearguard clashes at Nachod and Politz; 7 Dec D Braunau. **1745** Jan-May border security in the Waldenburg mountains; 4 Jun B Hohenfriedberg; 30 Sep B Soor (Grenadiers in Gren Bn Finck (4/24), heavy losses); 23 Nov ambush at Katholisch-Hennersdorf; Sk Gross-Strelitz. **1756-1763 Against Austria and others in the Seven Years War. 1756** 10-11 Sep before Pirna, Grenadiers in Gren Bn Grumbkow (17/24, 24/34); 1 Oct B Lobositz (grenadiers). **1757** 6 May B Prague, heavy losses; 6 May-20 1757 Jun S Prague, Grenadiers in Gren Bn Anhalt. Chef now GM Karl Christoph Freiherr von der Goltz (died in 1761 as GL). 24 May and 3 Jun sorties; 20 Jun rearguard action at Prague; 8 Aug reconnaissance to Loebau; 17 Aug cannonade at Hirschfelde; 7 Sep C Moys; 5 Nov B Rossbach (musketiers); 5 Dec B Leuthen (grenadiers); after the battle their Gren Bn Dieringshofen was made up of the grenadiers of IRs 24/21/27; 7-20 Dec S Breslau. **1758** 20 May-1 Jul before Olmuetz, Grenadiers in Gren Bn Naumeister (24/34); 27 Jul C Kramolna; 28 Nov C Schnellewalde. **1759** 23 Jul B Kay (musketiers) lost over 800 casualties; 12 Aug B Kunersdorf (musketiers); 27 Jul C Konradswaldau (grenadiers); 31 Aug cannonade at Johnsdorf (grenadiers); 11 Sep C Bernsdorf (grenadiers). **1760** Musketiers were at S Dresden. Grenadiers in Gren Bn Sobeck (24/34); 17 Jun C Landshut,

23 Jun C Landshut; the grenadiers were captured. 15 Aug B Liegnitz. 3 Nov B Torgau; the musketiers lost over 700 men. **1761** May-Jun in the camp at Glogau, Grenadiers in Gren Bn Rothenburg (24/34, 17/22); 20 Aug-11 Sep in the camp at Bunzelwitz; the grenadiers were at the following actions: 15 Sep C Gostyn; 30 Sep C Koeslin; 2 Oct C Spie; 22 Oct C Gollnow; 15 Nov cannonade at Greifenberg; 18-19 Nov advance to Belgard and clashes at Regenwalde and Petershagen; 30 Nov rearguard action at Naugard; 11 Dec C Drenow; 13-18 Dec rearguard action at Treptow-Stargard. From 1761-1763 the regiment was vacant. **1762** 6 Jul B Frieberg (musketiers); the grenadiers were at S Schweidnitz (4 Aug-10 Oct) and these actions: 20 Jul storm of Ohmsdorf castle; 21 Jul B Burkersdorf. 1763 Chef Oberst Bernhard Alexander von Dieringshofen; died in office as GM. 1776 Chef Oberst Prinz (Herzog from 1780) Leopold von Braunschweig, who was drowned in the Oder in 1785 as GM in a rescue attempt. **1778** 28 and 30 Sep rearguard action at Schweidnitz. **1779** 5 Feb C Brix. 1785 GM Gottlieb Ludwig von Beville; retired. 1791 Chef GM Karl Ludwig Sylvirus von Frankenberg; died in office **1794** In Poland; S Warsaw. 1795 Chef Oberst Johann Ludwig von Gruenberg; died as GM. 1799 Oberst August Wilhelm von Zenge; GM in 1806. **1806 Against France.** 14 Oct Auerstaedt. One battalion surrendered in Erfurt, the other in Pasewalk. The regiment was not re-raised.

25. Regiment zu Fuss.

Raised in 1713. Title in 1806 – **von Moellendorf.**

Berlin Inspectorate. Garrisons: Berlin, Neustadt-Eberswalde (three battalions).

Uniforms. Scarlet facings, yellow buttons, red stock. In 1806 officers wore two embroidered gold lace loops with open tassels under each lapel, three on each flap of the Brandenburg cuff, three on the pocket flap and two in the small of the back. The eight buttons on each lapel were evenly spaced. They had a narrow gold edging to the hat. NCOs also had the eight lapel buttons; they had two, smooth, gold lace loops (with one end pointed) under each lapel and two more in the small of the back. They had wide gold hat edging, three buttons on the cuff flap, but no lace decoration there. Privates had six white lace loops (in pairs), with pointed ends and blue stripes and tassels on each lapel, two under each; two on the cuff flap and two in the small of the back. White small clothes. Brass grenadier cap front plate and ornaments up to 1787, white headband, scarlet backcloth, white lace with three dark blue stripes. Pompon colours, from the centre out, red, yellow, white.

Colours. King's colour white with dark green centre and narrow yellow corner piles, black eagle, white scroll edged gold, gold decoration. Regimental colours dark green, white centre, black eagle, dark green scroll edged gold, gold decoration. Brown pike staffs.

Three colours were lost at the capitulation of Prenzlau, the other – of the 1st Bn - was destroyed by its bearer to prevent capture.

Campaign history. For full details of the battle history of this regiment, see those of IRs Nrs 1 and 19.

Raised in 1713 from garrison troops, initially in two independent battalions for GL Otto Freiherr von Schlaberndorf and GM Anton von Pannewitz. One of these units (Schlaberndorf's)had been part of the Kuestrin garrison and a Free Company von Driesen; the other was made up of two companies Free Corps von Pleiss; two von Memel and one von Friedrichsburg. Its Chef was GL Anton von Pannewitz. When he left in 1715 to take over IR Nr 13, von Schlaberndorf was given both battalions. **1715** S Stralsund. 1718 Chef Oberst Johann Christoph Graf von Wylich und Lottum. 1727 Chef Oberst Martin von Thile. 1728 Chef GM Friedrich August Graf von Rutowsky, from Saxon service; he was the illigitimate son of August the Strong (1756 FM and C-in-C of the Saxon troops in the camp at Pirna). Chef 1729 Oberst Christoph Wilhelm von Kalckstein (1747 GdI, died in 1759). **1740/1741** Grenadiers in Gren Bn Kleist (1/25). S Glogau. 10 Apr B Mollwitz, lost 247 men. **1742** Grenadiers in Gren Bn von der Trenck (25/27); they did not come into action. **1744/1745** Grenadiers in Gren Bn Finckenstein (19/25), except for the period 23-30 Aug 1744, when they were grouped 21/25. **1745** 4 Jun B Hohenfriedberg; 30 Sep B Soor; lost 264 men. **1756-1763 Against Austria and others in the Seven Years War.** Grenadiers in Gren Bn Ramin (19/25), then Hayden, Schwerin and Woldeck. **1756** 10-15 Sep Pirna. **1757** 9 May-20 Jun S Prague; 18 Jun B Kolin; 22 Nov assault on Breslau; 5 Dec B Leuthen; S Breslau. **1758** Olmuetz; 25 Aug B Zorndorf. **1759** In the camp at Schmottseiffen; 20 Nov C Maxen (grenadiers captured). **1760** GM Friedrich Ehrentreich von Ramin; 1793 FM; he died in 1816. The regiment was in Prinz Heinrich's corps; 3 Nov B Torgau, in Queiss' brigade. **1761** In the camp at Bunzelwitz, then off to Pommerania; 27 Sep C Spie at Colberg. **1762** In Silesia; 21 Jul B Buerkersdorf; S Schweidnitz. **1778/**

1779 Grenadiers in Gren Bn Loeben (19/25). 1782 Chef GL Wichard Joachim Heinrich von Moellendorf; **1806.** Auerstaedt 16 Oct. The regiment surrendered at Prenzlau and was not re-raised.

26. Regiment zu Fuss.
Raised in 1716. Title in 1806 – **Alt Larisch.**
Berlin Inspectorate. Garrisons: Berlin, Crossen (three battalions).

Uniforms. Brick red facings, yellow buttons, red stock. White small clothes; eight evenly-spaced buttons on the lapels. Originally officers wore embroidered gold lace loops with loose tassels to all buttons, but by 1792 these had been reduced to just two under each lapel, three on each Prussian cuff flap and packet flap and two in the small of the back. Their hats had a narrow gold edging. NCOs had the usual wide gold hat edging and wore two gold lace loops under each lapel and two in the small of the back. They had no gold lace to the cuffs. Privates wore plain yellow lace loops, pointed to the rear and with white tassels, in the same configuration as the NCOs. Brass grenadier cap front plate and ornaments up to 1787, white headband and back cloth, white lace with red edging. White pompon with red centre.

Colours. King's colour white with yellow centre and wavy, upright cross, black eagle, white scroll edged gold, gold decoration. Regimental colours yellow, white centre and wavy cross, black eagle, yellow scroll edged gold, gold decoration. Brown pike staffs.

The Leibfahne and one other were taken at the capitulation of Ratkau on 7 November; one other colour was captured on 14 Oct at Umpferstaedt, the last one at the capitulation of Erfurt.

Campaign history. For full details of the battle history of this regiment, see also those of IRs Nrs 13, 20 and 23.

Raised in 1714 from contingents of existing formations (Bn von Truchses and Bn Jung von Heyden) for GM Kurt Hildebrandt Freiherr von Loeben; 1721 GL; died in office. 1730 Oberst Alexander von Kleist; 1747 FM; died in office. From 1736-1738 the regiment was known as 'Alt-Kleist' to differentiate it from IR Nr 14, and from 1747 from IR Nr 27. 1740 Grenadiers in Gren Bn Wylich (23/26). **1741** 10 Apr B Mollwitz, lost almost 700 casualties. In 1742 the grenadiers were in garrison in Ottmachau. The main body of the regiment was in Neisse and Neustadt.

1744 Grenadiers in Gren Bn Geist (13/26) until the end of October. **1744/1745** Grenadiers in Gren Bn Lepel (20/26). 1745 4 Jun B Hohenfriedberg. **1749** Chef GM Dietrich Richard von Meyerinck; retired. **1756-1763 Against Austria and others in the Seven Years War. 1756** Grenadiers in Gren Bn Finck (13/26), then Kreckwitz, Humboldt, Billerbeck, Schwerin and Kalckstein throughout the war. The regiment was at the encirclement of the Saxon army at Pirna on 15 Oct. **1757** 6 May B Prague; 5 Nov B Rossbach; 5 Dec B Leuthen, as a reward for their conduct, 15 officers received the *Pour le Merite*. **1758** Chef GM Karl Heinrich von Wedell; later GL and Minister for War. 27 May-2 Jul S Olmuetz. 14 Oct B Hochkirch; the regiment was scattered by cavalry, only 150 men with three colours survived. **1759** In the camp at Schmottseiffen; 23 Jul B Kay (grenadiers); 12 Aug B Kunersdorf (grenadiers). **1760** GM Christian Bogislav von Linden; retired. 10 – 22 Jul S Dresden; 15 Aug B Liegnitz; 3 Nov B Torgau. **1761** Jun – Jul S Olmuetz; 11 Sep Dresden; **1762** 14 Oct C Brand (grenadiers); 29 Oct B Freiberg. 1764 Oberst Anton Abraham von Steinkeller; died in 1781 as GM and Commandant of Berlin. **1778/1779** 1778 GM Hans Christoph von Woldeck; died in office as GL. Grenadiers in Gren Bn Eberstein (23/26). 25 Oct C Jaegerndorf. **1779** 11 Jan C Moenick. From 1782 it was called 'Alt-Woldeck' to differentiate it from IR Nr 41. 1789 GM Philipp Albrecht von Schwerin; retired. The regiment was known as 'Jung-Schwerin' to differentiate it from IR Nr 52. **1794** In South Prussia and Poland. 1795 GM Karl Leopold von Larisch; from 1800 'Alt-Larisch' to differentiate it from IR Nr 53. In 1806 von Larisch was GL. **1806** Part of the regiment capitulated in Erfurt; part was in Bluecher's corps which surrendered after trying to take Luebeck; it was not re-raised.

27. Regiment zu Fuss.

Raised 1715. Title in 1806 – **von Tschammer.**

Magdeburg Inspectorate. Garrisons: Stendal (1st Bn and grenadiers), Gardelegen (2nd Bn), Burg (3rd Bn).

Uniforms. Poppy red facings with wavy white lace frame within the line of the buttons, yellow buttons, nine, evenly-spaced on each lapel; white small clothes, Prussian cuffs. Officers wore gold lace embroidery to the edgings of their facings and pocket flaps and a thin gold edging to the hat; above the button on the hat was a golden rosette. NCOs' lace was gold and was worn on lapels, cuffs and cuff flaps. The men had nine buttons on the lapels, two over each cuff and two in the small of the back. Their facings were edged in white, regimental lace in a zig-zag pattern, inside the lapel buttons. The regiment wore no buttonhole loops. Brass grenadier cap front plate and ornaments up to 1787, white headband and lace, poppy red backcloth, red pompon with yellow centre.

Colours. King's colour white with royal blue centre, black eagle, white scroll edged gold, gold decoration. A vertical, upright, wavy cross, the horizontal arms were red, the vertical arms were yellow. Regimental colours royal blue, white centre, black eagle, royal blue scroll edged gold, gold decoration. A vertical, upright, wavy cross, the horizontal arms were red, the vertical arms were yellow. Brown pike staffs.

All four colours were lost in the fighting at Luebeck on 6 Nov 1806.

Campaign history. For details of the battle history of this regiment, see also IRs Nrs 3, 9, 21, 22, 24 and 25.

Raised in 1715 as the 'Schwedisch-Pommersches Regiment' of the Prinz Leopold von Dessau (1742 FM). It included 400 Swedish prisoners of war; most officers and NCOs came from IRs Nr 3 and Nr 6. In 1747 the prince received his father's old regiment, the 'Alten-Dessauers Nr 3'. **1740/1741** Grenadiers in Gren Bn Bolstern (3/27). **1741** 7/8 May storm of Glogau. 10 Apr B Molwitz. **1742** Grenadiers in Gren Bn von der Trenck (25/27). 17 May B Chotusitz (musketiers) heavy losses. **1744** Grenadiers in Gren Bn Kahlbutz (24/27) S Prague (grenadiers). **1745** Grenadiers in Gren Bn Finck (24/27); from August Gren Bn Sydow (22/27); from November Gren Bn Kleist. 4 Jun B Hohenfriedberg; 15 Dec B Kesselsdorf; heavy losses. 1747 Chef GM Franz Ulrich von Kleist; mortally wounded at Lowositz. Until 1749 the regiment was known as 'Jung-Kleist' to differentiate it from IR Nr 26; from 1756 it became 'Alt-Kleist'. **1756 – 1763 Seven Years War. 1756.** Grenadiers in Gren Bn Lengefeld (21/27). 11-15 Oct Blockade of Pirna (grenadiers). 1 Oct B Lobositz (musketiers). After the battle the regiment all captains received the *Pour le Merite*. **1757** Chef GM Moritz Wilhelm von der Asseburg; retired. 9 May-20 Jun S Prague; Grenadiers in Gren Bn Dieringshofen (21/27); after the clash at Moys on 7 Sep it was combined with Gren Bn Anhalt (24/34). 22 Nov B Breslau (musketiers), heavy losses. 5 Dec B Leuthen. 8-21 Dec S Breslau (grenadiers); 28 Dec surrender of Liegnitz (grenadiers). **1758** Grenadiers in Gren Bn Dieringshofen (21/27). 31 Mar-18 Apr S Schweidnitz (grenadiers). 25 Aug B Zorndorf (musketiers), stabilised the line against Russian assaults; heavy losses. 14 Oct B Hochkirch. **1759** Chef GM Daniel Georg von Lindstaedt died in office. 26 Mar C Greiffenberg; the grenadiers were taken prisoner. 29 Oct C Pretsch (musketiers). **1760** Grenadiers in Gren Bn Kleist (21/27 and 37/40). No actions of note. **1761** Grenadiers (one company) in Gren Bn Kleist as in 1760; defence of Colberg including C Spie on 27 Sep. **1762** Grenadiers in Gren Bn Budberg (21/27). 7 Aug-9 Oct S Schweidnitz (1st Bn). 1764 Chef GM Peter Heinrich von Stojentin; died in office. 1776 GM Alexander Friedrich von Knobelsdorf; died in office as GFM. Grenadiers in Gren Bn Herzog von Holstein-Beck (21/27). **1778 / 1779** In the army of Prinz Heinrich in Bohemia. **1787** Expedition to Holland. 1 Oct C Halfwey; took a redoubt and four cannon. Next day they captured a 10-gun warship. **1793** On the lower Rhine. Sk Venlo; Sk Schwalm; Sk Roermond. Then joined the Austrian forces in Flanders. S Valanciennes, Sk Vicogne; Sk St Amand; 23 May B Famars. Sk Orchies; Sk Bouvines; 12 Jul capture of Conde. 27 Jul surrender of Valenciennes. In August it marched to Alsace; Sk St Imbert; 28-30 Nov B Kaiserslautern; on the right wing, they turned the battle into a victory. **1794** Sk Trippstadt; took three cannon. Sk Johanniskreuz. **1800** Oberst Friedrich Wilhelm Alexander von Tschammer; GM in 1806. **1806** The regiment was in Bluecher's corps and surrendered at the capitulation of Ratkau in November; it was not re-raised.

28. Regiment zu Fuss.
Raised 1723. Title in 1806 – **von Malschitzky.**
Upper Silesian Inspectorate. Garrisons: Brieg, Muensterberg (grenadiers).

Uniforms. Buff facings, white buttons, eight, evenly-spaced on each side of the chest; no lapels until 1787. No lace decoration, buff small clothes. Prussian cuffs. Officers had a narrow silver trim to their hats. NCOs wore silver lace to cuff, cuff flaps and hats. Tin grenadier cap front plate and ornaments up to 1787, dark blue headband, red backcloth, yellow pompon. The grenadier cap of 1806 had a buff headband edged white.

Colours. King's colour white with black centre, gold eagle, white scroll edged gold, gold decoration, narrow, light blue corner piles. Regimental colours black, white centre, black eagle, black scroll edged gold, gold decoration. Black pike staffs.

All four colours were lost at the capitulation of Magdeburg on 11 Nov 1806.

Campaign history. Raised in 1723 in Wesel from contingents of 'medium-sized men' from all other regiments, initially as a fusilier regiment for GM Konrad Wilhelm von der Mosel. 1733 Chef Oberst Friedrich Ludwig Graf zu Dohna (1747 FM); from 1747 the regiment was known as 'Alt-Dohna' to differentiate it from IR Nr 38. **1740** 14 Sep-20 Oct raid againdt Herstal; the grenadiers were in Gren Bn Lattorf-Hagen (13/28). **1742** Oberst Heinrich Karl Ludwig Herault de Hautcharmoy, who was mortally wounded as a GL in 1757 at Prague. Apr-Jun in Upper Silesia. **1744** 2-18 Sep S Prague; the grenadiers in Gren Bn Finck (28/38). 23 Sep capture of Tabor; 30 Sep capture of Budweiss; 25 Nov rearguard action at Prague; 11-13 Dec rearguard actions at Olbersdorf. **1745** Apr-Oct in Schweidnitz; from 4 July the commander of the Gren Bn was Ellert. 4 Jun B Hohenfriedberg (musketiers); 20 Oct C Hultschin (grenadiers). 30 Sep B Soor (musketiers). **1756-1763 Against Austria and others in the Seven Years War. 1756** in Cosel in Silesia; the grenadiers were in Gren Bn Kreytzen. No actions of note. **1757** Chef GM Lorenz Ernst von Muenchow; killed at Leuthen on 5 Dec 1757. 6 May B Prag (musketiers); 14 Aug C Landshut (grenadiers only; heavy losses); Aug-Dec in Neisse. The musketiers were in the 1st Austrian siege of Schweidnitz and were captured. **1758** 4 Apr: Chef GM von Kreytzen; died in office. The regiment was re-raised and was at the siege and storm of Schweidnitz on 16 Apr. S Olmuetz. 30 Jun C Domstaedl; heavy losses in protecting the supply convoy. **1759** 2 May: Chef GM Friedrich Ehrentreich von Ramin; in 1760 he received IR Nr 25. In the camp at Schmottseiffen. The grenadiers were in Glatz the whole year. **1759** 27 Jul C Conradswalde (grenadiers). **1760** GM Friedrich Wilhelm von Thiele; retired. 23 Jun C Landshut (grenadiers; very heavy losses). **1761** 8 Jul C Lubin; von Arnim gave temporary command of the Gren Bn to Podewils. 20 Aug-25 Sep in the camp at Bunzelwitz; 15 Sep C Gostyn; 22 Sep C Landsberg on the River Warthe; 30 Sep C Koerlin; 2 Oct C Spie; 2-19 Oct in the camp at Kolberg; 21 Oct C Zarnglaff. The grenadiers were captured; that winter only the musketiers were with the army. The grenadiers were reformed in the Gren Bn Hachenberg (Grsn Rs 5 and 10). **1762** The regiment had to be re-raised. From June the grenadiers were in Gren Bn de Chauvet (28/38); from 4 Aug command changed to Schaetzel. 14 Aug-10 Oct coverimg Schweidnitz; 16 Aug C Reichenbach 1770 GM Michael Konstantin von Kalinowa Zaremba; died in office as GL. **1778/ 1779** Grenadiers in Gren Bn Kamecke (28/32). 14 Sep C Jungbuchau. 1786 GL Ludwig Karl von Kalckstein; received IR Nr 5 and became Governor of Magdeburg as GL.

1789 Oberst August von Gentzkow; retired as GL. 1792 GM Friedrich Leopold von Ruits; he then received IR Nr 8. 1795 GM Karl Friedrich von Klinckowstroem (ennobled as Graf in 1798; retired as GL. 1799 Oberst Johann von Malschitzky (GM in 1806). **1794** In Poland. **1806.** Auerstaedt 16 Oct. The regiment was not re-raised.

29. Regiment zu Fuss.
Raised 1725. Title in 1806 – **von Treuenfels.**
Lower Silesian Inspectorate. Garrisons: Breslau.

Uniforms. Crimson facings, yellow buttons, red stock. Until 1787 no lapels; eight evenly-spaced buttons on each side of the chest (later on the lapels). Officers wore gold lace loops with loose tassels; two under each lapel, three on the cuff flap, three on the pocket flap and two in the small of the back. For NCOs the configuration of the tapes was as for the officers; they were plain gold as was their hat edging. No lace to the cuffs. Up to 1797 the other ranks wore 18 white lace loops with thin crimson worms at the edges; from then they had 10 lace loops with tassels; white, with a central light blue stripe and flanking crimson ones. They were set two below each lapel, two on each cuff flap and two in the small of the back. White small clothes. Prussian cuffs. Initially brass-fronted fusilier and grenadier cap front plates and ornaments up to 1787, red headband, white back cloth, dark blue lace edged red then white. Pompon colours, from the centre, dark blue, red, white.

Colours. King's colour white with royal blue centre, black eagle, white scroll edged gold, narrow, royal blue corner piles, gold decoration. Regimental colours royal blue, white centre, black eagle, royal blue scroll edged gold, gold decoration. Black pike staffs.

All four colours were lost at the capitulation of Pasewalk on 29 Oct 1806.

Campaign history. Raised in 1725 in Wesel from contingents of small men from regiments in East Prussia and Brandenburg as a fusilier regiment for Oberst Hans Christoph von Bardeleben (1726 GM; died in office). It was converted to a musketier regiment in 1741. 1736 Chef Oberst Georg Heinrich von Borck (1743 GL; died in office). Until 1741 the regiment was known as 'Jung-Borck' to differentiate it from IR Nr 22; from 1746 it became 'Alt-Borck' to differentiate it from IR Nr 20. **1740** 14

Sep-20 Oct raid against Herstal. **1741** Apr-Oct in the camp at Goettin, in the Gren Bn Itzenplitz (6/29 then 29/33). **1742** 21 May in the camp at Chotusitz in Gren Bn Byla (29/30). **1744** 1 Oct C Moldauthein in Gren Bn Kleist von Wuerttemberg (4/29); 7-14 Oct operations east of Tabor; now in Gren Bn Luck (12/29) 9-19 Nov defence of the Elbe River at Neu-Kolin; 30 Nov-6 Dec rearguard actions at Nachod and Friedland. **1745** 22 May C Landshut; 4 Jun B Hohenfriedberg; 22 Jun C Schmirschitz; 8 Oct Sk Wildschuetz; 1 Dec C Loewenberg. 1747 Chef GM Hans Caspar Ernst von Schultze, fatally wounded as GL at Breslau in 1757. **1756-1763 Against Austria and others in the Seven Years War. 1756** 20 Sep C Koenigslotha; 22 Sep C Schmirschitz. **1757** Chef GM Hans Caspar Ernst von Schultz; died of wounds received at Breslau. 18 Apr C Goldenoels; 26 Apr C Jung-Bunzlau; 6 May B Prague, then siege of the city until 20 Jun; 14 Jul D Reichstadt castle; 18 Jun B Kolin; terrible losses. 18-21 Jul rearguard actions at Boehm, Kamnitz and Kreibitz; 26 Sep cannonade of Barschdorf; 5 Nov B Rossbach; 22 Nov B Breslau; 5 Dec B Leuthen; 7-20 Dec S Breslau. **1758** GM Karl Heinrich von Wedell, who received IR Nr 26 in the same year. 1758 GM Karl Gottfried von Knobloch. 1 Jan-16 Apr S Schweidnitz; the regiment was reformed. 28 Sep C Ferhbellin; 15 Nov C Eilenburg. **1759** 15 Apr capture of Toeplitz and destruction of the magazines there; 16 Apr C Libochowitz; 11 May ambush at Himmelskron; 16 May capture of Bamberg; 31 May C Muenchberg; 12 Aug B Kunersdorf; heavy losses. Grenadiers now in Gren Bn Stwolinsky (29/31 and 9/10)16 Sep cannonade of Alt-Tanneberg; 21 Sep C Korbitz.21 Nov one battalion captured at Maxen. **1760** in Gren Bn Falkenhayn (29/31). Musketiers in Breslau. Grenadiers: 7 Jul C Klein-Welka; 13-22 Jul S Dresden; 19 Jul sortie at the Wilsdruff suburb; 15 Aug B Liegnitz; 17 Sep C Hochgiersdorf; 3 Nov B Torgau. After the battle the grenadiers were combined with those of 9/10. **1761** 20 Aug-11 Sep in the camp at Bunzelwitz; the grenadiers again in Bn Falkenhayn (29/31). 15 Nov cannonade at Greifenberg in Pomerania; 18-29 Nov thrust at Belgard and clashes at Regenwalde and Petershagen; 30 Nov rearguard action at Naugard; 11 Dec C Drenow; 12 Dec C Spie; 13-16 Dec rearguard actions at Treptow, Schwirsen and Guelzow. **1762** 11 Mar C Merseberg; 6 Jul C Adelsbach; 7 Jul C Friedland; 21 Jul B Burkersdorf; 4 Aug-10 Oct S Schweidnitz. 1764 Chef GM Johann Ferdinand von Stechow; died in office as GL. 1778 Chef GM Hans Ludwig von Flemming; later became

GL and Commandant of Breslau. 1782 Chef GM Balthasar Ludwig Christian von Wendessen; died in office in 1797 as GL and Governor of Warsaw. 1798 Chef Oberst Karl von Treuenfels; GM in 1806. **1806.** Jena 16 Oct. The regiment surrendered at the capitulation of Pasewalk and was not re-raised. Parts of the regiment went into the 2nd Westpreussischen IR Nr 7, Ser Nr 8 in the new regiments' section below.

30. Regiment zu Fuss.
Raised in 1728. Title in 1806: **vacant von Borcke.**
Pommeranian Inspectorate. Garrisons: Stettin, Colberg (three battalions). Remnants of the regiment in Colberg formed part of the new Colberg IR. Titles. 1 Jul 1813 - **Colbergisches IR Nr 10**; the number changed to 9 that same day. See Serial Nr 10 in the new regiments section below for history post 1807.

Uniforms. Initially red facings, no lapels, red stock, yellow buttons. Officers wore gold embroidered lace loops to all button holes. From 1787 buff facings, yellow buttons. All ranks wore eight evenly-spaced buttons on each lapel. Officers wore two embroidered silver lace loops under each lapel, three on each cuff flap and pocket flap and two in the small of the back. Their hats were edged in thin silver lace. NCOs wore plain silver buttonhole laces to the coat – as for officers - and a wide silver hat edging. No lace to the cuffs. Up to 1798 other ranks wore 22 plain white lace loops without tassels, thence white with a central red stripe flanked by dark blue ones; red, white and blue tassels. These were worn two under each lapel, two over each cuff and two in the small of the back. There were eight evenly-spaced buttons on each lapel. White small clothes. Initially brass-fronted fusilier caps, then bicorns and brass grenadier cap front plate and ornaments up to 1787, red headband, white back cloth, white lace edged dark blue. Pompon colours, form the centre, red, dark blue, white. From 1787 – 1799 the Kasket was worn.

Colours. King's colour white with black eagle and narrow corner piles, dark green centre, white scroll edged gold, gold decoration. Regimental colours dark green with white centre, black eagle and corner piles, dark green scroll edged gold, gold decoration. Black pike staffs.

All four colours were lost at the capitulation of Ratkau on 7 Nov 1806.

Campaign history. Raised in 1728 as a fusilier regiment for Oberst Martin von Thile from men capable of field service in the Kuestrin Garrison Regiment (* on 12 Feb 1717). It was converted into a musketier regiment in 1740. 1733 Chef Oberst Joachim Friedrich von Jeetze (1747 FM); from 1748 the regiment was known as 'Alt-Jeetze' to differentiate it from IR Nr 17. **1735** The regiment marched to the Rhine. **1741** 27-31 Oct S Neisse. **1742** 17 May B Chotusitz. In the 2nd Silesian War (1744-1745) the grenadiers were in the Gren Bn Lepel (7/30); from 26 Oct 1744 the title became Schoening (8/30). **1745** 14 Feb C Habelschwerd (2nd Bn); 4 Jun B Hohenfriedberg; 30 Sep B Soor (grenadiers); 15 Dec B Kesselsdorf. The regiment captured the village of Kesselsdorf, taking 20 cannon, four mortars, a colour and a pair of kettledrums. In recognition of this deed, the king awarded all field officers of the regiment the *Pour le Merite* and granted the regiment a new seal bearing the captured pieces. It bore the superscription: 'Bataille bei Kesselsdorf, den 15ten Dec. 1745.' 1752 Chef GM Christian Gottfried von Uchtlander. 1755 GM Berendt Sigismund von Blanckensee; in 1756 he received one of the Saxon regiments pressed into Prussian service. **1756-1763 Against Austria and others in the Seven Years War. 1756** Chef GM Hans Samuel von Pritz; died in office; succeeded that same year by GM Ernst Ludwig von Kannacher; retired. The grenadiers were initially in the Bn Kanitz and at Pirna 10 – 15 Oct; after the battle of Prag the battalion became 'Lubath', from 1762 Kanitz again. 1 Oct B Lobositz (musketiers) heavy losses. **1757** 6 May B Prague, distinguished. S Prague; 5 Nov B Rossbach (grenadiers); 22 Nov B Breslau and 5 Dec B Leuthen (musketiers) they then went to S Breslau (8-21 Dec). **1758** 14 Oct B Hochkirch; retook a lost battery and held it in the face of all enemy assaults until ordered repeatedly to withdraw. They lost half their strength. **1759** Chef GM Joachim Friedrich von Stutterheim; the regiment was known as 'Alt-Stutterheim' to differentiate it from IR Nr 20. Stutterheim received IR Nr 2. C St Sebastiansberg (grenadiers) they took the Austrian General Reinhard and 2,000 men prisoners. 12 Aug B Kunersdorf (grenadiers) took a Russian colour.13 Aug storm of Torgau; 21 Sep C Meissen. **1760** 19 Jul bombardment of Dresden (musketiers); 20 Aug C Strehla; 3 Nov B Torgau. Almost all officers in the regiment were wounded; despite heavy losses, they took took the Siptitzer heights at the third attempt. **1762** The regiment was reformed. 29 Oct B Freiberg. In 1768 Stutterheim transferred to IR Nr 2. 1768 Chef

Oberst Karl Franz Freiherr von Sobeck; died in office as GM. 1778 Chef Oberst Philipp Wolfgang Teuffel von Birckensee; resigned in 1781. 1780 Chef GM Georg August von Schoenfeldt; retired. **1792** Chef Oberst Georg Friedrich von Wegnern (killed as GM at Ensheim in 1793). The regiment was on the Rhine against France. Blockade of Longwy; Valmy. Sk Bockenheim. **1793** 19 Jun-23 Jul S Mainz; blockade of Landau. 1794 Chef GM Ernst Friedrich Wilhelm von Ruechel; in 1798 he received the Regiment-Garde. C Frankenthal; C Creuznach; B Kaiserslautern. **1798** Oberst Friedrich Adam von Borcke; died as GL in 1806. 1806 vacant. **1806** The regiment was in Bluecher's corps. The regiment itself was not re-raised, but the 3rd Bn (raised in 1788) was in Kolberg; in 1808 it became the 2nd Bn, Kolberg IR Nr 10. In June 1813 the number changed to 9. See Ser Nr 10 in the new regiments below.

31. Regiment zu Fuss.

Raised 1729. Title in 1806 – **von Kropff.**

Warsaw Inspectorate. Garrisons: Warsaw, Lencziz (three battalions).

On 21 Nov 1808 remnants of the regiment went to form the **1. Schlesisches IR Nr 10.** See Serial Nr 11 in the new regiments section below for subsequent history.

Uniforms. Rose red facings, yellow buttons, no lace decoration, Prussian cuffs; white small clothes. Initially no lapels; from 1787 lapels were worn; here were eight evenly-spaced buttons on each lapel. Officers wore gold lace embroidery to lapels, cuffs and pocket flaps and a thin gold edging to their hats. No embroidered buttonhole lace loops. NCOs had gold lace edging to cuffs, cuff flaps and hats. Initially brass-fronted fusilier caps; grenadiers wore brass-fronted grenadier cap front plate and ornaments; rose red headband and back cloth, rose red lace edged yellow then white. Pompon colours, from the centre, rose red, yellow, white. From 1787 – 1799 the Kasket was worn.

Colours. King's colour white with light blue centre and narrow corner piles, black eagle, white scroll edged gold, gold decoration. Regimental colours light blue, white centre and corner piles, black eagle, light blue scroll edged gold, gold decoration. Black pike staffs.

Three colours were lost at the capitulation of Schweidnitz on 6 Feb 1807; the fourth was hidden and could not be found again.

Campaign history. Raised in 1729 in Wesel as the last of the four fusilier regiments formed by the Soldier King. First Chef Oberst Friedrich Wilhelm von Dossow in 1729; he later received IR Nr 45 and became GFM. The men came from IRs 9, 10, 28 and 29. In 1742 it was converted to a musketier regiment. 1743 Chef Oberst Friedrich Wilhelm Marquis de Varenne; died in office. **1744** Chef Oberst Philipp Bogislav von Schwerin; he was promoted to GM and received IR Nr 13. The regiment was known as 'Schwarz-Schwerin' to differentiate it from IRs Nr 24 and 32. The grenadiers were in Gren Bn Jaeger (31/40). 2-18 Sep S Prague; 1 Oct C Moldauthein; 18 Oct-19 Nov security duties east of Prague; 19-22 Nov security duty on the River Elbe; 23 Nov withdrawal from Melnik; 2 Dec C Marschendorf. **1745** The grenadiers were in Gren Bn Lindstedt. 4 Jun B Hohenfriedeberg; 30 Sep B Soor; 16 Oct C Trautenbach-Schatzlar; Nov-Dec in Schweidnitz. 1746 Chef GM Johann Georg von Lestwitz (1762 GL, received IR Nr 32). **1756-1763 Against Austria and others in the Seven Years War.** The grenadiers were in Gren Bn Oestenreich (29/31); from 1760 it became Falkenhayn. No actions of note. **1757** 6 May B Prag; lost a third of their strength. S Prague (musketiers); The grenadiers fought at 18 Jun B Kolin, C Boehmisch-Leipa; 7 Sep C Moys or Goerlitz; 5 Dec B Leuthen; and C Kloster Wahlstadt; 22 Nov B Breslau; 5 Dec B Leuthen. 8-21 Dec S Breslau. **1758** The grenadiers were at the blockade of Schweidnitz (31 Mar-18 Apr), then in the Mark, 25 Sep C Fehrbellin (or Linum). **1759** 23 Jul B Kay; 12 Aug B Kunersdorf. The conduct of the regiment so pleased the king, that he gave all privates 8 Groschen. The grenadiers were at 8 Sep C Torgau; 14 Nov C Korbitz. **1760** S Dresden; 15 Aug B Liegnitz; 17 Sep C Hohengiersdorf (grenadiers); 3 Nov B Torgau, the regiment took the Siptitzer heights. **1761** No actions of note. **1762** 6 Jul C Adelsbach, threw the enemy off the heights at a loss of over 300 casualties. 7 Aug-9 Oct S Schweidnitz (grenadiers). 1763 Chef GL Bogislav Friedrich von Tauentzien (1791 GdI; died as Governor of Berlin). In the War of the Bavarian Succession (1778/1779) The grenadiers were in Gren Bn Kowalsky (29/31); for details see IR Nr 29. 1791 Chef GM Philipp Ernst von Borck; died in office. 1792 Chef Oberst Philipp Friedrich Lebrecht von Lattorf; became Governor of Kuestrin as GL. **1794** In Poland. 1800 Chef Oberst Bernhard Vollrad von Oldenburg; died in 1805 as GM. 1805 Chef GM Heinrich von Kropff. **1806** The main body of the regiment surrendered at the capitulation of Schweidnitz. The 3rd Bn (* 1788) was in Warsaw; together with other units, it formed the 1st Bn IR

Nr 11. In 1813 this number changed to 10. See Ser Nr 11 in the new regiments below.

32. Regiment zu Fuss.
Raised 1743. Title in 1806 – **Fuerst Friedrich Ludwig zu Hohenlohe-Ingelfingen.**
Lower Silesian Inspectorate. Garrison: Breslau.

Uniforms. At first the Swedish cuffs were in the coat colour and no lapels were worn; yellow buttons, no lace decoration, black stock, white small clothes. By 1787 buff facings, Prussian cuffs and lapels had been introduced. There were eight evenly-spaced buttons on each lapel, three on the cuff flap and two in the small of the back. Officers had a narrow gold edging to their hats. NCOs wore gold lace edging to cuffs, cuff flaps and hats. Initially brass-fronted fusilier caps; grenadiers wore brass-fronted grenadier cap front plate and ornaments; buff headband, red back cloth, lace, from the centre, red, white, dark blue, yellow. Pompon colours, from the centre, buff, dark blue, white, red. From 1787 –1799 the Kasket was worn.

Colours. King's colour white with light blue centre, black eagle, white scroll edged gold, gold decoration. Narrow red corner piles, upright, narrow, black wavy cross. Regimental colours light blue, white centre, black eagle, light blue scroll edged gold, gold decoration. Narrow red corner piles, upright, narrow, black wavy cross. White pike staffs.

All four colours were lost; those of the 1st Bn, probably at Jena; those of the 2nd Bn at the capitulation of Erfurt on 16 Oct 1806.

Campaign history. Raised in 1743 as a fusilier regiment from two garrison battalions in Minden and Geldern. The Geldern battalion later became the 1st Bn. In 1723 Chef was GL von Lilien; 1724 he was succeeded by Oberst von Roeseler and in 1728 GM von Kroecher took over. GM von Raders was first Chef of the Minden battalion in 1730; in 1731 Oberst von Beaufort took over. In 1743 Oberst Friedrich Julius von Schwerin became Chef of the regiment; it was known as 'Jung-Schwerin' to differentiate it from IR Nr 24 and 31. Schwerin died as GM. The regiment was converted to a musketier unit in 1745.

1744 Grenadiers in Gren Bn Hertzberg (32/33). 7 Dec advance to Oderberg; 20 Dec retreat to Ratibor; 24 Dec retreat to Cosel; on 31 Dec they crossed the Oder at Brieg. **1745** Musketiers in C Neustadt 22 May; grenadiers in Jan and Feb in Grottkau and Oppeln; Mar skirmishes with Hungarian light troops; 11 Apr C Gross-Strehlitz; 20 Apr C Konstadt; 4 Jun B Hohenfriedeberg; 31 Aug C Lewin, near Glatz; 9 Nov C Oppau. From 20 Aug-5 Sep the whole regiment was at S Cosel. 1747 Chef GM Joachim Christian von Tresckow (1763 GL); until 1754 'Jung-Tresckow' to differentiate from IR Nr 8. **1756-1763 Against Austria and others in the Seven Years War. 1756** In southern Silesia; no actions of note. **1757** 6 May B Prague and S Prague until 19 Jun; 7 Sep C Moys, very heavy losses. 22 Nov B Breslau again very heavy losses. **1758** 31 Mar-18 Apr S Schweidnitz (grenadiers); C Domstaedl (the whole regiment). Then in Breslau. **1759** The regiment was reduced to one battalion. 23 Jul B Kay in Pommerania. **1760** 23 Jun C Landshut (grenadiers very distinguished; almost wiped out). The regiment was in Schweidnitz. **1761** The regiment was captured in the fall of Schweidnitz and not re-raised for two years. **1763** Re-raised, using the ex-Saxon IR von Loen; new Chef GM von Horn. The grenadiers were initially in Gren Bn Kreytzen (28/32); from 1758 Arnim; 1762 de Chauvet/Schaetzel. 1763 Chef GL Johann Georg von Lestwitz (died in 1767). 1767 vacant von Lestwitz. 1770 Oberst Hans Christoph von Rothkirch; from 1782 'Alt-Rothkirch' to differentiate from IR Nr 11. Rothkirch became commandant of Neisse and died in 1785. **1777/1778** Grenadiers in Gren Bn Kamecke (28/32). **1779** 14 Jan C Zuckmantel. 1786 Chef Fuerst Friedrich Ludwig zu Hohenlohe-Ingelfingen (1806 GdI). **1792** On the Rhine and in Champagne. 30 Aug capture of Verdun; 20 Sep Valmy. **1793** 6 Jan C Hochheim; 1-7 Mar blockade of Koenigstein; C Stromberg and Bingen, together with Fus Bn Nr 10;19 Jun-23 Jul S Mainz with Fus Bn Nr 10; C Homburg; 26 Sep C Hornbach; C Eschweiler; Weissenburg lines; C Limburg; blockade of Landau. **1794** 23 May C Diedesheim, Forst and Wachterheim; took two colours. C St Martin; Plettersberg, Roth and Weier; 20 Sep B Kaiserslautern. **1806.** Jena 16 Oct. The regiment surrendered at the capitulations of Erfurt and Pasewalk and was not re-raised.

33. Regiment zu Fuss.
Raised 1740. Title in 1806 – **von Alvensleben.**
Upper Silesian Inspectorate. Garrisons: Glatz, Silberberg (three battalions).
On 21 Nov 1808 remnants of the regiment went to form part of the **Schlesisches IR Nr 12**. See Serial Nr 12 in the new regiments section below.

Uniforms. White facings and small clothes, yellow buttons, black stock, no lace decoration. Officers' hats had a narrow gold edging and they initially had six buttons, in pairs, on each lapel. In 1787 these were eight evenly-spaced buttons on each lapel, two under each pocket flap and two in the small of the back. Swedish cuffs with two buttons. NCOs wore gold lace edging to cuffs and hat. Initially brass-fronted fusilier caps; grenadiers wore brass-fronted grenadier cap front plate and ornaments; orange headband and back cloth, white lace with red, hollow diamond pattern, pompon colours, from the centre, red, orange, white. The grenadier cap of 1799 - 1806 had a light blue band edged yellow.

Colours. King's colour white with cornflower blue centre, black eagle, white scroll edged silver, silver decoration. Regimental colours cornflower blue, white centre, black eagle, cornflower blue scroll edged silver, silver decoration. White pike staffs.

All four colours were lost at the capitulation of Magdeburg on 11 Nov 1806. A colour that had been laid up in the depot was presented in 1814 to the newly-raised Fuesilier Bn of the Kaiser Franz Grenadier Regiment Nr 2.

Campaign history. Raised in 1734 by the princes of Anhalt at one battalion as their contingent to the Imperial army for the campaign on the Rhine. Following the peace of 1736, it went into Prussian service and was reinforced by King Friedrich, using men of the Magdeburg garrison to form a battalion and a grenadier company; Chef in 1734 Oberst von Schwandes; 1736 Chef Oberst von Wachholtz; died in office. 1736 Chef Oberst d'Arebaud; died in office. In 1740 it was converted to a fusilier regiment of two battalions for Oberst Andreas Johann de Persod. **1741** in the camp at Goettin; grenadiers in Gren Bn Lattorf (11/33), brigaded together with Gren Bn Itzenplitz (29/33?). 1742 grenadiers in Gren Bn Wiese (33/42) in Oppeln. 1743 Chef GM Samuel von Schlichting, who then received IR

Nr 2. 1743 Chef Oberst Asmus Ehrentreich von Bredow, who then received IR Nr 21. **1744** Chef Oberst Heinrich August Baron de la Motte Fouque (died in 1774 as GdI); grenadiers in Gren Bn Hertzberg (32/33). Dec retreat to Neisse. **1745** 11 Apr C Gross-Strehlitz; 20 Apr C Konstadt; 4 Jun B Hohenfriedberg (grenadiers); 31 Aug C Lewin; 9 Nov C Oppau. **1756-1763 Against Austria and others in the Seven Years War. 1756** Sep-Oct the advance to the Elbe; grenadiers in Gren Bn Nimschoefsky (33/42). No actions of note. **1757** 6 May B Prague (musketiers); lost a colour. The grenadiers were brigaded together with those of 12/39 in the following actions: 5 Jun C Gang; 13 Jun C Kuttenberg; 18 Jun B Kolin; 17-21 Jul actions on the retreat between Boehmisch-Kamnitz and Neudoerfel. From 22 Sep in they were in Glatz fortress. 22 Dec C Landshut. **1758** 6 Apr C Friedland; 20 May-1 Jul S Olmuetz (the whole regiment); the grenadiers were also at 17 Jun C Holitz; 14 Oct B Hochkirch. **1759** 10 Jul-16 Aug in the camp at Schmottseiffen; the musketiers saw no major actions; the grenadiers were at: 27 Aug advanced guard clash at Sorau; 2 Sep C on the retreat from Sorau; 1 Oct guarded the bridging operation at Meissen. **1760** On 23 Jun the musketiers were caught in the clash at Landshut and captured after fierce resistance. The grenadiers saw action as follows: 13-22 Jul S Dresden; 15 Aug B Liegnitz; 17 Sep C Hoch-Giersdorf; 2 Nov advanced guard clash at Schoena; 3 Nov B Torgau. After the battle the remaining grenadiers were brigaded with 8/46. **1761** 5-13 Apr the raid into the Voigtland; 22 Jul C Muensterberg; 20 Aug-25 Sep in the camp at Bunzelwitz. The regiment was captured and not reraised. **1762** The surviving grenadiers were at 21 Jul B Burkersdorf; grenadier battalion commander now Mosch. 4 Aug-10 Oct covering Schweidnitz. The musketiers were captured here and the regiment was not re-raised until 1763, when the men of the ex-Saxon regiment Prinz Maximilian (Prussian title Prinz Wilhelm) were brought in. 1774 Chef GM Georg Reinhold von Thadden; retired as GL. **1778** Grenadiers in Gren Bn Preuss (33/42). 15 Sep rearguard action at Hohenbrueck. **1779** 28 Feb cannonade of Neustadt in Upper Silesia. They were together with IR Nr 18 and were attacked by the Austrian General von Wallis with 12,000 men, but managed to escape after a hard fight. 1784 Chef GM Friedrich Wilhelm von Goetzen; Governor of Glatz in 1787, GL in 1790, died in office. **1794** Chef GM Theodor Philipp von Pfau (killed on 13 Jul 1794 at the Schaenzel); 1794 Chef GL Franz Andreas von Favrat; died in 1804 as GdI. The regiment was in Poland, at S Warsaw; on the Pilica

River and in South Prussia; C Pyzdry on the River Wartha; date unknown. The Poles were thrown back over the river. 1804 GM August Ludwig von Schlierstedt; was allowed to resign his office. **1805** Oberst Ludolf Friedrich von Alvensleben; GM in 1806. **1806.** Auerstaedt 16 Oct. The bulk of the regiment surrendered in the capitulations of Erfurt and Pasewalk and was not re-raised. The 3rd Bn was in Silberberg, which held out until the end of the war. Together with other units it formed the 2nd Silesian IR Nr 12. In 1813 this number became 11; see Ser Nr 12 in the new regiments section below.

34. Regiment zu Fuss.
Raised 1740. Title in 1806 – **Prinz Ferdinand von Preussen.**
The Mark Inspectorate. Garrisons: Ruppin, Templin (grenadiers), Nauen (three battalions).

Uniforms. Initially poppy red facings, Swedish cuffs, white buttons, yellow small clothes , black stock, six buttons, in pairs, on each lapel. By 1787 Brandenburg cuffs had been introduced. On each lapel were seven buttons; in 1797 this was changed to eight evenly-spaced buttons, three on each cuff flap. Officers wore eighteen embroidered silver lace loops with open tassels, two under each lapel, three on each cuff flap, three on each pocket flap and two in the small of the back. Their hats had a narrow silver edging. NCOs had a wide silver hat edging and wore only two silver lace loops with pointed ends under each lapel and two in the small of the back. They wore no silver lace to the cuff or cuff flaps. Privates wore the same configuration but their loops were white with pointed ends and white tassels.

Initially tin-fronted fusilier caps; grenadiers wore tin-fronted grenadier cap front plate and headband, dark blue back cloth, white lace. Pompon colours, from the centre, red, white, dark blue. From 1787 1799 the Kasket was worn.

Colours. King's colour white with pale blue centre, black eagle, white scroll edged silver,a pale blue, Maltese cross behind the central field, silver decoration. Regimental colours pale blue, white centre and Maltese cross, black eagle, pale blue scroll edged silver, silver decoration. An early version of the colours (up to 1763) had eight, narrow piles, pale blue on the king's colour, white on the regimental colours. Yellow pike staffs. The Colonel in Chief of this regiment was Grand Master of the Order of St John.

All four colours were lost at the capitulations of Erfurt and Pasewalk.

Campaign history. Raised in 1740 from the 2nd Bn of IR Nr 15; most officers came from the 6th and 15th Regiments. First Chef was Prinz Ferdinand von Preussen, youngest brother of the king, Friedrich II. Ferdinand was wounded in 1756 as GM at Prague, died in 1806 as GdI, still Chef of the regiment. His son, GL Prinz Louis (Ludwig) Ferdinand, was killed at Saalfeld on 10 Oct 1806. **1740** grenadiers in Gren Bn von Flanss (16/34). **1742** Grenadiers in Gren Bn von Kahlbutz (16/34); the whole regiment fought at B Chotusitz on 17 May. **1744** grenadiers in Gren Bn von Stange (34/Grsn R 5). **1745** 22 Mar C Landshut (grenadiers); 15 Dec B Kesselsdorf (musketiers). **Against Austria in the Seven Years War.** Grenadiers in Gren Bn von Grumbkow (24/34). **1756** 11-15 Oct capture of the Saxon army at Pirna; grenadiers were in 1 Oct B Lobositz. **1757** 9 Mar-20 Jun S Prague. 7 Sep C Moys or Goerlitz (grenadiers). 22 Nov B Breslau (musketiers); 5 Dec B Leuthen (the whole regiment). **1758** In Moravia, 27 May-2 Jul S Olmuetz; C Domstaedl 28 Jun (part of the regiment as supply convoy escort). **1759** In the camp at Schmottseiffen; no actions of note. **1760** 23 Jun C Landshut (grenadiers; very heavy losses). Musketiers were at S Dresden (July) and B Liegnitz on 15 Aug. The regiment suffered very heavy losses. The king was so well pleased with the regiment's conduct here that he awarded the *Pour le Merite* and 100 *Friedrichs d'or* to all captains of the regiment. **1761** The re-raised grenadiers were in Poland against the Russians; destruction of the great wagon train at Gostin on 15 Sep; they then went to Pommerania, where the musketiers were. The grenadiers fought at 19/20 Sep C Colberg and C Spie on 27 Sep. The musketiers were at Fehrbellin. In the war of the Bavarian Succession **(1777/1778)** the regiment was in the army of Prinz Heinrich; grenadiers in Gren Bn von Broesigke (12/34). **1793** On the Rhein; Sk Ginsheim; 19 Jun-23 Jul blockade of Mainz, at Mombach. Storm of the redoubt at Zahlbach; 17 and 20 Aug C Kettricher Hof. The regiment was the advanced guard in the thrust into the Vosges mountains; 13 Oct C Lembach (2nd Bn). **1794** 15 Jun C Lautern; 3 Jul C Trippstadt; Sk Johanniskreuz (1st Bn). **1806.** Auerstaedt 16 Oct. The regiment surrendered at the capitulation of Pasewalk and was not re-raised.

35. Regiment zu Fuss.

Raised 1740. Title in 1806 – **Prinz Heinrich von Preussen.**
Mark-Brandenburg Inspectorate. Garrisons: Koenigsberg in the New Mark, Pyritz (2nd Bn), Soldin (grenadiers), Cuestrin (1st and 2nd Bns).

Uniforms. Initially sulphur yellow facings, white small clothes, white buttons, black stock, Swedish cuffs, no lapels or lace decoration. By 1787 this had changed to Brandenburg cuffs and sulphur yellow lapels. Officers' hats, cuffs and collars were edged in narrow silver lace. They had eight embroidered silver loops with open tassels on and two under each lapel, three on each pocket flap and two in the small of the back. All ranks had eight buttons, evenly-spaced, on each lapel, but NCOs and men had no lace loops. NCOs wore plain silver lace to the hat, cuff and cuff flap. Initially they wore tin-fronted fusilier and grenadier caps and ornaments, yellow headband and backcloth, white lace with red pattern. The side and rear badges of the cap took the form of eagles. Pompon colours, from the centre, red, white, red. From 1787 – 1799 the Kasket was worn.

Colours. King's colour white with pale blue centre, black eagle, white scroll edged silver, silver decoration. Regimental colours pale blue, white centre, black eagle, pale blue scroll edged silver, silver decoration. White pike staffs.

Both colours of the 1st Bn were lost at the capitulation of Erfurt on 16 Oct 1806; those of the 2nd Bn at that of Magdeburg.

Campaign history. Raised in 1740 as a fusilier regiment in Potsdam from half of the Leibkompagnie of the old King's Regiment (Nr 6), grenadiers of the East Prussian grenadier battalions and new recruits for Prinz Heinrich (1726-1802), second brother to king Friedrich II. After the prince's death, the regiment remained vacant but continued to bear his name. **1741** During the First Silesian War the regiment was in Magdeburg fortress; the musketiers were at S Neisse (27-31 Oct). At the outbreak of the Second Silesian War (1744) the grenadiers went into the Gren Bn Finsck (35/36). **1744** 10-16 Sep S Prague. After the fall of Prague the grenadiers were together with those of IR Nr 39 under Oberst Joachim Christian von Tresckow. In November they were in the garrison of Prague and lost fewer men than other units in the subsequent retreat. 26 Nov C

Prague. 11-13 Dec C Olbersdorf. **1745** The regiment were garrison of Liegnitz; later (30 Nov) they formed part of the garrison of Leipzig. Grenadiers in Gren Bn Stranz; fought at B Soor on 30 Sep. **1756-1763 Against Austria and others in the Seven Years War. 1756** The grenadiers were together with those of IR Nr 36 under Major Friedrich August von Schenkendorff. At the capture of the Saxon army at Pirna (11-15 Oct). **1757** Both battalions were attacked by the Austrians in Jan and Feb in their winter quarters in Ostritz and Hirschfeld, but beat off the enemy, albeit with loss. 21 Apr B Reichenberg (the whole regiment) 6 May-20 Jun B and S Prague; 18 Jun B Kolin; the 1st Battalion scattered by Austrian cavalry. 22 Nov B Breslau (2nd Battalion and the grenadiers) took four cannon; 5 Dec B Leuthen. 8-21 Dec S Breslau. The grenadiers were with FM von Keith on the Moldau and not engaged at Prague or Kolin. Instead, the grenadiers were at: 17 Aug cannonade of Barschdorf; 29 Oct C Klettendorf an der Lohe; Breslau and Leuthen (took eight cannon). **1758** 27 May-2 Jul S Ollmuetz; 28 Jun C Domstaedl (grenadiers); 22 Sep Zehdenick; 28 Sep C Fehrbellin; 15 Oct ambush at Boitzenburg; 15 Nov C Eilenburg. **1759** In the camp at Schmottseiffen. 27 Feb capture of Erfurt; Grenadiers in Gren Bn Schwartz. They fought at: 7 May C Neundorf; 16 May C Bamberg; 12 Aug B Kunersdorf; the whole regiment; very distinguished in several assaults on the enemy batteries. Heavy losses; on one occasion the king rallied the regiment with one of its colours. Finally the regiment was scattered by Austrian cavalry. the grenadiers were also at: 16 Sep cannonade of Alt-Tanneberg; 21 Sep C Korbitz. Grenadiers in Gren Bn Stwolinsky (29/31/35/36). **1760** Grenadiers in Gren Bn Schwartz (35/36). 17 Sep C Hoch-Giersdorf; the grenadiers took some cannon. 2 Nov C Schoena (grenadiers only); 3 Nov B Torgau; in Queiss' brigade. The regiment led the assault and had heavy losses. **1761** The musketiers were in the camp at Bunzelwitz and saw no major actions in this year. The grenadiers were in Pommerania in Gren Bn Falkenhayn (29/31/35): 15 Nov C Greiffenberg; 18-29 Nov thrust at Belgard, clashes between Regenwalde and Petershagen; 30 Nov C Naugard; 11 Dec C Drenow; 12 Dec C Spie; 13-16 Dec rearguard actions at Treptow, Schwirschen and Guelzow. **1762** The grenadiers were at: 6 Jul C Adelsbach; 7 Jul C Friedland; 21 Jul B Burkersdorf. The whole regiment was at S Schweidnitz from 4 Aug-10 Oct. **1778** In the army of the king in Bohemia. Only the grenadiers saw action in Gren Bn Bruennow (35/46): 28 Jul-9 Sep advance into Bohemia to Niemes; from 10 Sep retreat into Saxony. **1779** Grenadiers only at C Zuckmantel on 14 Jan. **1793** On the Rhine: 17 and

20 Aug C Kettricher Hof; cannonade of Neu-Winzen; 14 Sep B Pirmasens; 28-30 Nov B Kaiserslautern; the regiment was on the left wing and decided the victory. Blockade of Landau. **1806.** Auerstaedt 16 Oct. The bulk of the regiment surrendered at the capitulations of Erfurt and Magdeburg and was not re-raised.

36. Regiment zu Fuss.
Raised 1740. Title in 1806 – **vacant von Puttkammer.**
Potsdam Inspectorate. Garrisons: Brandenburg, Coeslin (grenadiers).

Uniforms. Initially white facings and small clothes, Swedish cuffs, no lapels, yellow buttons, no lace decoration to coat; silver/white edging to the hats, black stock. From 1787 to 1797 lapels were introduced and buttons became white and Brandenburg cuffs were adopted. There was still no lace decoration to the coat except for NCOs who wore silver lace edging to cuff and flap. From 1797 officers wore embroidered silver lace loops with loose tassels; two under each lapel and in the small of the back, three on each cuff flap and pocket flap. Other ranks had no lace decoration. Initially brass-fronted fusilier and grenadier caps and ornaments, white headband and backcloth, white lace with three light blue stripes. Light blue pompon with white centre. From 1787 – 1799 the Kasket was worn.

Colours. King's colour white with pale purple centre, black eagle, white scroll edged gold, gold decoration. Regimental colours pale purple, white centre, black eagle, pale purple scroll edged gold, gold decoration. White pike staffs.

Two colours were lost at Auerstaedt, the other two were lost at the capitulation of Magdeburg.

Campaign history. Raised in 1740 in Potsdam as a fusilier regiment for Oberst Gustav Bogislav von Muenchow from garrison troops in Pommerania and Brandenburg. **1740** 7-8 Mar storm of Glogau (grenadiers). **1741** 10 Apr B Mollwitz (grenadiers). **1742** In Moravia; no major actions. **1744** In Bohemia; 10-16 Sep S Prague; lost many men in the retreat back to Prussia. C Moldauthein (grenadiers). **1745** 4 Jun B Hohenfriedberg; 30 Sep B Soor; led the attack and had heavy losses. **1756-1763 Against Austria and others in the Seven Years War. 1756** 1 Oct B Lobositz; distinguished. **1757** 21 Apr B Reichenberg, very distinguished.

18 Jun B Kolin; made seven charges against the enemy; at the end of the action there were only 97 men fit for duty. The regiment was reduced to one battalion. 22 Nov B Breslau; 5 Dec B Leuthen; the king attributed his victory to the regiment's conduct. 8-21 Dec S Breslau. **1758** 31 Mar-18 Apr blockade and surrender of Schweidnitz; 27 May-2 Jul S Olmuetz including C Littau. **1759** In the camp at Schmottseiffen. In von Finck's corps; 25 Sep ambush at Hoyerswerda; 29 Oct C Pretsch (Dohmitsch) and C Maxen (Dohna) on 20 Nov, where they were captured. Up to 1759 the grenadiers were in Gren Bn Schenckendorff; thence in Gren Bn Schwartz (35/36) and fought at Kunersdorf. **1760** One battalion was re-raised, but was again captured in Schweidnitz in Oct **1761.** The regiment was re-raised in 1763, using the men of the ex-Saxon regiment Prinz Lubomirsky (Prussian title von Haussen, later von Roebel). 1766 Chef Oberst Henning Alexander von Kleist; became Governor of Spandau as GL; died in office. **1778** The musketiers were in the army of the king. C Jung-Buchau; winter defence of Troppau. The grenadiers were brigaded together with those of IR Nr 35, in the army of Prinz Heinrich. 1780 Chef GM Peter Christoph von Zitzewitz; retired. 1785 Chef Oberst Wilhelm Magnus von Bruenneck (1786 GM); he received IR Nr 17. 1786 Chef GM Karl Albrecht Friedrich von Raumer; 1790 GL, received IR Nr 17. 1793 Chef Oberst Georg Henning von Puttkammer; retired in 1805. In 1806 the regiment was vacant von Puttkammer. The regiment was not mobilised for the Revolutionary campaigns in Poland or on the Rhine. **1806.** Auerstaedt 16 Oct. The regiment surrendered in Magdeburg and was not re-raised.

37. Regiment zu Fuss.

Raised 1740. Title in 1806 – **von Tschepe.**

South Prussian Inspectorate. Garrisons: Fraustadt, Lissa, Glogau (grenadiers), Zduny.

Uniforms. Initially crimson collar and Swedish cuffs, no lapels or lace decoration, yellow buttons, black stock, white small clothes. From 1787 to 1797 crimson lapels with six yellow buttons, in pairs, were worn. The facings were edged in white / silver lace until 1805 when this was removed. From 1805 officers wore silver embroidered lace loops with loose tassels; two under each lapel and in the small of the back, three on each cuff flap and pocket flap. Hat edging was white/silver. NCOs wore plain silver lace edging to the hat, cuff and cuff flap. There were eight evenly-spaced buttons on each lapel.

Initially brass-fronted fusilier and grenadier cap front plates and ornaments, red headband and back cloth, red lace edged white. Crimson pompon with white centre. From 1787 – 1799 the Kasket was worn.

Colours. King's colour white with lime green centre, black eagle, white scroll edged gold, gold decoration. Regimental colours lime green, white centre, black eagle, lime green scroll edged gold, gold decoration. White pike staffs.

All four colours were lost at the capitulation of Magdeburg on 11 Nov 1806.

Campaign history. The following actions are only those of the grenadiers from 1740-1778.

Raised in 1740 from men fit for field sevice of the Kolberg and Pillau Garrison Battalions, from the old King's Regiment (Nr 6) and from new recruits for Oberst Heinrich Tilio de Camus; he died in office. **1741** Chef Oberst Peter Ludwig du Moulin (1745 GL; retired as GdI). In Breslau. **1742** Grenadiers in Gren Bn Fouque (36/37/38); 19-26 Feb in Fulnek; 3 Mar C Walachisch-Meseritsch; 5 Mar-13 Apr in Kremsier; Grenadiers in Gren Bn Fouque (37/38); 14 Apr rearguard actions at Gross-Wisternitz and (25 Apr) Hartau-Kunzendorf. **1744** 23 Aug-7 Sep (Grenadiers in Gren Bn 37/39) advance to Prague; 8-17 Sep S Prague (musketiers); The Grenadiers were in Gren Bn 2/37 and were in the advance to Moldauthein; 18 Sep-6 Oct, then 9 Oct rearguard action at Moldauthein. Reorganization; Grenadiers in Gren Bn Geist (13/37). **1745** Jan-May border security in the Waldenburger mountains. 4 Jun B Hohenfriedbeg; 20 Aug-5 Sep S and capitulation of Cosel (musketiers). 30 Sep B Soor (grenadiers). **1755** Chef GM Heinrich Adolf von Kursell (fatally wounded at Zorndorf). **1756-1763 Against Austria and others in the Seven Years War. 1756** Grenadiers in Gren Bn Manteuffel (37/40); 20 Sep C Koenigslotha. 22 Sep C Jasena. The musketiers were in Silesia with von Schwerin's corps and saw much marching but little action. **1757** 6 May D Brandeis (grenadiers); 6 May B Prague (musketiers), lost 13 officers and over 660 men; 22 Jul C Zittau (musketiers). The grenadiers were at the following actions: 14 May C west of Kolin; 5 Jun C Gang; 13 Jun C Kuttenberg; 18 Jun B Kolin 17-21 Jul rearguard actions north of Boehmisch-Kamnitz;

17 Aug cannonade at Hirschfelde; 29 Aug C Lomnitz; 7 Sep C Moys; 22 Nov B Breslau; 5 Dec B Leuthen (grenadiers and 1st Bn) the 1st Bn took two guns; 6 Dec cannonade at Klein-Mochbern; 22 Dec C Landshut. The 2nd Bn was in the defence and fall of Schweidnitz and was captured. **1758** Chef GM August Wilhelm von Braun. 1st Bn was at S Schweidnitz (31 Mar-18 Apr). Grenadiers were at S Olmuetz, 16 Jun-1 Jul. 25 Aug B Zorndorf (musketiers); 14 Oct B Hochkirch (grenadiers); 30 Oct C Lauban (grenadiers). C Fehrbellin (musketier); 15 Nov C Eilenburg (musketiers). **1759** Grenadiers in Gren Bn Kleist; they fought in the following actions: 11 Aug ambush at Schreiberhau; 25 Oct C Vogelsang; 13 Nov cannonade at Nossen; 16 Nov C Oberhesslich; 19-22 Nov C Maxen, where they were captured. The musketiers were distinguished at Saalfeld on 26 Mar and B Kunersdorf on 12 Aug, where they lost 16 officers and 992 men. The grenadiers were part of von Finck's corps of 13,000 men and were captured at Maxen on 20 Nov. **1760** 1st Bn was in Fouque's corps at C Landshut on 23 Jun and were captured after a hard fight The grenadiers were in Gren Bn Kleist (37/40 combined with 21/27); 2 Jul advance to Maerkisch-Friedland 6 Aug C Hundsfeld; 19 Sep C Glogau. **1761** The regiment was in the camp at Bunzelwitz, at Breslau, then C Gostyn 15 Sep and C Coerlin on 30 Sep. The 2nd Bn was at C Treptow on 12 Sep and the grenadiers were in the following actions: 24 Aug-15 Nov in the camp at Kolberg; 19 Sep storm and defence of the Gruenen redoubt; 18-29 Nov thrust at Belgard, combats between Regenwalde and Petershagen; 30 Nov C Naugard; 11 Dec C Drenow; 12 Dec C Spie; 13-17 Dec actions on the retreat to Treptow (2nd Bn captured) and Stargard. **1762** Grenadiers in Gren Bn Kleist (37/40). 31 May-3 Aug in Upper Silesia; 11 Aug-10 Oct covering Schweidnitz. The musketiers fought at C Adelsbach, 6 Jul; C Friedland, 30 Aug and S Schweidnitz. 1770 Chef Oberst Johann Georg Wilhelm Freiherr von Keller (1785 GL; received IR Nr 8). **1778** Grenadiers in Gren Bn Goetz (37/38). 23 Nov C Weisskirch. 1st Bn Musketiers at Sks Staudnitz and Trautenbach. **1779** 28 Feb grenadiers at cannonade at Neustadt. 1785 Chef Oberst Adam Heinrich von Wolframsdorf (1686 GM; retired as GL). **1792** On the Rhine: capture of Longwy; capture of Verdun, where the grenadiers and the 2nd Bn stayed as garrison. The 1st Bn was at Valmy. **1793** Frankfurt / Main; 6 Jan C Hochheim; 19 Jun – 23 Jul S Mainz including the storm of Kostheim, took two cannon; Gustavsberg. Blockade of Landau (grenadiers and 1st Bn); 17 and 20 Aug Cs Kettricher Hof (2nd Bn); 14 Sep B Pirmasens (2nd Bn; took over 200 prisoners). **1794** Chef GM Johann Rudolf Hiller von Gaertringen; died in

1798. 8 Jan C Kreuznach (musketiers); the entire regiment was at Frankenthal on 3 Jan, C Kaiserslautern (23 May) and Sk Zweibruecken. 1799 Chef Oberst Johann Gustav von Stockhausen; died in 1804 as GM. 1804 Oberst Karl von Tschepe; GM in 1806. **1806.** Jena 16 Oct. The regiment surrendered at Magdeburg and was not re-raised. Parts of the regiment went into the 2nd Westpreussischen IR Nr 7, Ser Nr 8 in the new regiments' section below.

38. Regiment zu Fuss.
Raised 1740. Title in 1806 – **von Pelchrzim.**
Upper Silesian Inspectorate. Garrisons: Neisse, Cosel (three battalions).

Uniforms. Scarlet facings, yellow buttons, no lace decoration, Swedish cuffs, white small clothes, black stock. The regiment wore no lace loops on the coat; there were seven buttons on each lapel until 1787, when it changed to six (in three pairs) in 1798 there were eight, evenly-spaced buttons on each lapel, two on the Swedish cuff. Officers also had two under each pocket flap. NCOs had gold lace to the cuff. Initially brass-fronted fusilier caps; grenadiers wore brass-fronted grenadier cap front plate and ornaments, light blue headband and backcloth, white lace with three yellow stripes. Pompon colours, from the centre, yellow, white, yellow. From 1787 – 1799 the Kasket was worn; the grenadier cap of 1806 had a scarlet headband edged yellow.

Colours. King's colour white with scarlet centre, gold eagle, white scroll edged gold, gold decoration. Regimental colours scarlet, white centre, black eagle, scarlet scroll edged gold, gold decoration. White pike staffs.

All four colours were lost at the capitulation of Neisse on 16 Jun 1807.

Campaign history. Raised in 1740 in Berlin as a fusilier regiment from men fit for field duty in the Prussian garrison troops and new recruits for Oberst Wilhelm Alexander Graf zu Dohna (died in office as GL). 'Jung-Dohna' to differentiate it from IR Nr 28; from 1742 from IR Nr 44 and from 1748 from IR Nr 16. **1741** In Breslau. **1742** Grenadiers in Gren Bn Fouque (36/37/38; later 37/38). 19-26 Feb in Fulnek; 3 Mar C Walachisch-Meseritz; 5 Mar-13 Apr in Kremsier; 14 Apr C Gross-Wisternitz on the retreat; 25 Apr C Hartau-Kunzendorf. **1744** Grenadiers in Gren Bn Brandeis (28/38; from 18 Sep Gren Bn Finck). They fought at: 2-17 Sep

S Prague; 23 Sep capture of Tabor; 30 Sep capture of Budweiss; 25 Nov C Prague in the retreat; 11-13 Dec C Olbersdorf. **1745** Grenadiers in Gren Bn Finck (28/38; from 4 July Gren Bn Ellert). Apr-Oct in Schweidnitz; 20 Oct C Hultschin. The musketiers were at B Hohenfriedberg on 4 June. 1749 Chef GM Johann Christph von Brandeis; retired as GL. 1749 Chef GM Karl Anton Leopold Freiherr von Zastrow; retired as GL. **1756-1763 Against Austria and others in the Seven Years War. 1756** In Saxony at the surrender of the Saxon army at Pirna on 15 Oct. Grenadiers in Gren Bn Burgsdorff (38/43). 20 Sep C Koenigslotha; 22 Sep C Jasena; 29 Oct C Lewin on the retreat. **1757** Grenadiers: 18 Apr C Goldenoels; 6 May B Prague; 6 May-20 Jun S Prague; 2 Jul raid to Lobes; 17-19 Jul retreat from Boehmisch-Kamnitz to Zittau; 22 Nov B Breslau. The whole regiment fought at 6 May B Prague (very heavy losses) and 22 Nov B Breslau. The regiment was effectively destroyed and was re-raised in Berlin. **1758** The new regiment was from 1 Jan-16 Apr at S Schweidnitz. The musketiers were then in the defence of Dresden. The grenadiers also at: 21 May C Landshut; 13 Aug C Landsberg (Dechsel); 15 Aug C Kuestrin; 25 Aug B Zorndorf; 15 Nov C Eilenburg. **1759** The musketiers fought at B Kunersdorf on 12 Aug, they suffered such heavy losses that they were sent back to Schweidnitz to recruit. The grenadiers were at: 1 Jan capture of Dammgarten; 9 Jan-15 May S Stralsund; 28 Aug storm of Wittenberg; 31 Aug C Torgau; 5 Sep C Trachau; 8 Sep C Torgau; 13 Sep capture of Leipzig; 16 Sep cannonade of Alt-Tanneberg; 21 Sep C Korbitz; 7 Nov cannonade of Zoethain. 21 Nov one battalion was with von Finck's corps at Maxen and was captured. **1760** One musketier battalion was still in Schweidnitz and were captured there on 1Oct. Grenadiers: 13-22 Jul S Dresden; 20 Aug C Strehla; 26 Sep C Lichtenburg; 2 Oct C Dobien; 7 Oct C Berlin; 3 Nov B Torgau. the grenadiers fought at C Langensalza on 15 Feb, C Strehla on 20 Aug and were in the recapture of Wittenberg on 23 Oct and Torgau on 3 Nov (heavy losses). **1761** The musketiers had been captured at the fall of Schweidnitz; the regiment was not re-raised until 1763. The grenadiers marched into Thuringia and fought at C Langensalza on 15 Feb, where the enemy were defeated. **1762** Grenadiers: 15 May C Lampersdorf; 1 Jun C Kesselsdorf; 2 Aug C Teplitz; 29 Sep C Nassau an der Mulde; 15 Oct C Brand; 29 Oct B Freiberg. 1766 Chef Oberst Friedrich Gotthelf von Falkenhayn (from 1760-1762 he commanded Gren Bn 29/31). **1778** 23 Nov C Weisskirch; Grenadiers in Gren Bn Goetz (37/38).

1789 Grenadiers: 28 Feb cannonade of Neustadt. Here, together with IR Nr 18, they fought off an enemy corps of 12,000 men. 1781 Chef GM Heinrich Wilhelm von Anhalt; he received IR Nr 2. 1783 GM Eberhard von Hager; he became Director of the 1st Department of the Upper War College. 1790 Chef GM August Wilhelm Freiherr von Vietinghoff; died in office as GL. **1792** on the Rhine and in Champagne. Valmy; S Longwy, S Verdun. **1793** 6 Jan C Hochheim; 19 Jun – 23 Jul S Mainz; 26 Sep C Bliescastel; 27 Sep Sk Ensheim; 29 Sep Sk Bischmischheim; 17 Nov rearguard C Biesingen. Distinguished together with IR Nr 39. 28-30 Nov B Kaiserslautern. The grenadiers fought at Sk Wendelsheim; 28 Mar C Waldalgesheim; 27 Apr C Kostheim. 1796 Chef Oberst Christian August von der Marwitz; died in office as GL. 1800 Chef Oberst Karl von Pelchrzim (1806 GM). **1806** The regiment was in Neisse, Silesia and surrendered there on 16 June 1807. The regiment was not re-raised, but the 3rd Bn was in the successful defence of Cosel and in 1808 became part of the 2nd Silesian IR Nr 12. In 1813 this number changed to 11. See Ser Nr 12 in the new regiments section below.

39. Regiment zu Fuss.

Raised 1740. Title in 1806 – **von Zastrow.**

South Prussian Inspectorate. Garrisons: Posen, Kalisch (grenadiers and musketiers).

Uniforms. Initially Up to 1797 the regiment had buff facings and small clothes, black stock, white buttons, Swedish cuffs; no lace decoration to coat. In 1787 yellow buttons and blue Prussian cuffs were itnroduced, together with white small clothes. In 1798 facings changed to white and the cuffs became white; officers wore gold embroidered lace loops with loose tassels; two under each lapel, three on each cuff flap and pocket flap and two in the small of the back. NCOs had no loops, but plain gold lace edging to cuffs, cuff flaps and hat. There were eight, evenly-spaced buttons on the lapels; other ranks wore two above each cuff and two in the small of the back. Privates' lace loops (two under each lapel) were white with three crimson stripes, pointed and with white tassels. Initially tin-fronted fusilier caps; grenadiers wore tin-fronted grenadier cap front plate and ornaments, buff headband and backcloth, white lace. White pompon. From 1787 - 1799 the Kasket was worn. The band on the grenadier cap was light blue.

Colours. King's colour white with buff (or light yellow) centre, black eagle, white scroll edged silver, silver decoration. Regimental colours buff (or light yellow), white centre, black eagle, buff scroll edged silver, silver decoration. Yellow pike staffs.

All four colours were lost at the capitulation of Magdeburg on 11 Nov 1806.

Campaign history. Raised in 1740 as a fusilier regiment in Templin; most men came from a Braunschweig infantry regiment that had been in imperial service against the Turks from 1737-1739 and from men of the East Prussian garrison regiments. Chef in 1740 was Prinz Ferdinand von Braunschweig, commander of the guards. Later he became ruling Herzog (Duke) of Braunschweig and FM; from 1755 he was Chef of IR Nr 5. **1744** Chef Prinz Albrecht von Braunschweig; due to his youth he held no military rank. He was killed at the battle of Soor, 30 Sep 1745, leading the grenadiers in a charge against the Graner Koppe. Grenadiers in Gren Bn St Surin (37/39); from 8 Sep Treskow (39/43); from 18 Sep 35/39. The regiment served at S Prague (10-16 Sep). **1745** 30 Sep B Soor. Chef Prinz Friedrich Franz von Braunschweig (Oberst in 1751); killed as GM at Hochkirch 14 Oct 1758. From 1755 the regiment was known as 'Jung-Braunschweig' to differentiate it from IR Nr 5. From 1758-1763 the regiment was vacant. **1756-1763 Against Austria and others in the Seven Years War. 1756** In Pommerania; no actions. **1757** Grenadiers in Gren Bn Waldau (12/39); fought at B Kolin on 18 Jun. The musketiers were at B Prague 6 May; B Breslau 22 Nov; B Leuthen 5 Dec; took 14 cannon. 8-21 Dec S Breslau. **1758** Grenadiers in Gren Bn Pieverlink; 14 Oct B Hochkirch. The musketiers at S Olmuetz 27 May-2 Jul. **1759** Due to extremely heavy losses in 1758, the regiment saw no action this year. **1760** grenadiers in Gren Bn Stechow then Goerne; Jul S Dresden; 15 Aug B Liegnitz. The musketiers were at C Hohengiersdorf on 17 Sep; took 14 cannon and two howitzers. **1761** The regiment was in Pommerania; C Spie 12 Dec. **1762** 7 Aug-9 Oct S Schweidnitz (grenadiers). **1763** No actions of note. Chef Oberst Prinz Wilhelm von Braunschweig; he died whilst serving with the Russian army in Bessarabia. 1770 Chef GM Wichard Joachim Heinrich von Moellendorf; he later became Chef of IR Nr 25 (FM in 1806). **1778/1779** Grenadiers in Gren Bn Grolmann (24/39). The regiment was in the army of Prinz Heinrich; on 28 Sep 1778 the grenadiers fought in C Nickelsberg. **1779** 5 Feb C Brix (musketiers). 1783

Chef Oberst Christian Ludwig von Koenitz; retired as GL. From 1786 the regiment was known as 'Alt Koenitz' to differentiate it from IR Nr 17. **1792** On the Rhine; Longwy; 30 Aug capture of Verdun; 20 Sep Valmy. The grenadiers and 1st Bn, IR Nr 49 were attacked by over 6,000 French forces at Limburg/Lahn on 6 Nov, but held their ground for over three hours. 2 Dec storm of Frankfurt/Main (grenadiers and 2nd Bn). **1793** Chef Oberst Daniel Ludwig von Crousaz; retired as GL. 6 Jan C Hochheim (2nd Bn and grenadiers); 1-7 Mar blockade and capture of Koenigstein (1st Bn); 19 Jun-23 Jul S Mainz, including the storm of Kostheim. The regiment then went along the River Saar; C Saarbruecken; distinguished at C Biesingen on 17 Nov under GM von Kalckreuth, together with IR Nr 38. As a reward for their bravery here, the six most senior officers of the regiment were awarded the *Pour le Merite* and many NCOs and men received gallantry medals. 28-30 Nov B Kaiserslautern. **1794** 23 May C Kaiserslautern. 1800 Chef Oberst Friedrich Wilhelm von Zastrow (GM in 1806). **1806.** Jena 16 Oct. The regiment surrendered in Magdeburg and was not re-raised.

40. Regiment zu Fuss.
Raised 1741. Title in 1806 – **von Schimonsky.**
Lower Silesian Inspectorate. Garrisons: Schweidnitz, Striegau (grenadiers), Cosel (musketiers).

Uniforms. Initially rose red facings and small clothes, no lapels, Swedish cuffs, white buttons, black stock. Brandenburg cuffs. Officers wore three embroidered silver lace loops to each lapel, two under each lapel, two on the cuff, three on each pocket flap ans two in the small of the back. By 1787 Prussian cuffs had been introduced and officers wore seven silver loops on the lapel, otherwise the lace loops were as before. Small clothes were now white. In 1798 there were eight evenly-spaced buttons on each lapel. Officers now wore only two embroidered silver lace loops with loose tassels under each lapel, three on each pocket and cuff flap and two in the centre of the back. NCOs wore two plain, pointed silver lace tapes under each lapel and two in the small of the back; they had no lace edging to the cuff or cuff flap. Privates wore no lace loops. Hat edging was silver/white. Initially tin-fronted fusilier caps; grenadiers wore tin-fronted grenadier cap front plate and ornaments, rose red headband and back cloth, white lace with pink pattern. Pompon colours, from the centre, rose red, white, rose red. From 1787 - 1799 the Kasket was worn.

Colours. King's colour white with rose red centre, black eagle, white scroll edged gold, gold decoration. Instead of the usual laurel branches around the centre, it was surrounded by the chain of the Order of the Black Eagle, awarded for distinguished service in 1758. Regimental colours rose red, white centre, black eagle, rose red scroll edged gold, gold decorations. Centre surrounded by the chain of the Order of the Black Eagle. White pike staffs.

All four colours were lost at the capitulation of Magdeburg on 11 Nov 1806.

Campaign history. Raised in 1741 in Magdeburg from a regiment of three battalions of the duke of Sachsen-Eisenach which had been in imperial service against the Turks from 1732-1736. In 1740 it was reduced to one battalion and passed into Prussian service, where it was made up to a regiment with new recruits. The son of the duke refused to continue as Chef of the new fusilier regiment; instead, GL David Juergen von Grevenitz became Chef in 1741. 1743 Chef GM Georg Christoph von Kreytzen; died in office as GL. **1744** Grenadiers in Gren Bn Jaeger (31/40); S Prague 10-16 Sep. The musketiers were overwhelmed at Tabor on 23 Sep and Budweis on 30 Sep and captured. **1745** Grenadiers in Gren Bn Lindstedt (31/40); they fought at 4 Jun B Hohenfriedberg and 30 Sep B Soor. 1750 Chef Oberst Johann Friedrich von Kreytzen, brother of Georg; retired. The regiment was known as 'Alt-Kreytzen' to differentiate it from IR Nr 28; both generals were called Johann Friedrich. **1756-1763 Against Austria and others in the Seven Years War. 1756** Grenadiers in Grn Bn Manteuffel (37/40); no actions of note. **1757** 8 Apr the raid on the Austrian magazines at Jungbunzelau. The whole regiment fought on 6 May B Prague and 18 Jun B Kolin. The grenadiers were at C Moys (Goerlitz) 7 Sep; 12 Nov 2nd Bn captured at the fall of Schweidnitz; B Breslau 22 Nov; B Leuthen on 5 Dec. **1758** Musketiers were at Domstaedl 28 Jun; B Zorndorf 25 Aug; very distinguished, but lost two colours to Russian cavalry. Grenadiers were at S Olmuetz 27 May – 2 Jul; B Hochkirch 14 Oct (grenadiers captured). **1759** Chef GM Georg Karl Gottlob von der Gabelentz; in **1764** he became GL and Governor of Schweidnitz; died in office. The regiment was in the Pommeranian corps. The musketiers fought in B Kunersdorf 12 Aug; heavy losses. The grenadiers were in Gren Bn Kleist (37/40) and were captured at C Maxen on 20 Nov. **1760** Musketiers at C Weissen Hirsch near Dresden; heavy losses;

B Liegnitz 15 Aug. **1761** In the thrust into Poland then in the camp at Bunzelwitz. **1762** Musketiers at S and recapture of Schweidnitz 7 Aug-9 Oct. 1777 Chef GM Friedrich August Freiherr von Erlach (1786 GL); retired. **1763** No actions of note. **1778** In Bohemia with the main army. Musketiers at C Maehrisch-Ostrau 27 Oct. Grenadiers in Gren Bn Loewensprung (40/43). 15/16 Jul C Nachod; 16 Jul went to Upper Silesia with Stutterheim's detachment. 11 Aug ambushed enemy advanced guard at Glomnitz; 29 Oct C Maerisch-Ostrau. **1779** 11 Jan C Jaegerndorf. 1791 Chef GM Karl Ludwig von Pfuhl; retired. The regiment was known as 'Jung-Pfuhl' to differentiate it from IR Nr 46. **1794** In South Prussia. 1795 Chef GM Friedrich Wilhelm von Steinwehr; retired as GL. 1805 Chef Oberst Dietrich Leberecht von Schimonsky (GM in 1806). **1806.** Auerstaedt 16 Oct. The regiment surrendered in Magdeburg and was not re-raised. Parts of the regiment went into the 2nd Westpreussischen IR Nr 7, Ser Nr 8 in the new regiments' section below.

41. Regiment zu Fuss.
Raised 1741. Title in 1806 – **von Lettow.**
Westphalian Inspectorate. Garrisons: Minden, Herford (grenadiers), Embden (musketiers).

Uniforms. Initially raspberry facings, yellow buttons, six, in pairs, on each lapel; Swedish cuffs, black stock, buff small clothes. Officers wore embroidered golden loops; two under each lapel, two in the small of the back. NCOs wore the same configuration in plain gold braid with pointed ends and gold edging to the top and back of the cuffs; privates had yellow lace loops. In the first years the men wore brass-fronted fusilier and grenadier cap front plates and ornaments, raspberry headband, and back cloth, white lace with a central raspberry stripe. Pompon colours, from the centre, raspberry, white, raspberry. From 1787 - 1799 the Kasket was worn and white small clothes were introduced. Hat edging was gold/yellow. From 1798 there were eight buttons on each lapel.

Colours. King's colour white with light yellow centre, black eagle, white scroll edged silver, silver decoration. Regimental colours light yellow, white centre, black eagle, light yellow scroll edged silver, silver decoration. White pike staffs.

All four colours were lost at the capitulation of Hameln on 22 Nov 1806.

Campaign history. Raised in 1741 by taking over the Wuerttemberg Leibregiment Erbprinz (*1716) in Wesel. This regiment had been in imperial service up to 1739. The first Chef was Oberst Herzog August Wilhelm von Braunschweig-Bevern, but in the October he received IR Nr 7, so GM Johann Volprecht Freiherr von Riedesel became Chef. He had served in the Austrian army and retired as GL. **1744** In Berlin from August. Grenadiers in Gren Bn 41/44. **1745** Only the grenadiers saw action this year. Apr-Jul around Magdeburg; 31 Aug-15 Oct in the camp at Dieskau, Gren Bn Holstein-Beck. 23 Nov at Halle; 5 Dec capture of Torgau; 15 Dec B Kesselsdorf. 1746 Oberst Franz Karl Ludwig Graf Wied zu Neuwied; GL in 1762; died in office. **1756-1763 Against Austria and others in the Seven Years War. 1756** Grenadiers in Gren Bn Gemmingen (41/44) 10-13 Sep in the camp at Cotta; 18 Sep-24 Oct security duty at Aussig; 29 Oct rearguard actions at Hellendorf and Oelsen. **1757** Grenadiers: 25 Apr C near Schreckenstein; 30 Apr-9 May guarding the bakery in Welwarn; 6 May B Prague, lost over 1,000 men; 5 Jun C Gang; 13 Jun C Kuttenberg. The whole regiment fought on 18 Jun B Kolin. Grenadiers: 17-21 Jul rearguard actions north of Boehmisch-Kamnitz; Gren Bn Beneckendorff. 17 Aug cannonade of Hirschfelde; 7 Sep C Moys; after this action, the grenadiers were combined with Gren Bn Unruh (45/48/ Grsn R 9). The whole regiment fought on 22 Nov B Breslau and 5 Dec B Leuthen; the grenadiers were at 7-20 Dec S Breslau. **1758** Grenadiers: 31 Mar-8 Apr S Schweidnitz; 16 Apr storm of the Galgen Fort at Dresden in Gren Bn 41/44. 20 May-1 Jul S Olmuetz; the whole regiment. 7 Jul C Krenau (the whole regiment); 11 Oct C Jenkwitz (grenadiers); 25 Aug B Zorndorf (musketiers); 25 Sep C Fehrbellin (musketiers); 14 Oct B Hochkirch (grenadiers very distinguished). **1759** Grenadiers: 10 Jul 16 Aug in the camp at Schmottseiffen (Gren Bn 41/44); 29 Oct C Pretzsch; 13 Nov cannonade of Nossen; 16 Nov C Oberhesslich; 19-21 Nov C Maxen; the grenadier battalion was captured here. Musketiers: 12 Aug B Kunersdorf; heavy losses. **1760** Grenadiers in Pommerania: 14 May C Baehrenwalde; combined with 4/16 under Beneckendorff. 17 May C Belgard; 29 May D Koeslin; 3 Jun C Belgard; 19 Sep rearguard action between Brieg and Glogau. Musketiers: 15 Aug B Liegnitz; 3 Nov B Torgau. **1761** Grenadiers, marched this year from Pommerania to Silesia: 24 Aug-15 Nov in the camp at Kolberg; 19 Sep storm and defence of the Green Redoubt; 18-29 Nov thrust against Belgard; 20-25 Nov combats between Regenwalde and Petershagen; 11/12 Dec combats on the

Kreyerbach stream and at Spie; 13-17 Dec rearguard actions at Treptow and Stargard. Musketiers had no actions of note. **1762** Grenadiers in Gren Bn 41/44. 31 May-3 Aug operations in Upper Silesia. 11 Aug-10 Oct covering Schweidnitz, in Gren Bn Beneckendorff. Grenadiers: 16 Aug C Reichenbach. 1765 Chef Oberst Matthias Ludwig von Lossau; retired as GL. **1778/1779** The regiment was in Moellendorf's corps of the army of Prinz Heinrich in Bohemia. Grenadiers in Gren Bn Romberg Nr 3 (41/44). No actions. 1782 GM Alexander Friedrich von Woldeck; died in office as GL and Governor of Wesel. The regiment was known as 'Jung-Woldeck' to differentiate it from IR Nr 26. **1787** In the expedition to Holland. 1 Oct C Amstelveen. **1792** Oberst Karl Friedrich von Schladen; retired as GL. The regiment was on the Rhine and in Champagne; Valmy. **1793** Grenadiers: 17 and 20 Aug Kettricher Hof; 13 Oct C Lembach; Sk Roth; Sk Burweiler. Musketiers: failed storm of Bitsch fortress, 16 Nov. 1st Bn: B Pirmasens; 13 Oct C Lembach; Sk Roth; Sk Burweiler. 2nd Bn 19 Jun-23 Jul S Mainz including the failed storm of Fort Philippi, heavy losses. **1794** The whole regiment was at Sk Rupertsburg; 12-13 Jul C Schaenzel, heavy losses. Grenadiers fought at Hochspier; musketiers at B Kaiserslautern. 1804 GM Karl Ernst von Lettow. **1806** The regiment surrendered at Hameln on 22 November 1806 and was not re-raised.

42. Regiment zu Fuss.
Raised 1741. Title in 1806 **von Ploetz.**
Warsaw Inspectorate. Garrisons: Warsaw, Czenstochau (musketiers).

Uniforms. Orange facings, yellow buttons, no lace decoration, Swedish cuffs, six, evenly-spaced buttons on each lapel; white small clothes, black stock. Officers' hats had a narrow gold edging. NCOs wore gold braid to top and back of the cuff. Initially brass-fronted fusilier and grenadier caps and ornaments, orange headband and backcloth, white lace with orange central zig-zag and side stripes. White pompon with an orange ring. From 1787 the Kasket was worn. In 1798 there were eight buttons on each lapel. White pike staffs.

Colours. King's colour white with orange centre and narrow corner piles, black eagle, white scroll edged gold, gold decoration. Regimental colours orange, white centre and corner piles, black eagle, white scroll edged gold, gold embroidery. The fate of the colours in 1806-1807 is not clear.

Campaign history. Raised on 3 Nov 1741 from men of the Garrison Regiment of Brieg (part of them from the captured Austrian troops) as Fusilier Regiment. First Chef was GM Markgraf Heinrich von Brandenburg-Schwedt; he died in office in 1788. The grenadiers came from IR Nr 41 and of Wuerttemberg origin. As the Markgraf could not command the regiment himself, the commander, Oberstleutnant Martin Anton Freiherr von Puttkamer, took his place and the regiment thus had no Leibkompanie. 1745 25 Jan C Czarnowanz; 26 Jan C Oppeln; 6 Feb C Rosenberg; 27-30 Mar advance from Ratibor to Loslau; 31 Mar-17 Apr in Ratibor; from 26 Apr in Oppeln; from 31 May in Coeln; 27 Aug-5 Sep S Cosel; 28 Sep C Jaegerndorf; 20 Oct C Hultschin; 3-9 Dec advance into the Waldenburger mountains. 1750 Commander Oberst Johann Christian von Conrady. 1753 Commander Oberst Balthasar Rudolf von Schenckendorff; he received IR Nr 22. **1756-1763 Against Austria and others in the Seven Years War. 1756** Grenadiers in Gren Bn Nimschoefsky, later Mosch (33/42). No actions of note. **1757** 6 May B Prague then S Prague; 18 Jun (grenadiers) B Kolin, wiped out; refoermed with Saxon recruits. 22 Jul 2nd Bn captured in Zittau. **1758** 20 May – 1 Jul S Olmuetz. **1759** Raid on the Russian magazines at Posen, then in the camp at Schmottseiffen 10 Jul – 16 Aug. **1760** Oberst Heinrich Werner von Kleist. 23 Jul 1st Bn in B Landshut, very heavy losses. 2nd Bn in the area of Freiberg, no actions. Grenadiers: 19 Jul Dresden, 15 Aug Liegnitz, 17 Sep Hochgiersdorf, 3 Nov Torgau. **1761** In Pommerania; no major actions. **1762** 16 Aug C Peilau; 2nd Bn destroyed. 1764 Commander Major Heinrich Wilhelm von Lettow; he received IR Nr 46. 1776 Commander Oberstleutnant Stephan von Lichnowsky; he received IR Nr 23. 1778/1779 Grenadiers in Gren Bn Preuss (33/42). 1786 Commander Oberstleutnant Georg Bogislav von Koethen. 7 Feb 1789 Chef Oberst Nikolaus Reinhold von Hanenfeldt; retired as GL. 29 Dec 1794 Chef GM Kaspar Siegmund von Luettwitz; died in office. 5 Apr 1796 Chef Oberst Christian Franz Heinrich von Ploetz; GL in 1806. **1806.** The regiment was in l'Estocq's corps in eastern Prussia. The 3rd Musk Bn was captured at Czenczstochau. 1807. Parts of the regiment fought at Waltersdorf, 5 Feb; Wackern?; Braunsberg, 26 Feb; Steegen?; in the siege of Danzig and on the island of Ruegen? The regiment was not re-raised. On 21 Nov 1808 remnants of the regiment went to form the 1. Schlesisches IR Nr 10. See Serial Nr 11 in the new regiments section below for subsequent history.

43. Regiment zu Fuss.
Raised 1741. Title in 1806 – **von Strachwitz.**
Lower Silesian Inspectorate. Garrisons: Liegnitz, Striegau (grenadiers), Nimtsch (musketiers).

Uniforms. The first uniform had light orange facings, yellow buttons, Swedish cuffs, no lapels, but two rows of buttons on the chest, black stock, white small clothes, no lace decoration. White pike staffs. By 1787 the buttons were white buttons, still no lace decoration to the coat, Prussian cuffs, white small clothes. As of 1805, officers had embroidered silver lace loops; two under each lapel, three on each cuff flap and two in the small of the back. NCOs and men had no lace loops, NCOs had silver edging to cuff, cuff flap and hat. Initially tricorns until 1744, then brass-fronted fusilier or grenadier caps and ornaments, orange headband, white back cloth, white lace edged yellow. Light orange pompon with white centre. From 1787 – 1799 the Kasket was worn.

Colours. King's colour: all white ground, black eagle, blue-green scroll, gold decorations. Regimental colours: ground all blue green, black eagle, white scroll, gold decorations. White pike staffs.

All four colours were lost at the capitulation of Magdeburg on 11 Nov 1806.

Campaign history. For further details see IRs Nrs 38 and 40. Raised on 10 August 1741 by swearing in 600 of the town guard directly after the capture of Breslau from Austria to form the Breslau Garrison Regiment; Chef was the town commandant, GM Maximilien von Rampusch. 1742 Chef Oberst Philipp Ernst von Bardeleben. Converted to a fusilier regiment in 1744; the new Chef Oberst von Brandeis, who became Governor of Pillau in the same year. He was replaced by Oberst Joachim Ernst von Zimmernow, who died in Pardubitz on 22 Oct that year as a result of his wounds. **1744** Oberst Christian Ludwig von Kalsow (GM in 1756). The grenadiers covered the retreat from Prague, while the regiment was guarding the great magazine in Pardubitz from Austrian attacks, as on 9 November. **1756-1763 Against Austria and others in the Seven Years War. 1756** In Silesia. Grenadiers in Gren Bn Burgsdorff, later Heilsberg (38/43) 29 Oct Sk Lewin. **1757** GM Samuel Adolph von Kalckreuth; he went into Saxon service. Musketiers: 6 May B Prague, heavy losses; 9 May-20

Jun S Prague. In the retreat from Bohemia, the 1st Bn was cut off, surrounded and captured at Gabel on 15 Jul. The grenadiers fought at the defence of Zittau on 22 Jul. Grenadiers and 2nd Bn fought in B Breslau on 22 Nov. **1758** Oberst Joachim Leopold von Bredow; died in office. The regiment was in Saxony. Grenadiers: 31 Mar – 18 Apr S and capture of Schweidnitz; 25 Aug B Zorndorf, very heavy losses. Musketiers: 15 Oct C Freiberg; together with the 8th Hussars they attacked an enemy cavalry force of 2,000 men and killed or took almost all of them. **1759** Musketiers: at Passberg (St Sebastiansberg) and, 11 May, C Himmelskron in Franconia. 12 Aug B Kunersdorf; led the attack and captured a large battery at the beginning of the battle. Lost over 500 men. Grenadiers: 5 Sep C Dresden; 8 Sep C Torgau; 21 Sep C Gorbitz (Meissen). **1760** GM Christian Wilhelm von Zieten; the regiment was known as 'Jung-Zieten' to differentiate it from the hussar General von Zieten. Grenadiers: Jul S Dresden; 20 Aug C Strehla; 2 Oct Sk Wittenberg; 3 Nov B Torgau. **1761** In the camp at Bunzelwitz. Grenadiers: 15 Feb C Langensalza. **1762** Musketiers: 3 Jul C Adelsbach; distinguished, heavy losses. Grenadiers: 2 Aug C Toeplitz; 14 Oct C Freiberg. **1767** Oberst Doering Wilhelm von Krockow; received IR Nr 51. 1771 Oberst Karl Magnus von Schwerin; died in office in 1775. 1776 Oberst Leopold Ludwig Graf zu Anhalt (GL in 1786; died in office as GdI). **1778/1779** Grenadiers in Gren Bn Loewensprung (40/43). **1779** Musketiers: 14 Jan C Zuckmantel 1795 GM Alexander Leopold Graf von Wartensleben; he received IR Nr 59 (GL in1806). **1794** In Poland. Grenadiers and 1st Bn: 18 May C Skala; 6 Jun B Scelze; 15 Jun capture of Krackau. The whole regiment was at the blockade of Warsaw 27 Jul – 6 Sep, including: 19 Aug C Powontze (grenadiers) and C Wawbrecice (2nd Bn). The whole regiment went on the expedition to Thorn in October; the Musketiers: 10 Nov C Kapinos. 1803 GM Ludwig von Strachwitz. **1806.** Jena 16 Oct. The regiment surrendered in Magdeburg and was not reraised. Parts of the regiment went into the 2nd Westpreussischen IR Nr 7, Ser Nr 8 in the new regiments' section below.

44. Regiment zu Fuss.

Raised 1742. Title in 1806 – **von Hagken.**

Westphalian Inspectorate. Garrisons: Muenster, Soest (grenadiers), Drengsteinfurth and Ahlen (musketiers).

Uniforms. The original uniform had red facings, no lapels, two rows of eight yellow buttons on the chest, red button holes with white tassels, Prussian cuffs, buff small clothes. Initially brass-fronted fusilier and grenadiers caps and ornaments, red headband, buff back cloth, light blue lace edged black. Black pompon with light blue centre. From 1787 – 1799 the Kasket was worn.

From 1787 – 1797 they wore black facings, yellow buttons, 22 white lace loops, white small clothes, Prussian cuffs. In March 1798 the facings became buff, they lost the lapel laces, and the remaining buttonhole laces of the other ranks became white with three light blue stripes, pointed ends and white and light blue tassels. Officers had two embroidered gold lace loops with loose tassels below each lapel and in the small of the back, three on each cuff flap and under each pocket flap. Narrow gold hat edging. NCOs wore two plain gold lace loops with tassels under each lapel, in the small of the back and on the cuff flap. No gold lace to cuff edges. Privates had the same lace configuration as NCOs. Wide gold hat edging.

Colours. King's colour white with light blue centre, black eagle, white scroll edged gold, gold decorations. Regimental colours light blue, white centre, black eagle, light blue scroll edged gold, gold decoration. Brown pike staffs.

All four colours were lost at the capitulation of Hameln on 22 Nov 1806.

Campaign history. For further details see IR Nr 41, with whom the grenadiers were always combined.

Raised from 10 men each from every company in IR Nr 28, that were then transferred from Wesel to Silesia. This was a Fuesilier regiment. Chef was GM Graf zu Dohna; died in office as GFM. 1749 Chef Oberst Martin Eberhard von Jungkenn Muentzer von Mohrenstamm; GM in 1756; received Grsn R Nr 4. **1745** 15 Dec B Kesselsdorf (grenadiers) **1756-1763 Against Austria and others in the Seven Years War. 1756** Not mobilised. **1757** The musketiers were with the Anglo-Hanoverian army and involved in covering the retreat of the Duke of Cumberland. Grenadiers: 18 Jun B Kolin; 22 Nov B Breslau; 5 Dec B Leuthen. **1758** Grenadiers: 31 Mar – 18 Apr S Schweidnitz; 27 May – 2 Jul S Olmuetz; 14 Oct B Hochkirch. The musketiers were in Pommerania fighting the Swedes. **1759** Chef Oberst Rudolph August von Hoffmann; killed at Dresden. Musketiers: 15 Aug 2nd Bn surrendered to the Imperials at Leipzig. 4 Sep 1st Bn in the evacuation of Dresden. 5 Sep C Reichenberg; 8 Sep C Torgau. 21 Sep C Corbitz. Grenadiers 20 Nov C Maxen; captured. **1760** Chef GM Johann von Grant; became commandant of Neisse and died in office. Jul S Desden; 20 Aug C Strehla; 2 Oct C Wittenberg. The re-raised grenadiers were in Pommerania; defence of Coeslin; relief of Colberg. **1761** Grenadiers in Pommerania; defence of Colberg; 27 Sep C Spie. The musketiers were in Saxony; no major actions. **1762** 29 Oct B Freiberg. 1764 Chef Oberst Hans Christian von Brietzke; retired as GL. **1778** Chef Oberst Friedrich Wilhelm Ernst von Gaudi (1786 GM; died in office as GL and Commandant of Wesel). The regiment was in the army of Prinz Heinrich. **1787** In the expedition to Holland. **1790** The grenadiers were in Liege. 1789 Chef Oberst Franz Otto von Pirch; received IR Nr 8. 1791 Chef Oberst August Graf von Dohna; died in office as GL. 1792 Chef GM Franz Georg von Kunitzky; until now commander of the 1st Bn Garde; died in ofice. **1793** On the lower Rhine in von Knobelsdorff's corps; 3 Mar C Schwalmen (grenadiers, 1st and 3rd Bns). The 2nd Bn was at Venlo. The united regiment was at the recapture of Breda (30 Mar) and the clashes of Valenciennes, Vicogne in May and 23 May B Famars and the capture of Valenciennes (28 Jul) and Conde. They then marched to the upper Rhine and fought at Scheidt, Saarbrueck and 27 Sep C St Imbert and 28-30 Nov B Kaiserslautern. **1794** 23 May cannonade at Diedesheim; C Fenningen; 12-13 Jul Schaenzel and C Kesselsberg near Landau; cannonade of Mosbach; 17 Sep C Schorleberge; C Alseborn. 1799 Chef Oberst Ludwig von Strachwitz; he received IR Nr 43. 1803 Oberst Christian Alexander von Hagken (1806 GM). **1806** The regiment surrendered in Hameln and was not re-raised.

45. Regiment zu Fuss.

Raised 1743. Title in 1806 – **von Zweiffel.**

Franconian Inspectorate. Garrisons: Beireuth, Culmbach (grenadiers), Hof (musketiers).

Uniforms. At first the uniform had red facings, Prussian cuffs, yellow buttons, white small clothes, red stock. The coat had no lapels, but two rows of six buttons, in pairs, on the chest. Privates wore white lace button-holes with red and white tassels. NCOs wore no button-hole laces but had gold edging to the cuff and two gold butto-holes on the flap. Initially brass-fronted fusilier and grenadier caps and ornaments, dark blue headband, white back cloth, yellow lace, with dark blue and red stripes. Pompon colours, from the centre, yellow, dark blue, red. In 1792 the coat had dark blue lapels with seven buttons and yellow lace. Up to 1789 the regiment had dark blue lapels, crimson facings, yellow buttons, 8 white lace loops with three crimson stripes, white small clothes. In 1798 the facings changed to lemon yellow and the lapels were also in this colour. The distribution of the lace loops on the coat was as for the 44th Regiment. Those of the officers were embroidered gold lace with loose tassels; NCOs were plain gold lace with golden tassels; no gold lace edging to the cuffs. Privates' lace was white with three red lines and red and white tassels. From 1802 the facings changed to lemon yellow, Brandenburg cuffs were adopted and their were eight, evenly-spaced buttons on each lapel.

Colours. King's colour white with royal blue centre, black eagle, white scroll edged gold, gold decoration. Regimental colours royal blue, white centre, black eagle, royal blue scroll edged gold, gold decoration. Brown pike staffs.

One colour of the 1st Bn was lost at the capitulation of Erfurt on 16 Oct 1806; the other three at that of Magdeburg.

Campaign history. Raised on 1 August 1743 as a fusilier regiment from 10 men from each company of IR Nr 31 in Wesel. Chef GL von Dossow (FM in 1756); it was he who had previously commanded the 31st . Some of the officers came from other Prussian infantry regiments, others transferred in from foreign service. 1744 In Berlin as of August; grenadiers in Gren Bn Ingersleben (45/Grsn R 9/Grsn R 13). **1745** Oct/Nov in Brandenburg; 23 Nov at Halle; 30 Nov capture of Leipzig, thence garrison of the town. **1756-1763 Against Austria and others in the Seven Years War. 1756** In Wesel. **1757** Chef GL Erbprinz Friedrich von Hessen-Kassel (previously Chef of IR Nr 48; later Landgraf of Hessen-Kassel; 1785 FM). The musketiers were initially with the Duke of Cumberland's Anglo-Hanoverian army in the defence of Geldern, then back to Saxony. Grenadiers in Gren Bn Ingersleben (45/48/Grsn R 9) in Bohemia): 18 Apr C Goldenoels; 6 May B Prague, thence siege until 20 Jun. Grenadiers now in Gren Bn Unruh. 2 Jul raid to Lobes; 17-21 Jul rearguard actions north of Boehmisch-Kamnitz; 17 Aug cannonade at Hirschfelde; 7 Sep C Moys. After this action, the grenadiers were combined with 41/44. 22 Nov B Breslau; 5 Dec B Leuthen; 7-20 Dec S Breslau. **1758** Musketiers in the Saxon Corps. Grenadiers: 7 May-16 Jun guard duties at Maehrisch-Neustadt; grenadiers in Gren Bn 45/48/Grsn R 9. 16 Jun-1 Jul before Olmuetz; 14 Oct B Hochkirch. Musketiers: 28 Sep C Fehrbellin. **1759** Grenadiers: 30 Jun C Maifritzdorf; 23 and 27 Jul Cs Friedland in Silesia; 11 Sep C Bernsdorf. Musketiers: Sk Aue; 21 May C Reichenberg near Dresden; Sk Pretsch; 8 Sep C Torgau; 20 Nov C Maxen; the regiment was captured. **1760** Grenadiers: 18 Jun-26 Jul in Glatz fortress, then taken prisoner. The musketiers were reduced to a single battalion: 3 Nov B Torgau. **1761** Grenadiers in Pommerania against the Swedes; 29 Sep C Neuensund; took several hundred prisoners and three cannon. The whole regiment was in the defence of Colberg. **1762** The whole regiment was at C Reichenbach on 16 Aug. **1763** Regiment re-raised in Stettin; grenadiers in Gren Bn Nr 2. **1778-1779** In the army of Prinz Heinrich; no actions. The grenadiers in Gren Bn Meusel Nr 2 (45/48) 1786 Chef GM Johann Ludwig von Eckartsberg; retired. **1787** In the expedition to Holland. **1789** In the expedition to Liege. 1792 Chef Oberst Friedrich August von Grevenitz; received IR Nr 57. 1795 Chef Oberst Karl Philipp von Unruh; died in office as GL in 1805. 1805 Chef Oberst Johann Christian von Zweiffel (1806 GM). **1806.** Jena 16 Oct. The regiment surrendered and was not re-raised.

46. Regiment zu Fuss.

Raised 1743. Title in 1806 – **von Thile.**

Warsaw Inspectorate. Garrisons: Warsaw, Praga (musketiers).

On 21 Nov 1808 remnants of the regiment went to form the **1. Schlesisches IR Nr 10**. See Serial Nr 11 in the new regiments section below for subsequent history.

Uniforms. Originally the coat had black facings, Swedish cuffs, yellow buttons, buff small clothes. No lace decoration to the coat. Brass grenadier cap front plate and ornaments up to 1787, black headband, buff back cloth, black lace edged yellow and red. Then the Kasket with grenade badge was worn. Pompon colours, from the centre, black, yellow, red.

From 1787 – 1797 facings were black, yellow buttons, white small clothes, no lace loops, Swedish cuffs. In 1798 facings became scarlet, the cuffs changed to Brandenburg in style only in 1805. Officers wore golden aiguillettes and hat edging, but no buttonhole decoration. Eight evenly-spaced buttons on the lapels. NCOs wore gold lace edging to cuff, cuff flap and hat edging.

Colours. King's colour white with buff centre, black eagle, white scroll edged gold, gold decoration. Regimental colours buff, white centre, black eagle, buff scroll edged gold, gold decoration. White pike staffs.

All four colours were lost at the capitulation of Breslau on 5 Jan 1807.

Campaign history. Raised in 1743 in Brandenburg from new recruits from all over Germany, mainly from Wuerttemberg. Chef Erbprinz Karl Eugen von Wuerttemberg. Some of the officers came from foreign service. Until 1757 the regiment was known as 'Alt-Wuerttemberg'. **1745** *30 Sep B Soor (grenadiers); 15 Dec B Kesselsdorf, lost almost 600 men.* **1756-1763 The seven Years War. 1756** In Pommerania, no actions. **1757** Chef Oberst Johann Albrecht vob Buelow; died in office as GdI. C Reichenberg; 6 May B Prague, very distinguished, 50% losses. 15 Jul C Gabel; grenadiers and 2nd Bn captured. 22 Nov 1st Bn at B Breslau; 5 Dec B Leuthen. **1758** 31 Mar-18 Apr S Schweidnitz. 25 Aug B Zorndorf. Later at Torgau. **1759** The musketiers took part in the raid into Franconia; the whole regiment fought at Kunersdorf on 12 Aug; heavy losses. **1760** 23 Jun C Landshut (musketiers); most of the regiment was captured. Grenadiers: 15 Aug B Liegnitz; 17 Sep C Hohengiersdorf; 3 Nov B Torgau.

1761-1762 The regiment was recovering in Breslau; in 1762 it was re-raised at one battalion and served at S Schweidnitz 7 Aug-9 Oct. 14 Oct C Freiberg. 1776 Chef Oberst Heinrich Wilhelm von Lettow; retired as GM. **1778/1779** In the army of Prinz Heinrich; no actions. 1779 Chef GM von Pfuhl; retired as GdI and Governor of Spandau. **1794** In Poland; defence of the Vistula line from Thorn to Zakroczin; the grenadiers were in General von Schoenfeld's corps on the River Narew. 1795 Chef GM Alexander Heinrich von Thile; GL in 1806. **1806** The regiment was in Breslau fortress; surrendered there and was not re-raised. The 3rd Bn was in East Prussia; in 1808 it became part of the 1st Silesian IR Nr 11; in 1813 the number changed to 10. See Ser Nr 11 in the new regiments section below.

47. Regiment zu Fuss.
Raised 1743. Title in 1806 – **von Grawert.**
Upper Silesian Inspectorate. Garrisons: Glatz, Silberberg (musketiers).

Uniforms. Originally lemon yellow facings, yellow buttons, Swedish cuffs, no lace decoration, white small clothes, black stock. Brass fusilier and grenadier cap front plate and ornaments up to 1787, yellow headband and back cloth, red lace edged yellow. Then the Kasket with grenade badge was worn. Yellow pompon with red centre. Up to 1798 facings were lemon yellow, thence dark lemon yellow. Yellow buttons, no lace decoration, white small clothes. Officers wore gold aiguillettes; NCOs had gold lace to the Swedish cuff and to the hat.

Colours. King's colour white with lemon yellow centre, black eagle, white scroll edged gold, gold decoration. Regimental colours lemon yellow, white centre, black eagle, lemon yellow scroll edged gold, gold decoration. White staff pikes.

Three colours were lost at the capitulation of Magdeburg on 11 Nov, the other at that of Prenzlau on 28 Oct 1806.

Campaign history. Raised in 1743 for Prinz Georg Wilhelm von Hessen-Darmstadt as a fusilier regiment from a battalion raised by the house of Holstein-Gottorp out of imperial troops which had been stationed in Mecklenburg since 1734. Known as 'Jung-Darmstadt' to differentiate from IR Nr 12. **1745** Grenadiers in Gren Bn 46/47; Osten, then Kleist, then Aulack. 5 Dec capture of Torgau; 12 Dec capture of Meissen. Musketiers: B Kesselsdorf; capture of Dresden. **1747** Chef Oberst Karl Friedrich von

Derschau. 1752 GM Leopold Friedrich Ludwig von Weitersheim; he received a hitherto Saxon regiment. **1756-1763 Against Austria and others in the Seven Years War. 1756** Chef Oberst Kaspar Friedrich von Rohr; mortally wounded as GM at Leuthen. Grenadiers in Gren Bn Wangenheim. From 9 Sep in Dresden. Musketiers: capture of Dresden; 15 Oct capture of the Saxon army at Pirna. **1757** Grenadiers: 6 May B Prague; 12 May capture of the magazines in Schwartz-Kosteletz; 5 Jun C Gang; 13 Jun C Kuttenberg; 18 Jun B Kolin; 17-21 Jul rearguard actions north of Bohmisch-Kamnitz. The grenadiers were combined with Gren Bn Kahlden (Grsn Rs 3,4and NGR). 17 Aug cannonade at Hirschfelde; 26 Sep cannonade at Barschdorf; 22 Nov B Breslau; 5 Dec B Leuthen; 6-23 Dec S Breslau. The musketiers remained as garrison of Dresden. **1758** Chef GM Christoph Heinrich von Grabow. 1 Jan-21 Feb before Schweidnitz; 20 May-27 Jun before Olmuetz; Grenadiers in Gren Bn Carlowitz (47/ Grsn R 7); 30 Jun C Domstaedl; from August in Fouque's corps. 28 Nov ambush at Schnellewalde. Musketiers: the 1st Bn was in the expedition to Franconia; the whole regiment was at S Olmuetz 27 May-2 Jul and in the defence of the great convoy at C Domstaedl on 30 Jun. **1759** Grenadiers: 17 Apr C Troppau; 18 Jul C Lindenau; 23 Jul ambush at Friedland; 28 Jul C Friedland; 31 Aug cannonade at Johnsdorf; 11 Sep C Bernsdorf. Musketiers: B Kunersdorf; lost almost 600 men. The regiment was then attached to von Finck's corps and was captured at Maxen. **1760** Only one battalion and the grenadiers in action. Grenadiers: 7 Aug C Hundsfeld, in Gren Bn Bock; 19 Sep C between Brieg and Glogau. Musketiers: fought in Pommerania against the Swedes and against the Russians under General Tottleben in the protection of Berlin. **1761** 24 Aug-15 Nov in the camp at Colberg; 19 Sep storm and defence of the Green Redoubt; 18-29 Nov advance to Belgard; clashes between Regenwalde and Petershagen; 11/12 Dec C Drenow; 12 Dec C Spie; 13-17 Dec rearguard actions at Treptow and Stargard. Musketiers: 31 May-3 Aug operations in Upper Silesia. 11 Aug-10 Oct covering Schweidnitz. **1762** Musketiers: 3 Jan Sk Malchin; 30 Aug B Friedberg. 1764 Chef GM Prinz Johann Adolph von Nassau-Usingen. **1778** Chef GM Wenzelslaus Christoph von Lehwaldt (GL in 1786). Grenadiers in Gren Bn Restorff (30/47). 9-14 Aug advance to Leitmeritz; 27-31 Aug expedition to Melnik; from 21 Sep retreat to Saxony. 1788 GM Friedrich Wilhelm von Wangenheim; 1790 Chef GM Johann Karl Graf von Hertzberg; retired as GL. **1792** On the Rhine and in Champagne; 20 Sep Valmy. **1793** 1-7 Mar Blockade of Koenigstein; 28 Mar C Bingen; 13 Aug C Altstadt; 13 Aug C

Limbach, took 200 prisoners, two cannon and two colours. C Guntersblum near Mainz (grenadiers); Wiessenburg Lines (musketiers); 13 Oct C Herzogenhand (1st Bn); 16 Nov failed raid on Bitsch fortress (2nd Bn); 6 Dec cannonade of Zweibruecken; 22 Dec C Stoffelberg and Limbach; 24 Dec C Scheerhoehle, defended its post for four days against heavy French attacks. **1794** 23 May C Vogelweh. 1797 Oberst Julius August von Grawert (GL in 1806). **1806.** Jena 16 Oct. The regiment surrendered at Magdeburg and was not re-raised. The 3rd Bn was in Glatz fortress, which did not surrender. In 1808 it became part of the 2nd Silesian IR Nr 12; from 1813 Nr 11. See Ser Nr 12 in the new regiments section below.

48. Regiment zu Fuss.
Raised 1756. Title in 1806 – **Kurfuerst von Hessen-Kassel.**
Westphalian Inspectorate. Garrisons: Paderborn, Muehlhausen (grenadiers), Brackel (musketiers).

Uniforms. Poppy red facings, originally yellow buttons; these changed to white in 1798, buff small clothes, Prussian cuffs. Officers wore eight embroidered gold lace loops on and two under each lapel, two on the cuff flap,three on the pocket flap and two in the small of the back. Narrow gold hat edging. NCOs had six plain gold lace loops (even after 1798), in pairs, on the lapel, two under the lapen, two above the cuff and two in the small of the back. Wide gold hat edging. Privates had the same lace loop configuration as the NCOs but the laces were white, originally with white fringes; these changed to crimson in 1798. Brass fusilier and grenadier cap front plate and ornaments up to 1787, buff headband, poppy red back cloth, orange lace edged black and light blue. Pompon colours, from the centre, orange, black, light blue. This is the same cap as worn by Garnisons-R Nr 9. In 1801 buttons changed to white and there were six, in pairs, on each lapel.

Colours. King's colour white with crimson centre, black eagle, white scroll edged silver, narrow light blue corner piles, silver decoration. Regimental colours crimson, white centre, black eagle, crimson scroll edged silver, narrow light blue corner piles, silver decoration. Whit pike staffs.

All four colours were lost at the capitulation of Erfurt on 16 Oct 1806. The regiment was not re-raised.

Campaign history. When IR Nr 32 was raised in 1743, each company detached ten men who were used to form the basis of the 13th Garrison Regiment in Minden. Chef GM von Beaufort; died in office. 1743 Chef Oberst von Wobeser; died in office as GM. 1746 Chef Oberst von Wutgenau. 1755 Chef Oberst Friedrich Wilhelm von Salmuth. On 10 August 1756 in Wesel this battalion was increased to a fusilier regiment of two battalions by the addition of part of Fusilier Regiment Nr 41. First Chef was the Governor of Wesel, GL Friedrich, Erbprinz von Hessen-Kassel. For further details of campaigns see IR Nr 45.

1756-1763 Against Austria and others in the Seven Years War. 1756 Musketiers: In Wesel; cooperating with the Anglo-Hanoverian army of the Duke of Cumberland. Defence of Geldern; C Bielefeld in June, then transferred to Magdeburg. **1757** The musketiers were in Magdeburg. The grenadiers were in the king's army; 6 May B Prague; 5 Dec B Leuthen. **1758** Grenadiers: 27 May-2 Jul S Ollmutz; 14 Oct B Hochkirch. Musketiers: 28 Sep Fehrbellin; Nov-Dec defence of Dresden. **1759** Chef Oberst Friedrich Wilhelm von Salmuth (from Grsn Bn Nr 8); retired as GM. Erbprinz Friedrich had received IR Nr 45. The musketiers were in the expedition to Franconia, in several actions such as Aue. The 2nd Bn was then in the second defence of Dresden in September. The 1st Bn was in the recapture of Wittenberg on 27 Aug, C Dresden on 5 Sep, C Torgau on 8 Sep; 21 Sep C Gorbitz; 29 Oct C Pretsch. **1760** Musketiers: 20 Aug C Strehla; 2 Oct C Wittenberg. **1761** Grenadiers C Neuensund in Pommerania; captured several hundred Swedes and three cannon. **1762** Musketiers: 15 Oct C Brand (Freiberg), heavy losses. **1763** Chef GM Karl Friedrich von Beckwith; previously Chef of the Prussian Legion (the 'Legion Britannique' in the army of the Herzog Ferdinand von Braunschweig). 1766 Chef Oberst Martin Ludwig von Eichmann (GL in 1786). **1778/1779** In the army of Prinz Heinrich. **1787** In the expedition to Holland; 1 Oct C Oudekerken, Sk Wsuv, Sk Muiden. **1790** Grenadiers in the expedition to Liege. 1791 Chef GL Martin Ernst von Schlieffen; resigned his office to become Governor of Wesel. 1792 Chef GM Georg Bogislav von Koethen; retired. **1793** In the corps of the Herzog Friedrich August von Braunschweig-Oels (younger brother of the reigning Herzog) in the expedition into Flanders: March Sk Breda, Sk Brueggen, Sk Wasserburg; Roermond and Preussisch-Geldern; 1-2 May C Vicogne; 8 May C Hasnon; 23 May B Famars under General von Knobelsdorf, who took over from the Herzog in late March. In September they then marched to the upper Rhine; 27 Sep C St Imbert; 28-30 Nov B Kaiserslautern; C Moorlautern; Sk Saarbruecken. **1794** 3 Jul C Trippstadt. **1797** Chef GFM Landgraf Wilhelm von Hessen-Kassel; from 1803 Kurfuerst von Hessen-Kassel.

49. Regiment zu Fuss.
Raised 1741. Title in 1806 – **von Mueffling.**
Upper Silesian Inspectorate. Garrison: Neisse.

Uniforms. Originally the coat was all blue, no facings or lapels, brick red small clothes, black stock. From 1787 – 1797 facings were orange, lapels dark blue, buttons white, Brandenburg cuffs. In 1798 facings changed to white. All ranks wore eight, evenly-spaced buttons on the lapels. Officers wore embroidered silver lace loops with loose tassels; two under each lapel, three on each pocket and cuff flap and two in the small of the back; narrow silver hat edging. NCOs wore two plain silver loops under each lapel and two in the small of the back; no lace to the cuffs. The men wore lace loops in the same configuration as the NCOs, white with three light blue stripes, white and light blue tassels. Tin fusilier and grenadier cap front plate and ornaments up to 1787, red headband, dark blue backcloth, tin 'lace'. Light blue within red pompon. The Kasket with grenade badge was worn up to 1798. The 1798 grenadier cap had a light blue band edged white.

Colours. Initially the king's colour was white with white centre bearing an upright orange cross behind the black eagle, white scroll edged silver, a silver flaming grenade in each corner, silver decoration. The first regimental colours were all orange, with a white upright cross behind the black eagle, orange scroll edged silver, silver grenades and decorations.

In 1758 these changed to the conventional pattern (no corner piles or crosses); the centre of the king's colour chestnut brown, the regimental colour chestnut brown with white centre and chestnut brown scroll; all decoration silver. White pike staffs.

One colour was lost in the combat of Saalfeld on 10 Oct, the others at the capitulation of Magdeburg on 11 Nov 1806.

Campaign history. The following actions are only those of the grenadiers from 1740-1778.

Raised in November 1742 in Neisse as a regiment of ten companies of 'Pionniers' and two companies of 'Mineurs' from recruits from the mining areas of Upper Silesia and Moravia for work strengthening the fortifications. First Chef was GM Gerhard Kornelius von Walrave; removed from office for treason and imprisoned in the Star Redoubt in Magdeburg fortress until his death in 1773. **1747** Defence of Frauenberg castle; surrendered after a brave defence. **1748** Chef Oberst Philipp Loth von Sers (or Seers); resigned. **1756-1763 Against Austria and others in the Seven Years War. 1757** 22 Jul defence of Zittau (2nd Bn); 28 Dec recapture of Liegnitz. **1758** 31 Mar-18 Apr S Schweidnitz; 25 Aug B Zorndorf, very distinguished; stormed two batteries and took two howitzers and six ammunition wagons. 26 November 1758 converted to a fusilier regiment; Chef GM Christian Friedrich von Diericke; retired as GL. The two companies on miners were detached. There were no grenadiers in this regiment until 1787, when a four-company battalion was raised. Two companies had been added from Grsn R Nr 10. **1759** In Pommerania; 8-17 Jan S Demmin; 8-21 Jan S Anklam; 10 Apr capture of the Peenemuende Redoubt; 23 Jul B Kay; 12 Aug B Kunersdorf; distinguished, heavy losses. **1760** 3 Nov B Torgau; stormed the heights of Siptitzer and took several hundred prisoners. **1761** Under Prinz Heinrich. **1762** C Pretschendorf. 1770 Chef Oberst Karl August von Schwartz (GL in 1786); retired as GL and Governor of Neisse. **1788** On the raid into Bohemia; covered the withdrawal. Chef GM Otto Heinrich Friedrich von Borch; retired. **1792** On the Rhine and in Champagne; 20 Sep Valmy; 6 Nov C Limburg the 1st Bn, together with the grenadier battalion of IR Nr 39, defended their position against over 6,000 of the enemy for over three hours. 2 Dec storm of Frankfurt/Main. **1793** 6 Jan C Hochheim; took five cannon. 19 Jun-23 Jul S Mainz. The grenadier battalion assaulted Kostheim and took two cannon, but were driven out again. Some days later, the 1st Bn took Kostheim again, captured it and another cannon, but were again driven out. Some weeks later, the 2nd Bn again took the place. Together with three more cannon and over 300 prisoners. 14 Sep B Pirmasens; the 1st Bn took a cannon and over 100 prisoners. 13 Oct B Weissenburg lines (1st Bn); C Gleisweiler near Landau (musketiers); 28-30 Nov B Kaiserslautern the whole regiment; they took two cannon and many prisoners and much baggage. **1794** Chef GM Friedrich Wilhelm von Schoenfeldt; retired. The regiment was in all major actions of the year. In total the regiment had taken twelve enemy cannon; by order of the king, they were granted a new regimental seal showing their achievements. 1800 Chef Oberst Friedrich Wilhelm von Mueffling; GM in 1806.

1806. Saalfeld 10 Oct. Jena 16 Oct. The regiment surrendered in Magdeburg and was not re-raised.

50. Regiment zu Fuss.
Raised in 1768. Title in 1806 – **von Sanitz.**
Upper Silesian Inspectorate. Garrisons: Frankenstein, Muensterberg (grenadiers), Cosel (musketiers).

Uniforms. Peach blossom pink facings, white buttons (six, in pairs) on each lapel) Swedish cuffs, buff small clothes, black stock. Officers wore two embroidered silver lace loops under each lapel; NCOs loops were silver, privates were white. Tin fusilier and grenadier cap front plate and ornaments up to 1787, buff headband, peach blossom pink back cloth, rose red lace edged white. Pompon colours, from the centre, peach blossom pink, white, peach blossom pink.

From 1787 to 1797 small clothes were white; in 1798 facings changed to peach-blossom pink and there were eight, evenly spaced buttons on each lapel; Swedish cuffs. Officers wore embroidered silver loops under each lapel and two in the small of the back. Officers' facings were of crimson velvet. Narrow silver hat edging. NCOs had two plain silver loops under each lapel and two in the small of the back; no silver lace to the cuffs. Wide silver hat edging. Privates also had the same four loops but in white tape.

Colours. King's colour white with royal blue centre, black eagle, white scroll edged gold, scarlet corner piles, gold decoration. Regimental colours royal blue, white centre, black eagle, royal blue scroll edged gold, scarlet corner piles, gold decoration. Black pike staffs.

All four colours were lost at the capitulation of Magdeburg on 11 Nov 1806.

Campaign history. Raised in 1768 as a battalion of one grenadier and five musketier companies in Reichenbach/Silesia as fortress troops. The officers mostly came from Wuerttemberg and the Swiss Canton of Berne. First Chef GM Franz Ludwig von Rossieres from Sardinian service; he died as GM in office. 1778 Oberst Karl Ludwig Gottlieb von Troschke (GM in 1786); retired as GL. 1795 GL Georg von Steensen; he was appointed Governor of Neisse. It was expanded to a regiment in 1797. 1800 Oberst Karl Wilhelm von Sanitz; GM in 1806. **1806.** Jena 16 Oct. The regiment surrendered in Magdeburg and was not re-raised.

51. Regiment zu Fuss.

Raised 1773. Title in 1806 – **von Kauffberg.**

West Prussian Inspectorate. Garrisons: Danzig, Neugarten (musketiers).

Uniforms. Sulphur yellow facings, white buttons and small clothes, Swedish cuffs, black stock. From 1787 to 1797 the cuffs were of the Prussian style, thence Swedish again and the facings became dark lemon yellow. Tin fusilier and grenadier cap front plate and ornaments up to 1787, yellow headband and back cloth, dark yellow lace edged light blue. Light blue pompon with yellow centre.

Originally six buttons, in pairs, on each lapel; in 1798 this changed to eight, evenly-spaced. No lace decoration except in the parade uniform. This parade tunic had eight plain silver laces on the lapel, two below it, two each on the cuff, the pocket flap and in the small of the back. For daily wear, officers wore two embroidered silver laces under each lapel and two in the small of the back. Narrow silver lace hat edging. NCOs had no lace loops, but silver lace edging to the cuffs and the hat. Privates had no laces to the coat.

Colours. King's colour white with lemon yellow centre, black eagle, white scroll edged silver, narrow, pale blue corner piles. Regimental colours lemon yellow, white centre, black eagle, lemon yellow scroll edged silver, pale blue corner piles, silver decoration. Brown pike staffs.

All four colours were lost at the capitulation of Ratkau on 7 Nov 1806.

Campaign history. Raised in January 1773 in Marienburg from detachments mainly from west Prussian regiments but including foreigners. First Chef GM Otto Heinrich von Laxdehnen, who retired in the November. Next Chef 1773 GM Doering Wilhelm von Krockow (in 1786 GL and Graf) retired. **1778** The regiment was the advanced guard on the advance into Bohemia. 1789 Chef Oberst Ernst Friedrich Karl von Hanstein; died in office as GL. 1803 GM Friedrich August von Kauffberg. **1806** The regiment was in Bluecher's corps; fought at Luebeck, surrendered at Ratkau and was not re-raised. The 3rd Bn was in Danzig; in 1808 it became part of the 1st Westpreussisches IR Nr 6. See Ser Nr 7 in the new regiments' section below.

52. Regiment zu Fuss.
Raised on 1 Apr 1773. Title in 1806 **Reinhart.**
East Prussian Inspectorate. Garrisons: Rastenburg, Roessel, Angerburg (grenadiers), Lyck (musketiers).

Uniforms. Scarlet facings, white buttons, Swedish cuffs, no lace decoration for other ranks; eight, evenly-spaced buttons on the lapels. Black stock. Officers wore two embroidered silver laces with loose tassels under each lapel, on the cuff and in the small of the back and three on each cuff flap. NCOs had silver lace edging to cuffs and hats. Tin fusilier and grenadier cap front plate and ornaments up to 1787, scarlet headband, light blue back cloth, white lace with two red stripes. Pompon colours, from the centre, red, white, light blue, white. In 1787 Prussian cuffs were adopted, but these became Swedish again in 1798. In this year, the new uniform had eight, evenly-spaced buttons on each lapel.

Colours. King's colour white with dark green centre, black eagle, white scroll edged silver, upright, wavy scarlet cross, silver decoration. Regimental colours dark green, upright, wavy scarlet cross, , white centre, black eagle, dark green scroll edged silver, silver decoration. White pike staffs.

Campaign history. Raised on 1 April 1773 in Prussian Holland and Muelhausen. Most NCOs came from Prussian garrison regiments. First Chef GM Christian August von Lengefeld; received IR Nr 5, appointed Governor in Magdeburg and GL. **1778/1779** In the war of the Bavarian Succession the grenadiers were in Gren Bn Osorowsky (51/52). The regiment was in the advanced guard on the invasion of Bohemia. 14 Jan 1779 C Zuckmantel. 1785 Chef GM Wilhelm Friedrich Karl Graf von Schwerin; retired as GL. From 1789 'Alt-Schwerin' to differentiate from IR Nr 26. For more details of campaigns see IR Nr 51. **1794** In Poland. Rawka 6 Jun; Warsaw 27 Jul – 10 Aug; Socliaczew. The regiment was also present at Skala and Wola, losses unknown. 1795 Chef GM Joachim von Reinhart (GL in 1806). **1806.** The grenadiers combined with those of the IR 'von

Courbiere' Nr 58 as the Gren-Bn 'von Brauschitz. The 3rd Musketier-Bn was in the siege of Graudenz; the rest of the regiment, including the grenadiers, were besieged in Danzig. 1807. Chef GL Karl Friedrich von Hamberger. Actions during the siege of Danzig. C Dirschau 23 Feb; C Zigankenberg 16 Mar; C Scheidlitz 26 Mar. The sortie of 28/29 Apr.

The 3rd Bn was in Graudenz; in 1808 it became part of the 1. Westpreussisches IR Nr 6 on 7 Sep 1808. See Serial Nr 7, Grenadier-Regiment Graf Kleist von Nollendorf (1. Westpreussisches) Nr 6 for subsequent history. In 1808, parts of the regiment also went into the Kaiser Franz Garde-Grenadier-Regiment Nr 2, see Ser Nr 3 in the new regiments' section below.

53. Fusilier Regiment.
Raised in 1773. Title in 1806 – **Jung Larisch.**
West Prussian Inspectorate. Garrisons: Thorn, Prussian Stargard (grenadiers), Inowraclaw (musketiers).

Uniforms. Originally red Swedish cuffs, collar and shoulder strap, no lapels, white buttons and small clothes, black stock. Officers wore two embroidered silver lace loops under each lapel and two in the small of the back. Other ranks wore no lace loops; NCOs had silver lace to the cuff. Tin fusilier and grenadier cap front plate and ornaments up to 1787, light red headband and back cloth, white lace with two light red stripes. White pompon with light red centre. The grenadier cap of 1798 had a yellow band edged white.

From 1787 to 1798 facings were light blue, buttons white, cuffs Prussian; thence pale yellow facings and Prussian cuffs; yellow buttons (eight, evenly-spaced on each lapel). No lace decoration except for officers: two embroidered gold loops with loose tassels under the lapel, and in the small of the back, three on the cuff and the pocket flap. NCOs had gold lace edging to the cuff and the hat. White small clothes.

Colours. King's colour white with light green centre, black eagle, white scroll edged silver, narrow red corner piles, silver decoration. Regimental colours light green, white centre, black eagle, light green scroll edged silver, narrow red corner piles, silver decoration. White pike staffs.

The two colours of the 1st Bn were lost in combat at Schwartau on 6 Nov; that same day, the 2nd Bn was scattered between Luebeck and Mieslingen and both their colours were taken there.

Campaign history. Raised in April 1773 in Braunsberg, in the same way as for IRs 51 and 52. The following actions are only those of the grenadiers from 1773-1778. **1773** Chef Oberst Kaspar Fabian Gottlieb von Luck; retired as GL. **1778/1779** In the war of the Bavarian Succession the grenadiers were in Gren Bn Hausen (4/53). The regiment was in the advanced guard in the invasion. **1778** 5 Jul-15 Aug raid into Bohemia to Nachod; 16 Aug-5 Oct in the camp at Lewin, west of Glatz. 20 Nov C Kunzendorf. 1780 Chef Oberst Wilhelm Friedrich Freiherr von der Goltz; he received IR Nr 7. 1784 Chef Oberst Wilhelm Friedrich Karl Graf von Schwerin; he received IR Nr 52. 1785 Chef Oberst Karl Albrecht Friedrich von Raumer; he received Grsn R Nr 10. 1786 Chef GM Franz Andreas von Favrat; he received IR Nr 33 and became Governor of Glatz. **1794** Chef GM Albert Graf zu Anhalt; retired. The regiment was known as 'Jung-Graf Anhalt' to distinguish it from IR Nr 43. The regiment was in Poland, guarding the line of the Narew River; repelled repeated attempts by the insurgents to penetrate their line. The regiment was rewarded by being allowed to play the grenadiers' march. 1800 Chef Oberst Wilhelm Christoph von Larisch (the regiment was known as 'Jung Larisch' to distinguish it from IR Nr 26). **1806** The main body of the regiment was in Bluecher's corps and surrendered at Ratkow and was not re-raised. The 3rd Bn was in Graudenz; in 1808 it became part of the 1. Westpreussisches IR Nr 6 on 7 Sep 1808. See Serial Nr 7, Grenadier-Regiment Graf Kleist von Nollendorf (1. Westpreussisches) Nr 6, for subsequent history.

54. Fusilier Regiment.
Raised in 1773*. Title in 1806 – **von Natzmer.**
West Prussian Inspectorate. Garrisons: Graudenz (HQ and musketiers), Culm, Marienburg (grenadiers).

Uniforms. When raised the uniform had no lapels; pale purple facings on the collar and Swedish cuffs, yellow buttons, white small clothes, black stock. Officers wore two embroidered gold lace loops under the 'lapels'. Brass grenadier cap front plate and ornaments up to 1787, white headband, pale purple back cloth, yellow lace edged red. Pompon colours, from the centre, pale purple, yellow, pale purple. The brass fusilier and grenadier cap plate of 1798 had a white band edged in yellow tape edged red; pale purple cloth top, pompon from the centre: pale purple, yellow, pale purple.

From 1787 up to 1797 lapels were buff, but collar and Prussian cuffs were dark blue; white buttons and small clothes, no lace decoration. In 1798, lapels, collar and cuffs were all buff. Officers wore eighteen silver embroidered lace loops with long, loose tassels, arranged as for IR Nr 4. NCOs wore two silver lace loops, with pointed ends under each lapel, two in the small of the back; silver hat edging. No silver lace to cuffs. Privates wore white laces as for the NCOs but with white tassels.

Colours. King's colour white with black centre, gold eagle, white scroll edged gold, gold decoration. Regimental colours black, white centre, black eagle and scroll, the latter edged gold, gold decoration. White pike staffs.

All four colours were lost in the fighting in Luebeck on 6 Nov 1806.

Campaign history. Raised in Dec 1773 as for regiments 51-53. First Chef Oberst Albrecht Ehrentreich von Rohr; retired as GL. **1778/1779** In the war of the Bavarian Succession the grenadiers were in Gren Bn Franckenberg (54/55). The musketiers were in the advanced guard in the invasion of Bohemia; thence in the Upper Silesian corps of the Erbprinz von Braunschweig. 1784 Oberst Karl Kuno Ludwig von Klitzing; died in office. 1786 Chef GM Bogislav Ernst von Bonin; retired as GL. **1794** Chef GM Christian Friedrich von Mosch; became Commandant of Wesel. The regiment was in Poland; 18 May C Skala (grenadiers); one battalion at B Scelze on 6 Jun; it was rewarded for its actions by being allowed the play the grenadiers' march. It was also at the siege of Warsaw 27 Jul-6 Sep and threw the enemy out of Wolla. All three battalions stormed the redoubts at Wawrzice and Powonske at Warsaw. 1799 Oberst Hans Christoph von Natzmer (GM in 1806). **1806** The main body of the regiment was in Bluecher's corps; surrendered and was not re-raised. The 3rd Bn was in Graudenz; in 1808 it became part of the 2nd Westpreussischen IR Nr 7. See Ser Nr 8 in the new regiments' section below.

* Ten Saxon regiments were press-ganged into Prussian service at Pirna in October 1756; seven of these were disbanded within a year due to their heavy desertion rates. The IR Nr 54 remained in Prussian service until the end of the Seven Years War, when it was disbanded, the remaining men going to IR Nr 33. The original title of the regiment in 1756 was Sachsen-Gotha; in that same year it became von Saldern and in 1758 von Plotho. The uniform was faced white, white small clothes.

55. Fusilier Regiment.

Raised in 1774*. Title in 1806 – **von Manstein.**

West Prussian Inspectorate. Garrisons: Bromberg, Gnesen, Prussian Stargard (grenadiers), Graudenz (musketiers).

Uniforms. The first uniform had crimson facings to collar and Swedish cuffs, no lapels, yellow small clothes, black stock, white buttons, no lace decoration. NCOs wore silver lace to the cuffs. Tin fusilier and grenadier cap front plate and ornaments up to 1787, bright yellow headband, crimson back cloth, pink lace edged yellow, then white. Pompon colours, from the centre, white, yellow, crimson. Up to 1797 facings were light blue, yellow buttons, Prussian cuffs, no lace decoration. Thence crimson facings, white buttons, Brandenburg cuffs. Eight evenly-spaced buttons on the lapel. Officers wore gold embroidered lace loops with loose tassels; two each under the lapel and in the small of the back, three each on the cuff flap and pocket flap. NCOs had two plain gold laces with pointed rear ends under each lapel, two in the small of the back, gold hat edging, no lace to cuffs. Privates wore no laces.

Colours. King's colour white with light blue centre, black eagle, white scroll edged silver, upright, wavy yellow cross. Regimental colours light blue, white centre, black eagle, light blue scroll edged silver, upright, wavy yellow cross, silver decoration.

Three colours were lost in combat at Schwartau on 6 Nov; the other was destroyed by the bearer to prevent capture at the capitulation of Ratkau on 7 Nov 1806.

Campaign history. 1774 Raised in May in Mewe and Preussisch-Stargard from Polish recruits. Five officers were seconded from other Prussian regiments; the rest were foreigners. Chef Oberst Prinz Adolph von Hessen-Philippsthal; retired as GM. **1778/1779** In the war of the Bavarian Succession the grenadiers were in Gren Bn Franckenberg (54/55); with the four previous regiments in the advanced guard. **1778** 5 Jul-21 Sep raid into Bohemia, thence in the Upper Silesian corps of the Erbprinz von Braunschweig. 1780 Oberst Georg Ewald von Blumenthal; died in office as GM. 1784 Chef GM Melchior Sylvius von Koschenbahr; retired as GM. 1790 Chef GM Karl Eduard von Tiedemann; retired in 1792. 1792 Chef Chef GM Heinrich Graf von Bruehl; died on his way to take up his appointment. 1792 Chef Oberst Karl Friedrich von Holwede; retired. **1794** In Poland 6 Jun B Scelze (grenadiers and 2nd Bn); S Warsaw (the whole regiment) stormed four redoubts, took six cannon, two howitzers. 19 Aug C Powonsk (grenadiers and 2nd Bn); heavy losses. 23 Oct C Suchaczew the grenadiers, reduced to 300 men by previous actions, were attacked by a Polish corps of 4,000 men with ten cannon. They held their posts for six hours until relief arrived. On 23 Nov Prince Poniatowski with 8,000 men and 21 guns assaulted Suchaczew again. Despite being so outnumbered, the grenadiers beat off several attacks, leaving 800 men on the field. The battalion commander received a letter of thanks from the king and was promoted in the field. Many officers and soldiers received medals. The 2nd Bn was in an expedition into South Prussia and at the relief of Thorn in October. 1796 Chef GM Ernst Johann von Manstein; GL in 1806. **1806** The main body of the regiment was in Bluecher's corps; surrendered and was not re-raised. The 3rd Bn was in Graudenz; in 1808 it became part of the 2nd Westpreussischen IR Nr 7. See Ser Nr 8 in the new regiments' section below.

* Ten Saxon regiments were press-ganged into Prussian service at Pirna in October 1756; seven of these were disbanded within a year due to their heavy desertion rates. The IR Nr 55 remained in Prussian service until the end of the Seven Years War, when it was disbanded, the remaining men going to IR Nr 36. The original title of the regiment in 1756 was Lubomirsky; in that same year it became von Hauss and in 1760 von Roebel. The uniform was faced white, white small clothes. They fought at Kunersdorf and were captured at Maxen, then re-raised.

56. Regiment zu Fuss.
Raised in 1794*. Title in 1806 – **Graf von Tauentzien.**
Franconian Inspectorate. Garrisons: Neunkirchen / Baireuth, Culmbach (grenadiers), Crailsheim (musketiers).

Uniforms. From 1794 to 1797 the uniform was scarlet facings, white buttons, six (in pairs) to each lapel; no lace decoration, Brandenburg cuffs, white small clothes. Officers had thin silver hat edging and silver aiguillettes; no lace decoration. NCOs wore silver lace to cuffs, cuff flaps and hats. The grenadier cap of 1798 had scarlet band edged white. The regimental pompon colours were light blue over scarlet, divided by a black stripe. In 1798 the cut of the uniform was modernised, there were eight, evenly-spaced buttons on each lapel and Prussian cuffs were introduced.

Colours. Initially the regiment carried its old Bayreuth colours with diagonal black and white stripes. On the obverse was the crowned ducal cipher: CFCA (Christian Friedrich Carl Alexander) within green oak and palm leaves in the centre. In the four corners were the letters: M, Z, BA, B (Markgraf zu Brendenburg-Ansbach-Bayreuth), each in a green wreath, tied with a purple ribbon. In the centre of the reverse was thecrowned red eagle of Brandenburg; in the four corners: C, F, C, A.

In 1796 Prussian colours were presented. King's colour white with turquoise centre, wavy scarlet corner piles, black eagle, white scroll edged silver, silver decoration. Regimental colours turquoise with wavy scarlet corner piles, white centre, black eagle, turquoise scroll edged silver, silver decoration. White pike staffs.

All four colours were lost in the capitulation of Magdeburg on 11 Nov 1806. The regiment was not re-raised.

Campaign history. Raised in 1794 in Mainz from the infantry of Anspach-Bayreuth from the old regiments Voigt von Salzburg, von Reitzenstein and Grenadier Bn von Beust of the principality as IR Christian Ludwig Rudolph von Reitzenstein; he died in office. 1795 Chef Oberst Friedrich Gottlieb von Laurenz;died in office as GM in 1803. 1804 Chef GM Graf Friedrich Bogislav Emanuel von Tauentzien. The old princely regiments were in America in English pay from 1777 to 1783. **1777** in Clinton's expedition to the North River; the grenadiers were in the storm of Fort Montgomery. They also helped to clear the Delaware River. **1778** At Newport; the pursuit of the beaten foe; C Turkey Hill. **1780** C Springfield on the Jersey River. **1781** Under Lord Cornwallis in the fortified camp at Yorktown and were captured and interned in Virginia. **1788** Grenadier-Bn von Beust and IR von Reitzenstein went into Dutch service in Nijmegen. **1792** In Prussian service. **1793** In Holland; blockade of Breda; B Famars as the advanced guard of the Dutch army; assault on Muschin; defence of Lanoy in Flanders for twelve weeks; blockade of Maubeuge; defence of Bois de Tilleul. **1794** On the Rhine; no actions. **1806** 14 Oct Jena. The regiment surrendered in Magdeburg and was not re-raised.

* Ten Saxon regiments were press-ganged into Prussian service at Pirna in October 1756; seven of these were disbanded within a year due to their heavy desertion rates. The IR Nr 56 remained in Prussian service until the end of the Seven Years War, when it was disbanded, the remaining men going to IR Nr 32. The original title of the regiment in 1756 was Prinz August; in that same year it became von Loen, in 1758 von Kalkreuth then (in December) von Weitersheim, in December 1759 von Horn. The uniform was faced yellow, yellow small clothes. There was a mass desertion in March 1757; the regiment was subsequently re-raised.

57. Regiment zu Fuss.
Raised in 1794. Title in 1806 **von Grevenitz.**
Lower Silesian Inspectorate. Garrison: Glogau.

Uniforms. Pink facings, yellow buttons, Brandenburg cuffs (from 1798), eight evenly-spaced buttons on each lapel. Officers wore embroidered gold loops with loose tassels; two under each lapel, two in the small of the back, three each on the cuff flap and pocket flap. Narrow gold hat edging. NCOs had two plain gold loops under each lapel and two in the small of the back; gold lace edging to hat but none to cuffs. Privates wore the same configuration of lace loops as the NCOs, but they were white with three pink stripes, Prussian cuffs.

Colours. Presented in 1794. King's colour plain white with large black central crowned eagle, no wreath; silver decoration. Regimental colours plain pink with the same design.

The two colours of the 1st Bn were lost in the capitulation of Anklam on 1 Nov, those of the 2nd Bn at the capitulation of Hameln on 22 Nov 1806. The regiment was not re-raised.

Campaign history. Raised at one battalion in 1794 in Stettin from Polish prisoners and new recruits as the depot battalion von Weyher; increased to a regiment in 1795 by men of IR Nr 37. Chef GM von Grevenitz; retired as GL. **1806** Part of the regiment was at Jena and surrendered at Anklam on 1 November. Another battalion was in Hameln fortress and surrendered there on 22 November.

58. Regiment zu Fuss.
Raised on 20 Feb 1797. Title in 1806 **L'Homme de Courbiere.**
East Prussian Inspectorate. Garrisons: Goldapp, Gumbinnen, Angerburg (grenadiers), Olezko (musketiers).

Uniforms. Light yellow facings, white buttons, eight white lace loops with pointed ends and white tassels. Brandenburg cuffs. Officers had silver embroidered loops with tassels as for IR Nr 57; NCOs had plain silver loops as for IR Nr 57 as did the men. The grenadier cap had light yellow band edged white. The regimental pompon was light yellow within light blue.

Colours. Presented in 1797. King's colour white with light blue centre, lemon yellow corner piles, black eagle, white scroll edged silver, silver decoration. Regimental colours light blue, white centre, lemon yellow corner piles black eagle, light blue scroll edged silver, silver decoration. R. Redlin gives the colours as: kings – white with yellow centre and piles; regimental – light blue with yellow piles and centre.

Campaign history. Raised in 1797 with 24 native officers and 120 men of IR Nr 16 as cadre. Chef GL Guillaume Rene de L'Homme de Courbiere in Bartenstein and Preussisch-Eylau. On 20 Nov 1807 it absorbed men from the disbanded infantry regiments Nrs 29, 37, 40, 43, 54 and 55. On 7 Sep 1808 it became the **2. Westpreussisches IR (von Courbiere) Nr 7**. See Serial Nr 8 in the new regiments below for subsequent history. **1806/1807.** Besieged in Danzig. Christburg 19 Jan 1807; Dirschau 23 Feb; sorties from Danzig 26, 28/29 Apr. In 1808, parts of the regiment went into the Kaiser Franz Garde-Grenadier-Regiment Nr 2, see Ser Nr 3 in the new regiments' section below. The city surrendered on 24 May 1807; the regiment was not re-raised.

59. Regiment zu Fuss.
Raised in 1803. Title in 1806 **von Wartensleben.**
Franconian Inspectorate. Garrisons: Erfurt, Muehlhausen (grenadiers).

Uniforms. White facings, yellow buttons, in four pairs on each lapel; no lace decoration, Prussian cuffs. NCOs wore gold lace edging to hat, cuff and cuff flap and two plain gold lace loops under each lapel and in the small of the back.

Colours. Presented in 1803. King's colour white with pale blue centre and corner piles, black eagle, white scroll edged gold, gold decoration. Regimental colours pale blue, white centre and corner piles, black eagle, pale blue scroll edged gold, gold decoration.

Both colours of the 1st Bn were lost in the capitulation of Erfurt on 16 Oct; those of the 2nd Bn in that of Magdeburg on 11 Nov 1806.

Campaign history. Raised in Erfurt in 1803 from men of 2nd Bn, IR Nr 43, of the old Kur-Mainz IR von Knorr and of the town guards of Nordhausen and Muehlhausen in Thuringia **1806.** Auerstaedt 16 Oct. The regiment surrendered and was not re-raised.

60. Regiment zu Fuss.

Raised in 1802. Title in 1806 **von Chlebowsky.**

Warsaw Inspectorate. Garrisons: Warsaw, Kalisch (grenadiers).

On 21 Nov 1808 remnants of the regiment went to form the 1. Schlesisches IR Nr 10. See Serial Nr 13 in the new regiments section below for subsequent history.

Uniforms. Lemon yellow facings, yellow buttons, eight, evenly spaced, on each lapel, Brandenburg cuffs, white small clothes. Regimental hat pompons red over white. Officers had a narrow gold edging to the hat. NCOs wore plain gold lace edging to hat, cuff and cuff flap.

Colours. Designed but not presented in 1806. King's colour white with dark green centre, narrow pink corner piles, black eagle, white scroll edged silver, silver decoration. Regimental colours dark green with wide pink corner piles, white centre, black eagle, dark green scroll edged silver, silver decoration.

Campaign history. Raised in Warsaw in 1802 from the 3rd Battalion of IR Nr 44. 1806 The regiment surrendered and was not re-raised.

3

The Grenadier Battalions 1799-1807

By Order of 28 Feb 1799 (to be acted on on 1 June) the grenadier companies were to be formed into standing grenadier battalions as shown below. The grenadier companies of the following regiments were excluded from this scheme and stayed with their parent regiments: Nrs 3, 5, 18, 22, 36 and 46. The Fluegelgrenadier companies of the four guards battalions were to be concentrated together only in peacetime. If mobilised for war, they were to return to their parent regiments.

In the following list the pairs of numbers of the contributing regiments are shown divided by '/'.

1/13; 2/11; 3/21; 4/54; 5/20; 6/15; 7/30; 8/42; 9/44; 10/41; 12/34; 14/16; 17/51; 18/27; 19/25; 22/36; 23/26; 24/35; 28/50; 29/32; 31/46; 33/47; 37/57; 38/49; 39/44; 40/43; 45/56; 48/59; 52/58; 53/55.

From documents found in the Prussian War Ministry (destroyed in the Second World War) it seems that at least the first seven of these permanent grenadier battalions were presented with colours. The details are shown below.

1st Battalion. King's colour: white, light blue central disc, white scroll, black corner piles, gold decoration. Regimental colours: light blue, white central disc, light blue scroll, black corner piles, gold decoration.

2nd Battalion. King's colour: white, dark blue central disc, white scroll, darl orange corner piles, silver decoration. Regimental colours: dark blue, white central disc, dark blue scroll, dark orange corner piles, silver decoration.

3rd Battalion. King's colour: white, red central disc, white scroll, narrow red corner piles, narrow, upright cross, gold decoration. Regimental colours: red, white central disc, red scroll, narrow white corner piles, narrow white upright cross, gold decoration.

4th Battalion. King's colour: white, crimson central disc, white scroll, narrow upright light blue upright cross, silver decoration. Regimental colours: crimson, white central disc, crimson scroll, narrow upright light blue cross, silver decoration.

5th Battalion. King's colour: white, crimson central disc, white scroll, green corner piles, gold decoration. Regimental colours: crimson, white central disc, crimson scroll, green corner piles, gold decoration.

6th Battalion. King's colour: white, cornflower blue central disc, white scroll, narrow, cornflower blue corner piles and upright cross, silver decoration. Regimental colours: cornflower blue, white central disc, cornflower blue scroll, narrow white corner piles and upright cross, silver decoration.

7th Battalion. King's colour: white, dark blue central disc, white scroll, narrow crimson corner piles, silver decoration. Regimental colours: dark blue, white central disc, dark blue scroll, narrow crimson corner piles, silver decoration.

Some of these survived the catastrophe of 1806; they are shown below. On 14 Oct 1814 they were used to form the first two grenadier regiments of the new Prussian army.

1st Ostpreussisches Grenadier-Bataillon.

Raised on 28 Feb 1799 from two companies each from IR 'von Ruechel Nr 2' and IR 'von Schoening Nr 11'. For details of histories of these regiments see the original regimental histories.

Titles. 28 Feb 1799 - 'Grenadier-Bn von Below'; 1 Oct 1806 - 'Gren-Bn von Schlieffen'; 26 Aug 1807 - 'Gren-Bn von Kurowsky'; 7 Sep 1807 - '1. Ostpreussisches Gren-Bn'.

Campaign history. 1807. Preussisch Eylau, 8 Feb; Friedland 13 Feb; Koenigsberg 13 Feb; Heilsberg 22 Feb. **1813.** Gross-Goerschen 2 May; Bautzen 20/21 May; Hochkirch 4 Sep; Wartenburg 3 Oct; Moeckern 16

Oct; Freiburg 21 Oct; Hoerselberg 26 Oct. On 15 Nov the battalion was reduced to one weak company and was combined with the Schlesischen Gren-Bn. **1814.** Chateau-Thierry; Gue-a-Tremes 28 Feb; Paris. In Oct 1814 the remnants of the battalion became part of the **Kaiser Alexander Garde-Grenadier-Regiment Nr 1** (Die Kartoffelschaeler), see Serial Nr 2 in the new regiments section below.

2nd Ostpreussisches Grenadier-Bataillon.
Raised on 28 Feb 1799 from two companies each of IR von Besser Nr 14 and IR von Diericke Nr 16. For details of histories of these regiments see the original regimental histories.

Title. Grenadier-Bataillon von Fabecky; 7 Sep 1807 - '2. Ostpreussisches Grenadier-Bn'.

Colours: see Fuesilier-Bataillon of the Kaiser Aexander Garde-Grenadier-Regiment Nr 1.

Campaign history. 1807. Preussich Eylau 8 Feb; Heilsberg 22 Feb. **1813.** Halle/Saale 4 May; Luckau 4 Jun; Gross-Beeren 28 Aug; Dennewitz 6 Sep. Took 5 ammunition wagons and many prisoners. Wittenberg 25 Sep-4 Oct; Leipzig 18. **1814.** Lonhout 11 Jan; Antwerp 11 Jan; Soissons 26-28 Mar. In Oct 1814 the remnants of the battalion became part of the **Kaiser Alexander Garde-Grenadier-Regiment Nr 1**, see Serial Nr 2 in the new regiments section below. **Campaigns. 1815.** mobilised against France, no actions.

Pommerschen Grenadier-Bataillon Raised on 7 Nov 1801 as the Gren-Bn 'von Massow', from two grenadier companies each from the IR 'von Ruets' Nr 8, IR 'von Ploetz Nr 42 and from the Gren-Bn 'von Jung-Braun' (raised on 10 Oct 1805, from two companies each of IR 'von Kropff' Nr 31, 'von Thiele' Nr 46 and parts of the Freikorps 'von der Marwitz' (raised in 1806).

On 14 Oct 1814, this battalion became the 1st Battalion of the **Kaiser Franz Garde-Grenadier-Regiment Nr 2.** Serial Number 3 in the New Regiments section below.

Westpreussischen Grenadier-Bataillon Raised 21 Jan 1803 from the Gren-Bn 'von Brauschitz' and two companies each from IR 'von Reinhardt Nr 52 and IR 'von Courbiere Nr 58.

On 14 Oct 1814, this battalion became the 2nd Battalion of the **Kaiser Franz Garde-Grenadier-Regiment Nr 2.** Serial Number 3 in the New Regiments section below.

4

Feldjaegers

Raised in 1740.

Mark Brandenburg Inspectorate. Garrisons: Mittenwalde, Belitz, Zossen, Muencheberg, Treuenbriezen.

All details as for fusiliers of line infantry except as noted below. The shako had no top band and was decorated with dark green cords and a black plume for parades. Officers had black and silver cords.

Dark green tunic and turnbacks, red collar, shoulder straps, Swedish cuffs and piping to turnbacks, yellow buttons. Grey breeches, short, black gaiters, black leatherwork.

Campaign history. Raised in 1740 at a strength of 60 men; rapidly increased to 300 by dint of their good service. 1744 – Corps Feldjaeger zu Fuss; 1761 – Bataillon Fussjaeger; 1763 – reduced to two companies; 1773 – raised to five companies; 1778 – increased to six companies; 1 Jan 1784 – Fussjaeger-Corps of ten companies; 20 Jun 1786 - Regiment Fussjaeger; 23 Sep 1794 – absorbed the Ansbach-Bayreuth Jaeger-Bn of two companies; reformed in three battalions. They carried no colours. The corps fought as individual companies on several fronts in each campaign

1745. Against Austria; **1778-9.** Against Austria; **1792.** Against France, on the Rhine; two battalions were mobilised and saw action at Siereck,

Longwy, Volmerau, Ottange, Verdun, Montcheutin, Aux Froids, Fosses, Grand Pre, Chevieres, Valmy, Oberursel, Koenigstein; losses unknown.

1793. Weissenburg lines 13 Oct; Scheerhoehle 22 Dec. Also at Hochheim, Roermonde, Breda, Mainz, Kostheim, Waldalgesheim, Dreieckshausen, Mosbach, Bingen, Hoogstraten, Eich, Mosbach, St Amand, Vigogne, Neukirchen, Famars, Aux-Quatre-Cloches Abbey, d'Anchin, Auchy, Coutiches, Marchiennes, Bonans, Fuerstenmuehle, Lamponpont, Bouvines, Bretzenheim, Kostheim, Zahlbacher Muehle, Gleisweiler, Leimen, Bamberg, Sainghin, Hem, Altstadt, Limbach, Neukirchen, Woorsweiler Hof, Kettricher Hof, Leimen, Frisange, Menim, Limbach, Pirmasens, Wuerzbach, St Ingbert, Eschweiler, Saarbruecken, Weissenburg lines, Erlenbach, Bitsch, Landau, St Ingbert, Hauenstein, Burrweiler. **1794.** Weidenthal 23 May, Am Schorleberg 20 Jul, Duerkheim, Frankenthal, Alzey, Morschheim, Dannenfels, Neustadt/H, Bamelsberg, Galgenschanze, Kaiserslautern, Trippstadt, Kirrweiler, Weingarten, Fischbach, Edinghofen, Trippstadt, Steineck am Schmalenberg, Hoheneinoed, Johanniskreuz, am Platersberge, Hardt, Alsenborn, Neu-Leiningen, am Schorleberge, Kaiserslautern, Kreuzbach, St Johann. **1795.** On the River Ems and in Westfalia. **1796.** On the Weser in the Observation Corps. From 1792 – 1796 the corps lost 21 offrs and 468 men. **1806.** Four companies were with the Duke of Weimar's corps at Jena, three with Ruechel's corps at Jena, one with Le Coq in Westfalia. Schleiz 9 Oct; Saalfeld 10 Oct; Blankenburg 10 Oct; Jena 14; Altenzaun 26 Oct; Luebeck 6 Nov. 1806/7 – reformed from escaped POWs from the companies von Valentini and Werner in Danzig and Koenigsberg. A third company was raised on 17 Dec 1806; a fourth in spring 1807. When peace was declared there were nine companies and a detachment. **14 Nov 1808 – Jaeger-Bn Nr 1, Garde-Jaeger-Bn.**

1812 The Ostpreussisches Feld-Jaeger-Bn was mobilised with the 3rd Brigade of von Yorck's corps in Russia. **1813.** In Bluecher's Army of Silesia, 7th Brigade. Gross-Goerschen 2 May; Bautzen 20/21 May; Haynau 26 May. During the armistice, the 1st and 2nd Companies went to the Reserve Corps in the Army of Bohemia; the 3rd and 4th Companies remained in von Yorck's I Corps. Bunzlau 30 Aug; Hochkirch 4 Sep; Lilienstein 7 Sep; Koesen 21 Sep; Bischofswerder 22 Sep; Moeckern 16 Oct; Freiburg 21 Oct. **1814.** Muelheim 3 Jan; Chateau-Thierry 12 Feb; Paris 30 Mar. Took 12 guns. **1815.** No actions.

FELDJAEGER-CORPS ZU PFERDE (MOUNTED FELDJAEGERS)

Garrison Koepenick.

Uniform. Canary green coat, buff leather breeches, boots; white waistcoat, poppy red cuffs and collar, gold aiguillette. Officers wore eight embroidered gold loops to each lapel, two on each cuff, two on the pocket flap; four in the small of the back. A wide, scalloped gold hat edging. Jaegers' plain hats have plumes, white over black; green and silver hat cords and fist straps.

Raised in 1740 with 60 men including six 'Oberjaeger' (NCOs). In 1744 it was increased by 112 men; mostly foresters. Their duties in peace and war are to act as couriers. Chef in 1740 was GM Graf von Hacke. The incumbent of this post was usually 1st General Adjudant to the king. 1750 GM von Buddenbrock. 1754 Oberst von Ingersleben; killed at Breslau. 1757 Oberst von Wobersnow; killed at Kay. 1759 Oberst von Krusemark; as GM he became Chef of the Regiment Gensd'armes. 1768 GM von Anhalt; became Chef of IR Nr 33. 1784 Oberst von Hanstein; became Chef of IR Nr 51. 1787 Oberst von Geusau; promoted to GM and transferred to the 3rd Department of the War Office. In 1806 he was GL and QMG. 1790 Oberst von Bischofswerder; retired in 1798 as GL. 1798 Oberst von Zastrow; in 1800 he became Chef of IR Nr 39. 1801 Oberst von Roeckritz; GM in 1806.

5

The Garrison Regiments 1718-1787

The uniforms of these regiments were of line infantry cut, but single-breasted (with two rows of six buttons-in pairs- on the chest) and all regiments had dark blue collars and black stocks. Apart from the 2nd and 6th Regiments, who wore white waistcoat and breeches, they wore dark blue small clothes. The turnbacks were red for all regiments. As with the line regiments, these units also had grenadier companies, which were brigaded together with those of other units in time of war. Their grenadiers wore the typical mitre cap as is proven by surviving examples illustrated in Siegfried Fiedler's publication. Strangely enough, Dr Hans Bleckwenn does not show them as being worn in his book on this army, but Menzel does.

As will be seen, they frequently went on campaign. In 1787 these regiments were disbanded, the men going to form many of the new Fuesilier Regiments.

Up to 1763 the garrison regiments carried their own designs of colours. These bore a large, crowned king's cipher (FR) in golden palm leaves in the centre, a small golden grenade in the centre of each side and a large golden grenade in each corner. The central field and the surrounds were the same colour for each regiment. Regiments 1 – 4 had narrow black corner piles, the other regiments had plain colours.

From 1763 onwards, the central cipher was replaced by the usual black eagle and the corner grenades by the small, crowned royal cipher within a wreath as on the colours of the line regiments. All pikes were white.

1st Garrison Regiment. Location 1782-1788: Anaburg, Gumbinnen, Rastenburg, Roessel in East Prussia.

Red piping to the flap of the Prussian cuff, white buttons. Tin grenadier cap front plate and ornaments, red headband, dark blue back cloth, white lace with two flanking red worms and a central yellow worm. Pompon colours red within medium green.

Colours. Up to 1763, the 'FR' pattern central cipher with gold decoration; king's colour all white, regimental colour all violet. As of 1763: king's colour white, with narrow black corner piles, light blue central disc, black eagle, white scroll, gold decoration. Regimental colour royal blue, white central disc, black eagle, royal blue scroll, gold decoration.

Campaign History. Raised 1n 1718 in Memel at two companies, three in 1719 and five in 1721. Chef Oberst Karl de Brion. **1728** Oberst Ludwig de l'Hospital (GM in 1740. **1744/1745** Oct-Apr on the Wartha pass; grenadiers in Gren Bn Kleist von Jung-Schwerin (16/Grsn). 4 Jun B Hohenfriedberg; 30 Sep B Soor; 16 Oct C Schatzlar and Trautenbach.). **1755** Oberst Christoph Georg von Luck. **1757** Oberst Werner Friedrich von Puttkamer 'Alt Puttkamer' to differentiate from Grsn R Nr 7. 29 Aug C Puschdorf forest; grenadiers in Gren Bn Lossau (Grsn Rs 1/11). 30 Aug B Gross-Jaegersdorf. **1758** 7-14 Jan S Peenemuende redoubts; 18 Jan-18 Jun S Stralsund; 25 Aug B Zorndorf; 3 Oct cannonade at Passkrug near Pyritz; 15 Nov C Eilenburg. **1759** 1 Jan capture of Dammgarten; 9 Jan-15 May S Stralsund; 23 Jul B Kay; 12 Aug B Kunersdorf; 21 Sep C Korbitz. **1760** 13-22 Jul S Dresden; 19 Jul C Wilsdruff suburb; 20 Aug C Strehla; 7 Oct C Mariendorf/Berlin; 3 Nov B Torgau. **1761** 15 Feb C Langensalza; 2 Apr C Schwarza; 2 Jul C near Marienberg; 2 Aug C Teplitz; 14 Oct C Hartmannsdorf and at Rathswalde. 15 Oct C Brand; 29 Oct B Freiberg. **1772** Oberst Friedrich Sylvius von Hallmann. **1786** Oberst Karl Ernst von Bose. In 1787 four companies were given to Fus Bn von Thile Nr 3. The regiment was disbanded in **1788**. **1778/1779** no actions.

2. Garrison Regiment. Location 1782-1788: Barten, Domnau, Fischhausen, Pillau, Preussisch-Holland in East Prussia. Red piping to the flaps of the Prussian cuffs, white buttons. Tin grenadier cap front plate and ornaments, red headband, dark blue back cloth, white lace with two red flanking worms and a central dark blue worm. Light green pompon.

Colours. Prior to 1763, the 'FR' colours were white for the king's, light blue for the regimental colours. Then the king's colour was white, with narrow black corner piles, lemon yellow central disc, black eagle, white scroll, gold decoration. Regimental colour lemon yellow, white central disc, black eagle, lemon yellow scroll, gold decoration. **Campaign History.** Raised in 1718 in Pillau at five companies under Oberst Georg Bogislav von Wobeser. **1723** Chef Oberst Kaspar Dietlof von Winterfeld. **1725** Oberst Peter de Sers. **1727** Oberst Johann de Natalis. **1742** Obstlt Georg Werner von der Schulenburg. **1743** Oberst Christoph Ernst von Roeder. **1754** Oberst Hans Sigismund von Sydow; from 1759 'Alt-Sydow' to differentiate from Grsn R Nr 5. **1756** Not mobilised. **1757** 30 Aug B Gross-Jaegersdorf; they were in the second line and so confused by the battle that they fired into their own first line, allowing the Russians to launch a counter-attack. **1758 / 1759** In Schweidnitz; the grenadiers fought at Zorndorf in 1758 and Kay and Kunersdorf in 1759. **1760 - 1762** 1st Bn and grenadiers in Prinz Heinrich's army in Saxony and Silesia; they fought at Torgau 3 Nov 1760. 2nd – 4th Bns were in Breslau, Miltitsch, Neisse, and Trachenberg. **1762** 1st Bn and grenadiers at Freiberg on 29 Aug. **1773** Chef Oberst Georg Wolf von Tuempling. **1777** Chef Oberst Georg Lorenz von Pirch. 1778/1779 Grenadiers in Gren Bn von der Hardt Nr 4 (Grsn Rs 1 and 2). 1787 The grenadier company went to form the new Fusilier Bn Nr 6; see there for later history.

3. Garrison Regiment. Location 1782-1788: Colberg, Treuenbrietzen in Pommerania. All parts of the coat were dark blue, except the turnbacks and the red piping to the flaps of the Prussian cuffs; buttons gold.

Colours. These were of the style issued under Frederick the Great, with the central cipher 'FR' instead of the eagle and scroll. King's colour white, with narrow black corner piles, gold decoration. Regimental colour light blue, gold decoration.

Campaign History. Raised in 1718 in Kolberg under Oberst Sigismund von Sack. **1740** Chef GL Georg Rudolph von Glaubitz, then Oberst Volrath Hellermann (ennobled in 1743). **1744** Grenadiers in Gren Bn Byla (Grsn Rs NGR/3/4). 2-18 Sep S Prague; 26 Nov C Prague; 6 Dec a chain of rearguard actions began at Drum; 11 Dec Pankratz, 12 Dec Olbersdorf, 14 Dec Hermsdorf. **1745** Jan-Mar frontier security at Friedland / Silesia Apr-Jun in Liegnitz; 11 Jul C Neustadt; 27/28 Jul C Ziegenhals; 7 Aug C

Tillow nr Friedland; 26 Aug 5 Sep S Cosel; 28 Sep C Jaegerndorf; 20 Oct C Hulschin. **1756** Chef Oberst Georg Arnold von Grolmann. Aug-Dec in Pommerania; grenadiers in Gren Bn Kahlden (NGR/3/4). **1757** Grenadiers: 12 Mar raid on Busch-Ullersdorf; 21 Apr C Reichenbach; 6 May B Prague, thence S Prague until 2 Jun; 18 Jun B Kolin. After this action, on 1 Aug, the grenadier battalion was combined into a single unit with those of IR Nr 47/Grsn R 7. 17-21 Jul rearguard actions north of Boehmisch-Kamnitz; 17 Aug cannonade at Hirschfelde; 26 Sep cannonade at Barschdorf; 22 Nov B Breslau; 5 Dec B Leuthen; 7-20 Dec S Breslau. **1758** 1 Jan-21 Feb blockade of Schweidnitz; 1 May-30 Jun before Olmuetz; grenadiers in Gren Bn Wangenheim (NGR/Grsn Rs 3 and 4)7 Jul rearguard action at Kroenau; 11/12 Jul rearguard actions at Holitz; 14 Oct B Hochkirch. **1759** 17 Apr C Troppau; 18 Jul C Lindenau; 27 Jul C Konradswaldau; 11 Sep C Bensdorf. **1760** Grenadiers: 6 Aug C Hundsfeld; 2 Nov C Schoena; 3 Nov B Torgau. After this action, the grenadiers were in Gren Bn Carlowitz (NGR/Grsn Rs 3 and 4). The 1st Bn was captured in Torgau on 27 Sep. **1761** no actions. **1762** 27 Jun C Konstappel. **1763** Chef Oberst August Friedrich Marschall von Biberstein. The regiment absorbed men of the *Freikorps* von Schony**1767** Chef Oberst Ernst Friedrich von der Heyden. **1772** Chef Oberst Detlof Freiherr von Vietinghoff. **1778/1779** no actions. Disbanded in **1788**; the men went to the depot battalions of IRs Nrs 1 and 19.

4. Garrison Regiment. Location 1782-1788: Aken, Koenner, Loebejun, Treuenbrietzen, in the area of the modern German state of Thueringia. All parts of the coat were dark blue, except the turnbacks and the piping to the Prussian cuffs which were red; buttons gold.

Colours. These were of the style issued under Frederick the Great, with the central cipher 'FR' instead of the eagle and scroll. King's colour white, with narrow black corner piles, gold decoration. Regimental colour royal blue, gold decoration.

Campaign History. Raised in 1736 in Aken on the Elbe from a battalion in the service of the princes of Anhalt of one grenadier and five musketier companies; completed by drafts from the Magdeburg garrison and the IR Anhalt-Dessau. First Chef Oberst Kaspar Jochen von Wachholtz. For campaign details see Grsn R Nr 3. **1737** Oberst Peter d'Arbaud. **1739** Oberst Andreas Johann de Persod. **1740** GM Adam von Weyher, died on 10 Dec

1744; the regiment remained vacant until 1746. **1746** Oberst Jakob Heinrich von Grape. **1756 – 1757** Not mobilised. **1758** In Prinz Heinrich's corps in Saxony; no actions. **1759** GM Martin Eberhard von Jungken, known as Muenster von Mohrenstamm; previously Chef of IR Nr 44. 21 Aug as garrison in Wittenberg. Surrendered the fortress to the Imperials; then, reinforced by Kleist's corps, retook it. Took Torgau in a night attack. **1760** Oberst Ewald Georg von Lettow. The regiment was captured in Torgau fortress on 27 Sep. The regiment was not re-raised until the end of the war. **1763** Obstlt Friedrich Franz Ernst Freiherr von Plotho, previously Chef of a Saxon regiment. **1766** Oberst Franz Thoro von Groscreuz. **1769** Oberst Ludwig von Gohr. **1772** Oberst Peter Georg von Puttkamer. **1775** Oberst Heinrich Ludwig von Gotter. **1782** Oberst Valentin Friedrich von Ruechel. **1784** GM Otto Kasimir von Meerscheidt, known as Huellessem. **1788** The regiment was disbanded; the men went to the depot battalions of IRs 9 and 10, which, as 3rd Bns of those regiments, surrendered in Hameln.

5. Garrison Regiment. Location 1782-1788: Glogau, Beeskow, Drossen, Krossen, Zuellichau in Silesia. Black Prussian cuffs, flaps piped red, yellow buttons. Brass grenadier cap front plate and ornaments, white headband and back cloth, white lace with black edges. Black pompon centre, edged white with thin black stripes.

Colours. These were of the style issued under Frederick the Great, with the central cipher 'FR' instead of the eagle and scroll. King's colour plain white, gold decoration. Regimental colour black, gold decoration.

Campaign History. Raised in 1741 in Glogau fortress at two battalions and two grenadier companies. First Chef GM Christoph Friedrich von Thuemen. **1743** Chef Oberst Friedrich Julius von Muetzschefahl (GM in 1756). **1744** Grenadiers in Gren Bn Kahlbutz (24/Grsn R 5). 27 Aug capture of Teschen castle; Sep-3 Dec in Leitmeritz; Grenadiers in Gren Bn Stange (34/Grsn R 5). 6 Dec rearguard action at Drum. **1745** Jan-May on border security at Friedland / Silesia; 22 May C Landeshut; 28 May-Jun in Liegnitz fortress; 30 Sep B Soor; Nov-Dec in Schweidnitz fortress. **1756** Grenadiers in Gren Bn Rath (Grsn Rs 5 and 10). Sep-Oct raids into eastern Bohemia. **1757** Grenadiers in Cosel fortress. Three musketier battalions were captured in Schweidnitz on 12 Nov. **1758** 1st and 2nd Bns 16 Feb advance to Graetz; 18 Feb C Troppau; 19 Feb rearguard action at

Troppau; 1 May-27 Jun before Olmuetz; 30 Jun C Domstaedl; grenadiers combined with Gren Bn Carlowitz (47/Grsn R 7). 2 Jul-9Aug retreat to Landshut; Aug-Nov in Fouque's corps in Silesia; 14 Nov C Cosel. **1759** Chef Oberst Gustav Adolf von Sydow; 'Jung-Sydow' to differentiate it from Grsn R Nr 2. 1st and 2nd Bns: 23 Jul 1st C Friedland / Silesia; 28 Jul 2nd C Friedland / Silesia; 12 Aug B Kunersdorf. 11 Sep C Bernsdorf. **1760** Grenadiers in Gren Bn Koschenbahr Nr 5 (Grsn Rs 5 and 10). 17 Jun 1st C Landeshut; 23 Jun 2nd C Landeshut. 1st and 2nd Bns in Prinz Heinrich's corps. **1761** May-Jul at Glogau; 15 Aug C Gross-Wandriss; 20 Aug-11 Sep in the camp at Bunzelwitz; 14 Sep C Kobylin; 30 Sep C Koerlin; 2 Oct C Spie; 20 Oct C Zargelow; 22 Oct C Gollnow; 20-25 Nov actions between Regenwalde and Petershagen; 13-17 Dec rearguard actions at Stargard and Treptow. **1762** 1st and 2nd Bns in Saxony; Jun-Oct in Upper Silesia. 3rd and 4th Bns in garrisons in Silesia. **1763** Oberst Christian Friedrich von Berner. **1770** Chef Oberst Johann Benjamin von Hasslocher. **1771** Oberst Adam Friedrich von Arnstaedt. **1778** Chef Oberst Paul von Natalis, who commanded a grenadier battalion 1760-1763. 14 Sep rearguard action at Jung-Buchau. **1788** The regiment was disbanded, the men going to the depot battalions of the IRs Nrs 3, 5, 20, 21, 27, 36 and 41, all of which surrendered in 1806.

6. Garrison Regiment. Location 1782-1788: Brieg, Cosel, Namslau in Silesia. Red Prussian cuffs, red piping to the cuff flaps, white buttons and small clothes. Tin grenadier cap plate and ornaments, red band and back cloth, white lace with red pattern, white pompon with a red ring.

Colours. These were of the style issued under Frederick the Great, with the central cipher 'FR' instead of the eagle and scroll. King's colour plain white, silver decoration. Regimental colour turquoise, silver decoration.

Campaign History. Raised in 1741 in Brieg after the battle of Mollwitz, as the 'Brieg'sches Garnisons-Regiment' for Oberst Kaspar Heinrich von Stechow, using 400 of the Austrian garrison and 800 released Prussian prisoners of war. The two grenadier companies came from the Wuerttemberg regiment taken into Prussian service as IR Nr 41. Form October 1741 the Chef was GM Heinrich von Saldern. **1744** Grenadiers in Gren Bn Vinne (Grsn Rs 6 and 10). Sep-Dec in Jaegerndorf; 23 Dec Sk Cosel; thence garrison in Cosel. **1745** Chef GM Johann Georg von Lehmann. From April in Markgraf Karl's corps. 22 May C Bratsch; 3

Jun-14 Aug garrison of Schweidnitz; Aug-Dec border security at Crossen on the Oder. **1750** GL Heinrich Guenther Gottfried von Bosse. **1753** Chef GM Christian Friedrich von Lattorff. **1756** Grenadiers in Gren Bn Ploetz Nr 6 (Grsn Rs 6 and 8). 20 Sep-29 Oct raids into eastern Bohemia. Musketiers not mobilised. **1757** Grenadiers: 6 May B Prague; thence garrison of the place until 20 Jun. 22 Nov B Breslau; 5 Dec B Leuthen; 7-20 Dec S Breslau. Musketiers in garrison. **1758** Grenadiers in Gren Bn Rohr (composition as before). 3 May-2 Jul before Olmuetz; 23 Jun C Littau; 7 Jul rearguard action at Kroenau; 11/12 Jul rearguard action at Holitz; 25 Aug B Zorndorf; 14 Oct B Hochkirch. After this battle, the grenadiers were combined with those of IRs Nrs 8 and 14. 4th Bn in Moravia; S Olmuetz. **1759** Grenadiers in Gren Bn vor dem Busche (6 and 8). 12 Aug B Kunersdorf; 17-29 Aug in the camp at Fuerstenwalde; thence protecting Glogau between Freystadt and Herrnstadt until 26 Oct; then to Landeshut. **1760** Jul-Aug in Prinz Heinrich's army in the relief of Breslau. 25 Aug-8 Oct in Goltz's corps guarding Glogau; 29 Aug-3 Sep raid on Bunzlau and Loewenberg; 10-15 Oct relief of Berlin; Oct-Nov in Goltz's corps in Silesia, relief of Cosel, raid on Landshut. **1761** Jun-Dec before Kolberg; 18-29 Nov thrust at Belgard, actions between Regenwalde and Petershagen; 30 Nov rearguard action at Naugard; 11/12 Dec C Kreyerbach nr Drenow; 13-18 Dec rearguard actions at Treptow and Stargard. **1762** Chef Oberst Gerhard Alexander Freiherr von Sass. 13 May-30 Jul minor skirmishes in Upper Silesia; 11 Aug-10 Oct covering Schweidnitz. **1778** Grenadiers in Gren Bn Nr 6, Gillern (Grsn Rs 6 and 8). 14 Feb C Wartha; 19 Feb C Neuheide. **1788** The regiment was disbanded, the men distributed to the depot battalions of IRs Nrs 28, 29, 31, 32, 38 and 43 and to Fues Bn Nr 8. All these units disappeared in 1806 except the 3rd Bn of IR Nr 31 in East Prussia. This became part of 1st Schlesischen IR Nr 11 (as of 1813 Nr 10). See Ser Nr 11 in the new regiments' section below.

7. Garrison Regiment. Location 1782-1788: Angermuende, Bernau, Neustadt-Eberswalde, Cuestrin, Reppen, Spandau, Straussberg, Templin in Brandenburg. Crimson Prussian cuffs and piping to the cuff flaps, yellow buttons. Brass grenadier cap plate and ornaments, crimson band and back cloth, royal blue lace, white pompon.

Colours. These were of the style issued under Frederick the Great, with the central cipher 'FR' instead of the eagle and scroll. King's colour plain white, gold decoration. Regimental colour crimson, gold decoration.

Campaign History. For details of the campaign actions, see IRs Nrs 36, 43, 47. Raised in 1741 in Stettin from shorter recruits, Austrian prisoners of war and volunteers of the Stettin Land-Regiment, with two battalions and two grenadier companies. The grenadiers came from the princely Schwarzburg grenadier company, which had been in the service of Mecklenburg. Chef GM Karl Wilhelm von Bredow. **1746** Chef Oberst Hans Christoph von Jeetze. **1754** Chef Oberst Christian Henning von Lange. In 1744/1745 the grenadiers were in Gren Bn Brandes (43/Grsn 7). **1756-1763** the grenadiers were in Gren Bn (47/Grsn 7) Wangenheim, then Carlowitz, then Buddenbrock, then Bock. The musketier battalions were distributed as follows: 1st Bn – Glogau, 2nd – Breslau, 3rd and 4th – Berlin. **1757** 3rd and 5th Bns moved to the fortresses on the Elbe. 5th Bn raised in Dresden; 6th Bn raised in Berlin.16 Oct Austrian corps of General Hadik raided Berlin; the outnumbered 4th Bn was destroyed in the fighting. 22 Nov 2nd Bn released from Breslau at its fall to the Austrians. **1758** 3rd and 5th Bns in the Saxon corps; 6th Bn sent to reinforce the garrison of Cuestrin. **1759** 3rd and 5th Bns again in the Saxon corps; both captured on 4 Sep at the surrender of Dresden. **1760** Chef GM Joachim Christian Friedrich von Itzenplitz. 9 Oct the battalions in Berlin were captured by the Russians and came home only in 1762. 3rd and 5th Bns in the defence of Leipzig, then in Magdeburg. **1761 - 1762** No actions of note. **1766** GM Christoph Friedrich von Rentzel; he received IR Nr 23; Chef GM Christian Ernst von Puttkamer (ex-Chef of IR Nr 23); 'Jung-Puttkammer' to differentiate from Grsn R Nr 1. **1771** Chef Oberst Georg Lorenz von Kowalsky. **1778-1779** the grenadiers were in Gren Bn Scholten (36/Grsn 7). **1788** The regiment was disbanded; the men were distributed among the depot battalions of IRs Nrs 13, 18, 19, 23, 25, 26, 36 and 46. These units all vanished in 1806 except the 3rd Bn IR Nr 46, which in 1808 became part of the 1st Schlesisches IR Nr 11. In 1813 this became Nr 10. See Grenadier R Koenig Friedrich Wilhelm II (1st Schlesisches) Nr 10', Ser Nr 11 in the new regiments below. The two grenadier companies went to form part of Fusilier Bn Nr 2, which surrendered at Ratkow.

8. Garrison Regiment. Location 1782-1788: Glatz, Reichenbach in Silesia. Black Prussian cuffs, red piping to cuff flaps, white buttons. Tin cap plate and ornaments, black band and back cloth, white lace, pink pompon.

Colours. These were of the style issued under Frederick the Great, with the central cipher 'FR' instead of the eagle and scroll. King's colour plain white, silver decoration. Regimental colour royal blue, silver decoration.

Campaign History. For details of the campaign actions see IR Nr 42 and Grsn R Nr 6. Raised in 1741 in Muenchberg at one battalion from imperial volunteers. Chef Oberst Karl Friedrich von der Recke. **1745** Chef Oberst Kurt Lebrecht Freiherr von Leoben. **1744/1745** The grenadiers were in Gren Bn von der Goltz (42/Grsn 8). **1746** Chef Oberst Kaspar Friedrich von Knobelsdorff. **1748** Chef Oberst Georg Ernst von Nettelhorst. **1753** From this date, the grenadiers were in Gren Bn Nr 6 (Grsn Rs 6 and 8). **1756-1762** The grenadiers were in Gren Bn (Grsn Rs 6 and 8) Ploetz, then Rohr, then vor dem Busche. **1756** Musketiers not mobilised. **1757** Chef Oberst Friedrich Freiherr Quadt von Wickeradt. 1st Bn defeated at C Landshut where they had been sent to combat the Austrian invaders on 12 Aug. **1758, 1759** No actions. **1760** 13,000 Austrians under General Draskovich blockaded Glatz in which the regiment was. As it was composed mostly of Austrian ex-prisoners of war and deserters, it put up little resistance. The Austrians broke in on 26 July; the Prussian commandant surrendered with 3,200 men. The regiment was not re-raised until 18 Mar 1763, when Oberst Francois de le Noble (previously Chef of a Free Regiment) became the new Chef of the four-battalion regiment. **1772** Chef Oberst Karl Benedikt von Bremer. **1778** Chef Oberst Sigismund August von Berrenhauer; he received Grsn R Nr 11. **1778/1779** The grenadiers were in Gren Bn Gillern (Grsn Rs 6 and 8). **1782** Chef Oberst Ulrich von Heuking. **1787** The grenadiers went into Fusilier Bn Nr 9; see there for future actions. **1788** The two musketier battalions were disbanded and distributed among the depot battalions of IRs Nrs 12, 24, 34, 35, 37 and 43; all these battalions vanished in 1806.

9. Garrison Regiment. Location 1782-1788: Geldern in the area west of Wesel fortress on the lower Rhine. Black Swedish cuffs, buttons gold. Brass grenadier cap plate and headband badges, buff headband, red backing, tapes decoration: orange central stripe, flanked by black then light blue lines. Pompon orange centre, then black and light blue. This same cap was worn by IR Nr 48.

Colours. These were of the style issued under Frederick the Great, with the central cipher 'FR' instead of the eagle and scroll. King's colour plain white, gold decoration. Regimental colour orange, gold decoration.

Campaign History. Raised in 1743 in Geldern as a replacement for the Infantry Battalion Kroecher (the new IR Nr 32) which had moved off to Silesia. 50 men of the old battalion formed the cadre of the new regiment, the rest were new recruits. Chef GM Georg Volrath von Kroecher. **1744** grenadiers were sent to Minden, where they were joined by two companies of IR Nr 45 and one company from Grsn R Nr 13 to form a battalion; this became the Gren Bn Ingersleben next year. **1748** Chef GL Ernst August de la Chevallerie Baron de la Motte, ex-Chef of IR Nr 17. The Gren Bn Ingersleben became Gren Bn Nr 2. **1757** D Geldern; after the capitulation of 23 Aug, they were free to march out to Magdeburg, but suffered heavily from desertion on the way there. The regiment was re-established in Magdeburg. Then moved to Berlin and took part in the defence against Hadik's raid. **1756-1760** The grenadiers were in Gren Bn Nr 2 was in Glatz; when that fortress fell, they were captured and the battalion was not reformed until 1763, when it became the Pommersches Provinzial Gren Bn Ingersleben. **1763** Chef GM Konstantin Natanael von Salenmon, ex-Chef of a free regiment. In this year, the regiment took in selected men of the Freiregiment Salenmon. **1776** grenadier company sent to IR Nr 48. **1778-1779** grenadiers in Gren Bn Meusel (45/48). For details of campaign actions see these regiments. **1787** the grenadiers went back to IR Nr 48; in 1806 they fought in Gren Bn Krafft (48/59) which surrendered in Erfurt and Magdeburg. **1788** the regiment was disbanded; the men were distributed among the depot battalions of IRs Nrs 44 and 48, which surrendered in Hameln in 1806.

10. Garrison Regiment. Location 1782-1788: Neisse, Nimptsch, Ottmachau, Patschkau in Silesia. Black Swedish cuffs, white buttons. Tin cap plate and ornaments, black band and back cloth, yellow lace and pompon.

Colours. These were of the style issued under Frederick the Great, with the central cipher 'FR' instead of the eagle and scroll. King's colour plain white, silver decoration. Regimental colour sky blue, silver decoration.

Campaign History. Raised in 1743 in Muencheberg from shorter members of IR Nr 47. Chef Oberst Georg Ewald von Puttkamer; he then received Grsn R Nr 11. **1744** Chef GM Arnold Goswin von Rettberg. Grenadiers in Gren Bn Vinne (Grsn Rs 6 and 10). **1747** Chef Oberst Busso Christian von Blanckensee. **1753** Grenadiers in Gren Bn Nr 5 (Grsn Rs 5 and 10). **1756** The regiment was increased by two new battalions to four battalions. Throughout the Seven Years War the musketiers of the regiment weres in Neisse fortress. Grenadiers in Gren Bn Nr 5, Rath, then Koschenbahr, then Hachenberg. **1765** Chef Oberst Christoph von der Muelbe. **1778/1779** Grenadiers in Gren Bn Nr 5, Lentzke. **1780** Chef Oberst Heinrich Gottlieb von Koenitz; he received IR Nr 17. **1786** Chef GM Karl Albrecht Friedrich von Raumer, ex-Chef of IR Nr 53. He then retired and was succeeded by Oberst Friedrich von Oven. **1787** Grenadiers transferred to the new Fus Bn Nr 10, which surrendered after the battle of Jena. **1788** Disbanded; the men were distributed among the depot battalions of IRs Nrs 33, 37, 40, 42, 47, 49 and Fus Bn Nr 13.

11. Garrison Regiment. Location 1782-1788: Allenstein, Guffstein, Heiligenberg, Heilsberg, Wartenburg, Wormditt in East Prussia. Crimson Prussian cuffs, red piping to the cuff flaps, white buttons. Tin cap plate and ornaments, crimson band, back cloth and pompon, plain white tape.

Colours. These were of the style issued under Frederick the Great, with the central cipher 'FR' instead of the eagle and scroll. King's colour plain white, silver decoration. Regimental colour bright red, silver decoration.

Campaign History. For details of the campaign actions see I R Nr 2, Grsn Rs Nrs 1 and 2. Raised in 1744 in Angerburg, East Prussia at two musketier battalions and two grenadier companies from drafts from Grsn Rs Nrs 1 and 2 and new recruits. Chef Oberst Georg Ewald von Puttkamer. **1748** Chef Oberst Franz Christoph von Manteuffel. **1756** increased by two battalions. **1757** B Gross-Jaegersdorf. **1758** S Schweidnitz then installed as garrison of the fortress. **1759** In Schweidnitz. **1760** Chef Oberst Henning Christian von Mellin. In Fouque's corps; one battalion was captured at Landeshut and not re-raised. The three battalions were then transferred to Neisse fortress, where they stayed for the rest of the war. **1769** Chef Oberst Karl Ludwig von Ingersleben. **1772** A new 4th Bn was raised and stationed in Muelhausen in the newly acquired West Prussia. **1782** Chef Oberst Sigismund August von Berrenhauer. **1788** Disbanded; the men went to the depot battalions of IRs Nrs 7, 51, 53, 54, 55 and the 11. Fuesilier-Bn.

12. Garrison Battalion. Location 1782-1788: Emden in Ostfriesland. Black Swedish cuffs, yellow buttons. This battalion had no grenadiers.

Colours. These were of the style issued under Frederick the Great, with the central cipher 'FR' instead of the eagle and scroll. King's colour plain white, gold decorations. Regimental colour light blue, gold decorations.

Campaign History. Raised in 1744 in East Fresia (Aurich, Emden and Greetsiel) as the 'Ostfrieslaendische Garnison'. There were no grenadiers. Chef Oberst Ernst Georg vion Kalckreuth. **1763** Chef Oberst Guillaume Rene de L'Homme de Courbiere, ex Chef of a Free Regiment. **1787** Chef Obstlt Philipp Lambert de Bauvrye. **1757** On 3 July the battalion was captured by the French. **1759** Exchanged; it then served as a replacement depot. **1788** Disbanded; the men went to the depot battalions of IRs Nrs 10, 45 and 48.

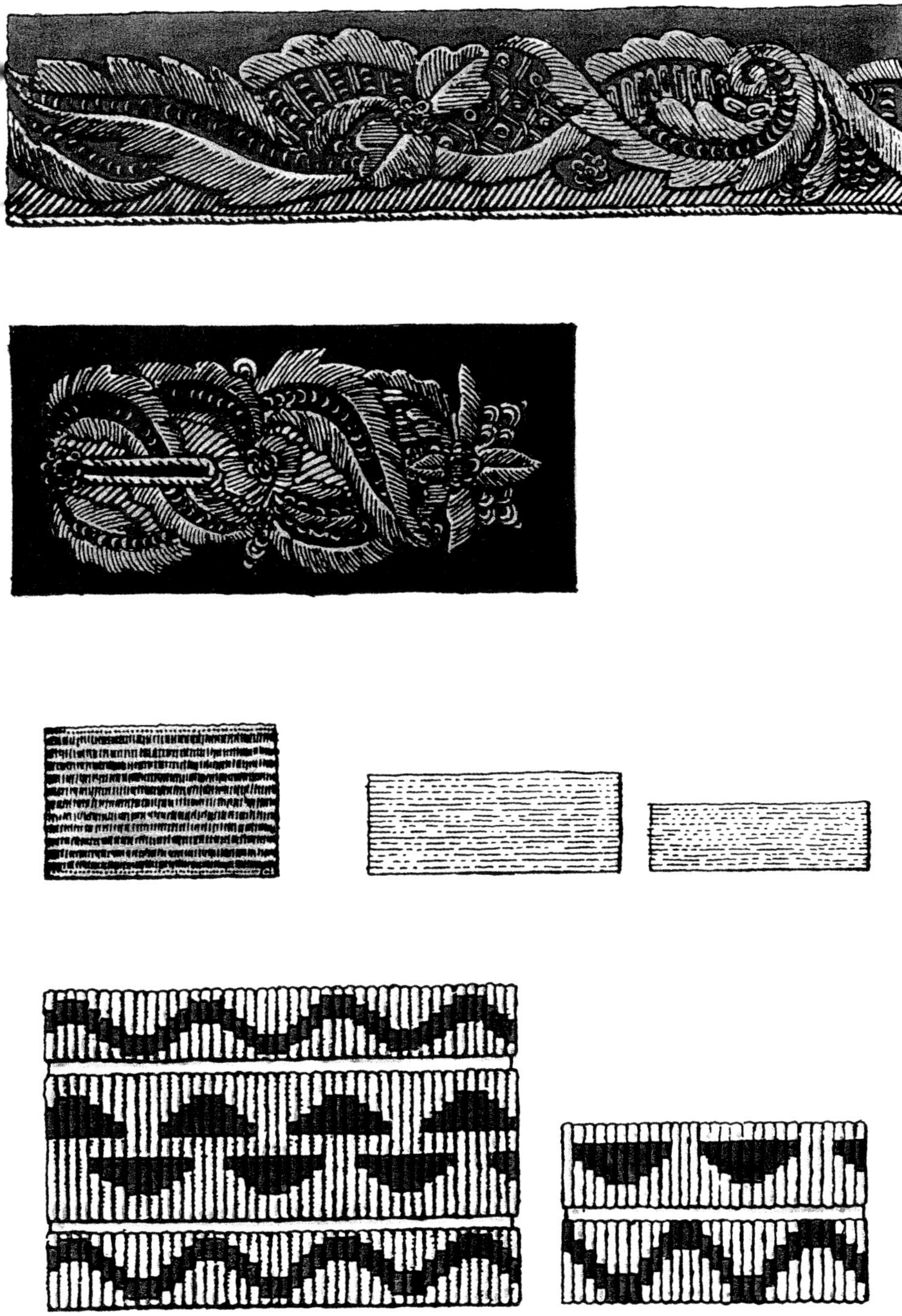

Lace of IR Nr 1. Top to bottom. Officers: silver on red to the facings; silver on blue for the buttonholes. Left. NCOs: plain silver. Right. Privates: white, large and small. Drummers: red on white. Painted by Alfred Gay and presented to the author.

Lace of IR Nr 2. Top to bottom. Officers: gold on red. NCOs: gold. Privates: crimson with white tassels. Drummers black and white on crimson. Painted by Alfred Gay and presented to the author.

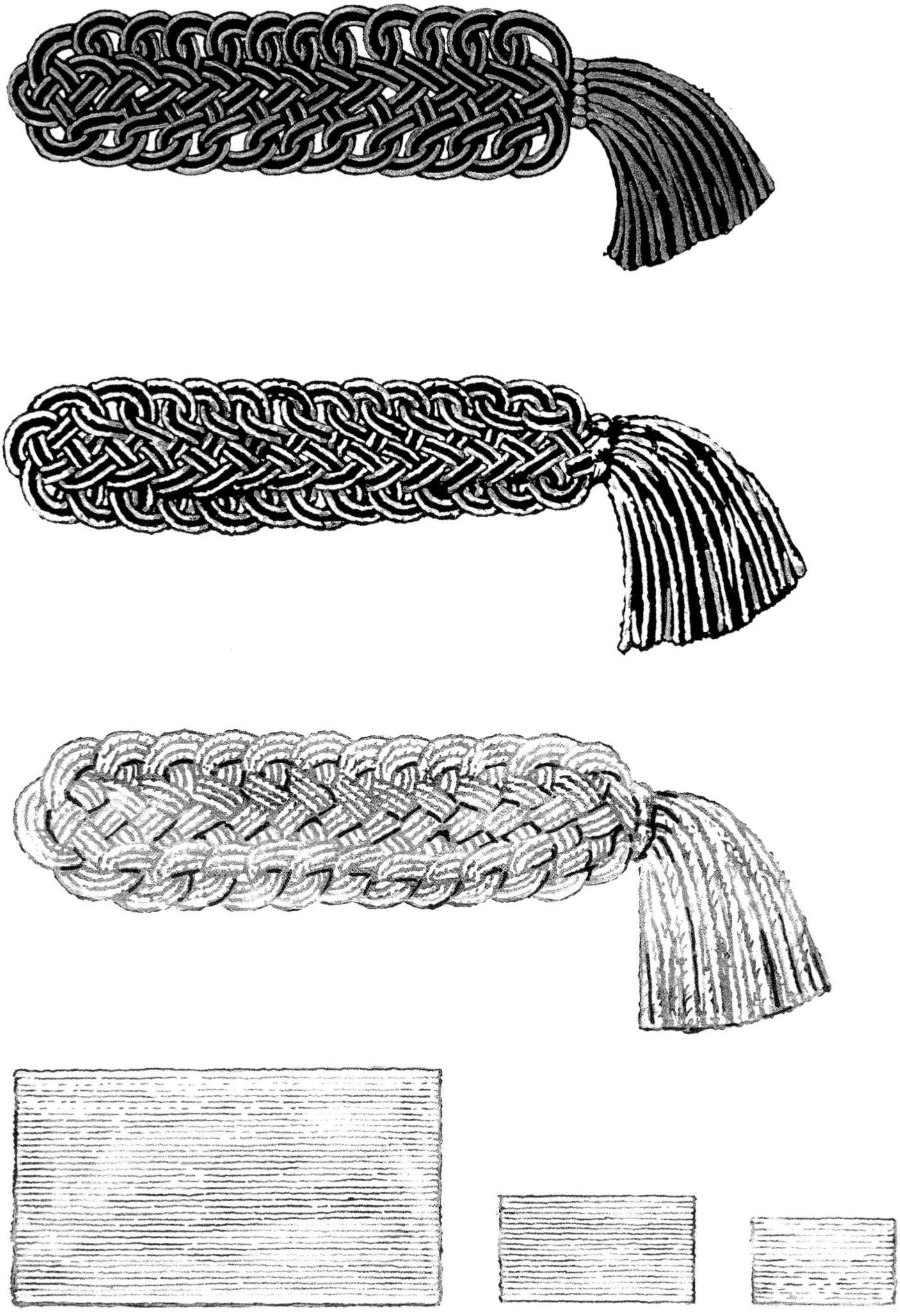

Lace of IR Nr 3. Top to bottom. Officers: none. NCOs: black and silver. Privates: black and white. Drummers: white (in three sizes). Painted by Alfred Gay and presented to the author.

Lace of IR Nr 4. Officers, NCOs and Privates not shown. Drummers: alternate red and blue on white. Painted by Alfred Gay and presented to the author.

Lace of IR Nr 5. Top to bottom. Officers (large and small); NCOs: gold, plain, no tassels. Privates: orange with white tassels. Drummers: light blue within yellow diamonds on white. Painted by Alfred Gay and presented to the author.

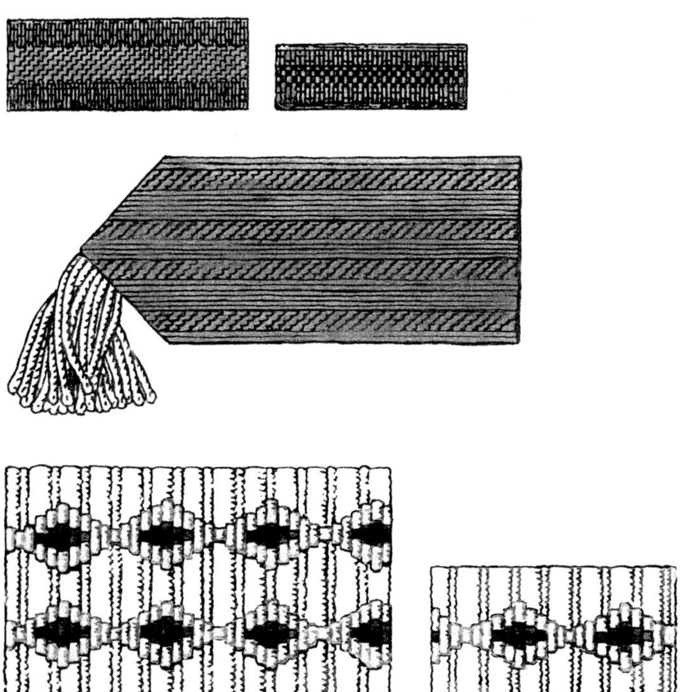

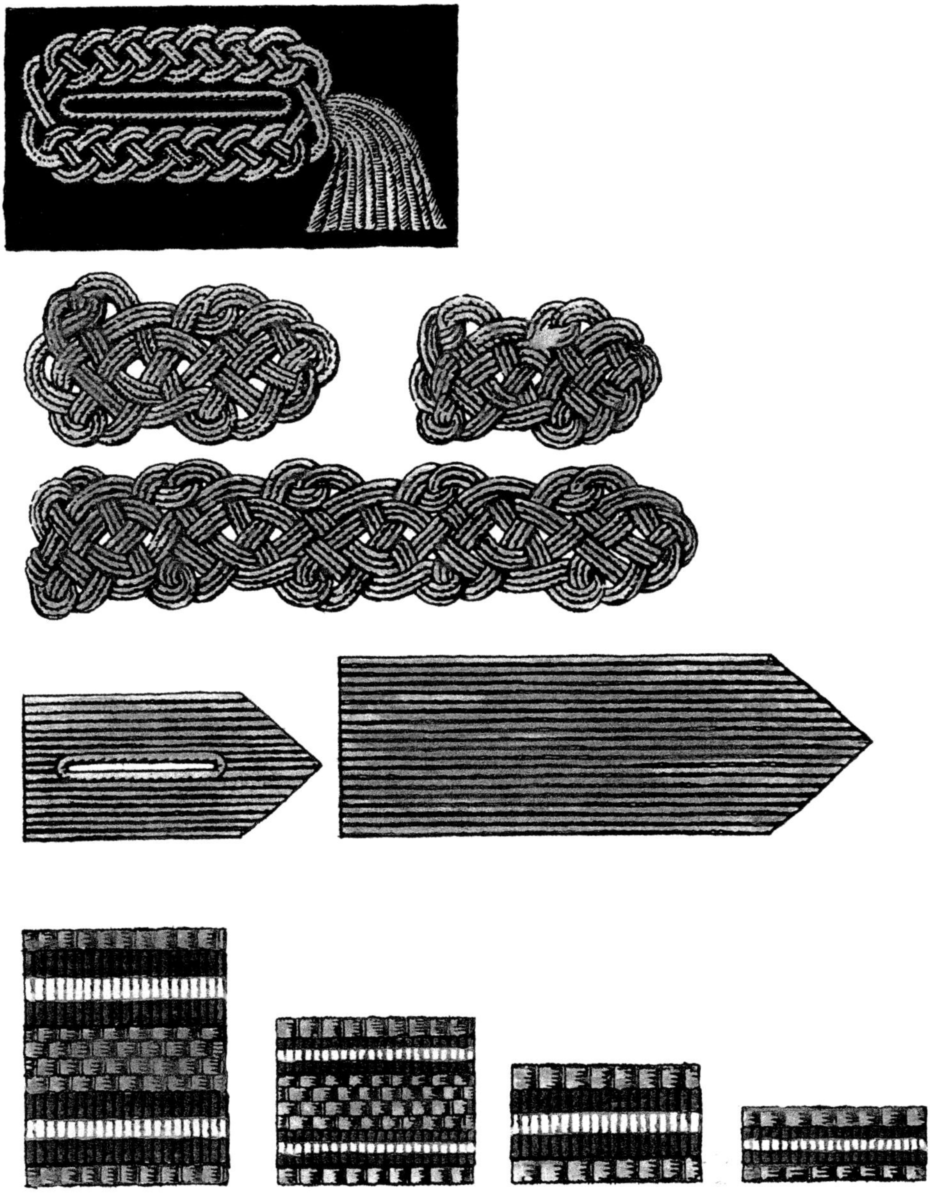

Lace of IR Nr 6. Top to bottom. Officers and NCOs: gold. Privates; gold. Drummers: yellow within red on gold. Painted by Alfred Gay and presented to the author.

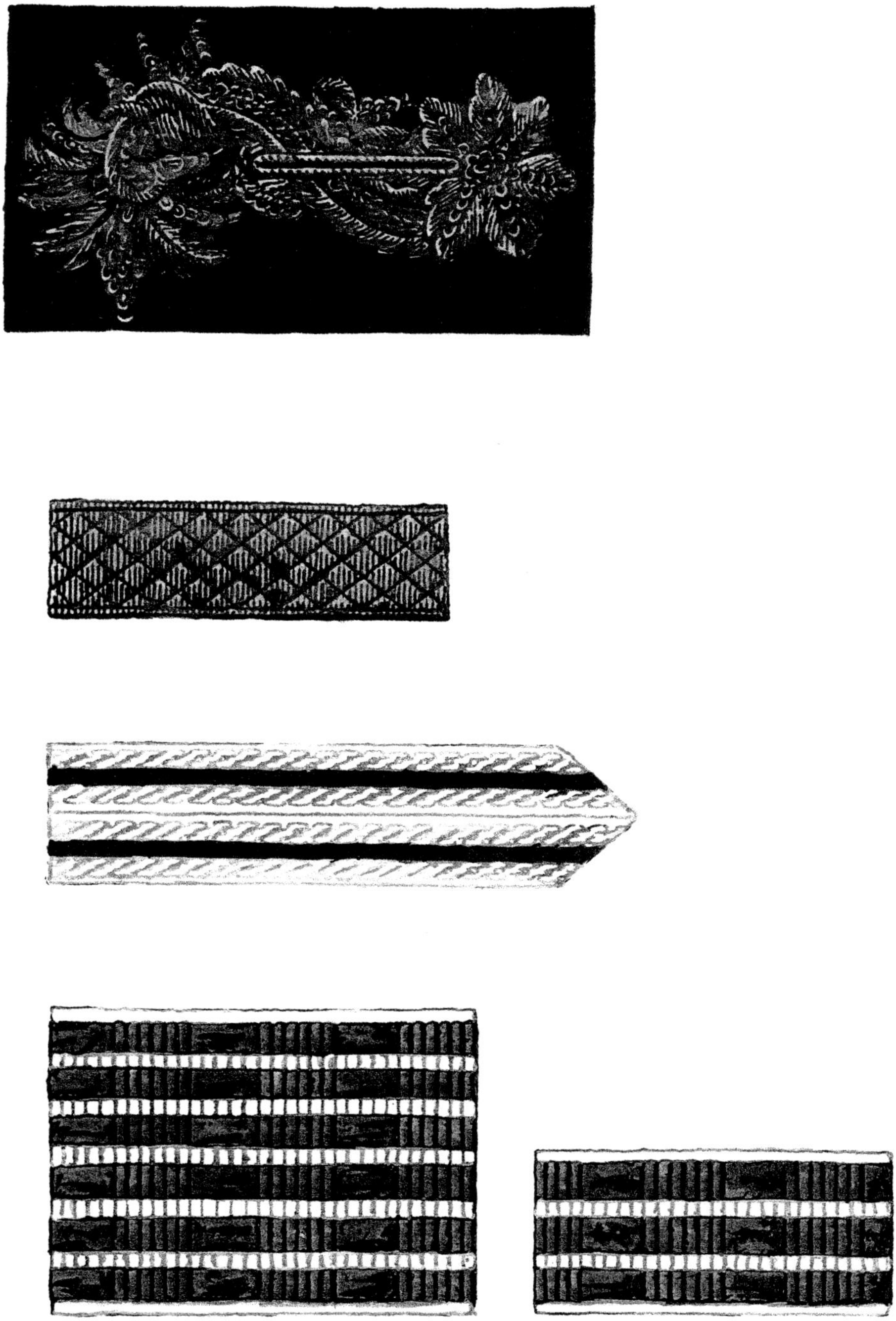

Lace of IR Nr 8. Top to bottom. Officers; gold embroidered. NCOs: plain gold. Privates: light blue on white. Drummers: alternate red and green bars with white stripes. Painted by Alfred Gay and presented to the author.

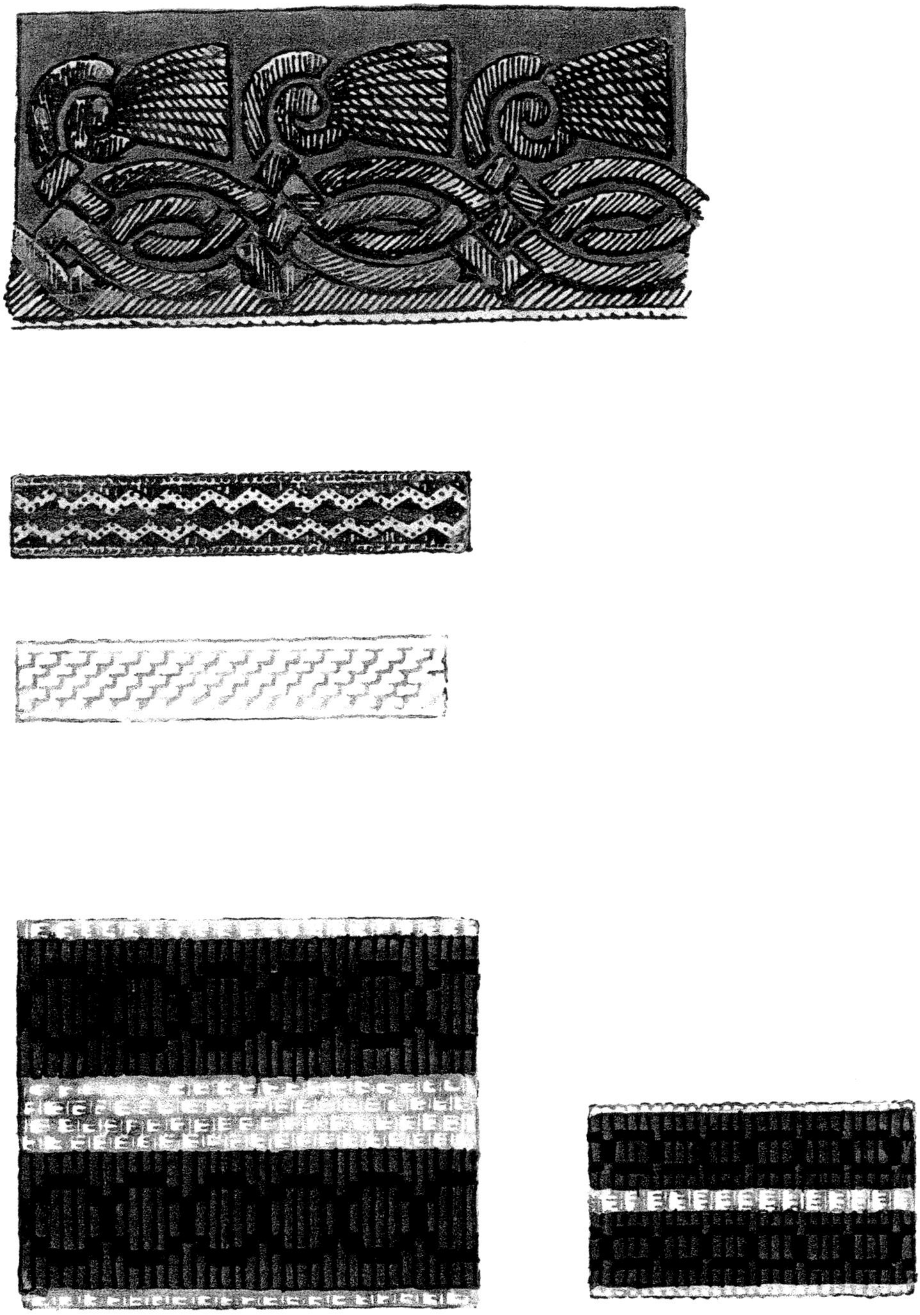

Lace of IR Nr 9. Top to bottom. Officers: gold embroidered. NCOs: plain gold. Privates: white. Drummers: black chains on red with white stripes. Painted by Alfred Gay and presented to the author.

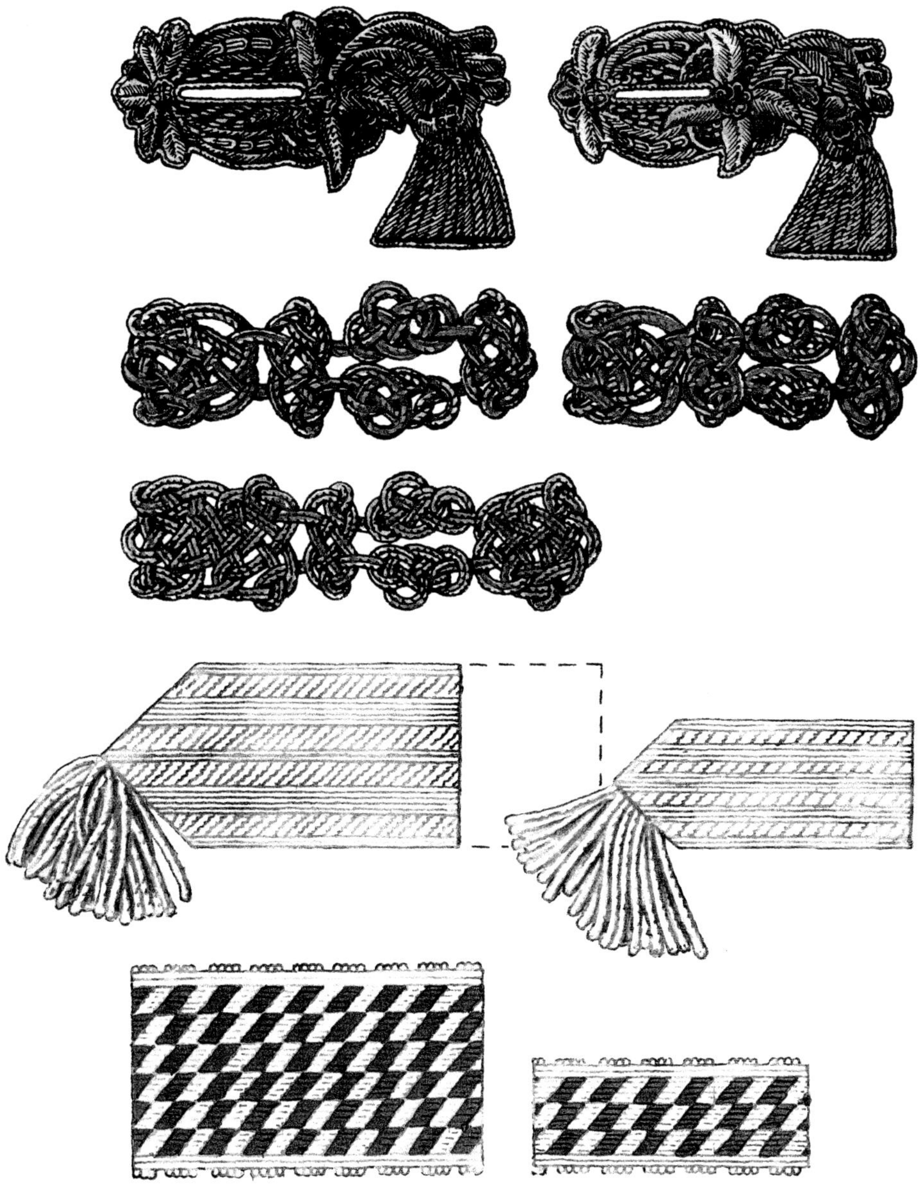

Lace of IR Nr 10. Top to bottom. Officers: silver embroidered (large and small). NCOs: silver. Privates: plain white. Drummers: red and white. Painted by Alfred Gay and presented to the author.

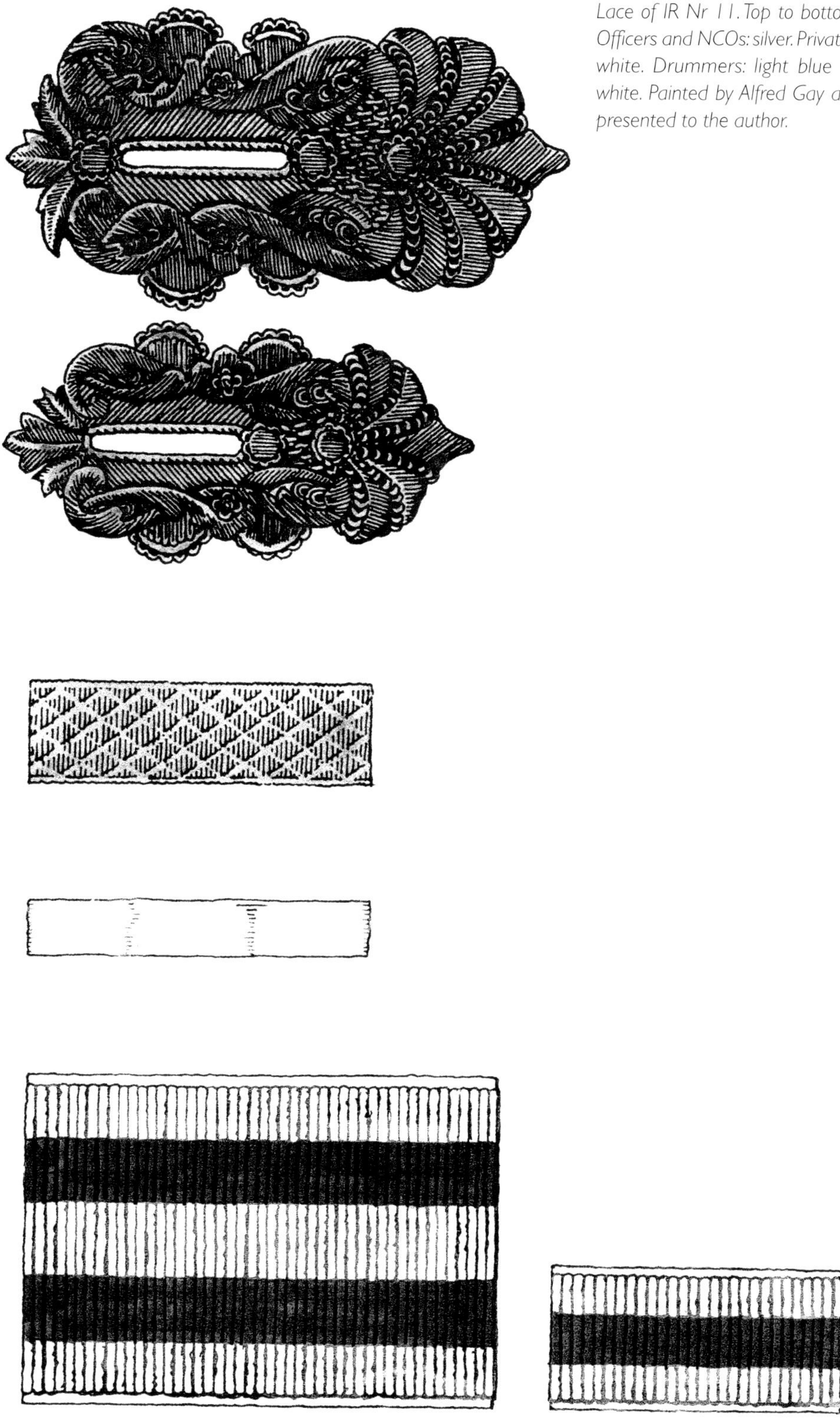

Lace of IR Nr 11. Top to bottom. Officers and NCOs: silver. Privates: white. Drummers: light blue on white. Painted by Alfred Gay and presented to the author.

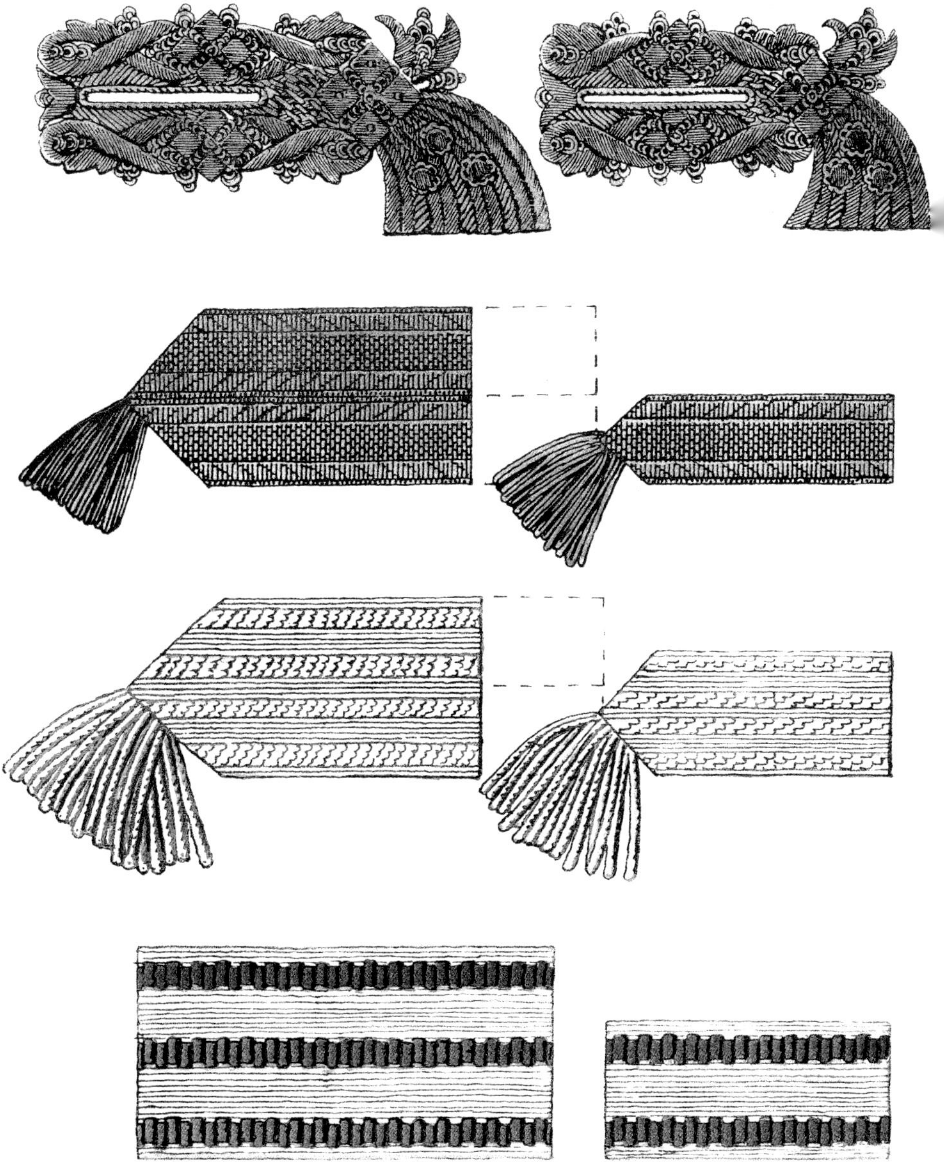

Lace of IR Nr 12. Top to bottom. Officers: gold, large and small. NCOs: gold. Privates: white. Drummers: three red stripes on white. Painted by Alfred Gay and presented to the author.

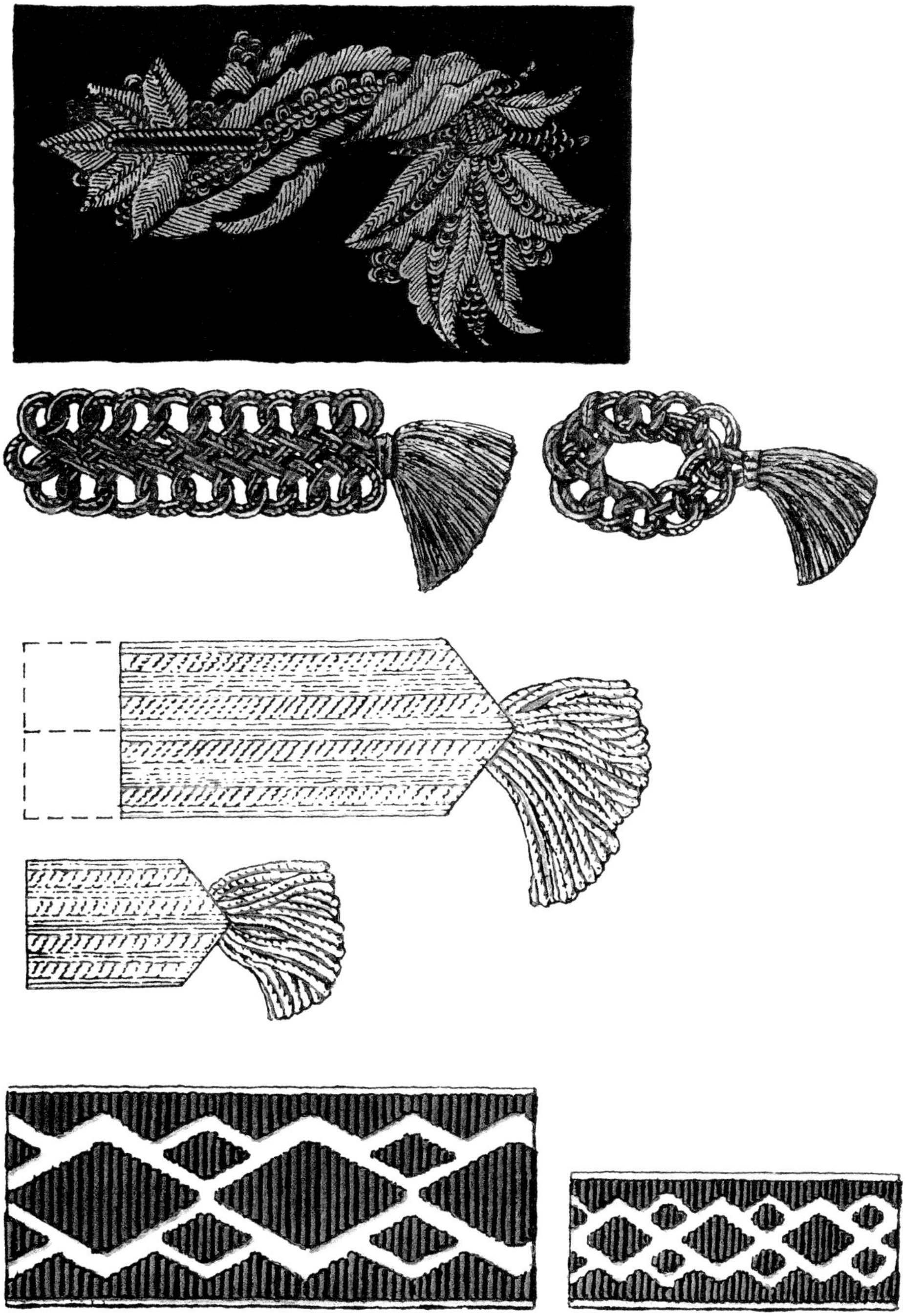

Lace of IR Nr 13. Top to bottom. Officers: silver embroidered. NCOs: silver. Privates: white. Drummers: red and white. Painted by Alfred Gay and presented to the author.

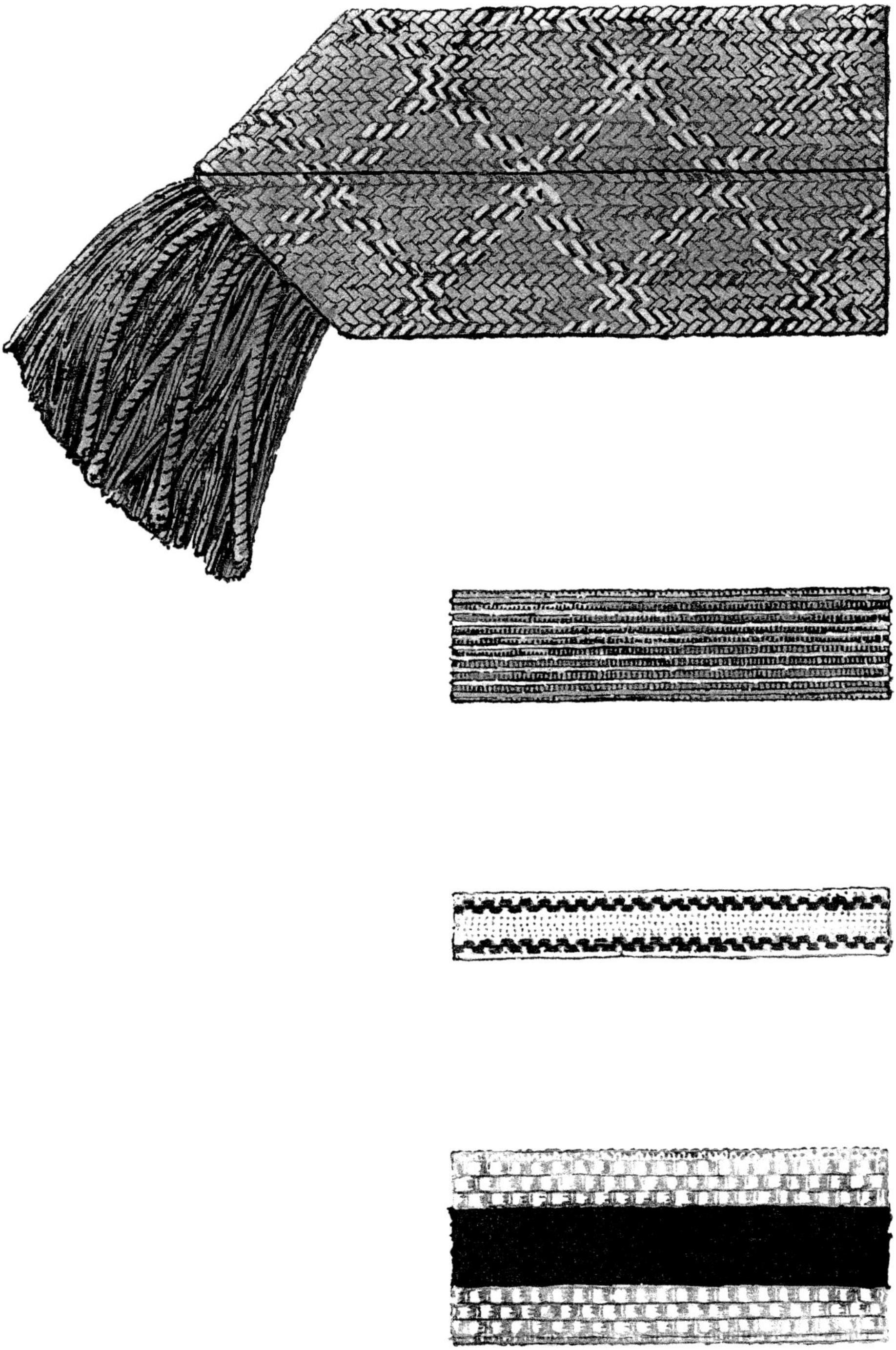

Lace of IR Nr 14. Top to bottom. Officers: gold. NCOs: gold with narrow orange stripes at the edges. Privates: two orange lines on white. Drummers: red within white. Painted by Alfred Gay and presented to the author.

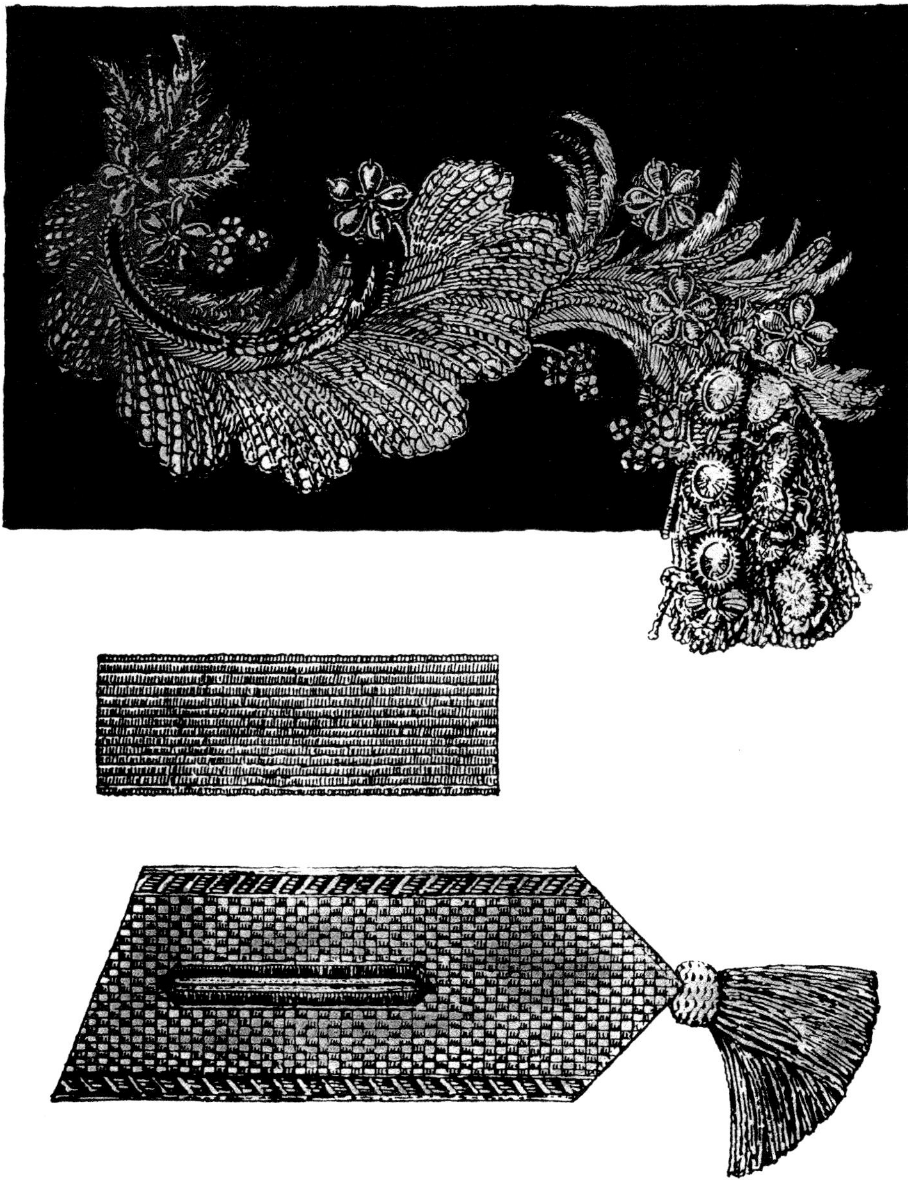

Lace of IR Nr 15, 1st Battalion. Top to bottom. Officers, NCOs and Privates: silver. Drummers as for 2nd and 3rd Battalions. Painted by Alfred Gay and presented to the author.

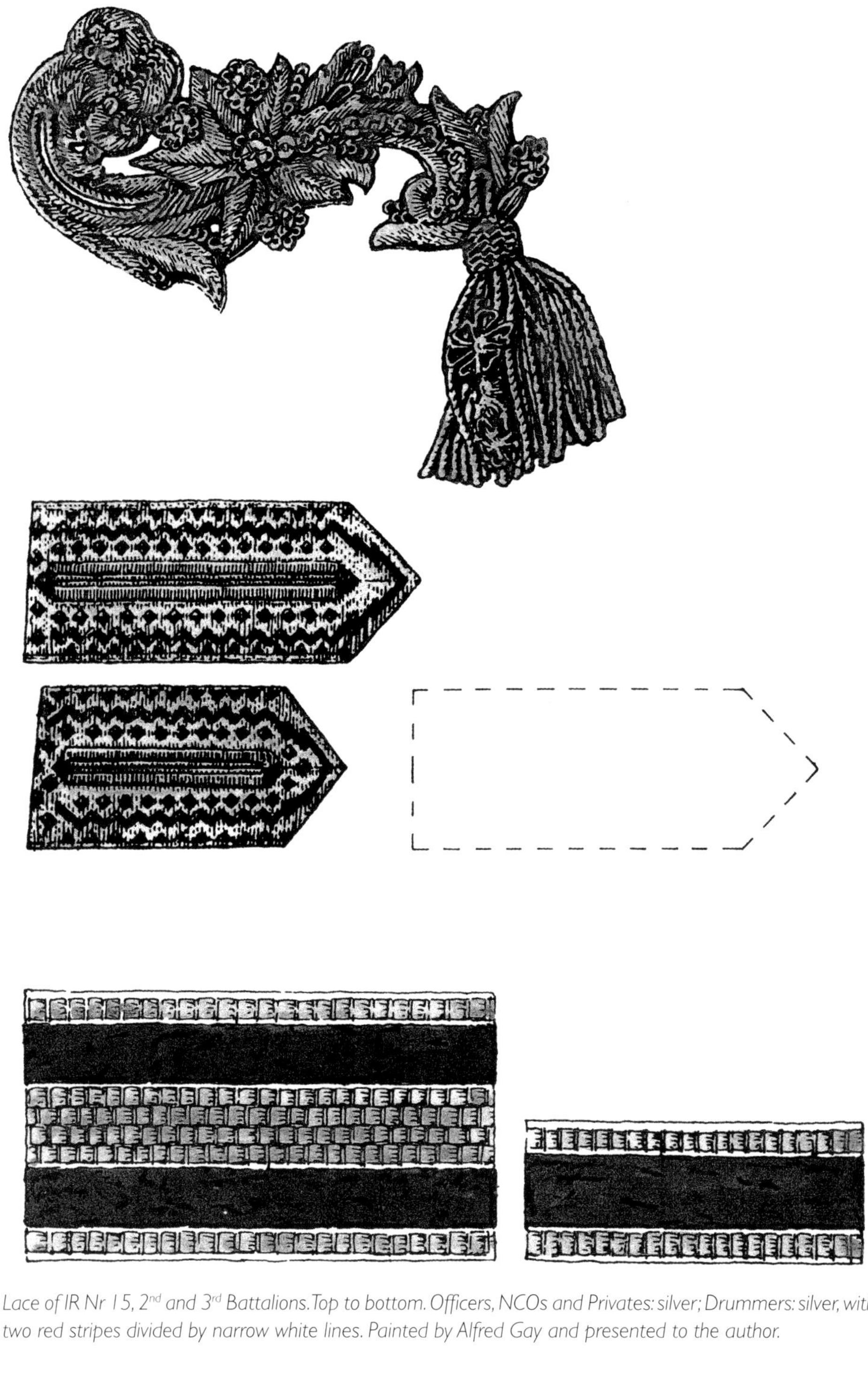

Lace of IR Nr 15, 2nd and 3rd Battalions. Top to bottom. Officers, NCOs and Privates: silver; Drummers: silver, with two red stripes divided by narrow white lines. Painted by Alfred Gay and presented to the author.

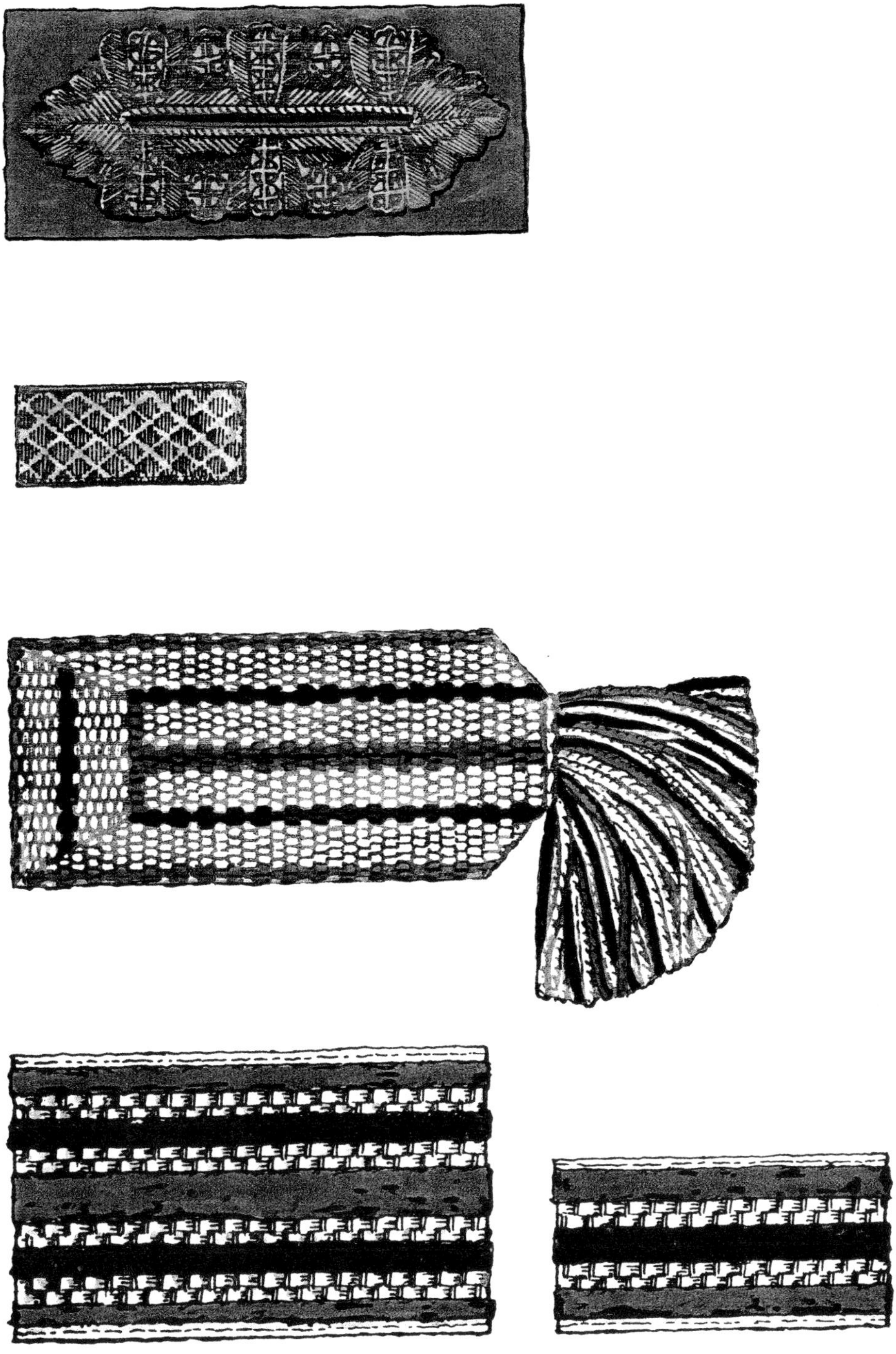

Lace of IR Nr 16. Top to bottom. Officers: gold embroidered. NCOs: 'plain' patterned gold. Privates: white with a central black stripe and red edgings; tassels. Drummers: white with red and black stripes. Painted by Alfred Gay and presented to the author.

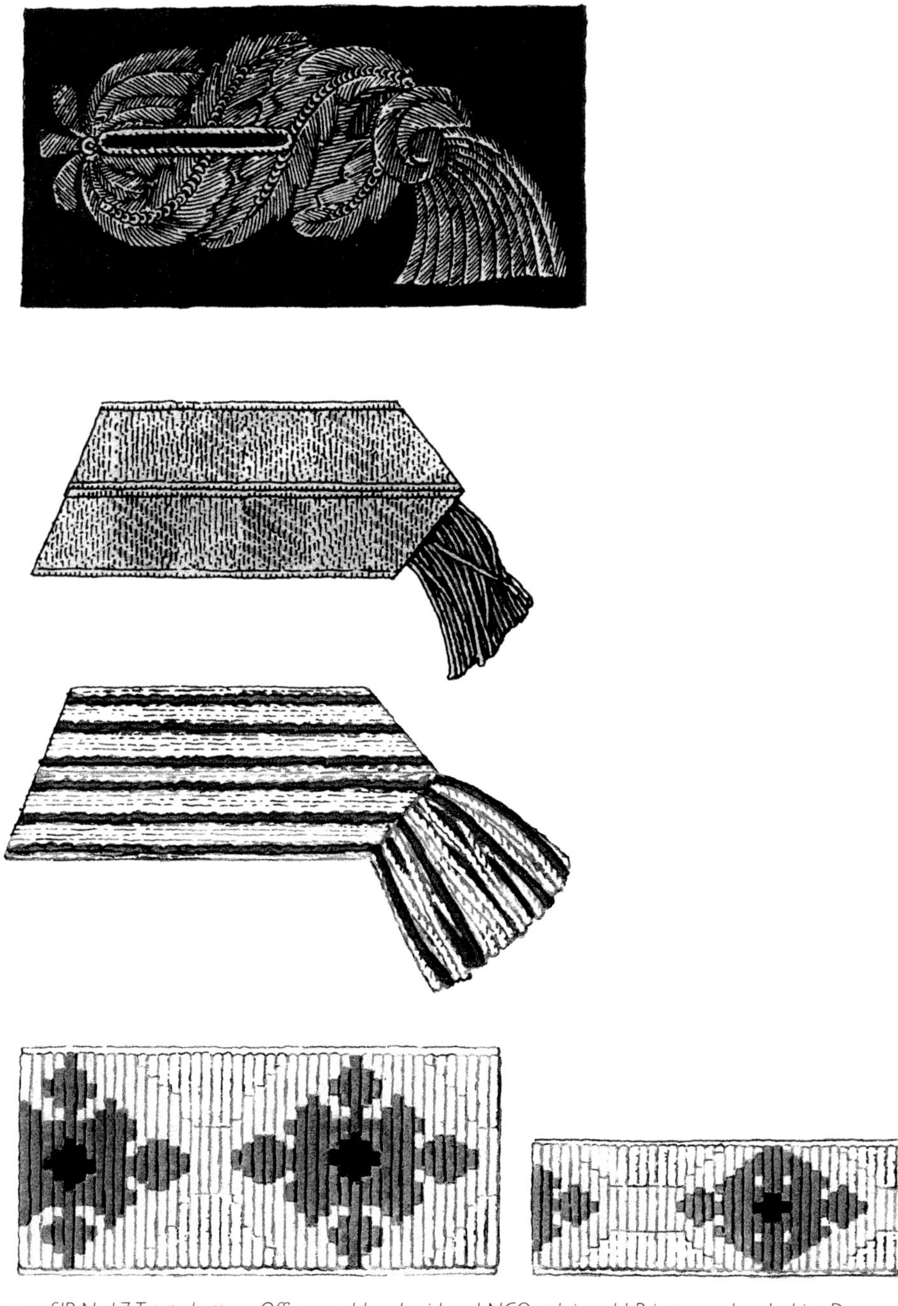

Lace of IR Nr 17. Top to bottom. Officers: gold embroidered. NCOs: plain gold. Privates: red and white. Drummers: white with light blue crosses having black diamond centres. Painted by Alfred Gay and presented to the author.

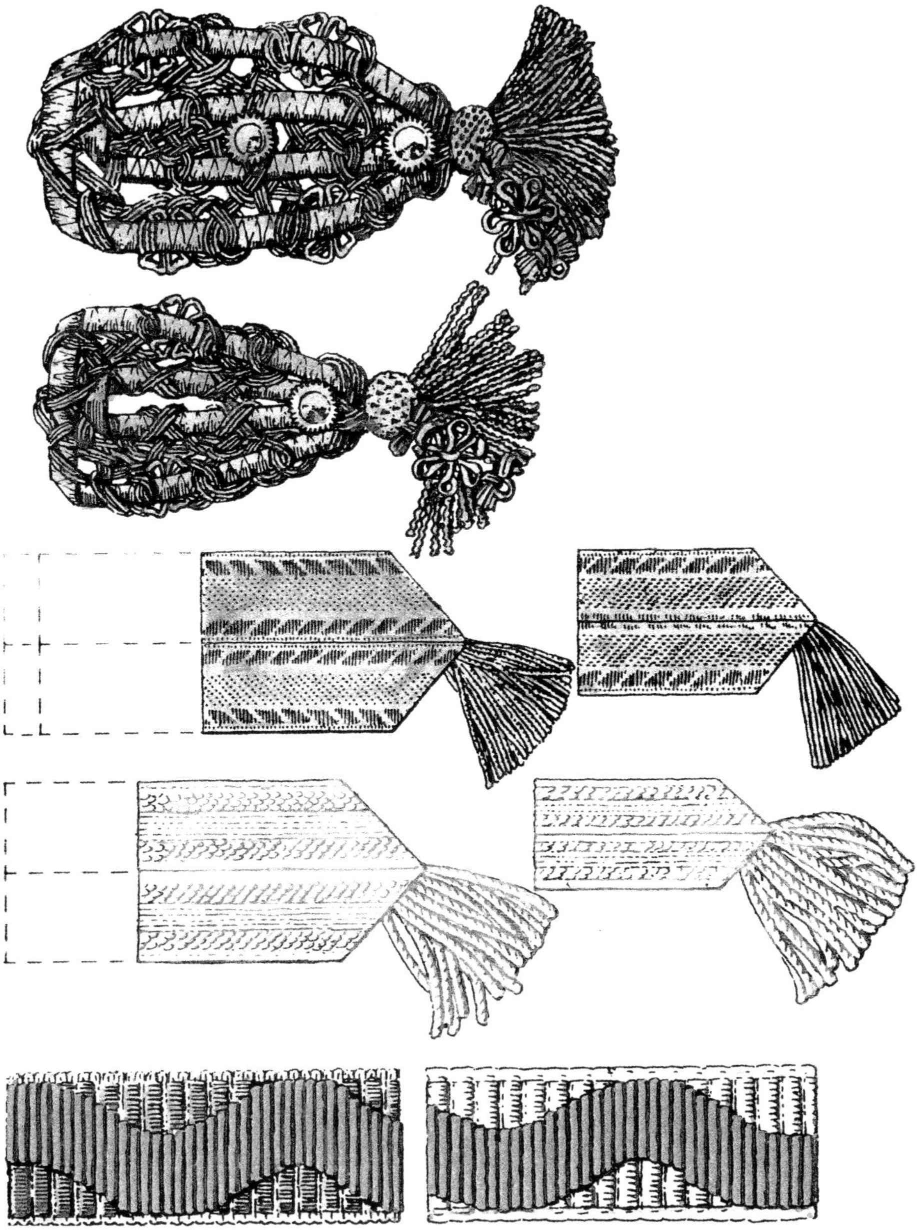

Lace of IR Nr 18. Top to bottom. Officers: silver embroidered. NCOs: plain silver. Privates: white. Drummers: left – wavy pink line on silver; right – pink on white. Painted by Alfred Gay and presented to the author.

Lace of IR Nr 19. Top to bottom. Officers; gold embroidered. NCOs: plain gold. Privates: (large and small) orange pattern on white, with orange and white fringes. Drummers: orange and white. Painted by Alfred Gay and presented to the author.

Musketier, IR Nr 12. A reconstruction. Dark blue coat, red facings, yellow buttons and small clothes. By kind permission of the Wehrgeschichtliches Museum, Rastatt.

Grenadier cap plate, IR Nr 4. Brass, red and royal blue pompon, yellow backing, red within royal blue lace. The central rear emblem is three brass grenades. By kind permission of the Wehrgeschichtliches Museum, Rastatt.

Grenadier cap plate, IR Nr 6, the 'Grenadier-Garde'. Brass plate and headband, red backing, gold lace. The central rear emblem is three grenades; the pompon is missing; it was red within white. By kind permission of the Wehrgeschichtliches Museum, Rastatt.

Grenadier cap plate, IR Nr 12. Brass plate, black eagle, light red band, royal blue backing, brass 'lace', royal blue within light red pompon. The central rear emblem is a crowned brass cartouche bearing the black Prussian eagle and flanked by flaming grenades. By kind permission of the Wehrgeschichtliches Museum, Rastatt.

Grenadier cap plate, IR Nr 18. Tin plate, black eagle. It seems that this regiment also wore another version, in which the eagle had gilt crown, beak, sword hilt, talons and lightning bolt. Pink band and backing, white tape, green-within red-within white pompon. This regiment wore several different versions of cap plates. The central rear emblem was a large, tin grenade. By kind permission of the Wehrgeschichtliches Museum, Rastatt.

Grenadier cap plate, IR Nr 21. Brass plate, red band, buff backing, black tape edged white, black, white and red pompon. The rear central emblem was a crowned cartouche bearing an eagle and flanked by flaming grenades. This was similar to that of IR Nr 12, except that here the eagle is brass coloured. By kind permission of the Wehrgeschichtliches Museum, Rastatt.

Grenadier cap plate, IR Nr 35. Tin front plate, lemon yellow band and backing, white tape with red pattern. The pompon was red within white within red, not as shown here. The central rear emblem was an eagle over crossed cannon barrels and trophies. By kind permission of the Wehrgeschichtliches Museum, Rastatt.

Fusilier cap plate, IR Nr 33. Brass plate, orange band and backing. By kind permission of the Wehrgeschichtliches Museum, Rastatt.

Grenadier cap plate, IR Nr 50. Tin plate, buff band, pink backing, pink and white tape, pink-white-pink pompon. The central rear emblem was a sunburst under a grenade. By kind permission of the Wehrgeschichtliches Museum, Rastatt.

Grenadier cap plate, IR Nr 55. Tin plate, yellow band, pink backing; white tape with a central pink stripe flanked by yellow stripes. Pom pon white within yellow within pink. The central rear emblem was a grenade over trophies of arms. By kind permission of the Wehrgeschichtliches Museum, Rastatt.

Bombardiers cap plate, artillery. The backing to the brass plate is black as is the headband. The central rear emblem is four crossed cannon barrels behind a pair of kettle drums. Author's collection; Rastatt museum.

Officer's Kuerass with gilt decoration. Author's collection; Rastatt museum.

A rare print of Bluecher in hussar uniform. Author's collection.

Grenadier cap of the Grenadier-Garde 1806 with cut feather plume. Author's collection; Rastatt museum.

Tin cap plate (IR Nr 35) and officer's gorget. Author's collection; Rastatt museum.

General Hans David von Yorck in 1814. Among his decorations are the Iron Cross, Pour le Merite, *Order of Maria Theresia, the Order of St George and the star of the Order of the Black Eagle. Author's collection.*

Field Marshal von Bluecher, Fuerst von Wahlstatt and General Yorck Graf von Wartenburg in 1815. Author's collection.

General B. F. E. von Tauentzien, Graf von Wittenberg and General F. W. Freiherr von Buelow, Graf von Dennewitz. Author's collection.

13 Garrison Battalion. Location 1782-1788: Minden in Westphalia. Raised in 1743 in Minden from new recruits and 50 men of the IR Nr 32. Red Swedish cuffs, yellow buttons. According to Fiedler, the grenadiers of this battalion wore the same cap as Fuesilier-R Nr 48 i.e. brass front plate and headband emblems, red bag. Tape: orange central stripe, flanked by black then light blue stripes. Pompon from the centre: orange, black, light blue. Chef Oberst Joachim Wrocislaus von Wobeser. **1746** Chef Oberst Hans Leonhard von Wutginau. **1755** Chef Oberst Friedrich Wilhelm von Salmuth. **1756** Converted to Fusilier R Nr 48 on 10 August.

Neues Garrison Regiment. Location 1782-1788: , Cuestrin, Dratzig See, Moers, Spandau, Peitz near Kottbuss, Regenstein, Stettin. Raspberry facings, yellow buttons. Brass grenadier cap plate and head band, red backcloth, yellow lace; pompon raspberry within white. This cap was apparently also worn by the grenadiers of the 3rd and 4th Garnisons-Regimentern. Raised in 1742 from various independent companies in Kuestrin, Moers, Peitz, Regenstein/Harz, Spandau, Stettin and Tempelburg. **1763** Disbanded except for the two grenadier companies, which had been in Charlottenburg from 1742. **1744/1745** Gren Bn Byla Nr 1. **1756-1763** Gren Bn Kahlden etc. **1778/1779** Gren Bn Oldenburg. For details of campaign actions see Grsn R Nr 3.

6

The Fusilier Battalions 1787-1806

King Friedrich Wilhelm II reintroduced light infantry, specifically trained to fight in extended order and armed with shorter, lighter weapons than the musketiers, in 1787. The first light troops of this type in the Prussian army had been raised in 1756, at the outbreak of the Seven Years War. At the first partition of Poland, the infantry regiments Laxdehn Nr 51, Lengfeld Nr 52, Luck Nr 53, Rohr Nr 54 and Hessen-Philippstal Nr 55 were raised. In 1768 a battalion was raised which initially bore the name of Oberst von Rossiere, but then became the regiment von Troschke Nr 50. As with the line infantry, each battalion had 40 sharpshooters (10 in each company) armed with rifled weapons. These battalions were trained to skirmish in open order.

Being light infantry, the fusiliers carried no colours. Their coats were, from 1788, mid-green in colour, with facings shown on collar, lapels and plain round cuffs. Coats were to the knee. They wore no distinguishing lace decorations. The small clothes, hat edging and belts were white. The front of the hat was decorated with the Prussian eagle badge in the button colour. In 1798 the coat became dark green, belts and hat edging were blackened, green trousers replaced the white, the green turnbacks were edged in the facing colour and were shortened to hip length. The battalion hat tufts were all replaced by short black, worsted plumes, and only the front of the hat was turned up. Short boots replaced the gaiters. The facings were now changed to reflect the new brigading and were extended to include the shoulder straps. Badges of rank were as for the line infantry.

Drummers and hornists wore coloured lace decoration to lapels, collars and swallows nests. This varied in colour and pattern according to the battalion of the wearer. Officers' facings were of velvet.

Everything changed again in 1802/3. The shako was introduced; it had a white top band with a pompon in the battalion colour. The coat now bore facings on lapels, collar, cuffs and turnback edging for all battalions. The sabre belt was now worn around the waist. White breeches were again introduced and worn in short, black gaiters. The collapse of the Prussian army broke the thread of continuity of the surviving troops.

With the army reorganization of 1808, the fusilier battalions received the same dark blue coats as the line troops, and wore the same facings and buttons as the regiments to which they were now attached. Belts remained black, grey breeches and black gaiters were worn. The shako bore the star for the Guards regiment, all other battalions had a black-within white cockade and a regimental button and loop from the cockade up to the black-within-white pompon.

Within each territorial brigade, regimental seniority was shown by the colour of the shoulder straps:- 1st – white, 2nd – red; 3rd – yellow; 4th - light blue.

The FUSILIER Battalion Histories

1. Fusilier-Bataillon. Kurmark Fusilier Brigade in 1798; Magdeburg Fusilier Brigade in 1802. Garrison Hildesheim.

Uniforms. Light green facings, yellow buttons. Hat tuft green with an orange ring. In 1798 this changed to crimson collar, shoulder straps and cuffs with yellow buttons. The lapels were in the coat colour. The shako pompon was red in 1802.

Campaign history. Raised from four companies of the IR von Leipziger Nr 3 and named von Langelair.**1787** The expedition to Holland; C Amstelveen. 1789 Chef von Schenke; retired. **1792** Valmy; Sk Otange; Sk Fontoi; Sk Briquenai. **1793** Chef von Wedel; received command of IR Nr 41. Sk Darweiler; 15 Mar Sk Stromberg; 27 Mar C Waldalgesheim (Weiler); Sk Wendelsheim; 30 Mar Sk Alzey; Sk Carlsberg; 13 Aug C Limbach; Sks Neunkirchen / Pleisse; 14 Sep C Spiessen; 26 Sep C Blieskastel; 27 Sep C St Imbert; Bebelsheim; 17 Nov C Biesingen; 28 Nov C Vogelweh. 28-30 Nov B Kaiserslautern. **1794** 8 Jan C Kreuznach.

1800 Chef von Carlowitz; became commander of IR Nr 20 in 1805 as Oberst. 1806 Chef von Keyserlingk. **1806.** In the Duke of Weimar's corps; missed the battles of Jena and Auerstaedt; joined up with Bluecher's corps and fought at Altenzaun and Luebeck. Surrendered at Ratkau; disbanded.

2. Fusilier-Bataillon. Kurmark Fusilier Brigade in 1798. Magdeburg Fusilier Brigade in 1802.

Uniforms. Pink facings, white buttons. Hat tuft green with a white ring. In 1798 this changed to crimson collar, lapels shoulder straps and cuffs with yellow buttons. The shako pompon was white in 1802. Garrison Burg.

Campaign history. Raised in 1788 from two companies of IR von Leipziger Nr 3 and the grenadier companies of Grsn R Nr 7 (in Gren Bn von Borke Nr 1). **1787** Chef von Renouard; became Oberst and commander of IR Nr 56. Expedition to Holland; 17 Sep capture of Gorkum; capture of Dordrecht. **1792** Sk Sirck; Sk Fontois; 30 Aug capture of Verdun; 14 Sep C Grandpre; Chalons; 6 Nov C Limburg / Lahn; 2 Dec C Ursel. **1793** Chef von Bila; in 1806 GM. 6 Jan C Hochheim; 28 Mar C Bingen; 30 Mar C Alsheim; Nov blockade of Landau. **1794** 23 May C Diedesheim; together with the 10th Fusiliers they drove the enemy back with a bayonet charge. 13 Jul C Edighofen; 20 Sep B Kaiserslautern. **1806.** In the Duke of Weimar's corps; missed the battles of Jena and Auerstaedt; joined up with Bluecher's corps and fought at Altenzaun and Luebeck. Surrendered at Ratkau; disbanded.

3. Fusilier-Bataillon. 1st East Prussian Fusilier Brigade in 1798. Garrison Bialystock.

Uniforms. White facings, yellow buttons. Hat tuft white with a green ring and orange top. In 1798 this changed to violet lapels, yellow buttons; collar and cuffs were in the coat colour. The shako pompon was yellow in 1802.

Campaigns. Raised on 22 May 1787 from four companies of the Grsn R Nr 1. Titles – 1787 - von Thile; became commander of IR Nr 46 as GM. 6 Feb 1795 Chef – von Tilly; became commander of IR Nr 8. 1797 Chef von Wackenitz. **1794.** In Poland in GL von Wolcky's corps. Skirmishes at

Modlin, Warsaw, Mieszkowice, Stanislawow, Dobrylas, Skwa, Slonsarze, Margowiecki, Dobrylas 12 Sep, Prsym, Mieszkowice 29 Sep, Muszewska, Longo, Mieszkowice, Muszewska 1 Oct, Novy-Dwor; in Gen von Schoenfeld's corps, which crossed the River Narew and defeated the enemy. C Sava; under GL von Guenther. **1806.** No action. **1807.** 5 Feb C Waltersdorf – heavy losses; the remnants were

captured, but were released some days later. All the Polish recruits deserted; 30 officers and NCOs withdrew to Tilsit. The unit also fought at Preussisch-Eylau, Braunsberg, Danzig, Zagern, Koenigsberg and Baumwalde. Losses not known. In 1808 the battalion became the Fus Bn of the **1st Westpreussisches IR Nr 6**; later **Gren R Graf Kleist von Nollendorf (1st Westpr) Nr 6**; see Ser Nr 7 in the new regiments' section below.

4. Fusilier-Bataillon. 2nd Warsaw Fusilier Brigade in 1798. Garrison Wrazlaweck.

Uniforms. Light blue facings, yellow buttons. Hat tuft green with a yellow ring. In 1798 this changed to light green facings, yellow buttons. The shako pompon was red in 1802.

Campaign history. Raised from two companies of Grsn R Nr 2 and two of the Leichte IR von Arnould; Chef Obstlt von Lieberoth; died as Oberst. **1794** In Poland, maintaining peace in Thorn. Various companies were detached at times. All the sharpshooters were sent off to chase the Polish corps of Gen Dembrowski. 1795 Chef Major von Greiffenberg; Oberst in 1806. **1806.** Auerstaedt 16 Oct. joined up with Bluecher's corps and fought at Altenzaun and Luebeck. Surrendered at Ratkau; disbanded.

5. Fusilier-Bataillon. Kurmark Fusilier Brigade in 1798; Magdeburg Fusilier Brigade in 1802. Garrison Hildesheim.

Uniforms. Dark green facings, yellow buttons. Hat tuft white with a red top. In 1798 this changed to crimson lapels, collar and cuffs were in the coat colour with yellow buttons. The shako pompon was yellow in 1802.

Campaign history. Raised from four companies of Gren Bn von Borcke Nr 1; Chef Major von Borcke; became commander of Fus Bn Nr 20. **1794** to the Rhine; no actions of note. 1797 Chef Major Graf von Wedel;

Oberst in 1806. **1806** In the Duke of Weimar's corps; missed the battles of Jena and Auerstaedt; joined up with Bluecher's corps and fought at Altenzaun and Luebeck. Surrendered at Ratkau; disbanded.

6. Fusilier-Bataillon. 1st East Prussian Fusilier Brigade in 1798. Garrison Stallupoehnen.

Uniforms. Orange facings, yellow buttons. Hat tuft green with a red ring and orange tip. In 1798 this changed to violet facings, yellow buttons. The shako pompon was white in 1802.

Campaign history. Raised in 1787 from the grenadiers of Grsn Rs Nrs 1 and 2 (in Gren Bn von Holstein-Beck Nr 4. Chef in 1787 Oberst Herzog von Holstein-Beck; he became Chef of an infantry regiment. 1789 Chef Major von Rembow; in 1806 GM. **1794** In Poland; C Piontnitza / Narew, where it defeated an enemy force six times its own number; 10 Jul C Collno (two companies); threw the enemy out of the town and took 90 prisoners, a howitzer and two small cannon. 18 Jul C Demnicki; a company made a feint attack which paved the way for victory. **1806** in eastern Prussia; fought in l'Estocq's corps; 14 Jun Koenigsberg. In 1808 this battalion went into the **2nd Ostpr IR Nr 3; in 1914 Grenadierregiment Koenig Friedrich Wilhelm I (2. Ostpr) Nr 3.** See 11. Regiment zu Fuss above.

7. Fusilier-Bataillon. South Prussian Fusilier Brigade in 1798. Garrison Creutzburg.

Uniforms. Pink facings, white buttons. In 1798 this changed to light blue facings, white buttons. The shako pompon was red in 1802.

Campaign history. Raised in 1787 from four companies of the Grsn R von Heucking Nr 8; Chef Oberst von Schmidhenner; became Commandant of Glatz. 1788 Chef Oberst von Schultz; died as GM. **1794** In Poland; 18 May C Skala; guarded the magazines in Pilika; Czarnowiec and Pinczow. Then engaged in disarming rebel troops coming out of Warsaw. 1800 Chef Major von Rosen; Obstlt in 1806. **1806.** Schleitz 9 Oct; Jena 14 Oct, destroyed in the battle.

8. Fusilier-Bataillon. South Prussian Fusilier Brigade in 1798. Garrison Sieradz.

Uniforms. Light green facings, white buttons. In 1798 this changed to light blue collar and cuffs, green lapels, white buttons. The shako pompon was white in 1802.

Campaign history. Raised in 1787 from Grsn R von Sass Nr 6; Chef Oberst von Plueskow; transferred to IR Nr 47. 1790 Chef Major von Hanff; became Commander of Fort Preussen at Neisse. **1794** Chef Major von Eisenhardt; died in office that same year. In Poland; 6 Jun C Scelze (two companies); 15 Jun capture of Krakau; 26 Oct C Magniszewo (two companies). Chef Major von Kloch; Oberst in 1806. **1806** 14 Oct Auerstaedt; most captured there. Some joined up with Bluechers corps, fought at Luebeck and surrendered at Ratkau.

9. Fusilier-Bataillon. 1st Warsaw Fusilier Brigade in 1798. Garrison: originally Patschkau in Silesia; in 1794 it was moved to Pultusk in New East Prussia.

Uniforms. Buff facings, white buttons, Hat tuft crimson. In 1798 this changed to light green lapels, collar and cuffs in the coat colour, white buttons. The shako pompon was red in 1802.

Campaign history. Raised from the old Gren Bn von Gillern Nr 6. Title changes: 15 Feb 1789 Chef Oberst von Gillern; he became Chef of IR Nr 16. 1789 Chef Major von Diebitsch (Jung-Diebitsch); died in office. 26 Apr 1789 Chef Major von Prosch; retired as Obstlt. 1797 Chef Major von Ledebur; transferred as Oberst to command the 3rd Musketier Bn of IR Nr Schoening Nr . 9 Feb 1802 Chef Major Borell du Vernay; Oberst in 1806. 13 Sep 1807 - 'von Danielewitz'. 1794. 18 May C Skala; 6 Jun C Rawka took 4 guns; 15 Jun capture of Krakau; Sk Modlin; Sk Blonie; Sk Marymont; C Nowy Dwor. 1806/1807. The battalion reached Graudenz without being engaged. It was combined with remnants of Fues Bn von Knorr Nr 12 as Fues Bn von Danieliewicz. 1808 21 Nov: It became part of **1st Schleschises IR Nr 11**; in 1813 Nr 10. See Ser Nr 11 in the new regiments' section below.

10. Fusilier-Bataillon. Upper Silesian Fusilier Brigade in 1798.

Uniforms. Buff facings, white buttons. Hat tuft black with vertical white stripes. In 1798 this changed to black facings, white buttons. The shako pompon was white in 1802.

Campaign history. Raised in 1787 when the grenadiers of the Grsn R Nr 10 were transferred to the new Fusilier Bn; Chef Major von Diebitsch; retired as Oberst. 1791 Major von Forcade; resigned. 1792 Major von Martini; retired. 1793 Major Fuerst zu Anhalt-Pless; transferred to Hus R von Schimmelpfennig Nr 6 as Chef. 1802 Chef Obstlt von Erichsen; Oberst in 1806. **1792** On the Rhine and in Champagne. 2 Sep Capture of Verdun. **1793** 6 Jan C Hochheim took two cannon; 30 Mar C Alzheim; 19 Jun-23 Jul S Mainz including C Kastel; 14 Sep B Pirmasens; Sk Ober-Steinbach; Sk Busenberg; Sk Neustadt. **1794** Tuerkheim; Neppenheim; Obernheim; 23 May C Diedesheim; Fenningen; Felmersheim; B Kaiserslautern; Enkebach; Hochspeier; 20 Sep Kaiserslautern. **1806** 14 Oct Jena; the battalion was destroyed at Umpferstaedt after the battle.

11. Fusilier-Bataillon. 1st East Prussian Fusilier Brigade in 1798. Garrison Memel.

Uniforms. White facings and buttons. In 1798 this changed to violet collar and cuffs, yellow buttons; the lapels were in the coat colour. The *Stammliste* gives the facings as being light green in 1806. The shako pompon was red in 1802 and 1806.

Campaign history. Raised in 1787 from men of the Grsn R von Bernhauer Nr 11. Chef Obstlt von Dessauniers; became GM and Commandant of Glogau. 1795 Major von John; resigned. 4 Apr 1795 Major von Eichler; retired. 1798 Major von Thuemen; became Commander of IR Nr 1. 17 Jan 1805 Major von Bergen. **1806** the battalion was in East Prussia and fought in l'Estocq's corps. **1807** 10 Jun B Heilsberg; 14 Jun Koenigsberg. **1808** It became the fusilier battaion of the **1st Ostpreussische IR Nr 1; later the Grenadier-R Kronprinz (1st Ostpreussisches) Nr 1.** See 2. Regiment zu Fuss above.

12. Fusilier-Bataillon. 1st Warsaw Fusilier Brigade in 1798. Garrison Bielsk.

Uniforms. Orange facings, white buttons. Hat tuft green with an orange tip. In 1798 this changed to light blue lapels, the collar and cuffs were in the coat colour, white buttons. The shako pompon was yellow in 1802.

Campaign history. Raised in 1787 from the Gren Bn von Baehr Nr 11; Chef Oberst von Baehr; retired as GM. 1788 Oberst Albert Graf zu Anhalt; became GM and Chef of IR Nr 53. 1795 Major von Eiche; retired as Oberst. 1805 Major von Knorr. **1794** In Poland; no actions of note. **1806** 17 Oct Halle; the battalion then joined Bluecher's corps, fought at Luebeck and most surrendered at Ratkau. 30 men joined up with Fus Bn Nr 9 and were incorporated into that unit. See there for later history.

13. Fusilier-Bataillon. Lower Silesian Fusilier Brigade in 1798. Garrison Jauer.

Uniforms. Buff-brown facings, white buttons. Hat tuft white with a green ring and yellow tip. In 1798 the uniform had black collar and cuffs, lapels in the coat colour, yellow buttons. The shako pompon was yellow in 1802.

Campaign history. Raised in 1787 from two companies of the Leichtes IR von Chaumontet, which itself had been raised only the previous year; in 1788 it took in two companies of the disbanded Grsn R Nr 11. Chef Major von Thadden; died in office as Oberst. 1794 Chef Major von Nordeck zu Rabenau (known as Rabenau); Oberst in 1806. **1792** On the Rhine. **1793** 6 Jan C Hochheim; took two cannon. 1-7 Mar bombardment of Koenigstein; 28 Mar C Bingen; 13 Aug C Limbach and Altstadt; Sk Hombach and Schwegem; 16 Nov Failed storm of Bitsch fortress (a detachment of volunteers); 6 Dec C Zweibruecken; blockade of Landau; 27 Dec C Anweiler valley. **1794** 20 Sep C Moorlautern: took two colours. **1806.** Saalfeld 10 Oct. Jena 16 Oct. The battalion was destroyed at Umpferstaedt after the battle.

14. Fuesilier-Bataillon. Lower Silesian Fusilier Brigade in 1798. Garrison Bunzlau.

Uniforms. Black facings, yellow buttons. Hat tuft yellow with a black ring and green tip. In 1798 the uniform had black facings, yellow buttons. The shako pompon was white in 1802.

Campaign history. Raised on 1 Jun 1787 as von Poellnitz from part of the Leichtes IR von Chaumontet. Chef Oberst von Pollitz; later GM and Inspector of the Warsaw Fusilier Brigade. 15 Apr 1794 Chef Oberstlt von Pelet; GM in 1806. 29 May 1807 Chef Major Schuler von Senden **1794.** 18 May C Skala; 6 Jun B Rawka (or Scelze); for their bravery here they were awarded the right to play the Grenadier March; Sk Nowi Miesto; 27 Jul-6 Sep S Warsaw including actions at Gorce, Wawkzyszew and the Poworsker suburbs. **1806.** Saalfeld 10 Oct; Jena 14 Oct; Zehdenick 26 Oct. The battalion made its way back to Danzig, took part in the defence of the fortress. **1807.** Parts of the battalion were in Pillau, where they were combined into Fues Bn Nr 15. On 21 Nov **1808** remnants of the battalion went to form the fusilier battalion of the **Grenadier-Regiment Koenig Friedrich Wilhelm II (1. Schlesisches) Nr 10**, see Serial Nr 11 in the new regiments section below.

15. Fusilier-Bataillon. Lower Silesian Fusilier Brigade in 1798. Garrison Loewenberg.

Uniforms. Buff-brown facings, yellow buttons. Hat tuft yellow with a green tip. In 1798 the uniform had black lapels, yellow buttons; collar and cuffs in the coat colour. The shako pompon was red in 1802.

Campaign history. Raised on 1 Jun 1787 as von Schurf from part of the Leichtes IR von Chaumontet. 1787 Chef Major von Schurf; became commander of the 3rd Bn, IR Nr 35. 1788 Chef Major von Forcade; took over Fus Bn Nr 10. 1791 Chef Major von Ruhle; Oberst in 1806. **1793** In Poland. **1794** Followed Madalinski's corps to the Pilica River, then took part in the blockade and surrender of Czenstochau fortress. 18 May C

Skala; Sk Bistry in South Prussia. **1806.** Saalfeld 10 Oct. Jena 16 Oct. It then withdrew into Danzig and took part in the defence of the fortress. On 21 Nov 1808 remnants of the battalion went to form the fusilier battalion of the **Grenadier-Regiment 'Koenig Friedrich Wilhelm II' (1. Schlesisches) Nr 10,** see Serial Nr 11 in the new regiments section below.

16. Fusilier-Bataillon. 2nd Warsaw Fusilier Brigade in 1798. Garrison Petrikau.

Uniforms. Black facings, white buttons. Hat tuft black with a green tip. In 1798 this changed to light blue lapels, collar and cuffs in the coat colour, yellow buttons. The shako pompon was yellow in 1802.

Campaign history. Raised in 1787 from part of the Leichtes IR von Arnould (raised itself in the previous year); Chef Oberst von Ruempel; retired. 1789 Chef Major von Oswald; GM in 1806. **1794** In Poland. Initially engaged in the pursuit of Madalinsky's brigade along the River Pilica; 18 May C Skala; 6 Jun B Scelze; 15 Jun capture of Krakau; 28 Jul C Wola near Warsaw; took a colour and many prisoners. **1806.** Auerstaedt 16 Oct; escaped to join Bluecher's corps, fought at Luebeck and surrendered at Ratkau.

17. Fusilier-Bataillon. 1st Warsaw Fusilier Brigade in 1798. Garrison Plock.

Uniforms. Light blue facings, white buttons. Hat tuft light blue with a white ring and green tip. The shako pompon was white in 1802.

Campaign history. Raised in 1787 from part of the Leichtes IR von Arnould; Chef Obstlt Freiherr von Wahlbrunn; died in office. 1788 Major von Hinrichs; GM in 1806. **1794** On border guard along the River Pilica; in July they joined the king's army and fought at Blonien. S Warsaw, then marched into South Prussia with Oberst Szekeli's column; fought at Laduschin and Bromberg. **1806** 17 Oct Halle; they escaped to join Bluecher's corps, fought at Luebeck and surrendered at Ratkau.

18. Fusilier-Bataillon. Magdeburg Fusilier Brigade in 1798; Westfalian Fusilier Brigade in 1802. Garrison Lippstadt.

Uniforms. Crimson facings, white buttons. Hat tuft green with a white ring. In 1798 this changed to crimson lapels and shoulder straps with white buttons. The collar and cuffs were in the coat colour. The shako pompon was red in 1802.

Campaign history. Raised in 1787 from part of the Leichtes IR von Mueller (raised itself in the previous year); Chef Obstlt von Wilhelmini; died in office. 1788 Chef Major von Mueffling; transferred as Obstlt to IR Nr 12; in 1806 he was GM and Chef of IR Nr 49. **1792** On the Rhine and in Champagne. 20 Sep Valmy. **1793** 28 Mar C Waldalgesheim and Bingen; Sk Ensheim; Sk Gundersheim; 15 Apr Carlsberg; Homburg; 13 Oct C Weissenburg Lines. **1794** Chef Major von Holzschuer; died in office as Oberst. 23 May C Weidenthal; Edinghofen; Roth; 13 Jun C Edesheim; Sk Weiher; Sk Edinghofen; 13 Jul C Schaenzel; Sk Eselsfurth; 20 Sep B Kaiserslautern. 1800 Chef Major von Sobbe; Oberst in 1806. **1806.** In Ruechel's corps. Jena 16 Oct; surrendered at Magdeburg on 20 Oct.

19. Fusilier-Bataillon. Magdeburg Fusilier Brigade in 1798; Westfalian Fusilier Brigade in 1802. Garrison Muenster.

Uniforms. Crimson facings, yellow buttons. Hat tuft white with a green ring and orange tip. In 1798 this changed to crimson collar, lapels and cuffs with white buttons. The shako pompon was white in 1802.

Campaign history. Raised in 1787 from part of the Leichtes IR von von Mueller; Chef Major von Ernest; GM in 1806. **1792** On the Rhine and in Champagne; 15 Sep C Montcheutin; 20 Sep Valmy; Sk Pillon; Sk Moosbach; Sk Biebrich. **1793** C Alzey; 15 Apr C Carlsberg, 13 Aug C Limbach and Altstadt; Sk Limbach; 14 Sep C Zweibruecken; 13 Oct C Herzogenhand; Sk Bockweiler; Sk Neu-Hornbach. **1794** Sk Welschstein; Sk Moorsheim; Sk Kirkheim-Poland; 3 Jul C Trippstadt; Sk Leimen; Heltersberg; Sk Johanniskreuz; Creuznach. **1806.** In Wobeser's corps. Jena 16 Oct; destroyed at the bridge on the Ilm. The remnants surrendered at Magdeburg.

20. Fusilier-Bataillon. Magdeburg Fusilier Brigade in 1798; Westfalian Fusilier Brigade in 1802. Garrison Muenster.

Uniforms. Dark green facings, white buttons. Hat tuft white. In 1798 this changed to crimson collar, and cuffs, green lapels with white buttons. The shako pompon was yellow in 1802.

Campaign history. Raised in 1787 from part of the Leichtes IR von von Mueller; Chef Major von Legat; transferred to command 3rd Bn IR Nr 32. 1797 Chef Oberst von Borcke; became Chef of IR Nr 32; died as GL. 1798 Chef Major von Invernois; Oberst in 1806. **1792** On the Rhine and in Champagne; C Briquenai. **1793** 19 Jun-23 Jul S Mainz (1 company); 13 Aug C Altstadt and Limburg; 20 Sep B Kaiserslautern. **1806** The battalion was in Ostfriesland at the start of the campaign. On 4 Nov they joined Bluecher's corps; fought at Luebeck and surrendered at Ratkau.

21. Fusilier-Bataillon. 2nd East Prussian Fusilier Brigade in 1798. Garrison Heilsberg.

Uniforms. Black lapels, dark green collar, cuffs in the coat colour, white buttons. In 1798 this changed to violet facings, white buttons. In 1806 facings were light green. The shako pompon was white in 1802.

Campaign history. Raised in 1795 from new recruits; Chef Major von Stutterheim; Oberst in 1806. **1806** The battalion was in eastern Prussia. **1807** 10 Jun B Heilsberg; 14 Jun C Koenigsberg. **1808** it became the Fusilier Bn of the 3. Ostpreussisches IR Nr 4. See **Grenadier-Regiment Koenig Friedrich der Grosse (3. Ostpreuss.) Nr 4,** Ser Nr 5 in the new regiments' section below.

22. Fusilier-Bataillon. Upper Silesian Fusilier Brigade in 1798. Garrison Neumark.

Uniforms. In 1798 the uniform was black lapels, collar and cuffs in the coat colour, white buttons. The shako pompon was yellow in 1802.

Campaign history. Raised in 1797 from new recruits; Chef Major Baron zu Putlitz; transferred to command 3rd Bn IR von Grevenitz. **1800** Chef Major von Boguslawsky; Obstlt in 1806. **1806** 14 Oct Jena, destroyed.

23. Fusilier-Bataillon. 2nd East Prussian Fusilier Brigade in 1798. Garrison Johannisburg.

Uniforms. In 1798 the uniform was violet collar and cuffs, yellow buttons; lapels were in the coat colour. Lapels, collar and cuffs were light green in 1806. The shako pompon was red in 1802.

Campaign history. Raised in 1797 from new recruits; Fus Bns 3 and 12 provided the cadres. Chef Major Johann David Ludwig von Yorck; transferred to command the Jaeger-Regiment as Oberst; in 1806 he was its Chef. In 1813 he was a Feld-Marschall. 1800 Chef Major von Schachtmeyer. **1806** The battalion fought in eastern Prussia, in l'Estocq's corps. **1807** 5 Feb C Waltersdorf. **1808** The battalion became the Fusilier Bn of the 4th Ostpreussisches IR. This became the **Grenadier-Regiment Koenig Friedrich I (4. Ostpreuss.) Nr 5.** See Ser Nr 6 in the new regiments' section below.

24. Fusilier-Bataillon. 2nd East Prussian Fusilier Brigade in 1798. Garrison Soldau.

Uniforms. In 1798 the uniform was violet lapels, green collar and cuffs, yellow buttons. The shako pompon was yellow in 1802. By 1806 the facings were light green, the buttons white.

Campaign history. Raised in 1797 from new recruits; the cadre was provided by Fus Bns 12, 16 and 17. Chef Major Friedrich Wilhelm von Buelow; Oberst in 1806; GL in 1813. **1806.** The battalion fought in eastern Prussia in l'Estocq's corps. **1807** 5 Feb C Waltersdorf. **1808** The battalion became the Fusilier Bn of the 4th Ostpreussisches IR. This became the **Grenadier-Regiment Koenig Friedrich I (4. Ostpreuss.) Nr 5.** See Ser Nr 6 in the new regiments' section below.

WE HOPE THAT YOU ENJOY THIS BOOK...and that it will occupy a proud place in your library. We would like to keep you informed about other publications from Schiffer Books. Please return this card with your requests and comments. **(Please print clearly in ink.)**
Note: We don't share our mailing list with anyone.

Title of Book Purchased ____________________

❑ Purchased at: ____________________ ❑ received as a gift

Comments or ideas for books you would like to see us publish: ____________________

____

____

Your Name: ____________________

Address ____________________

City ____________ State ______ Zip ________ Country ________

E-mail Address ____________________

Please provide your email address to receive announcements of new releases

❑ Please send me a **free** *Schiffer Antiques, Collectibles, & the Arts*
❑ Please send me a **free** *Schiffer Woodcarving, Woodworking, and Crafts Catalog*
❑ Please send me a **free** *Schiffer Military, Aviation, and Automotive History Catalog*
❑ Please send me a **free** *Schiffer Lifestyle, Design, and Body, Mind, & Spirit Catalog*

K: user\do\wp\basic\bouceback

Printed in China

7

The Invalides

Garde-Invaliden
Founded by King Friedrich Wilhelm I in 1730 and located in Werder near Potsdam. It took in those soldiers of the Garde and Grenadiergarde who were no longer fit for field service. Those invalids of the 1st Bn Garde and of the Garde du Corps had their own establishment. It consisted of a captain, two sergeant-majors, 31 NCOs 477 privates and 14 musicians. They receive full allowances, but no uniforms are issued. In time of peace they mount guard around Potsdam. In time of war, they guard the royal palaces in Potsdam and other locations.

Invaliden-Corps in Berlin.
Uniform Dark blue coat, facings and breeches; officers have a narrow gold hat edging. This institute was built in Berlin in 1748; all regiments were to send their wounded and poverty-stricken members to it. They formed three companies each of 200 men. They received the same uniforms and allowances as the field regiments, free lodgings, firewood and light. Those with capable hands and feet perform security duties for the institute.

Invaliden-Corps in the Provinces.
King Friedrich Wilhelm II founded similar institutions, beginning in 1788, in Czenstochau, Habelschwert, Konitz, Labes, Langenzenn, Mansfeld, Mewe, Neustaedtel, Patschkau, Peine, Swienemuende, Tapiau, Trebbin, Weichselmuende and Ziegenhals. The uniform was a dark blue coat, crimson cuffs, white buttons and small clothes. Officers have a thin silver hat edging, soldiers white. Inmates received full allowances.

By Order of 20 Feb 1809 all existing regimental invalide hospitals were concentrated into 25 Provincial Invalide Companies.

8

Infantry Regiments Raised after the Disaster of 1806

The titles shown in brackets are those which they bore in 1900.

Regiments Which Later Became Grenadier Regiments

1. Zweites Garde-Regiment zu Fuss.
Raised on 19 Jun 1813 from existing units as shown below.
1st Battalion. Raised on 14 May 1811 as the Normal-Infanterie-Bataillon from men from all existing infantry regiments.

Campaign history. 1813 Against France. 2 May Gross-Goerschen; 20/21 May Bautzen.

2nd Battalion. Raised on 14 Feb 1788 as depot battalion of IR von der Goltz Nr 7; 8 Jan 1796-it became the 3rd Musketier Bn (see there for campaign history); 20 Aug 1808 it became the 1st Bn, 2nd Pommersches-IR, which became the Colberg IR on 26 Aug 1808. On 12 Mar 1812 it was mobilised for the Russian campaign as the 2nd Bn, 3rd Mobile IR until 7 Mar 1813.

Campaign history. 1807 5/6 Jan Sk Wollin; 7 Mar-8 Jul D Colberg, with sorties at Wolfsberg on 17/18 May, 15, 16 and 17 Jun. **1812** 19 Jul Eckau; 7 Aug Wolgund; 27 Sep Eckau; 29/30 Sep Meseton; 18 Oct Wolgund. **1813** 17 Apr Wittenberg; 28 Apr Halle; 2 May Gross-Goerschen; 5 May Colditz; 20/21 May Bautzen; 24 May Siegersdorf.

Fuesilier Battalion.
Raised in Dec 1806 as von Schills Freikorps, which received royal approval on 13 Jan 1807. By Nov 1807 it had three battalions. On 20 Aug 1808 it became the'Leichtes-Bn von Schill' of the 2nd Pommeranian IR. On 26 Aug 1808 this regiment became the Leib-Infanterie-Regiment. 1900 - Fuesilier Bn, Leib-IR. (2. Garde Regiment zu Fuss).

Campaign history. 1806 - 1807. Defence of Colberg. **1809.** The Leib-Kompagnie took part in von Schill's raid through north Germany. **1812.** In Russia with X Corps as part of the 4th Combined IR; Garossenkrug. Samsonkrug, Klawe, Dahlenkirchen. The Leib-Kompagnie rejoined the parent regiment on 17 Mar 1813. **1813.** Gross-Goerschen 2 May, Colditz, Koenigswartha, Bautzen. Siegersdorf. **1814.** Paris 30 Mar. **1815.** No actions.

2. Kaiser Alexander Garde-Grenadier-Regiment Nr 1 (Die Kartoffelschaeler).
Raised on 14 Oct 1814 from:
2a. Leib-Grenadier-Bataillon. (raised on 27 Dec 1806 in Colberg from escaped prisoners of war as the 'Grenadier-Bn von Waldenfels'; which became the 'Leib-Gren-Bn' of the Leib-Infanterie-Regt on 26 Aug 1808.

Titles. In 1807 it was known as the 'Grenadier-Bn von Waldenfels'; on 27 Nov 1808 the title changed to 'von Buelow'; on 4 Jun 1813 to 'von der Goltz'; on 10 Jul 1813 to 'von Carlowitz'.

Campaign history. 1813. In von Yorck's I Corps. Gross-Goerschen 2 May, Bautzen 20/21 May, Hochkirch 4 Sep, Bischofswerda 22 Sep, Moeckern 16 Oct, Friedburg. **1814.** Chateau-Thierry. After this battle, the regiment was combined with the Westpreussischen Gren-Bn. Triport 23 Mar, Paris; The regiment took 10 guns.

2b. 'Ostpreussischen Grenadier-Bn Nr 1' (in 1806 this was the 'Grenadier-Bn von Schlichting' or 'von Below'. This latter unit had been raised on 28 Feb 1799 from two companies each of the grenadiers of the IRs 'von Ruechel Nr 2' and 'von Schoening Nr 11'. See the Grenadier Battalions for actions.

2c. The fusilier battalion was raised from the 'Ostpreussisches Grenadier-Bn Nr 2 (in 1806 this was the 'Gren-Bn von Fabieckij') and from two companies each of IR 'von Besser Nr 14' and 'IR von Dierecke Nr 16'. See the grenadier Battalions for actions.

Titles. 14 Oct 1814 – 1. Grenadier-Regiment; 19 Oct 1814 – Gren-Regt Kaiser Alexander Nr 1; 27 Nov 1819 – Kaiser Alexander Garde-Grenadier-Regt Nr 1.

Colours.
1st Battalion: they use the Retirierfahne of the Leib-Gren-Bn of the Leib-Inf-Regt, which was presented to this battalion on 28 May 1812. On 15 Jun 1815 the EK (Iron Cross) 1813 model was added to the finial. The ground of the colour was black with narrow white corner radials, orange medallions, gold embroidery. Under the central medallion 'COLBERG 1807' in gold on a blue label. From 14 Oct 1814 the 1st Bn carried this colour.

2nd Battalion: they used the Retirierfahne of the 2nd Bn, 1. Ostpreussischen IR, which had been presented in 1753. On 22 Mar 1812 this colour was presented in Graudenz to the 1. Ostpreussischen Gren-Bn, which had been raised **on 2 Aug 1789**. On 15 Jun 1815 the EK 1813 was added to the finial. The colour was black with white medallions, gold decorations, black scroll, gold 'FR' in the finial. From 14 Oct 1814 the 2nd Bn carried this colour.

Fuesilier-Bn: they carried the Retirierfahne of the 1. Bn, 3.Ostpreussischen IR, which had been presented to the 2.OstpreussischenGren-Bn (raised on **28 Feb 1789**). This colour was presented to the Fuesilier-Bn at Berendt on 16 May 1812. This unit became the Fuesilier-Bn of the 1. Gren-Regt on 14 Oct 1814. Black ground, white corner radials, orange medallions, gold decorations, 'FWR' in the finial. From 14 Oct 1814 the Fuesilier Bn carried this colour. On 15 Jan 1815 the EK was added to the finial.

3. Kaiser Franz Garde-Grenadier-Regiment Nr 2. ('Franzer', 'Bluthunde').

Raised 14 Oct 1814:

1st Bn – from the Pommerschen Gren-Bn (raised on 7 Nov 1801 from the Gren-Bn 'von Massow', from two grenadier companies each from the IR 'von Ruets' Nr 8, IR 'von Ploetz Nr 42 and from the Gren-Bn 'von Jung-Braun' (raised on 10 Oct 1805, from two companies each of IR 'von Kropff' Nr 31, 'von Thiele' Nr 46 and parts of the Freikorps 'von der Marwitz' (raised in 1806).

2nd Bn – from the Westpreussischen Gren-Bn (raised 21 Jan 1803); from the Gren-Bn 'von Brauschitz' and two companies each from IR 'von Reinhardt Nr 52 and IR 'von Courbiere Nr 58.

Fuesilier-Bn – from the Schlesischen Gren-Bn, the Gren-Bn 'von Losthin', from the 4th and 5th Bns of IR 'von Alvensleben'Nr 33, and the 3rd and 4th Bns of IR 'von Grawert Nr 47. All these battalions had been raised in 1807. For details of histories of these regiments see the original regimental histories.

Title in 1814: 2. Grenadier-Regiment. On 25 Dec 1814 this became: 'Kaiser Franz Garde-Grenadier-Regiment Nr 2'.

Colours. By AKO of 28 Mar 1812, the battalions carried the following colours:

1st Bn. The Pommersches Gren-Bn was presented with the *Retirierfahne* of the 2nd Bn, Pommerschen IR, which was originally presented to the IR 'von Hacke' Nr 8 on 25 Jun 1772.

Black ground, white flame cross, gold grenades, white medallions with the usual eagle and wreath, black scroll with gold letters 'PRO GLORIA ET PATRIA'. Gold embroidery, 'FWR' in the corner medallions and the finial; on 15 Jun 1815 the EK was added to the finial.

2nd Bn. The *Retirierfahne* presented on 25 Feb 1798 to the 1st Bn, 2nd Westpreussisches IR: light blue with yellow corner piles and centre, light blue scroll, silver embroidery, white staff. Old IR Nr 58?

Fues-Bn. From 14 Oct 1814 the battalion carried the *Retirierfahne* presented in 1779 to the IR von Alvensleben Nr 33, which was then carried by the 1st Bn, 2nd Westpreussisches IR. It was cornflower blue with a white centre, cornflower blue scroll, silver decoration, white staff.

Campaign history. 1807. Kanth 14 May, Adelsbach, the siege of Glatz, 20-24 Jun. **1813.** In 1st Brigade, I Corps. Gross-Goerschen 2 May; Bautzen 20/21 May; Katzbach 26 Aug; Wartenburg 3 Oct; Moeckern 16 Oct; Freiburg 21 Oct. **1814.** Chateau-Thierry 11/12 Feb. After this action, they were combined with the Ostpreussische Gren-Bn. Paris 30 Mar.

4. Grenadier-Regiment Kronprinz (1. Ostpreussisches)

Titles: 1793 - 'von Brueneck' Nr 2; 17 Aug 1805 – I.R. 'von Ruechel Nr 2; 11 Jul 1807 – vacant von Ruechel Nr 2; 7 Sep 1808 – 1.Ostpreussisches IR. Colours. This regiment lost no colours in 1806/1807. For details of history and colours of this regiment see the original Infantry Regiment Nr 2.

4a. Campaigns of the Fusilier-Bataillon 'von Bergen' Nr 11.
Raised on 1 Jun 1787 from the disbanded Garnison-Regiment Nr 11.

Titles. 4 Apr 1795 – von Eichler; 17 Jan 1805 – von Bergen; 1 Jan 1808 – Fuesilier-Bataillon 'Kronprinz'.

5. 7 Sep 1808 – 3.Ostpreussisches IR Nr 4. Raised on 20 Nov 1807 from I.R.Nr 14; see there for early history.

Title in 1901: Grenadier-Regiment Koenig Friedrich der Grosse (3. Ostpreussisches) Nr 4. Page 245, Vol I.

Colours. New style; King's colour was white, black corner piles, orange centre, black eagle, light blue scroll, gold decorations, black staff. Regimental colour black, white corner piles, orange cantre, black eagle, light blue scroll, gold decorations, black staff.

1812. The 1st Bn was mobilised as part of the 2nd Combined IR together with the 1st Bn, and Fuesiliers of IR Nr 5. They fought at Eckau and Lautsch-Krug. **1813.** The regiment formed part of von Buelow's III Corps. Magdeburg 13 Apr; Halle 2 May The regiment took 3 guns, 3 ammunition wagons. Luckau 4 June; Gross-Beeren 23 Aug. Took 2 guns.

Dennewitz 6 Sep Took 4 guns. Leipzig 18/19 Oct. After the battle, the 2nd Bn could muster only 3 officers, 5 NCOs and 100 R&F. **1814.** Hoogstraten, Antwerp and Laon; losses unknown. **1815.** The regiment saw no action.

6. Grenadier-Regiment Koenig Friedrich I (4. Ostpreussisches Nr 5). Raised on 11 Mar 1689.

Titles: 7 Feb 1789 – IR 'von Gillern' Nr 16; 20 May 1792 -'von Hausen'; 1 Oct 1799 - 'von Dierické'; **7 Sep 1808 – 4. Westpreussisches IR 'von Dierické'.**

Colours. Old style. King's colour white with vertical, upright wavy cross, white centre, gold decorations, brown staff. Regimental colour orange with vertical wavy cross, white centre, orange scroll, gold decoration, brown staff.

Campaign history. 1806. The grenadiers of the regiment were combined with those of the IR 'von Besser' Nr 4 as the Grenadier-Bn 'von Fabecky'. During 1806 and 1807 this battalion fought at Soldau, Gross-Schiemanen, Waltersdorf, Wackern, Preussisch-Eylau, Heilsberg, Braunsberg, Gollau and Koenigsberg as part of L'Estocq's corps. **1807.** 10 Mar – 25 May: the regiment was in the defence of Danzig.

1812. The 1st and Fuesilier Bns were mobilised with the 1st Bn, 3rd Ostpreussisches IR as the '2nd Combined IR. Eckau 19 Jul, the regiment recovered a lost gun. Dahlenkirchen 28 Aug; Bauske Dahlenkirchen 15 Oct; Dahlenkirchen 16 Nov. **1813.** The regiment was in the 4th Brigade of Buelow's III Corps; the grenadiers were combined with those of the 3rd Ostpreussischen-IR, to form the 2nd Ostpr. Gren-Bn in the 3rd Brigade of the same corps. Vehlitz 5 Apr; (Fues. Bn only); Koenigsborn 22 Apr (Fues-Bn only); Spandau 1 – 27 Apr (1st and 2nd Bns only); Halle 4 May (Grenadiers only); Luckau 4 Jun (The whole regiment was engaged); Schoenwalde 20 Aug (Only the Fues-Bn was engaged); Wittstock 22 Aug (Fues-Bn only); Gross-Beeren 23 Aug (Grenadiers and 2nd Bn only); the regiment took two guns. Dennewitz 6 Sep the regiment took 4 guns, 6 ammunition wagons and 5,000 POW. Wittenberg 10 Oct; Coswig ? Oct (Fues-Bn only); Leipzig 16 - 18 Oct; the regiment took 23 offrs and 1,000

R&F prisoner. Arnheim 30 Nov (Fues-Bn only; they took a gun); Gorkum 20 Dec 1813 – 21 Feb 1814. **1814.** Laon 9/10 Mar; Soissons 2 Mar (1st Bn only). **1815.** The regiment saw no action.

7. Grenadier-Regiment Graf Kleist von Nollendorf (1. Westpreussisches) Nr 6.
Raised on 14 Oct 1772.

Titles: 1798 – IR 'Alt-Schwerin' Nr 52; 19 May 1795 - 'von Reinhart'; 30 Jan 1807 - 'von Hamberger'; **7 Sep 1808 – 1. Westpreussisches IR Nr 6.**

Colours. Old style. King's colour white with madder red flame cross, dark green centre, white scroll, silver decoration, white staff. Regimental colour dark green with madder red flame cross, dark green scroll, silver decoration, white staff.

Campaign history. 1812. The 1st Bn was mobilised and combined with the 1st and Fues Bn of the 2nd Westpreussisches IR as the '5th Komb. Feld-Regiment'. Oley 27 Sep; Graefenthal 28 Sep. **1813.** Gross-Goerschen 2 May (2nd, 3rd and Fues Bns only); Weissig 19 May (1st Bn. Only); Bautzen 20 May (1st Bn only(; Bautzen 20 May (2nd, 3rd , Fues Bns only); Dresden 26 Aug (Fues Bn only); Leubnitz 28 Aug (Fues Bn only); Haeselich 28 Aug (1st, 2nd Bns only); Kulm 30 Aug (Fues Bn only); the battalion retook a lost gun. Zinnwald 11 Sep; Leipzig 16 – 18 Oct (the whole regiment fought here). **1814.** Beauval 28 Feb; Laon 9 Mar (Fues Bn only); they took 8 guns on the windmill hill. Montils 26 Mar (Fues Bn only); Paris 30 Mar. **1815.** In the I Corps. Gilly 15 Jun; Ligny 16 Jun; Villers-Cotterets 28 Jun; the regiment took 18 guns, 24 ammunition wagons and a field smithy.

8. Grenadier-Regiment 'Koenig Wilhelm I' (2. Westpreussisches Nr 7).
Raised on 20 Feb 1797 as the IR Nr 58 'von Courbiere'. **On 7 Sep 1808 it became the 2. Westpreussisches IR (von Courbiere) Nr 7.**
On 20 Nov 1807 it absorbed men from the disbanded infantry regiments Nrs 29, 37, 40, 43, 54 and 55.

Colours. Old style. King's colour white with narrow yellow corner piles and centre, white scroll, silver decoration, white staff. Regimental colour light blue, yellow corner piles and centre, light blue scroll, silver decoration, white staff.

Campaign history. 1812. The 1st and Fues Bns were mobilised with the 1st Bn, 1st Westpreussisches IR as the Komb. Feld-Regiment Nr 5. Riga 6 Aug (Fues Bn only); Krebsenkrug 15 Aug (Fues Bn only); St Olay 22 Aug (Fues Bn only); Staroi-Krug 26 Sep (Fues Bn); Mittau 28 Sep (1st Bn only); Kiopen 29 Sep (1st Bn only); Schlockhof 30 Sep (Fues Bn only); Garossenkrug 1 Oct; Katherinenhof 17 Oct; Dolbing 29 Oct (Fues Bn only); Samsonskrug 29 Oct (Fues Bn only); Thomsdorf 17 Nov (1st Bn only). **1813.** The 2nd Bn formed a combined regiment together with the 2nd and Fues Bns of the 1st Westpreussisches IR in von Klux's 9th Niederschlesische brigade, II Corps; the 1st and Fues Bns stayed in von Horn's brigade of Yorck's I Corps until 30 May 1813. Bernburg 13 Apr (1st Bn only); Alsleben 16 Apr (Fues Bn only); Gross-Goerschen 2 May (2nd Bn only); Bautzen 20/21 May; Colditz 5 May (Fues Bn only); Bischofswerda 20 May (Grenadiers only); Dresden 26 Aug (the whole regiment); Kulm 30 Aug (1st and 2nd Bns only); Hellendorf 5 Sep (Fues Bn only); Pirna 8 Sep (Fues Bn only); Hellendorf 15 Sep (Fues Bn only); Leipzig 16 and 18 Oct the whole regiment was present; Ilversgehofen 5 Nov.

1814. Etoges 14 Feb; Gue-a-Tremes 28 Feb (Fues Bn only); Laon 9 Mar (Fues Bn only); Claye 28 Mar (Fues Bn only). **1815.** Lambusart 15 Jun; Ligny 16 Jun.

9. Leib-Grenadier-Regiment 'Koenig Friedrich Wilhelm III' (1. Brandenburgisches) Nr 8.

Raised on 7 Jun 1808 from men of the 2. Pommersches Res Inf Bn, 3. Neumaerkisches Res Inf Bn, Leichtes Bn von Schill, the grenadier Bn 'vac von Weissenfels', Gren Bn 'von den Osten',, IR 'von Kuhnheim' Nr 1, IR 'von Kleist' Nr 5, IR 'von Arnim' Nr 13, IR 'von Tschammer' Nr 27, IR 'Prinz Ferdinand' Nr 34.

The 'Leib-Grenadier-Bn' became the 'Kaiser Alexander Garde-Grenadier-Regiment Nr 1' on 14 Oct 1818.

Titles. 26 Aug 1808 – Leib-IR, 7 Sep 1808 – 1. Brandenburgisches IR Nr 9, **14 Sep 1808 – Leib IR Nr 9, 1 Jul 1813 – Leib IR Nr 8.**

Campaign history. 1807. Siege of Colberg. During the siege the unit fought at Stargard, Naugard, 1st Sellnow, 1st Grenadier-Schanze, 2nd Sellnow, Maikufle (18 Jun), 2nd Grenadier-Schanze (16 Jul). **1809.** The Leib-Kompagnie went on von Schill's raid with 4 offrs, 12 NCOs and 104 R&F; 3 offrs, 2 NCOs and 11 R&F from the rest of the regiment joined them. **1812**. 1st Eckau 19 Jul; the regiment took a colour. 2nd Eckau 27 Sep; Lautsch-Krug 30 Sep; Garossen-Krug 1 Oct. **1813.** March-out strength; 1st, 2nd Musketier-Bns, Fues-Bn, Grenadiers: 95 offrs, 3,000 ORs, 700 Fr-Jg in four companies. The regiment was in Colonel von Steinmetz's brigade of von Yorck's I Corps. Gross-Goerschen 2 May; Koenigswartha (Weissig). Following this battle, the 1st and 2nd Bns were combined and the Fues Bn was combined with the Fues Bn of the 1st Ostpreussisches IR into one battalion. Bautzen 20/21 May; Loewenberg 21 Aug; Bunzlau 30 Aug. On 1 Sep the regiment was transferred to von Zeppelin's brigade. Hochkirch 4 Sep (Grenadiers only); Bishofswerda 22 Sep (Grenadiers only); Wartenburg 3 Oct (2nd Bn and Fues Bn only); the regiment took 1 gun and 3 ammunition wagons. Moeckern 16 Oct. The regiment took 4 guns. Gohlis 18 Oct; Freiburg 21 Oct. **1814.** La Chasse 3 Feb; the regiment claimed that it took a colour of the *59e Ligne*; this is most unlikely. Chateau-Thierry 12 Feb; Brienne 28 Jan; Laon 9 Mar; the regiment took a cannon, a howitzer and an ammunition wagon. Berry-au-Bac 14 Mar (Fues-Bn only); Trisport 27 Mar; Paris 30 Mar. **1815.** In von Thielmann's III Corps. Ligny 16 Jun; Wavre 18 Jun; Versailles 1Jul.

10. Colbergisches Grenadier-Regiment 'Graf Geisenau' (2. Pommersches) Nr 9.

Titles. 7 Jun 1808 – IR 'von Borcke' Nr 30; **1 Jul 1813 - 'Colbergisches IR Nr 10; the number changed to 9 that same day.**

Colours. New pattern; king's colour white with black corner piles, orange centre, light blue scroll and light blue 'COLBERG 1807' patch edged gold, black staff. Regimental colour black with white corner piles, otherwise as above.

Campaign history. 1807. The regiment was formed in Colberg from the 3. Musketier-Bns of the IRs 'von Owstein' Nr 7, 'von Borcke' Nr 30 and the Fues-Bn 'von Moeller.'* The siege lasted from 20 Mar – 2 Jul and losses were over 50% of the initial 6,000-strong garrison.

1812. The 1st Bn of the regiment was mobilised together with the 2nd and Fues Bns of the 1. Pommersches IR to form the 3rd Combined Field Regiment in von Yorck's corps in Courland, Russia. The march-out strength was 20 offrs and 700 ORs. Eckau 19 Jun, Schlock 5 Aug, St Annen 7 Aug, Eckau, 27 Aug; losses not known. **1813.** 1 Jul 1813 - **'Colbergisches IR Nr 10'; the number changed to 9 that same day.** In von Kleist's II Corps; this formed part of the army of Bohemia after the armistice. Sk Wittenberg . Sk Alt Damm 7 Apr. Sk Gross-Goerschen 2 May. Sk Colditz 5 May. B Bautzen 20/21 May. On 2 Jun the regiment was transferred to von Buelow's III Corps in the Army of the North. B Dennewitz 6 Sep. B Leipzig 17-19 Oct. Sk Doesburg 23 Nov; the Fues-Bn took a gun. Sk Zuetphen 24 Nov. C Arnheim 25-28 Nov. C Arnheim; the 2nd Bn took 2 guns. **1814.** C Wyneghem 13 Jan; the regiment took 2 guns. C 's - Hertogenbosch 26 Jan; the regiment took 1 gun and 7 colours. C Wyneghem 9 Mar. C Soissons 23 Mar. C Compiegne 1 Apr. **1815.** B Ligny 16 Jun. C Namur 20 Jun. The regiment also saw action at the sieges of Landrecy, Philippeville and Givet and suffered very slight losses.

11. Grenadier-Regiment 'Koenig Friedrich Wilhelm II' (1. Schlesisches) Nr 10.

Raised on 21 Nov 1808: 1st Bn from the men of the disbanded regiments 'von Kropff ' Nr 31, 'von Ploetz' Nr 42, 'von Thile' Nr 46, 'von Chlebowski' Nr 60 and the Silesian Reserve-Bn 'von Welling'.

The 2nd Bn was raised from the Fues-Bns 'von Danielewitz' Nr 9, 'von Knorr' Nr 12 and 'von Hinrichs' Nr 17.

The Fues-Bn was raised from the old Fues-Bns 'Schuler von Senden Nr 14, 'von Ruehle' Nr 15 and two companies of the National-Jaeger-Korps (which was raised itself in 1807).

The grenadiers were raised from two companies of the Gren-Bn 'von Losthin'.

*Captain von Moeller collescted together the depots of the Fuesilier Bns Nrs 1, 5,18, 19 and 20, which were in Hildesheim in October 1806. He then led them to Colberg, where, on 9 November 1806, together with ex-prisoners of war, he raised this unit.

Titles. 20 Nov 1808 – **1. Schlesisches I R Nr 11; 1 Jul 1813 – 1. Schlesisches IR Nr 10.**

Colours. On 9 May 1809 the regiment was presented with one Ordinairfahne of the disbanded IR 'von Pelzchrim' Nr 38; red, white centre, gold decoration, red scroll; the cipher 'FR' in the corners and in the finial; white staff. On 14 Feb 1814, this colour was buried at the battle of Vauchamp and not recovered.

Campaign history. The regiment was formed in 1808 from the 3rd Bn, IR Nr 31 and other units.

1812. The 2nd Bn was mobilised together with the 2nd and Fues-Bns of the 2. Schlesisches IR to form the 6. Kombiniertes IR in von Yorck's corps in Courland, Russia. They fought at Klivenhof, St Annen, Kiopen and Riga. **1813.** Until the armistice, the 2nd Bn fought in von Yorck's I Corps with the 6. Komb.IR. Gross-Goerschen (Incl 2nd Bn) 2 May; Bautzen (Incl 2nd Bn) 20/21 May; Dresden 26/27 Aug (In the 11th Brigade); Kulm 30 Aug; Leipzig 16 – 18 Oct; Erfurt ? Oct. The grenadiers of the regiment were with the Army of Silesia; and fought in the following engagements: Gross-Goerschen; Bautzen; Katzbach 30 Aug; Moeckern 16 Oct; Halle / Saale.

1814. Vauchamps 14 Feb The main body of the regiment, less the grenadiers; in von Ziethen's 11th Brigade of von Kleist's II Corps. Only few of the 11th Brigade, and about half of the 12th Brigade, survived; they were combined into the 10th Brigade. The remnants of the 3rd Musk Bn were combined with those of the 10th Res IR to form the 'Bn von Huellesheim'. Those of the Fues Bn were combined with the fusiliers of the 2. Schlesisches IR and those of the 10th and 11th Res IRs to form the 'Fues Bn von Haas'. The regiment was not reformed until after the end of the war.

The fusiliers were in action at Laon on 9 Mar and took part in the capture of 19 guns and 2 howitzers.

Grenadiers; in von Yorck's I Corps. Chateau-Thierry 12 Feb; after this battle, they combined with the Ostpreussisches Gren Bn. Paris 30 Mar. **1815.** In von Haake's 13th Brigade, von Buelow's IV Corps. Waterloo 18 Jun. The regiment took a colour of the Young Guard, a cannon and a howitzer. Aubervillers 30 Jun.

12. Grenadier-Regiment 'Koenig Friedrich III' (2. Schlesisches) Nr 11.

1st and 2nd Battalions. Raised on 21 Nov 1808; from men of the 3rd Musk Bn, 4th and 5th Bns of the IR 'von Alvensleben' Nr 33 (raised in 1740), 3rd Musk Bn, of the IR 'von Pelzchrim Nr 38 (also raised in 1740), 3rd and 4th Bns, IR 'von Grawert Nr 47 (also raised in 1740) and the Land-Reserve (National-) Bn 'von Glan' (raised in 1807).

Fuesilier Battalion. Raised from the following units which were raised during the 1806/1807 war: Kompanie 'von Richtmeister', 'von der Berswordt', 'von Polczinski', 'von Offency', 'von Sell', 'von Freyberg', 'von Vaersth' and parts of the light companies 'von Clausewitz', 'von Ingenheim', 'von Rekowski' and 'von Stengel'.

Grenadier Battalion. Raised from two companies of the 'Braunen Grenadier-Bataillon von Losthin' (raised on 4 Jun 1807) which, together with the grenadiers of the 1. Schlesisches IR, formed the 'Schlesisches Grenadier-Bataillon'. On 14 Oct 1814 this became the Fuesilier Bn of the later Kaiser Franz Garde-Grenadier-Regiment Nr 2.

Titles. 21 Nov 1808 – **Schlesisches IR Nr 12; 1 Jul 1813 – Schlesisches IR Nr 11.**

Colours. On 9 May 1809 the regiment was presented with four regimental colours of the disbanded IR 'von Alvensleben' Nr 33. They were cornflower blue with white medallions, black eagle, silver decorations and grenades, cornflower blue scroll with silver PRO GLORIA ET PATRIA, the crowned cipher 'FR' in the corners and in the finial, white staff.

On 15 Jun 1815 the Fusilier Battalion received its colour in Paris. It was black, with orange medallions, gold decorations and the cipher 'FWR' in the corners and in the finial.

Campaign history. 1812. The 2nd and Fusilier battalions were mobilised, together with the 1st Bn, 1. Schlesisches IR Nr 10 as the 'Komb. Feld-R Nr 6'. Eckau 19 Jun; Dahlenkirchen 22 Aug; St Annen ? (2nd Bn only); Plehlen 26 Sep (Fus Bn only); Eckau 27 Sep (Fus Bn only); Riopen, Kosakenkrug 29 Sep (the whole regiment was present); Schlockhof 30 Sep; Garossenkrug 1 Oct. 1813. The regiment was in von Yorck's I Corps. Alvensleben 6 Apr; Gross-Goerschen 2 May; the regiment recovered a lost cannon. Colditz 5 May; Koenigswartha 19 May; Bautzen 20/21 May;

Waldau 24 May; Dresden 26/27 Aug; Kulm 30. The Fusiliers took 4 guns. Hellendorf 5 Sep; Wartenburg 15 Sep; Tellnitz 17 Sep (Fus Bn only); Leipzig (Probstheida and Markkleeberg) 16 – 18 Oct; the regiment took 20 guns. Erfurt ? **1814.** Etoges 14 Feb. After this action, the 1st and 2nd Bns were combined with the 3nd Bn, IR Nr 11 into a single battalion. The Fus Bn was combined with the Fuslilers of IR Nr 11 into one battalion. Gue-a-Tremes 28 Feb; La Claves 27/28 Mar; Laon 9/10 Mar; the regiment recovered a lost Prussian gun and took 3 French pieces. Paris 30 Mar. **1815.** The regiment was in von Ryssel's brigade of von Buelow's IV Corps. Waterloo 18 Jun. The regiment was also in action at St Denis and at Aubervilliers on 1 Jul, and at Steins on 29 Jul.

13. Grenadier-Regiment 'Prinz Carl von Preussen (2. Brandenburg'sches) Nr 12. Nickname: the Kolonialregiment. **Page 327, Vol I.**

Raised on 1 Jul 1813. The 1st Bn from the 1st (Res) Bn, Leib-IR; the 2nd Bn from the 2. (Res) Bn, Leib-IR; the Fus Bn was raised from the 3rd Musk Bn of the 1st Westpr. IR.

Titles. 1 Jul 1813 – Brandenburg'sches IR Nr 12; 5 Nov 1816 – **12. IR (2nd Brandenburg'sches).**

Colours. These were ordered on 3 Jun 1814, but not presented until 8 Oct 1815.

Campaign history. 1813. 3rd Musk Bn, 1. Westpr. IR. The regiment was in Huehnerbein's brigade of von Yorck's I Corps.

Gross-Goerschen 2 May; Bautzen 20/21 May; Loewenberg 19/20 Aug; Goldberg 23 Aug (1st Bn only); Katzbach 26 Aug; Bunzlau 30 Aug; Bischofswerda 22 Sep (1st Bn only); Goedau 22 Sep (1st Bn only); Moeckern 16 Oct; the whole regiment was engaged. The regiment took 16 guns. After this action, the two musketier battalions were combined into one; the fusiliers were combined with the 1st Fus Bn of Res IR Nr 12. Freiburg 21 Oct; Hoerselberge 26 Oct. **1814.** Epernay 22 Feb (the regiment less the Fusiliers); Mery-sur-Seine (Fus Bn only); 22 Feb; Gue-a-Tremes 28 Feb; Laon 9 Mar; Paris 30 Mar.

1815. The regiment was in von Steinmetz's 1st Brigade of von Ziethen's I Corps. Gosseliers 25 Jun; Ligny 16 Jun; Waterloo 18 Jun; Issy (Paris) 2/3 Jul.

The New Line Infantry Regiments

Infanterie-Regiment 'Herwarth von Bittenfeld (1. Westfaelisches) Nr 13.

Raised on 1 Jul 1813; 1st Bn – from the 3rd Musketier-Reserve-Bn of the 1.Ostpreussisches IR. 2nd Bn – from the 1. Ostpreussisches Reserve-Bn. Fusilier Bn – from the 1. Littauisches Reserve-Fuesilier-Bn. During 1813 a 4th Bn was raised from men of the 2. Ostpreussisches Res-Musk-Bn. On 3 Jun 1814 it was disbanded and the men distributed among the senior battalions.

Titles. 1 Jul 1813 – 1. Reserve-IR; 25 Mar 1815 – **13. IR.**

Colours. Granted by AKO of 3 Jun 1814; presented on 18 Oct 1815. Black cloth, narrow white corner 'spokes', orange medallions, black eagle, gold decorations, light blue scroll, FWR ciphers in the corner medallions and the finial.

Campaign history. 1813. In Hirschfeld's brigade of von Buelow's III Corps, Army of the North.

Luckau 4 Jun (1. and 2. Ostpr IRs); (1. Litt. IR); Stettin (3. Musk Bn); Koenigsborn 21 Aug (1. and 2. Ostpr IRs); Hagelberg 27 Aug (the whole regiment). The regiment took 4 guns and 7 ammunition wagons. Dessau 12 Oct. In von Tauentzien's brigade. Sieges of Wittenberg and Magdeburg. **1814.** No action. **1815.** In von Ziethen's I Corps. The 1st Bn and the Fusiliers were at the siege of Landau; the 2nd Bn was at the siege of Mainz.

15. Infanterie-Regiment 'Graf Schwerin' (3. Pommersches) Nr 14.

Raised on 1 Jul 1813; 1st Bn – from the 3rd Musk Bn of the 1. Pommersches IR (raised itself on 6 Jul 1811). 2nd Bn – from the 1. Res-Bn, of the same regiment, which was raised on 20 Dec 1812 as the 'Pommersches Res-Bn Nr 3'. The Fus-Bn was raised from the Res-Fues-Bn of the 1.Pommersches IR (which itself was raised on 20 Dec 1812 as the '4. Res-Bn' of the regiment with the title of 'Pomm. Miliz-Bn Nr 6'.

Titles. 1 Jul 1813 – '2. Reserve-IR'; 25 Mar 1813 – **14. IR**; 5 Nov 1816 – **14. IR (3. Pommersches).**

Colours. Awarded by AKO of 3 Jun 1814; presented on 3 Sep 1815 in Givet. Black cloth with narrow white corner 'spokes', orange medallions, gold decoration, light blue scroll, black eagle, FWR ciphers in the corner medallions and in the finial.

Campaigns. 1813. In von Borstell's brigade of von Buelow's III Corps, Army of the North. Damm 15 Apr (3rd Musk Bn and 1. Res Bn); Hoyerswerda 26 Apr (Res Fues Bn only); Gross-Beeren and Schmielkendorf 23 Aug; Thiessen 3 Sep; Woltersdorf 5 Sep; Dennewitz 6 Sep; the regiment took a howitzer. Leipzig 19 Oct. The Fus Bn stormed the outer Grimma gate and took 2 generals, 46 offrs, 2,670 men prisoner and took 28 guns. Neuss 2/3 Dec; the regiment took the eagle of the *150e Ligne*. At a parade in Duesseldorf on 5 Dec 1813, the 1st and Fusilier battalions wore red plumes and the 2nd Battalion wore yellow plumes, which they had captured in the depot in Neuss. **1814.** Hoogstraten 11 Jan; Lier 31 Jan; Oudenarde 5 Mar; Soissons 6 Apr. 1815. In the 7th Brigade, von Pirch's II Corps. Ligny 16 Jun; Namur 20 Jun; Maubeuge 29 Jun – 13 Jul, and Givet 3 Sep.

16. Infanterie-Regiment 'Prinz Heinrich der Niederlaende' (2. Westfaelisches) Nr 15. Raised on 1 Jul 1813.
1st Bn – from the 3rd Ostpreussisches-Musketier-Reserve-Bataillon of the 2nd Ostpreussisches IR.
2nd Bn – from the 4th Ostpreussisches-Musketier-Reserve-Bataillon of the 2nd Ostpreussisches IR.
3rd Bn – from the 2nd Littauisches Fuesilier-Reserve-Bataillon of the 2nd Ostpreussisches IR.

Titles. 1 Jul 1813 – 3. Reserve-IR; 25 Mar 1815 – 15. IR (4. Schlesisches); 11 Jul 1815 – 15. IR 'Graf Buelow von Dennewitz'.

Colours. Awarded by AKO of 3 Jun 1814; presented on 24 Sep 1815 in Vendome. Black cloth with narrow white corner 'spokes', orange medallions, gold decoration, light blue scroll, FWR ciphers in the corner medallions and in the finial.

Campaign history. 1813. The regiment was in von Buelow's III Corps, Army of the North and served at the siege of Stettin, the relief of Pillau and the clash at Luckau early in the year. Blankenfelde 23 Aug; Zahne 5 Sep; Dennewitz 6 Sep; the regiment took a colour of the 7th Wuerttemberg IR. Dessau 12 Oct; Sieges of Torgau 18 Oct – 26 Dec and Wittenberg through Dec 1813. Blockade of Magdeburg up to 14 May 1814. **1815.** In von Hiller's brigade of von Buelow's IV Corps. Waterloo 18 Jun; they took 3 guns and ammunition wagons. At Genappes on 19 Jun they took 80 abandoned guns, 3,000 POW, Napoleon's military chest and a colour each of the *25e* and *29e Ligne*.

17. Infanterie-Regiment 'Freiherr von Sparr' (3. Westfaelisches) Nr 16.

Raised on 1 Jul 1813:
1st Bn from the 3rd Musk-Bn, 3rd Ostpreussisches IR.
2nd Bn from the 1st Res-Bn, 3rd Ostpreussisches IR.
3rd Bn from the 2nd Res-Bn, 3rd Ostpreussisches IR.

Titles. 1 Jul 1813 – 4th Res IR; 25 Mar 1815 – **16th IR**; 5 Nov 1816 – **16th IR (4th Westfaelisches).**

Colours. Awarded by AKO of 3 Jun 1814, 28 Sep 1814 and 15 Jun 1815; presented on 24 Sep 1815 in Alencon-Mortagne. Black cloth with narrow white corner 'spokes'; orange medallions, gold decorations, light blue scroll, FWR ciphers in the corner medallions and in the finial.

Campaigns. 1813. In the Prinz Ludwig von Hessen-Homburg's brigade, von Buelow's III Corps.

Halle/Saale 2 May; the 3rd Musketier Bn took 2 cannon, 1 howitzer and 3 ammunition wagons. Callau 3 Jun; Luckau 4 Jun. Gross-Beeren 23 Aug; the regiment took a gun. Dennewitz 6 Sep; the regiment took a colour of the 2nd Wuerttemberg IR and captured 20 officers and 700 men. Wittenberg 24 Sep; Leipzig 18 Oct; Arnheim 30 Nov; the regiment took a gun. Gorkum; 14 Dec. **1814.** In Zielinski's brigade, III Corps. Gorkum 22 Jan; Soissons 3 Mar; Laon 8-10 Mar; the fusiliers re-took a lost gun. **1815.** Saw no action.

The 1st Westprussian IR storming the villages of Kaja, Rahna, Klein- and Gross-Goerschen on 2 May 1813. A painting by Richard Knoetel Author's collection.

The death of Prinz Louis Ferdinand at Saalfeld on 10 October 1806. Author's collection.

Friedrich Wilhelm Graf von Goetzen, one of the leadir Prussian reformers after 1806. Author's collection.

Below: *Field Marshal von Bluecher at the crossing of th Rhine at Kaub by the Army of Silesia on New Year's Da 1814. The famous castle on the rocks in the middle c the Rhine can be seen in the background. Author collection.*

Prince Friedrich Ludwig von Hohenlohe-Ingelfingen. This general served in the Auatrian army against Prussia in the Seven Years War, then transferred to their service. In 1806 he commanded the left wing of the army; sick and over-influenced by Colonel von Massenbach (his Chief of Staff) he was defeated at Jena and surrendered at the capitulation of Prenzlau on 28 October 1806. After two years as a prisoner of war, he retired. Author's collection.

General Ernst Wilhelm Friedrich von Ruechel; often scapegoated for causing the Prussian defeat at Jena by the late arrival of his corps on the field. Author's collection.

Field Marshal Wichard Joachim Heinrich Graf von Moellendorf. Born in 1724, he made his reputation in all the wars up through 1794. In 1806 he acted in an advisory role to the king at Auerstaedt, where he was wounded. He died in 1816. Author's collection.

General Field Marshal Friedrich Adolf Graf Kalckreuth 1737 - 1818. Joining the Prussian army in 1752, he was a competent officer whose career was stunted by his differences with Prince Heinrich, brother to the king. His reputation was made at the siege of Mainz in 1793. In 1806 he covered the retereat from Auerstaedt, then commanded the defence of Danzig until 26 May 1807. Author's collection.

Major von Gneisenau (1760 – 1831), as young defender of Colberg in 1806. Gneisenau had a tempestuous youth and had to flee into Austrian service from Erfurt in 1779 to escape being put on trial for complicity in a murder during a brawl. In 1782 he left Austrian service under another cloud, having caught his commander trying to rape a girl. He served in British pay in the War of American Independence and entered Prussian service in 1786. At Jena in 1806 he was a staff officer of engineers, but came to fame as the successful defender of Colberg up to 2 July 1807. He was a strong ally of Scharnhorst's reforms and became Bluecher's Chief of Staff after Scharnhorst's being mortally wounded at Luetzen on 2 May 1813. He is famous for having ordered the beaten Prussian army to march to join Wellington at Waterloo when Bluecher was incapacitated by a bad fall from his horse at Ligny. Author's collection.

Napoleon passes the famous Brandenburg Gate in Berlin on 27 October 1806. Author's collection.

Queen (Auguste Wilhelmine Amalie) Louise of Prussia in 1806. Naploeon called her 'the only man in Prussia'. She died in 1810. Author's collection.

Karl Wilhelm Ferdinand Duke of Brunswick, 1735 - 1806; veteran commander of the Prussian army, mortally wounded at Auerstaedt on 14 October 1806. He was married to the daughter of Frederick, Prince of Wales and began military life in 1757 under the Duke of Cumberland and later under his uncle, Frederick of Brunswick. His leberal sentiments were so well known, that he was offered (and declined) the command of the French Revolutionary armies in 1792. His uncertain convictions led to him being scared off by the French at Valmy, after which he resigned. Queen Louisa persuaded him to take up command again in 1806. Author's collection.

Grenadier cap plate, apparently IR Nr 2. Brass plate, red band, buff backing, white tape. Author's collection; Rastatt museum.

Prince Louis Ferdinand of Prussia, killed in action at Saalfeld on 10 October 1806. Author's collection.

Cap of the Mineur-Korps, Seven Years War. Tin front plate, peach pink band and top, white tape and pompon. Museum Fuer Deutsche Geschichte, Berlin; author's collection.

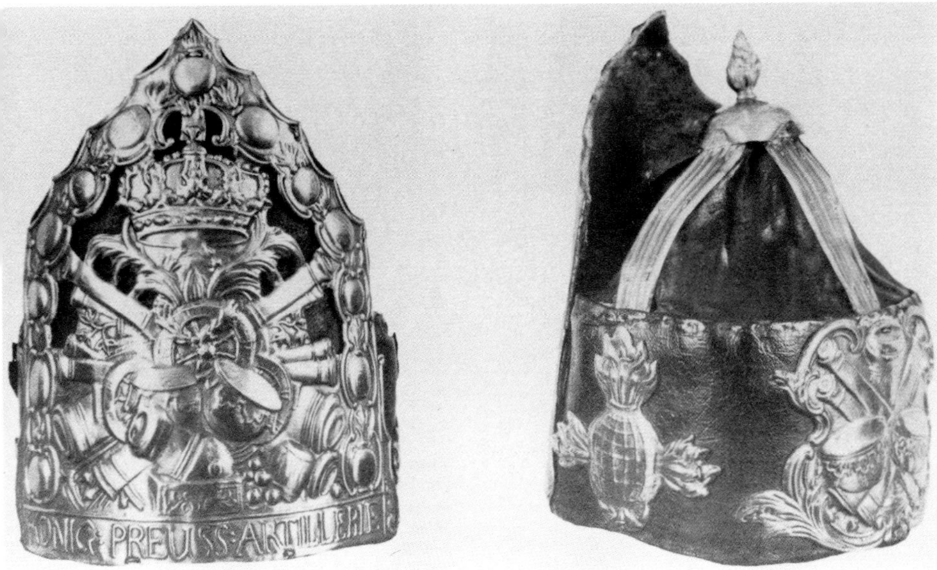

Cap of an artillery bombardier, Seven Years War. Brass front plate and emblems, black band and top. Museum Fuer Deutsche Geschichte, Berlin; author's collection.

The regimental seal of Kuerassier-Regiment von Reitzenstein Nr 7, commemorating the capture of cannon from the French at Pirmasens on 14 September 1793. From the Stammliste of 1806.

Cap of a grenadier, IR Nr 21, Seven Years War. Brass front plate, red band and buff top, white tape with black central stripe, pompon black within white within black within red. Museum Fuer Deutsche Geschichte, Berlin; author's collection.

Cap of a fusilier, IR Nr 49, Seven Years War. Tin front plate, red band, royal blue top, white tape and pompon. Museum Fuer Deutsche Geschichte, Berlin; author's collection.

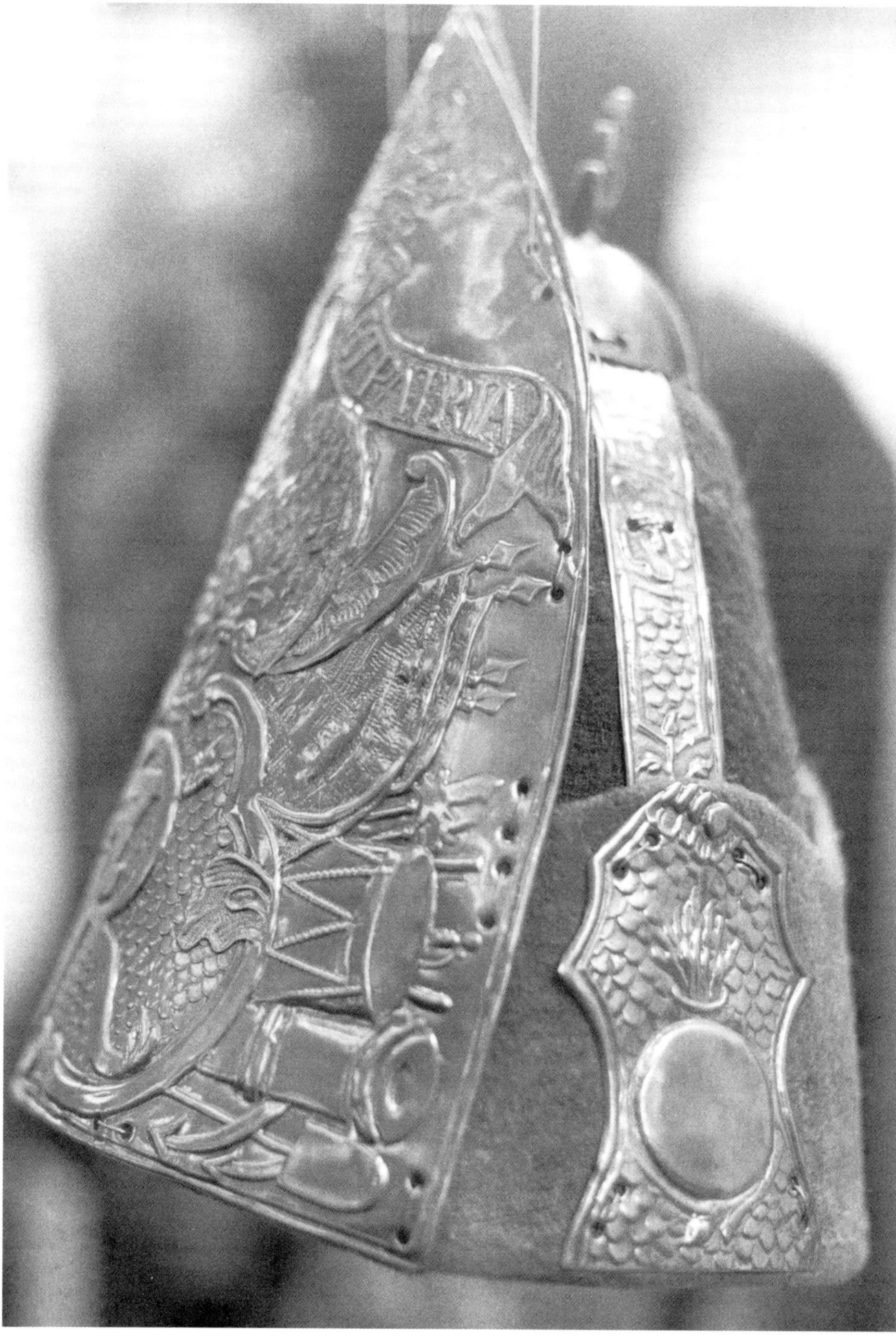

Cap of a fusilier, IR Nr 49, Seven Years War. Tin front plate, red band, royal blue top, white tape and pompon. Museum Fuer Deutsche Geschichte, Berlin; author's collection.

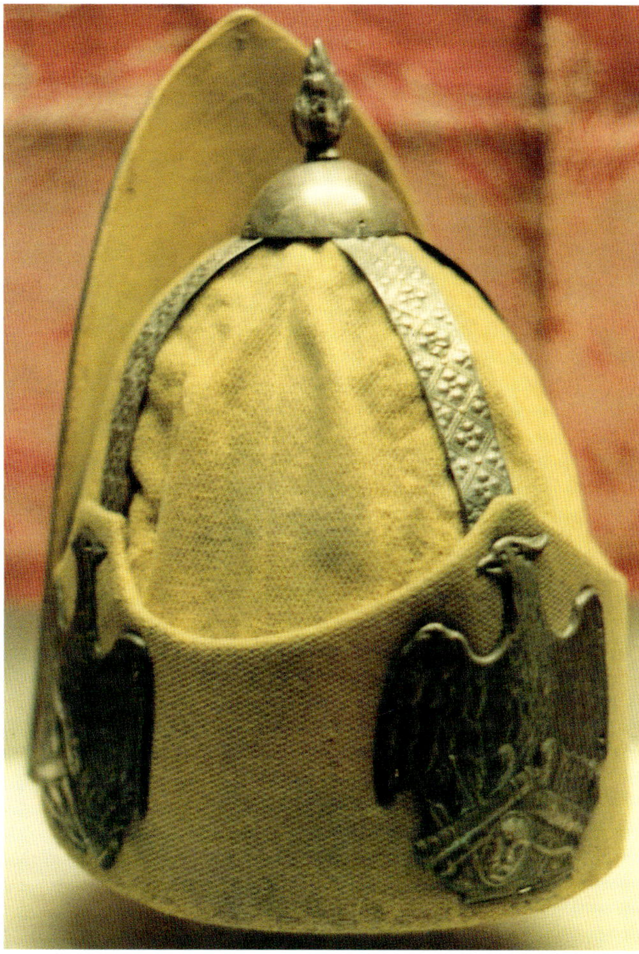

ear view of a fusilier's cap, IR Ir 35, Seven Year's War, Rastatt Auseum, author's collection.

Officer's gorget, Seven Years War, Rastatt Museum, author's collection.

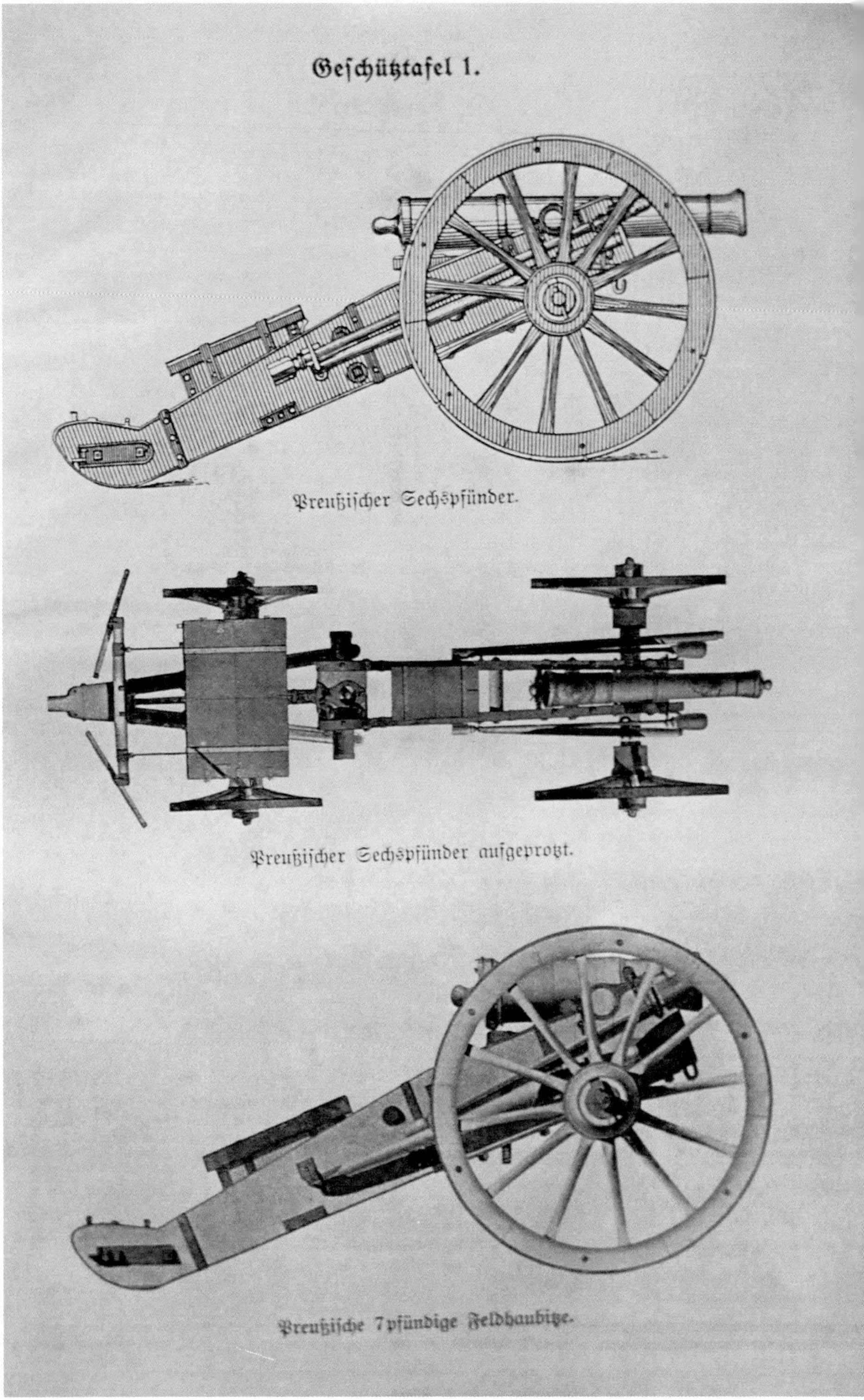

Prussian artillery pieces of the 1813 - 1815 period. Author's collection.

Badges worn on the Kasket from 1788-1799. Top: fusiliers. Bottom row, left to right: grenadier, musketier, artillery. The badges were in the button colour. Kling.

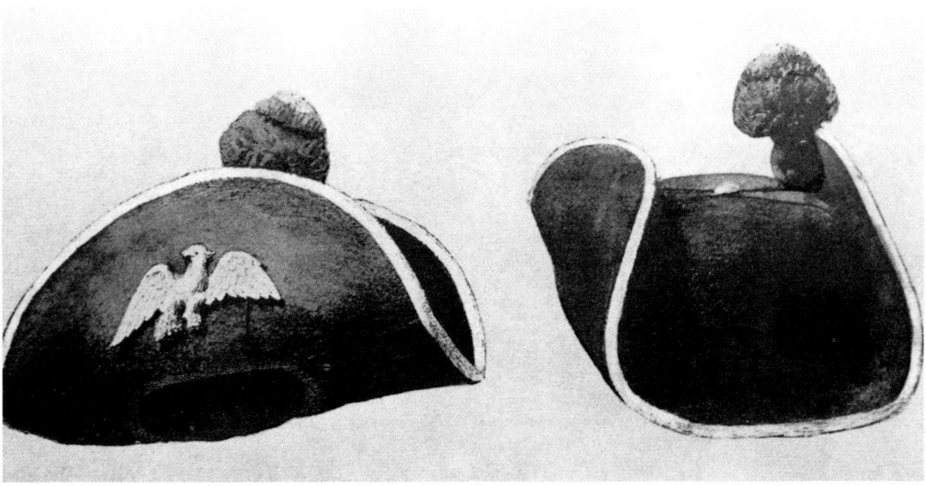

Two views of the Kasket worn from 1788 – 1799. Kling.

Members of the battered Prussian army give thanks for victory after the battle of Leipzig.

Gerhard David von Scharnhorst, the Hanoverian-born officer so active in leading the reforms of the Prussian army after 1806. He is in the uniform worn in early 1813 as Bluecher's Chief of Staff. Author's collection.

The gigantic monument to the fallen in the battle of Leipzig.

Prussian troops storming the outer Grimma gate at Leipzig on 19 October 1813.

Naïve representation of an officer and trooper from the 1st Dragoons (Prittwitz) in 1806 from a series by Ramm. Author's collection.

Naïve representation of an officer and trooper from the 4th Kuerassiers (Wagenfeld) in 1806 from a series by Ramm. Author's collection.

Guidons and standards carried by the 1st Leib-Husaren Nr 5. Top: Regimental guidon carried from 1742 – 1763; black, silver embroidery, red wolf's tooth edging. Centre: two yellow guidons with silver embroidery, taken from the French Hussar Regiment Polleretzky in the clash at Stoecken-Drebben on 23 Feb 1758. Bottom. Two dark blue standards with gold embroidery, taken by the regiment at the battle of Krefeld from the Cuirassier Regiment Bellefonds on 23 Jun 1758. Author's collection.

The regimental seal of 30th Infantry Regiment, von Borke, awarded for recognition of their capture of 20 cannon, 4 mortars, a pair of kettle drums and a colour from the Austrians in the battle of Kesselsdorf in 1745. From the Stammliste of 1806.

The regimental seal of 49th Infantry Regiment, von Mueffling, awarded to commemoration of their capture of 12 cannon from the French in the campaigns of 1793 - 1794. From the Stammliste of 1806.

The regimental seal of the 5th Dragoons 'Koenigin', presented to commemorate their achievements in taking four cannon and 67 colours from the Austrians in the battle of Hohenfriedberg, during the campaign of 1745. From the Stammliste of 1806.

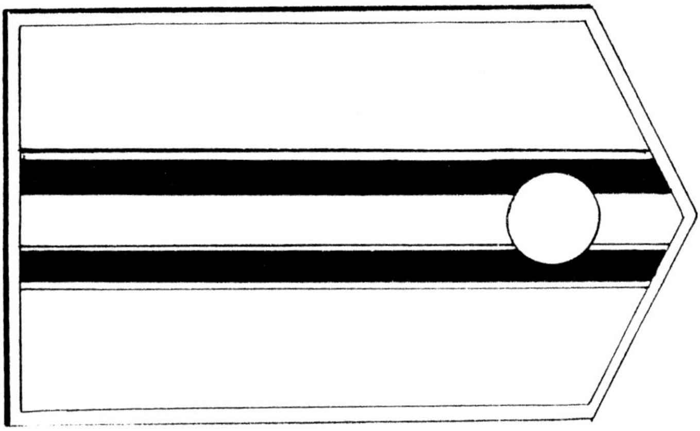

The new Prussian officers' shoulder strap rank badges introduced in 1808. Top to bottom: junior lieutenant, lieutenant and staff captain; centre – captain; right – field officer. The outer piping of the strap is red, the field is in the regimental facing colour and the lace is black and silver.

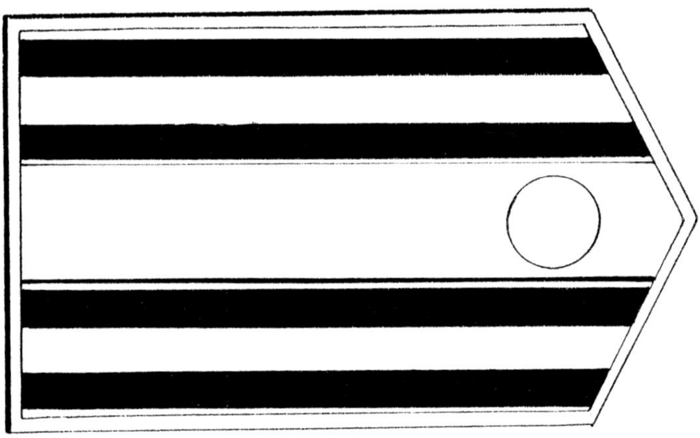

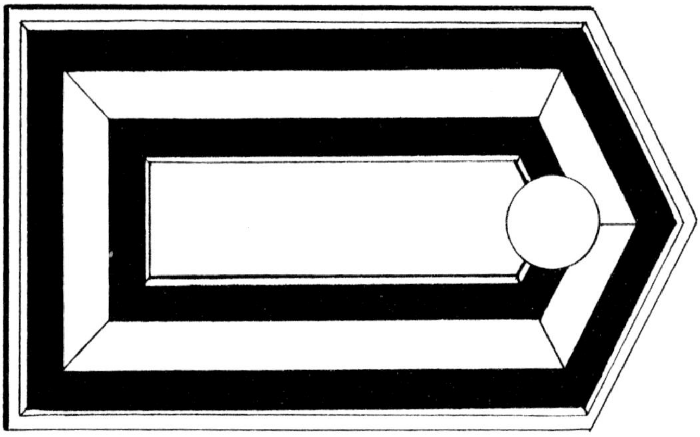

Prussian Heavy Cavalry 1792. Augsburger Bilder; author's collection.

Prussian Light Cavalry 1792; hussars and Bosniaken. Augsburger Bilder; author's collection.

Prussian Line Infantry 1792. Augsburger Bilder; author's collection.

Prussian Light Infantry 1792. Augsburger Bilder; author's collection.

Prussian artillery 1792. Augsburger Bilder; author's collection.

Favrat's Freicorps 1763. Left to right: grenadier officer, grenadier, Jaeger, hussar and dragoon. A Knoetel plate; author's collection.

Gschray's Freicorps. Left to right: dragoon, infantry private, infantry officer. A Knoetel plate; author's collection.

NCOs of the Kuerassier Regiments Wagenfeld (left) and Koenigin 1806. From Kling's great work; author's collection.

Offizier im Parade-Anzug

vom

Königl. Preuß. Infanterie-Regiment von Arnim 1806.

(Nr. 13 der Stammliste).

Das Esponton, der Ringkragen, die silbergestickte Parade-Uniform und das Achselband blieben beim Ausmarsch ins Feld 1806 in der Garnison zurück. (Siehe § 47, 58 und § 59).

Zu Seite 24

Officer in parade dress, IR von Arnim Nr 13, 1806. From Kling's great work; author's collection.

Grenadier and Musketier, IR von Holstein-Beck, Nr 11, 1794. From Kling's great work; author's collection.

18. Infanterie-Regiment 'Graf Barfuss' (4. Westfaelisches) Nr 17.
Raised on 1 Jul 1813;
1st Bn – from the 3rd Musk Bn, 4th Ostpreussisches IR.
2nd Bn – from the 1st Res Bn, 4th Ostpreussisches IR.
3rd Bn (Fues Bn from 1815) – from the 2nd Res Bn, 4th Ostpreussisches IR.
4th Bn(disbanded on 12 Dec 1813) – from the 3rd Lithuanian Res Fus Bn.

Titles. 1 Jul 1813 – 5. Reserve-IR; 25 Mar 1815 – **17. IR**; 5 Nov 1816 – **17. IR (4. Westfaelisches).**

Colours. Authorised by AKOs of 3 Jun 1814, 28 Sep 1814 and 15 Jun 1815; presented on 11 Nov 1815 in Minden. Black cloth, narrow white corner 'spokes', orange medallions, gold decoration, light blue scroll; the cipher 'F WR' in the corner medallions and in the finial.

All units were in von Buelow's III Corps in 1813.

Campaign history. 3rd Musk Bn, 4th Ostpreussisches IR. **1813.** Spandau **18a.** 1-21 Apr; Luckau 4 Jun. **18b. Campaigns. 1st and 2nd Res Bns, 4th Ostpreussisches IR. 1813.** Stettin Mar-Nov, blockade losses unknown. **18c. Campaigns. 3rd Lithuanian Res Fus Bn.**

1813. Luckau 4 Jun; Wilmersdorf, Wittstock and Gross-Beeren 21-23 Aug; Musketier Berdick, of the 1st Bn, took 4 guns, but men of other regiments stole them from him. Dennewitz 6 Sep; the regiment took a gun. Coswig (Wittenberg) 12 Oct; Arnheim 30 Nov; the regiment took 3 guns. **1814.** Hoogstraten and Merxem 11-13 Jan; Doerne 1 Feb; Laon 9 Mar; Fere-Champenoise 25 Mar. allied infantry scarcely came into action here this day. The regiment took two caronnades that then stood in front of the Zeughaus in Berlin until 1945. The regiment was not mobilised in 1815.

19. Infanterie-Regiment 'von Grolmann' (1. Posensches) Nr 18.
Raised on 1 Jul 1813: 1st Bn – from the 1st Reserve-Bn, 1st Westpreussisches IR.
2nd Bn – from the 4th Reserve-Bn, 1st Schlesisches IR.
Fus Bn – from the 2nd Reserve-Bn, 1st Westpreussisches IR.

Titles. 1 Jul 1813 – 6. Reserve-IR; 25 Mar 1815 – **18. IR (1.Westfaelisches)**; 5 Nov 1816 – **18. IR (3. Westpreussisches)**; 10 Mar 1823 – **18. IR**; 4 Jul 1860 – **1. Posensches IR (Nr 18).**

Colours. Awarded by AKOs of 3 Jun 1814, 28 Sep 1814 and 15 Jun 1815; presented on 24 Sep 1815.

Black cloth, narrow white corner 'spokes', orange medallions, gold decoration, light blue scroll, 'FWR' ciphers in the corner medallions and in the finial.

Campaign history. 1813. 1st and 2nd Reserve-Bns. Siege of Glogau total loss 2 offrs, 17 ORs. 2nd Res Bn – Neukirch total loss 41 men. When the regiment was formed on 1 Jul, it went to von Kluex's brigade in von Kleist's II Corps. Dresden 26/27 Aug and Kulm 30 Aug; Leipzig 16 Oct; Leipzig 18 Oct; Erfurt ? The 2nd Bn then stayed in Erfurt until 16 May 1814.

1814. Claye-Soully 28 Mar; Paris 30 Mar; the regiment took 2 guns and a howitzer. **1815.** In von Losthin's 15th Brigade, von Buelow's IV Corps. Waterloo 18 Jun; the regiment had the highest losses of any Prussian regiment engaged here this day.

20. Infanterie-Regiment 'von Courbiere' (2. Posensches) Nr 19.
Raised on 1 Jul 1813: 1st Bn from the 3rd Musk Bn, 2nd Westpreussisches IR.
2nd Bn from the 1st Res Bn, 2nd Westpreussisches IR.
Fus Bn from the 2nd Res Bn, 2nd Westpreussisches IR.

Titles. 1 Jul 1813 – 7. Res IR; 25 Mar 1815 – **19. IR (4. Westpreussisches)**; 5 Nov 1816 – **19. IR (4. Westpreussisches)**.

Colours. Awarded by AKOs of 3 Jun and 28 Sep 1814 and 15 Jun 1815; presented on 7 Oct 1815

Black cloth, narrow white corner 'spokes', orange medallions, gold decoration, light blue scroll, 'FWR' ciphers in the corner medallions and in the finial.

Campaign history. 1813. Gross-Goerschen 2 May (3rd Musk Bn, 2. Westpr IR only); Bautzen 20/21 May (3rd Musk Bn, 2. Westpr IR only). In July the regiment went to von Kleist's II Corps; Dresden 26/27 Aug; Kulm 30 Aug (1st and 2nd Bns only); Hellendorf 8 Sep and Pirna 9 Sep. From 25 Sep the 2nd Bn was on lines of communication policing duties. Leipzig 16 – 19 Oct (1st and Fus Bns only); Erfurt ? The 1st and 2nd Res Bns were at the siege of Glogau late in the year; losses unknown.

1814. The regiment was in von Pirch I's brigade of the II Corps. Sarrechamps 14 Feb The 3rd Musk Bn was destroyed in defence of this farm, losing 6 offrs and 180 men killed. Vauchamps 14 Feb. The 1st Bn had to cover the allied withdrawal; only 80 escaped. Following this disaster, all Kleist's infantry was combined into one weak brigade. Beauval 28 Feb; Laon 9/10 Mar. After this action, the 2nd Bn was combined with the 2nd Bn, 2nd Westpr. IR, until after the battle of Paris. **1815.** In von Ziethen's I Corps. Ligny 16 Jun (1st and 3rd Musk Bns only); Wavre 18 Jun (1st and 3rd Musk Bns only).

21. Infanterie-Regiment Graf Tauentzien von Wittenberg (3. Brandenburgisches) Nr 20.

Raised on 1 Jul 1813:

1st Bn - from the 2. Res Bn, 1. Pommersches IR.

2nd Bn - from the 3. Res Bn, 1. Pommersches IR.

Fus Bn - from the 3. Res Bn, Colberg'sches IR.

Titles. 1 Jul 1813 – 8. Reserve-IR. 25 Mar 1815 – **20. IR**. 5 Nov 1816 – **20. IR (3. Brandenburg'sches)**.

Colours. Awarded by AKOs of 3 Jun and 28 Sep 1814 and 15 Jun 1815; presented on 28 Aug 1816

Black cloth, narrow white corner 'spokes', orange medallions, gold decoration, light blue scroll, 'FWR' ciphers in the corner medallions and in the finial.

Campaign history. 1813. The parent units were in von Tauentzien's IV Corps and at the sieges of Stettin, Wittenberg and Magdeburg; losses unknown. **1815.** In von Thielmann's III Corps at Landau and Bitsch; losses unknown.

22. Infanterie-Regiment 'von Borcke' (4. Pommersches) Nr 21.
Raised on 1 Jul 1813:
1st Bn – from the 1. Reserve Bn, Colberg'sches IR.
2nd Bn – from the 2. Reserve Bn, Colberg'sches IR.
3rd Bn – from the 4. Reserve Bn, Colberg'sches IR.

Titles. 1 Jul 1813 – 9. Reserve IR; 25 Mar 1815 – **21.IR**; 5 Nov 1816 – **21. IR (4. Pommersches)**.

Colours. Awarded by AKOs of 3 Jun and 28 Sep 1814 and 15 Jun 1815; presented on ?

Black cloth, narrow white corner 'spokes', orange medallions, gold decoration, light blue scroll, 'FWR' ciphers in the corner medallions and in the finial.

Campaign history. 1813. The Colberg IR was at the siege of Stettin. On 1 July the new regiment went to von Buelow's III Corps, Army of the North. Gross-Beeren 23 Aug; Dennewitz 6 Sep; the regiment took 4 guns and a howitzer, but could not bring the 4 guns off the field. Wittenberg ? Leipzig 18 Oct (1st Bn only); Arnheim 30 Nov; the regiment took 2 guns. Port of St Andreas 14 Dec (Fusiliers only). 1814. Antwerp 11-13 Jan (1st and Fusilier Bns only). 's-Hertogenbosch 26 Jan (Fusiliers only); Laon 9 Mar (1st and Fusilier Bns only). Soissons 16 – 30 Mar; Compiegne 1 Apr. **1815**. The regiment was in von Langen's 8th Brigade in von Pirch I's II Corps. Ligny 16 Jun; Wavre 18 Jun.

23. Infanterie-Regiment 'Keith' (1. Oberschlesisches) Nr 22.
Raised on 1 Jul 1813:
1st Bn – from the 1st Reserve Bn, 1st Schlesisches IR (this unit was the 2nd Bn until 23 May 1815).
2nd Bn – from the 2nd Reserve Bn, 1st Schlesisches IR (this unit was the Fus Bn until 23 May 1815).
Fus Bn – from the 3rd Musk Bn, 1st Schlesisches IR (this unit was the 1st Bn until 23 May 1815).

Titles. 1 Jul 1813 – 10. Reserve IR; 25 Mar 1815 – **22. IR (1. Rheinisches)**; 5 Nov 1816 – **22. IR (3. Schlesisches)**; 4 Jul 1860 – **1. Oberschlesisches IR (Nr 22).**

Colours. Awarded by AKOs of 3 Jun and 28 Sep 1814 and 15 Jun 1815; presented on 3 Oct 1815 in Givet. Black cloth, narrow white corner 'spokes', orange medallions, gold decoration, light blue scroll, 'FWR' ciphers in the corner medallions and in the finial.

Campaign history. 1813. The 1st and 2nd Reserve Bns were at the siege of Glogau; Gross-Goerschen 2 May; Bautzen 20/21 May. The regiment was in von Jagow's 11th Brigade of von Kleist's II Corps. Dresden 26/27 Aug; Kulm 30 Aug; the regiment took a French colour, 8 guns and 1,200 prisoners. Dohna and Pirna 8 Sep (1st Bn only); Leipzig 16 – 18 Oct; Erfurt ? (2nd and 3rd Bns only). **1814.** Vauchamps 14 Feb; the 1st and 2nd Bns were destroyed; the 11th Brigade was reorganised into one weak battalion. It fought at Laon, Athies, Pontavert, Fismes, Claye-Soully and Paris; losses unknown. **1815.** In the 8th Brigade, von Pirch I's II Corps. Ligny 16 Jun; Namur 20 Jun. The regiment was also present at the sieges of Maubeuge, Landrecies, Rocroy ands Givet; losses not known.

24. Infanterie-Regiment 'von Winterfeldt'
(2. Oberschlesisches) Nr 23.

Raised on 1 Jul 1813:
1st Bn – from the 1st Reserve Bn, 2nd Schlesisches IR (this unit was the 2nd Bn until 3 Mar 1815).
2nd Bn – from the 2nd Reserve Bn, 2nd Schlesisches IR (this unit was the Fus Bn until 3 Mar 1815).
Fus Bn – from the 3rd Musk Bn, 2nd Schlesisches IR (this unit was the 1st Bn until 3 Mar 1815).

Titles. 1 Jul 1813 – 11. Reserve IR; 25 Mar 1815 – **23. IR (2. Rheinisches)**; 5 Nov 1816 – **23. IR (4. Schlesisches)**; 10 Mar 1823 – **23. IR**; 4 Jul 1860 – **2. Oberschlesisches IR (Nr 23).**

Colours. Awarded by AKOs of 3 Jun and 28 Sep 1814 and 15 Jun 1815; presented on 3 Sep 1815 in Paris. Black cloth, narrow white corner 'spokes', orange medallions, gold decoration, light blue scroll, 'FWR' ciphers in the corner medallions and in the finial.

Campaign history. 1813. The 2. Schlesisches IR was at the siege of Glogau in the corps of General Schuler von Senden.

Gross-Goerschen 2 May (3. Musk Bn, with von Yorck's I Corps). Bautzen 20/21 May; Dresden 26/27 Aug. In 12th Brigade, von Kleist's II Corps. Kulm 30 Aug; the regiment captured two French battalions with 1,586 prisoners. Hellendorf 5 Sep; Heldenau 6 Sep; Hellendorf 15 Sep (1. and 2. Bns only); Leipzig 16 – 18 Oct. Together with the 2.Westpreussisches IR, the regiment took 2 guns.

1814. Etoges 14 Feb. Due to the devastating losses in other regiments, the 2nd Bn was combined with the 1st and 2nd Bns of the 2. Schlesisches IR to form the 'Bataillon Graf Reichenbach'. The Fus Bn was combined with that of the 2. Schlesisches IR to form the 'Fus Bn von Haas'. As of 1 Feb, the 1st Bn was in Vitry as garrison. Gue-a-Tremes 28 Feb (both the combined battalions); Laon 9 Mar (Fus Bn only); the battalion took 22 guns. Claye-Soully 28 Mar (Bn von Reichenbach only); Paris 30 Mar; both battalions were present. **1815.** In von Lungen's 8th Brigade of von Pirch I's II Corps. Ligny 16; Wavre 18 Jun; Namur 20 Jun; Philippeville 22 Jul – 9 Aug; Givet 22 Aug – 22 Sep.

25. Infanterie-Regiment 'Grossherzog Friedrich Franz II von Mecklenburg-Schwerin' (4. Brandenburg'sches) Nr 24.

Raised on 1 Jul 1813 from:

1st Bn from the 4th Reserve Bn of the Leib-IR.

2nd Bn from the 5th Reserve Bn of the Leib-IR (this later became the Fusilier Bn).

Fusilier Bn from the 3rd Reserve Bn of the 3rd Ostpreussisches IR (this later became the 2. Bn).

Titles. 1 Jul 1813 – 12. Reserve IR; 25 Mar 1815 – **24. IR**; 5 Nov 1816 – **24. IR (4. Brandenburg'sches).**

Colours. Awarded by AKOs of 3 Jun and 28 Sep 1814 and 15 Jun 1815; presented on 3 Sep 1815 in Paris. Black cloth, narrow white corner 'spokes', orange medallions, gold decoration, light blue scroll, 'FWR' ciphers in the corner medallions and in the finial.

Campaign history. 1813. The 4. Res Bn of the Leib IR fought at Luckau on 3 Jun; they were in von Huehnerbein's brigade on von Yorck's I Corps.

Goldberg (1. Bn only); Katzbach 30 Sep; Hochkirch 4 Sep (Fus Bn only); Lang-Wolnsdorf 15 Sep (2. Bn only); Gross-Hartla 20 Sep (2. Bn only); Bischofswerda 21 Sep (2. Bn only); Moeckern 16 Oct; the regiment took 3 guns. After this action, the 1.Bn (only 170 ORs strong) was combined with the Fus Bn, 2. Brandenburg'sches IR; the 2nd Bn (126 ORs) was combined with the Fus Bn of the regiment (253 ORs). There were only 17 Fr-Jg left. Freiburg 21 Oct; Hoerselberge 26 Oct. **1814.** St Johann 7 Jan (2. Bn only); St Avold 10 Jan (2. Bn only); they took a gun. Mery-sur-Seine? 21 Jan (Fus Bn only); Gue-a-Tremes 28 Feb; Laon 9 Mar; they took 3 guns. Paris 30 Mar.

1815. In von Steimnetz's brigade, I Corps. During the campaign, 194 recruits from the Rheinland deserted. Ligny 16 Jun; Waterloo 18 Jun; Sevres 2 Jul; Issy 2 Jul.

26. Infanterie-Regiment 'von Luetzow' (1. Rheinisches) Nr 25.
Raised on 18 Feb 1813 as the 'Royal Prussian Frei-Corps'.

Titles. 18 Feb 1813 – Koeniglich Preussisches Freikorps; 19 Jan 1814 – **I R von Luetzow**; 25 Mar 1815 – **25. IR**; 4 Jul 1860 – **Rheinisches IR (Nr 25).**

Colours. Awarded by AKOs of 3 Jun and 28 Sep 1814, 15 Jun 1815 and 21 Jan 1816; presented on 21 Apr 1816 in Erfurt. The presentation had been delayed by an inquiry into the conduct of the regiment at Ligny. Black cloth, narrow white corner 'spokes', orange medallions, gold decoration, light blue scroll, 'FWR' ciphers in the corner medallions and in the finial.

Campaign history. 1813. The corps was in von Wallmoden's commandand attached to von Buelow's III Corps, Army of the North. A girl, Eleonora Prochaska, served in the Freikorps as 'Jaeger Renz'; she was killed at the Goehrde on 16 Sep. Lauenburg 17 – 19 Aug ; Gadebusch 20 Aug Lt Theodor Koerner and Volunteer Graf Hardegg were killed here. Guedow, Moelln 4/5 Sep; Boitzenburg 16 Sep; Goehrde 16 Sep; Zarentin 18 Sep; Weisser Hirsch; Hohendorf 10 Oct; Bremen 12 – 15 Oct. **1814.** Juelich 21 Mar; Ardennes 16 May? **1815.** In von Pirch I's II Corps; 330 of the men were Rheinlanders from the Duchy of Cleve (ex Confederation of the Rhine) and of very poor loyalty. Ligny 16 Jun; Waterloo 18

Jun. The Fusiliers took part in the pursuit to Genappes, together with the Fus Bn, IR Nr 15 and took a case of Napoleon's medals. Maubeuge 26 Jun; Marienbourg 30 Jun – 7 Aug.

27. Infanterie-Regiment 'Fuerst Leopold von Anhalt-Dessau' (1. Magdeburg'sches) Nr 26.

Raised:

1st Bn – 12 Mar 1813 as the 'Auslaender-Bn 'Major von Reuss';formed of released POWs of 1812 and Westfalians; taken into the Prussian army on 5 Jul 1813.

2nd Bn – 26 Apr 1813 from volunteers, POW and deserters of old west Prussian provinces.

Fus Bn – 21 Jun 1813 from volunteers and deserters of old west Prussian provinces.

Titles. 12 Mar – **Auslaender-Bn 'von Reuss'**; 5 Jul 1813 – **Elb-IR**; 25 Mar 1815 – **26. IR**; 5 Nov 1816 – **26. IR (1. Magdeburg'sches)**.

Colours. Awarded by AKOs of 3 Jun and 28 Sep 1814, 15 Jun 1815; presented on 8 Oct 1815 in Vervet. Black cloth, narrow white corner 'spokes', orange medallions, gold decoration, light blue scroll, 'FWR' ciphers in the corner medallions and in the finial.

Campaign history. 1813. Initially only the 1. Bn, in von Boyen's brigade, von Buelow's III Corps, Army of the North.

On 26 Apr the 2nd Bn was formed almost exclusively of POW.

After the end of the Armistice, the regiment went to von Thuemen's 4th Brigade. Wietstock 22 Aug Gross-Beeren 23 Aug (2.Bn only); Luebnitz, Hagelberg 27 Aug (1.Bn only, in von Hirschfeld's division); Dennewitz 6 Sep; they shared in the capture of a gun. Wittenberg 13 Sep – 12 Oct. Until the end of Jan 1814 the regiment stayed in garrison in Minden and Muenster. The Fr-Jg detachment was attached to the Fusilier Bn. **1814.** In the corps of the Duke of Weimar. Antwerp 14 Feb – 16 Apr.

1815. In von Krafft's 6th Brigade, von Pirch I's II Corps. Ligny 16 Jun; Namur 20 Jun; the regiment took 2 guns. The regiment was also at the sieges of Landrecies, Rocroy, Philippeville and Givet.

28. Infanterie-Regiment 'Prinz Louis-Ferdinand von Preussen' (2. Magdeburg'sches) Nr 27.

Raised on 7 Mar 1815:

1st Bn - from the the 1. Schlesisches Ersatzbataillon

2nd Bn – from the 9th and 10th Companies of the Auslaendische Jaeger-Bn 'von Reiche', of

the 11th and 12th Companies of von Hellwig's Frei-Korps, from the Reserve-Bn of the Elb-

IR.

Titles. 7 Mar 1815 – 27. IR; 5 Nov 1816 – **2. Magdeburg'sches IR (Nr 27)**; 10 Mar 1823 – **27. IR**; 4 Jul 1860 – **2. Magdeburg'sches IR (Nr 27).**

Colours. Awarded by AKO of 3 Oct 1815; black cloth, narrow white corner 'spokes', orange medallions, gold decoration, light blue scroll, 'FWR' ciphers in the corner medallions and in the finial.

28a. Campaigns of the Auslaender Jaeger-Bataillon 'von Reiche'.

1813. In General Graf von Wallmoden's Corps. Vellahn and Ortskrug 20 Aug; Moelln 4 Sep; Buechen 5 Sep; Ghoerde 16 Sep; Hohensdorf 23 Sep; Winsen an der Luhe 26 and 30 Sep; Bremen 13 Oct; Rothenburg 13 Oct. On I Nov the battalion was transferred to von Buelow's Corps. Zuetphen 24 Nov; Aalst 14 Dec; Hedell and St Andree 15 Dec. 1814. **Hoogstraten 11 Jan; Gindrich and Herungen 12 Feb and 1 Apr; Tegels 9 Apr.**

28 b Campaigns of von Hellwig's Freikorps.

Strength at the start of the campaign – 550 all ranks in three companies plus a Fr-Jg detachment.

They served in von Buelow's III Corps, Army of the North. They fought at Hoyerswerda,

Schoenweide, Holzdorf, Wartenburg, Soemmerda, Wustwetzel, Lonhaut, Peer, Lovenjaut, Lessines, Ypres, Menin, Oudenarde, Harlebeck, Menin, Belleghem and Courtrai. Their total combat losses were 5 offrs and 220 ORs.

28 c Campaigns of the Reserve Bn, Elb-IR (as the 1. Westfaelisches LIR).

The regiment was organized into two battalions each of one grenadier, one voltigeur and four fusilier companies in the usual French manner, with a strength of 2,270 all ranks.

1809. The regiment was engaged in the chase after von Schill's renegade regiment.

Dodendorf 5 May The regiment was destroyed; 14 offrs and 400 ORs were captured together with both colours. Delmenhorst 6 Sep Fighting the Duke of Brunswick's Black Corps. Losses not known.

1812. Defence of Danzig; total loss – 16 offrs, 488 ORs. The regiment entered Prussian service as the Reserve Bn, Elb-IR, with 14 offrs and 400 ORs. **1815.** In the 12th Brigade of von Thielmann's III Corps. Ligny 16 Jun; Wavre 18 Jun; Saarlouis 13 Nov.

29. Infanterie-Regiment 'von Goeben' (2. Rheinisches) Nr 28.

Raised on 5 Dec 1813 from men of the old 1. IR of the ex-Grand Duchy of Berg.

Titles. 5 Dec 1813 – 28. IR; 5 Nov 1816 – **28. IR (2. Rheinisches).**

Colours. Awarded by AKOs of 3 Oct and 12 Dec 1815; presented ? Black cloth, narrow white corner 'spokes', orange medallions, gold decoration, light blue scroll, 'FWR' ciphers in the corner medallions and in the finial.

Campaign history. 1815. The regiment served in von Pirch II's 2. Brigade of von Ziethen's I Corps; march-out strength – 78 offrs, 180 NCOs, 2,232 R&F, plus a Fr-Jg detachment of ca 200 men.

Couillet 15 Jun total losses were 15 offrs, 680 ORs. The fusilier battalion was almost destroyed by a cavalry charge; it was then combined with the fusilier battalion of the 2. Westfalian LWR.

Ligny 16 Jun; after this action the two musketier battalions were combined into one. Issy 3 Jul.

30. Infanterie-Regiment 'von Horn' (3. Rheinisches) Nr 29.

Raised on 5 Dec 1813

1. Bn – from the Grenadier-Bn of the ex-Grand Duchy of Berg.
2. and Fusilier Bns – from the 2. LIR of the ex-Grand Duchy of Berg.

Titles. 5 Dec 1813 – 29. IR; 5 Nov 1816 – **29. IR (3. Rheinisches).**

Colours. Awarded by AKOs of 3 Oct and 12 Dec 1815; presented on 14 Apr 1816 in Koblenz. Black cloth, narrow white corner 'spokes', orange medallions, gold decoration, light blue scroll, 'FWR' ciphers in the corner medallions and in the finial.

Campaign history. 1815. In von Jagow's 3rd Brigade of von Ziethen's I Corps. Gosselies 15 Jun; Ligny 16 Jun.

31. Infanterie-Regiment 'Graf Werder' (4. Rheinisches) Nr 30.
Raised on 6 Sep 1812 as part of the **Russian-German Legion**. On 25 Mar 1815 it became the **30. IR**.

At this time, it was formed of the 1st, 2nd and 7th Bns of the Legion. The 1st Bn consisted mainly of Prussian POWs, the 2nd and 7th of Saxons, Coburgers and Dutch. They totalled 49 offrs, 110 NCOs and 365 R&F. Later the regiment was reinforced by the 1st, 4th, 5th, 10th, 11th and 13th Ersatz-Bns.

Titles. 5 Nov 1816 – **30. IR (4. Rheinisches).**

Colours. Awarded by AKOs of 3 Oct and 12 Dec 1815; presented on 9 May 1816 in Danzig. Black cloth, narrow white corner 'spokes', orange medallions, gold decoration, light blue scroll, 'FWR' ciphers in the corner medallions and in the finial.

Campaign history. 1815. In von Ziethen's I Corps; march-out strength – 62 offrs, 187 NCOs, each battalion had a Fr-Jg detachment of 6 NCOs and 140 R&F, which only joined the regiment in July of the year.

Ligny 16 Jun; Wavre 18 Jun.

32. Infanterie-Regiment 'Graf Bose' (1. Thuering'sches) Nr 31.
Raised on 6 Sep 1812 from the 2.IR of the Russo-German Legion:
1st Bn – from the 3.Bn; (Confederation of the Rhine).
2nd Bn – from the 4.Bn; (Confederation of the Rhine).
Fus Bn - from the 6. Bn (Saxon IR 'Prinz Maximilien).
In 1815 the regiment was reinforced by NCOs and men of the disbanded 2., 4., 6., 8., 9., 11. and 12. Ersatz-Bataillone.

Titles. 6 Sep 1812 – 31. IR; 5 Nov 1816 – 31. IR (3. Magdeburg'sches); 4 Jul 1860 – **1. Thuering'sches IR (Nr 31).**

Colours. Awarded by AKOs of 3 Oct and 12 Dec 1815; presented on 21 Apr 1816 in Erfurt. Black cloth, narrow white corner 'spokes', orange medallions, gold decoration, light blue scroll, 'FWR' ciphers in the corner medallions and in the finial.

Campaign History. 1813. In the corps of General Graf Wallmoden-Gimborn. Camin-Vellahn 21 Aug; Boden 4 Dec; Sehestadt. **1814.** Strength on 1 Mar 1814 was 1,749 NCOs and men; after the regiment entered Holland, most of its Dutch members deserted. Neuland 9 Feb. **1815.** In von Thielmann's III Corps. Ligny 16 Jun; Wavre 19 Jun.

33. 2. Thuering'sches Infanterie-Regiment Nr 32.
Raised on 7 Mar 1815 from:
1st Bn – Elb-IR.
2nd Bn – Westfalian LW.
3rd Bn – Saxon LW.

Titles. 7 Mar 1815 – **32. IR**; 5 Nov 1816 – **32. IR (4. Magdeburisches)**; 4 Jul 1860 – **2. Thuering'sches IR (Nr 32).**

Colours. Awarded by AKOs of 26 Jun 1825; presented on? Black cloth, wide white corner 'spokes', orange medallions, gold decoration, light blue scroll, 'FWR' ciphers in the corner medallions and in the finial.

Campaigns. None until 1830.

34. Fuesilier-Regiment 'Graf Roon' (Ostpreussisches) Nr 33. Raised on 1 Oct 1815 in ex-Swedish Pommerania from the Swedish regiments 'Koeniglich Schwedisches Regiment', 'Leib-Regiment der Koenigin' and the 'Regiment von Engelbrechten'.

Campaigns of the Leib-Regiment der Koenigin.
1735 – 1736: War of the Polish Succession. 1757 – 1762: Seven Years War.

1788 – 1790: The Finnish War. 1790: on the fleet as marines. 1805 – 1806: one combined battalion in the field against Prussia. 1806 – 1807: War against France. 1808: on the fleet as marines against Russia. 1809: against Russia; two battalions. March-out strength: 19 offrs, 31 NCOs. 488 R&F. campaign losses – 9 offrs, 282 ORs. 1812: the regiment was disarmed and interned by the French in Stralsund. 1813: the regiment was re-raised at a strength of one battalion and fought against the French in the Army of the North under Bernadotte (Crown Prince of Sweden). Present at, but not engaged in the battles of Gross-Beeren and Dennewitz; present at Leipzig on 18 Oct. Fought at Dessau.

1814: present at the sieges of Juelich and Maastricht.

Campaigns of the Regiment von Engelbrechten.
1757 – 1762: Seven Years War; one combined battalion. 1788 – 1790; Finnish War. 1790: on the shallows fleet as marine infantry. Fought in the sea battles of Reval, Friedrichshamm, Kaskila, Vli-Urpala, Kronstadt, Koivisto, Wyborg bay, Svenska Sound. Total campaign loss: 719 NCOs and men killed and wounded; POW 5 offrs, 4 NCOs, 149 R&F of whom one third died. 1806 – 1807: war against France. 1809: war against Russia. Original strength: 19 offrs, 32 NCOs, 456 R&F. The regiment fought at Djeknaboda, Soefvar, Ytterboda, Ratan. Total campaign losses: 5 offrs, 6 NCOs, 131 R&F. 1812: the regiment was disarmed and interned by the French in Stralsund. 1813: the regiment was re-raised at a strength of one battalion and fought against the French in the Army of the North under Bernadotte (Crown Prince of Sweden). Present at, but not engaged in, the battles of Gross-Beeren and Dennewitz; present at Leipzig on 18 Oct.

1814: present at the sieges of Juelich and Maastricht.

9

The Freikorps

The Freikorps of the Seven Years War

So desperate was the Prussian king for light infantry and light cavalry units to counter the excellent light troops of the Austrians, that he turned to the device of raising *Freikorps*. In effect he sold patents for inventive adventurers to recruit forces outside the normal military establishment for money. These formations were subject to their own disciplinary codes and attracted soldiers of fortune from many different states. The commander of each designed their own uniforms and laid down which weapons were to be carried. The Prussian state funded them.

These units sometimes took the form of a 'legion' having infantry and cavalry components under the same commander. Many of those attracted to these regiments were old hands at military life, rootless nomads who often had served in, and deserted from, several different armies. Many recruited from the Austrian, French and Imperial prisoners of war. Their reputations were frequently blackened by outrages against the Prussian civil population which they had been hired to protect. The combat performance and reliability of these troops was 'patchy' at best.

The best of the bunch was the *Freikorps* raised by Friedrich Wilhelm von Kleist (known as 'Green Kleist' from their uniforms.

Three such commissions were given on 18 August 1756 to a Dutch Colonel d'Angelelli, a Saxon Lieutenant-Colonel Mayr and a Lieutenant-Colonel le Noble from the Palatinate. These were shortly followed by a Lieutenant-Colonel Kalben. In all, there were twenty-three such formations raised.

Unless otherwise stated, all units wore dark blue coats with light blue facings, waistcoats and breeches.

Freibataillon Nr 1. Raised in 1756; title Le Noble. Actions: Landshut (captured); exchanged and increased to two battalions; fought at B Freiberg 29 Oct 1762. Uniform: dark green coat, yellow collar, cuffs, waistcoat and breeches (some sources show light green facings, waistcoat and breeches), buff belts, buttons and button-hole laces. Austrian-style black cap with front flap bearing the crowned 'FR' in white; tall black gaiters. At least some of the men were armed with rifles.

Freibataillon Nr 2. Raised in 1756; title Mayr; 1759 Collignon; 1760 Courbiere. Actions: Landshut (captured); exchanged and re-raised; a 2nd Bn was raised in 1760. They were captured by the Russians in 1761, exchanged and fought at Freiberg on 29 Oct 1762.

Freibataillon Nr 3. Raised in 1756 title Kalben; 1758 Salenmon. Uniform: coat and facings dark blue, light blue waistcoat and breeches. Actions: captured at Maxen, exchanged and fought at Torgau in November 1760, fought at Freiberg on 29 Oct 1762.

Freibataillon Nr 4. Raised in 1756 title d'Angelelli; 1760 Chossignon. Actions: B Breslau (distinguished), B Hochkirch (distinguished). Captured at Landshut, exchanged and a 2nd Bn raised, very heavy losses against the French at Nordhausen on 27 Mar 1761; exchanged.

Freibataillon Nr 5. Raised in 1757 title Chossignon, 1758 Montjou. Actions: B Bautzen on 6 Sep 1757; exchanged. Absorbed into *Freibataillon Nr 7* in 1759.

Freibataillon Nr 6. Raised in 1757 from French prisoners of war taken at Rossbach. Title Rapin; 1759 von Luederitz. Actions: Landshut (captured); exchanged; captured at Berlin and exchanged again. A 2nd Bn was raised in 1760; fought at Freiberg on 29 Oct 1762. Uniform: Tricorn, dark blue coat with light blue round cuffs, dark blue collar with red tabs to the ends, each with a small yellow button. No lapels, but two rows each of six yellow buttons; red turnbacks. Light blue waistcoat and breeches, long black gaiters.

Freibataillon Nr 7. Raised in 1758 title Wunsch. Absorbed *Freibataillon Nr 5* in 1759. Actions: Berlin (2nd Bn captured by the Russians) exchanged. Uniform: dark green coat, waistcoat and breeches, red collar, lapels, cuffs and turnbacks, natural brown leather belts. Tricorns. The Jaegers were armed with rifles.

Freibataillon Nr 8. Raised in 1758 title Du Verger; 1759 Quintus Icilius. Uniform: tricorn with light blue pompons; dark blue coat, light blue facings, waistcoat and breeches, yellow buttons and button-hole lace, red turnbacks. White belts, high black gaiters. The Jaeger company had dark green coats. Actions: B Hochkirch. A 2nd Bn was raised in 1760 and a 3rd in 1761.

Freibataillon Nr 9. Raised in 1758, title Haerdt. A drummer is shown wearing a tricorn edged white and with light blue pompons, dark blue coat with light blue collar, lapels, round cuffs all edged white; white buttons and button-hole lace, dark blue swallows nests edged and decorated in white; red turnbacks. White belts, brass drum with red and white striped hoops, black gaiters. Actions: not known.

Freihusaren Lubomirsky (Nr 10). Raised in 1758 title Lubomirsky. Uniform: green dolman, red pelisse, fur busby. This unit existed only a few months. Actions: not known.

Freicorps Nr 11. Raised in 1758 title *Corps franc des Volontaires de Prusse*. It originally had a battalion of infantry and a squadron of dragoons. Infantry uniform: green coat, and waistcoat, red collar and cuffs, white belts, buttons, waistcoat and breeches. Dragoon uniform: dark green coat and breeches, red collar, cuffs, turnbacks and shoulder strap, yellow buttons, high jack boots. Dark green, square saddle cloth and pointed holster covers, both edged in wide yellow braid having two thin red stripes. Black harness, steel fittings; Prussian dragoon sword with brass hilt in black sheath. A 2nd Bn was raised in late 1762. Actions: not known.

Freibataillon Nr 12. Raised in 1759 title *von Kleist'sches Freicorps*. In 1759 it consisted only of a hussar regiment; in 1760 a *Frei-Dragoner-*

Regiment was raised; in 1761 the corps was expanded by a unit of *Kroaten* and one of *Jaegers*. Uniform: Hussars: red dolman and breeches, green pelisse, white buttons, fur and frogging, brown fur busby with red bag, red and white plume and yellow cords. Yellow and white waist sash, buff belts. The red sabretasche was edged in green and bore the crowned cipher 'FR' Dragoons: green coat, facings, waistcoat and breeches, fur busby, yellow boots. *Kroaten:* green coat, waistcoat and breeches, black felt tricorn. Jaeger: green coat and waistcoat, red collar and cuffs, buff breeches. The Croats were dressed in all-black, Hungarian-style uniforms with white buttons and lace frogging. The men wore black felt Mirlitons with red cords, red waist sash, white belts. An officer is shown with a black coat with green, pointed cuffs edged silver, silver buttons and button-hole lace; Tricorn edged in silver scalloped lace. Moss-green waistcoat and breeches decorated in silver, silver and black waist sash, silver cartouche bandolier edged red, yellow boots. He is armed with a short musket and a sabre with gilt hilt and Prussian silver and black *portepee*. Actions: on the raid into Franconia; B Kunersdorf on 12 Aug 1759; B Torgau (hussars and dragoons) 3 Nov 1760; B Freiberg 29 Oct 1762 (*Kroaten* and Jaegers).

Freicorps Schoney (Nr 13). Raised in 1760, title Schoney. It consisted of a battalion of Hungarian grenadiers and a squadron of *Frei-Husaren*. Uniform: grenadiers: Austrian-style grenadier cap with brass front plate, bearing the black Prussian eagle, black fur trim and headband, red bag with yellow piping and light blue tassel. Dark blue coat, light blue collar and cuffs, red turnbacks; yellow edging to collar, yellow frogging to the chest (seven bars), yellow 'bear's claws to the pointed cuffs. Light blue waistcoat and breeches trimmed yellow, red waist sash, short boots, white belts. The hussars had light blue dolmans and dark blue pelisses. Actions: not known.

Freicorps Nr 14. Raised in 1760; title *Frei-Dragoner-Regiment von Glasenapp*. Actions and uniforms: not known.

Freibataillon Nr 15. Raised in 1760; title: *Volontaires d'Ostfriese* or *Freibataillon Jeney*. Actions: B Freiberg 29 Oct 1762. Uniforms not known.

Freibataillon Nr 16. Raised in 1760; title von Schack. Actions: B Freiberg, 29 Oct 1762.

Freibataillon Nr 17. Raised in 1761; title: *Freibataillon von Heer* or *Schweizer-Bataillon.* Actions: B Freiberg, 29 Oct 1762.

Freibataillon Nr 18. Raised in 1761; title: Bequignolles. Actions: not known.

Freibataillon Nr 19. Raised in 1761. Title: La Badie or *Royal Etranger.* Actions: not known; this unit was quickly disbanded due to its high rate of desertion.

Freicorps Nr 20. Raised in 1761; title: Gschray. It consisted of a *Freibataillon* and a *Freidragoner-Eskadron.* Dragoon uniform: light blue coat, black lapels and cuffs, yellow buttons, button-hole lace, waistcoat and breeches, white belts. The dragoons wore bicorns with yellow pompons, white gauntlets, high, cuffed boots. The infantry wore tricorns with yellow pompons, dark blue coats with dark blue facings, yellow buttons and button-hole lace, red turnbacks. A yellow fringed shoulder strap was worn on the left shoulder. Yellow waistcoat and breeches, white belts, high black gaiters. Actions: Nordhausen, 27 Mar 1761; Gschray and many of his corps were captured.

Freicorps Nr 21. Raised in 1761. Title: *Tartarisches Ulanencorps Krczowsky.* Uniform: red dolman, brown, baggy trousers, white turban. Actions: not known; this unit never reached its intended full strength.

Freicorps Nr 22. Raised in 1761. Title: *Schwarze Brigade von Favrat.* Uniforms: Grenadiers – black coat, red cuffs and lapels, straw waistcoat and breeches. Dragoons: as for the grenadiers. Jaegers: black coat, green lapels, cuffs, waistcoat and breeches. Hussars: yellow dolman, black pelisse, Mirliton cap. Actions: not known; there was never more than the grenadier battalion raised.

Freibataillon von Below. Raised in May 1760. Uniform unknown; this was an improvised force, captured at Landshut, 20 Jun 1761.

The Freikorps of the War of the Bavarian Succession

Despite the lessons of the Seven Years War, Frederick had disbanded all the light infantry units in 1763; consequently they had to be re-raised in 1778.

The following *Freicorps* were raised during this brief war.

1. ***Freiregiment*** **Graf Hordt.** Raised on 23 Feb 1778 in Oranienburg from volunteers from Mecklenburg and Swedish Pommerania; commander Lt-Col Johannes Kuempel. By July 1778 it had joined the 2nd Army with two battalions. Uniform unknown.

2. ***Freibataillon*** **von Pollitz.** Raised on 23 Feb 1778 in Calbe and Stassfurt at one battalion; in July 1778 it joined the 2nd Army. Uniform unknown.

3. ***Freibatallon*** **von Steinmetz.** Raised on 23 Feb 1778 in Namslau; commander Major von Steinmetz (killed at Comeise on 24 Nov 1778); Uniform dark blue coat, light blue collar, Swedish cuffs, waistcoat and breeches, white buttons and plumes to bicorns, long black gaiters. White plumes; for NCOs with black tips. Officers had wide, scalloped silver lace edging to their bicorns, white plumes with black bases, black cockade, silver epaulettes.

4. ***Freiregiment*** **Salenmon.** Raised on 10 Mar 1778 in Wesel as part of Garrison R Nr 9. It did not take the field. Uniform as for Steinmetz, with dark blue coat, light blue collar, Swedish cuffs, waistcoat and breeches, yellow buttons; NCOs wore seven gold button-hole laces to each breast, a gold hat edging and boots. officers had wide scalloped gold lace edging to the hat.

5. ***Freiregiment*** **l'Homme de Courbiere.** Raised on 10 Mar 1778 in Wesel; did not take the field. Uniform dark blue coat, black facings, white small clothes.

6. ***Freiregiment*** **von Muenster.** Raised on 21 Mar 1778 in Soest at two battalions and initially attached to the 2nd Army. By the end of the year, the 1st Bn had been transferred to the 1st Army. Uniform dark blue coat, light blue collar, Swedish cuffs and small clothes, white buttons, long black gaiters. NCOs had silver button-holes to the chest, silver hat edging and boots; officers as for Steinmetz.

7. ***Freibataillon*** **von Rentz.** Raised on 21 Mar 1778 in Geldern; never completed, did not take the field. Uniform unknown.

8. ***Freiregiment*** **von Stein.** Raised on 4 Apr 1778 in Halberstadt; commander Colonel von Colong; by Aug 1778 it had joined the 1st Army. Uniform unknown.

9. ***Freibataillon*** **von Bremer.** Raised on 13 Apr 1778 in Brieg; from 7 Aug 1778 it was titled von Freystedt; no other details known.

10. ***Freibataillon*** **von Schlichting.** Raised on 6 Dec 1778 in Wartenberg in Silesia. Uniform dark blue, light blue lapels and cuffs, white buttons and small clothes. Officers had silver embroidery to lapels and cuffs.

11. ***Freijaeger-Kompagnie*** **von Spitznas.** Raised on 4 May 1778 in Cleve and Wetzlar at 180 men to reinforce the Jaeger Battalion with the 1st Army; it joined that battalion in Feb 1779. Uniform light green as for the Jaeger, red collar and Swedish cuffs, white buttons and small clothes. NCOs and men wore black, fusilier-type caps with the white, crowned 'FR' on the front flap. NCOs had silver lace edging to the cuff and the cap front flap and black tips to their white plumes. Officers had wide, scalloped silver lace edging to their tricorns and black bases to their white plumes; they wore boots.

12. ***Bosniakon-Eskadron*** **Moczydowsky.** Raised in 1778 in Wartenberg; no other details known.

13. ***Tartarenkorps*** **von Schill.** Raised on 4 Jan 1779 in Solsau, Willenberg and Gumbinnen. Uniform: yellow square-topped '*Konfoederatka*', white plume, black fur headband; dark blue Kurtka with yellow collar, lapels, Swedish cuffs, buttons and small, fringed epaulettes, black stock. Dark blue Polish trousers, trimmed at the bottoms with elaborate yellow leather cuffing. The men wore hussar-style barrel sashes, officers had the usual black and silver waist sashes. Dark blue Polish saddle cloths with yellow wolfs-tooth edging. The front rank bore lances with blue and yellow fanions.

14. Bischoffswerder'sche Frei-Jaeger-Corps. This corps was unusual in that it was raised in Dresden, capital of Saxony, the officers were appointed by Frederick and all ranks took oaths of loyalty to both the king of Prussia and the Elector of Saxony, Friedrich August. Raised on 21 Sep 1778; Chef was Major Johann Rudolf von Bischoffswerder; it had four companies with 17 officers, 48 NCOs, five surgeons, four hornists and 600 men. Uniform: officers wore the conventional tricorn, but NCOs and men wore what could best be described as a Royal Canadian Mounted Police hat with a vertcal black leather front plate, pointed at the top and bearing the cipher 'FA' under an electoral cap. NCOs' front plates were of brass. On the left hand side was a white plume. The coat, turnbacks and breeches were green, the collar, Swedish cuffs and waistcoats buff; belts and long gaiters black. The cartridge pouch lid bore 'FA' under a cap.

The Freikorps of the War of 1806-1807

Due to the wholesale destruction of the army in 1806, there was plenty of scope for active individuals to raise 'private armes', particularly in the besieged fortresses in 1807. Due to the shattered state of Prussia's finances and the shortage of clothing stocks, it is very likely that the colourful uniforms attributed to some of these corps remained wishful thinking; the men would have worn what they could lay their hands on.

1. *Freikorps von Krokow.* Raised by AKO of 27 Dec 1806 in Danzig at one battalion and a squadron of cavalry. The uniform was a Roman-type helmet with a combe and cut, black horsehair crest; the body of the helmet was given a leopardskin effect, the black front plate bore the skull and bones in white. The front of the combe was decorated with a lion's head. The coat was green with black collar, Swedish cuffs, lapels and edging to the green turnbacks and yellow buttons; belts were black; the sabre bandolier bore an oval bress plate with the black Prussian eagle. Grey trousers with black leather booting and yellow buttons up the sides were worn.

2. ***Freikorps von der Marwitz.*** Raised by AKO of 27 Dec 1806 in Danzig

3. ***Grenadier-Bataillon Schmeling*** Raised in Danzig by AKO of 27 Dec 1806 from members of the garrison of the city

4. ***Fuesilier-Bataillon Danielewitz.*** Raised in Danzig from the remnants of the Fusilier Battalions Borel, Hinrichs and Knorr

5. ***Bataillon-Garde.*** Raised in November 1806 in Graudenz from remnants of the 1st Bn of the Regiment Garde, the Grenadier-Garde-Bn, IR Koenig Nr 18. It was then transferred to Memel where, on 27 Jun 1807 it received the title shown.

6. ***Grenadier-Bataillon von Waldenfels.*** Raised in Colberg from escaped prisoners of war and from volunteers

7. ***Freikorps von Schill.*** Raised in Colberg from escaped prisoners of war and from volunteers; seven companies strong. Many men wore the uniforms of their old regiments; belts were white.

8. ***Fuesilier-Bataillon Moeller.*** *No details known.*

9. ***Grenadier-Bataillon Losthin.*** Raised in Silesia and called the 'Brown Grenadiers', this unit wore bicorns, blue tunics, brown collar, Brandenburg cuffs, lapels and breeches, red turnbacks and yellow buttons.

10. ***Freikorps von Kalkreuth.*** Uniform was a shako with yellow pompon, loop, and button, Prussian cockade; blue tunic, yellow buttons, green collar and Swedish cuffs piped red, red turnbacks, black belts and grey overalls.

11. ***Grenadier-Jaeger-Kompagnie von Sell.*** Uniform was a shako with yellow loop and button, Prussian cockade for officers (other ranks wore the brass eagle) and a green 'sausage' crest running from thee rear base of the shako to the top centre of the front. Officers had small gilt eagles to the sides of the shako, connected by gold chains. This practice became general for Prussian infantry officers in the new uniform regulations. The dark green tunic had black collar, Swedish cuffs and lapels, yellow buttons, red turnbacks, grey overalls or breeches, black gaiters. At least some of the men were armed with rifles.

There were several other such units which were disbanded again when the fortresses of Brieg, Neisse and Schweidnitz fell.

On 15 May 1807, the king approved the following uniform for the defenders of Colberg: dark blue coat, white lining, collar and cuffs, yellow buttons, white waistcoats and breeches. Bicorn with gold lace loop and button; black leather belt with a brass plate bearing the crest of the town. They carried a sabre of the same pattern as the *Bergwerks-und Hutten-Departement* and wore no epaulettes or Portepees.

The Freikorps of the Wars of Liberation

In early 1813 there was a distinct upsurge of patriotic spirit in Prussia, following von Yorck's Convention of Tauroggen with the Russians and the prospect of taking revenge for all the humiliations of the year of 1806 – 1812. Scharnhorst advised the king to beware of allowing persons of doubtful character or motivation to raise these volunteer formations; the king agreed. However, there were so many applications from the upper classes and the academics, who would not have taken service in the regular army for a number of reasons, that it was decided to raise a certain number of units to harness the national spirit instead of stifling it.

The following units were raised:

1, von Luetzow's Royal Prussian *Freikorps*. Raised on 18 Feb 1813 by the retired officers von Luetzow, von Helden (aka Sarnowski) and von Petersdorff, all of whom had been in Schill's corps in 1806 / 7. The volunteers were mostly foreigners. There were three battalions each of four companies (the 2nd Bn had a company of Tyrolean riflemen), three squadrons of hussars, one each of mounted rifles and lancers. A company of artillery was also raised. The infantry uniform was a shako with black, drooping horschair plume and cords, brass 'loop' and button. The plain black coat was in the '*Litewka*' style, a double-breasted frock coat with red piping to collar, shoulder straps and Polish cuffs. Trousers, gaiters, belts were also black, buttons yellow. Prussian badges of rank and items of equipment were worn. The Tyrolean rifle company wore a pike grey tunic with green collar, cuffs, Polish cuffs, lapels and turnbacks, white buttons, pike grey trousers with double, green side stripes and the Tyrolean hat with the right hand brim extended and turned up; a green plune was worn. The hussars were also all in black, black glass buttons; the black overalls had a red stripe and yellow buttons down the sides. The mounted rifles wore the black *Litewka,* yellow buttons, red piping and hussar overalls and an infantry shako. The lancers again wore black with red piping, yellow buttons, hussar breeches and the shako. Harness and saddle furniture was black.

2, *Auslaender-Bataillon* von Reuss. Raised in Berlin 21 Mar 1813 from deserters from foreign armies. The battalion wore the uniform of the Leibregiment but with (initially) red shoulder straps, later yellow.

3, *Auslaendische Jaeger-Bataillon* von Reiche. In Berlin on 10 Mar 1813; mainly from Westphalians of the *Grande Armée.* Uniform was a Prussian shako with brass chinscales, green *Litewka* with two rows of yellow buttons, red collar and cuffs, light green shoulder straps piped red, grey breeches in boots; black belts. It saw no action prior to the armistice and on 11 Aug it was sent to join von Wallmoden's corps in northwestern Germany. Raised on 29 Apr. Uniform of the infantry was a shako with steel chinscales, a green tunic with black collar, cuffs, shoulder straps, belts, white buttons, grey overalls (white in summer). The cavalry wore English uniforms consisting of brown Colpacks, red dolman and pelisse with white buttons and lace, grey overalls.

4, von Hellwig's *Freikorps.* Major von Hellwig, of the 2nd Silesian Hussars, operating in Bluecher's advanced cavalry, came to his commander's attention due to his skill and daring. On the night of the 12 / 13 April, he ambushed a Bavarian unit in Langensalza with his 90 hussars and took two howitzers, three guns, three ammunition wagons and 32 horses. Five nights later he captured a Westphalian detachment of two officers, 32 men and 50 horses in Wanfried. He applied to lead his own *Streifkorps* in the enemy rear areas and on 29 April the king approved his request. He was allowed to take the 3rd and 4th Squadrons of his regiment on 11 May and a few days later the Jaeger Detachment of the Combined Silesian Hussars (29 men in all) was added. Hellwig was also given permission to recruit more volunteers if he could. In the second half of May 1813, after the battle of Gross-Goerschen, Hellwig, who had been operating in the enemy's rear areas, began to recruit an infantry unit from volunteers among enemy prisoners and deserters, mainly Westfalians and other Germans. His success was such that he soon had three squadrons of hussars as well. Infantry uniform was to be a shako with tin chinscales, green tunic with black collar, cuffs, shoulder straps and belts; white buttons, white or grey trousers. French weapons were taken from the battlefield. The hussars wore English uniforms: brown Colpacks with a blue bag, red dolman and pelisse with white lace and buttons. On 12 July Hellwig's formation was attached to Colonel von Oppen's Reserve Cavalry in von Buelow's III Corps. By the end of November, Hellwig's infantry numbered 400 men in three companies.

5, Schill's Husaren. In March 1813 Major von Schill of the 2nd Silesian Hussars (brother of the famous Ferdinand von Schill, killed by Dutch troops in Stralsund in 1809) received permission to take 100 men of his regiment to Hamburg to recruit a corps of volunteers of infantry and cavalry. The Prussian state would meet all costs incurred. The choice of Hamburg was unfortunate as the rich Hanseatic city, just evacuated by the French, was busy raising its own volunteer formations. Money was also short; in May, von Schill took out personal loans of 700 Thalers to help pay the upkeep of the corps.

Due to the lack of progress of the recruitment, it was decided on 4 August, that Schill's command should be attached to von Hellwig's in the III Corps in the Army of the North and the men sent back to their parent regiments. However, on 7 August the Crownprince of Sweden (commanding the Army of the North) attached von Schill to Vegesack's division and the corps remained in action until the end of the 1814 campaign.

6, The Thueringian Battalion. On 13 April 1813 a detachment of 16 hussars from Major Franz von Bluecher's* command surprised the Ducal Saxon Battalion in the neighbourhood of Hof and challenged them to come over to the allied side, which they promptly did. On 24 April this battalion had 12 officers, 22 NCOs, two surgeons, ten musicians, 274 men and four women. The king of Prussia took the unit into his service as 'The 1st Thuringian Battalion'; it was sent to join the forces blockading Glogau. On 17 June it was transferred to Schweidnitz, where its ranks were filled up with German deserters and prisoners from the *Grande Armée*. Two officers and 60 men of the Herzog von Coburg's contingent were captured in September and added to the corps. The uniforms were a complete mix of the blue, faced red, of the Ducal Battalion, green faced yellow of the Weimar part of the original battalion and white English garments. The unit was disbanded on 23 November 1813 when the Herzog von Wiemar promised to raise a battalion in his duchy to fight for the allies.

* - Son of the field marshal and commander of the Combined Silesian Hussars.

7, Minor Volunteer Units. On 14 Feb 1813 the Prussian government published an order that required all trained foresters and gamekeepers under 50 years of age, who lived within a radius of eight *Meilen* from a fortress, to report the the fortress commander for special duties. By mid-May there was such a unit in Liegnitz of three officers and 209 men, who performed 'excellent service and killed a lot of the enemy.'

Similar units were raised in Glatz, Schweidnitz and Silberberg; that in Schweidnitz was badly led and had a poor reputation. Attempts were made in the Riesengebirge, along the Bohemian border, to raise similar units (*Landwehrschuetzen*) from poachers. This scheme had limited success as the poachers threatened to burn down the houses of anyone who reported them to the authorities.

Efforts were also made to raise units from among the Silesian miners and one battalion of 400 men, known as the 'Milewskischen Bergleute' was attached to Major von Boltenstern's *Streifkorps*. Some other similar units were raised.

The city of Koenigsberg raised and paid for the 'Koenigsberger or Ostpreussische National-Jaeger-Detachement' of one officer, 2 NCOs, five hornists and 172 Jaegers. It was attached to von Buelow's III Corps.

8, Freiwilliges Lausitzer Jaegerkorps. In late autumn 1813 this unit was raised in the Lausitz region in Saxony by Captain von Lyncker. Bluecher reported that it had grown to the size of a battalion by 9 October. When the Saxon army went over to the allies on 19 October, the Prussian king ordered that this unit too should be transferred to the command of General von Thielmann, who was in charge of the reorganization of that army.

9, Schivelbein'sches Kavallerie-Detachement. The county of Schivelbein had rasied a cavalry unit of 40 men by June 1813; it was sent to von Yorck's I Corps, but found to be of such dubious quality that it was disbanded.

10

The Foot Artillery

From 1785 there were four field artillery regiments:

1st. Chefs – 1785 Oberst Johann Wilhelm von Dittmar; 1792 Oberst Johann Friedrich von Meerkatz.

2nd Chefs – 1785 Oberst Adam Heinrich von Pritzelwitz; 1787 Oberst Christoph Carl Friedrich von Bardeleben; 1794 GM Karl Philipp von Anhalt; 1795 Oberst Karl Friedrich von Block; 1797 Oberst Johann Wilhelm von Lentken; 1803 Oberst Georg Friedrich Wilhelm von Schoenermark.

3rd Chefs – 1786 Oberst Christian Friedrich von Moller; 1795 Oberst Georg Friedrich von Tempelhoff.

4th Chefs – 1785 Oberst Johann Friedrich von Meerkatz; 1792 Oberst Alexander August von der Lochau. 1801 Oberst Gottfried Ludwig von Hartmann. Chef of the regiment was Oberst Heinrich Ernst von Hueser.

Uniforms. Dark blue tunic of line infantry cut, black facings, red turnbacks, yellow buttons, Brandenburg cuffs, white small clothes and belts, red stock. Privates wore no lace decoration; NCOs wore 22 gold lace buttonholes. Until 1757 bombardiers of this corps wore fusilier caps; brass fronted, with black headband and cap head piece. Across the base of the frontplate was: 'KONIG:PREUSS:ARTILLERIE' . To the sides of the headband was a brass, three-flamed grenade, at the back were crossed cannon barrels and a pair of kettle drums. It seems doubtful that this headgear was ever in general use, although it seems that bombardiers wore it.

When the Kasket was introduced in 1787, it bore a brass grenade badge having three flames. It was worn until 1798. On the body of the grenade was the cipher 'FWR'. The same badge was worn on the pouch lid. The pompon was yellow over red over black. Prior to and after the kasket, the men's tricorns were decorated with a button and a black loop. The artillery carried straight-bladed swords.

Foot Artillery Post 1808. As for musketiers except as noted below. Yellow top band to shako, brass, three-flamed grenade badge; black collars and cuffs edged red. Shoulder straps in the provincial colour.

Black belts.

11

The Pontoniers

This corps was part of the artillery. Uniform of infantry cut, dark blue coat, dark blue lapels, red collar, Prussian cuffs, shoulder strap and turnbacks, yellow buttons, white belts and small clothes. The bicorn was decorated with a button, black loop and light blue over red over yellow pompon. The kasket bore the royal cipher 'FWR' in brass.

Pontoniers post 1808.
As for the foot artillery.

12

The Foot Artillery and Supply Train

Dark blue, single breasted, long skirted coats, with dark blue collars, red cuffs, yellow buttons, red turnbacks, buff breeches, high boots. The bicorn was decorated with a button and black loop

Members of the supply train wore the same uniform but with white buttons and light blue collars

13

Horse Artillery

Frederick the Great raised the first units of this arm in 1759. In 1796 there were seven such companies in the 2nd Battalion of the 4th Field Artillery Regiment; in this year the Horse Artillery Regiment became independent. Chef of the regiment was Oberst Heinrich Ernst von Hueser. By 1805 there were ten companies.

Uniforms as for the foot artillery but the coat was light blue and of dragoon cut. The bicorn was decorated with a button and yellow loop; white plumes were worn. The light blue turnbacks of the mens' tunics were edged in wide black lace, piped red on both sides. Light cavalry sabres in black leather sheaths with brass fittings were carried. The saddlecloth was light blue, edged with a wide black lace piped yellow. The rear corner bore the crowned royal cipher. In 1806 some companies were already wearing the new shako, with white top band and cords, black and white cockade, three-flamed grenade badge.

Horse Artillery Post 1808. As for the foot artillery but with dragoon shako, white plumes, yellow cords, brass chinscales. Dark blue turnbacks edged in red-black-red lace. Dragoon boots.

14

Horse Artillery Train

Dark blue, single breasted, dragoon-style tunics with yellow buttons and red turnbacks, buff breeches, high boots. The bicorn was decorated with a button, black loop and cockade and a white plume.

15

Garrison Artillery

Uniforms as for the foot artillery but with black stocks.

Artillery Material 1740 - 1800

The transformation of the artillery from a civilian guild into a purely military began during the reign of King Frederick William I; it was this monarch who re-classified field guns into four main categories: 3, 6, 12 and 24 pounders and this system was imposed as from 1715.

The bronze used for cannon barrels was an expensive material and old, worn out items were melted down and re-used.

A similar scheme was imposed on the mortars; there were to be bronze pieces at 10, 25, 50 and 75 pounds, the light, wrought-iron Coehorn weapon and a few heavy, stone items.

Howitzers came into general use early, but in about 1740 the 18 pounder became the standard piece.

General von Linger recorded in 1722 that the Prussian army then had 722 bronze cannon, 1,425 iron cannon, 171 bronze mortars, 128 iron mortars, 28 bronze howitzers and 27 of iron. This total of 2,501 guns represented an increase of 500 pieces over the inventory of 1712.

When Frederick II came to the throne, the Prussian artillery was at much the same state of evolution as that of any other European nation.

The following table is based on Gohlke pages 92-93:

Cannon

	Barrel					Projectile		Chamber	Charge	Team
	Calibre cm *	Length calibres	Length cm	Weight per pound of shot	Weight kg	Calibre cm	Weight kg		kg	horses
3 Pdr bronze 1740 (H)	7,32	16	115	150-165	210-231	7,19	1,4	conical	0,47	3-4
3 Pdr bronze 1740 (H)	7,32	16	115	100-165	140-147	7,19	1,4	cylindrical	0,34	4
3 Pdr bronze 1746-50 (L)	7,32	20	144	148	207	7,19	1,4	conical	0,47-0,58	4
3 Pdr bronze 1746 (B)	7,32	22	160	193-200	270-280	7,19	1,4	none	0,7	4
3 Pdr bronze 1754-1768 (D)	7,32	18	130	147	206	7,19	1,4	none	0,47-0,58	4
3 Pdr bronze 1771 (D)	7,32	20	144	165-188	462-526	7,19	1,4	none	0,47-0,58	4
3 Pdr bronze 1774 (H)	7,32	20	144	?	?	7,19	1,4	none	0,47-0,58	4
6 Pdr bronze 1740 (H)	9,42	16	145	150-165	420-462	9,06	2,8	conical	0,9	6
6 Pdr bronze 1740 (H)	9,42	16	145	150-165	210-231	9,06	2,8	cylindrical	0,7	3
6 Pdr bronze 1754 (D)	9,42	16	145	96	268	9,06	2,8	conical	0,9	3
6 Pdr bronze 1760-66 heavy (H)	9,42	26	235	322-330	902-925	9,06	2,8	none	1,4	6
6 Pdr bronze 1762 heavy (D)	9,42	22	199	257	719	9,06	2,8	none	1,4	6

12 Pdr bronze 1754-59										
light (?)	11,9	14	160	71	397	11,4	5,6	none	1,4-1,6	4
12 Pdr bronze 1740 (H)	11,9	16	182	150-165	340-462	11,4	5,6	conical	1,8	4
12 Pdr bronze 1740 (H)	11,9	16	182	100-105	560-588	11,4	5,6	cylindrical	1,4	4
12 Pdr bronze 1744 (L)	11,9	16	182	87	487	11,4	5,6	conical	1,63	4
12 Pdr bronze 1754 (D)	11,9	14	160	64	358	11,4	5,6	conical	1,4-1,6	4
12 Pdr bronze 1754 (?)	11,9	24	274	263	1472	11,4	5,6	none	2,3-2,8	8
12 Pdr bronze 1758 (?)	11,9	16	182	142	795	11,4	5,6	none	2,3-2,8	8
12 Pdr bronze 1759 (A)										
medium	11,9	18	205	170	952	11,4	5,6	none	1,8	8
12 Pdr bronze 1761 (Br)										
medium	11,9	22	251	266	1492	11,4	5,6	none	2,3	12
12 Pdr bronze 17?? (?)										
siege and fortress	11,9	24	274	?	1697	11,4	5,6	none	2,8	none
24 Pdr bronze 1740 (H)	14,99	16	230	150-165	1680-1848	14,39	11,2	none	3,2	6
24 Pdr bronze 1740 (H)	14,99	16	230	100-105	1120-1176	14,39	11,2	none	2,8	4
24 Pdr bronze 1744 (?)	14,99	16	230	150-165	1680-1848	14,39	11,2	none	1,8	4
24 Pdr bronze 1754 (?)	14,99	16	230	150-165	1680-1848	14,39	8,14 Gran	none	4,7	8

* Designers: A = Austria, B = Beauvry, Br = Brummer, D = Dieskau, H = Holtzmann, L = Linger

Howitzers

	Barrel					Projectile		Chamber	Charge	Team
	Calibre cm *	Length calibres	Length cm	Weight per pound of shot	Weight kg	Calibre cm	Weight kg	cm	kg	horses
7 Pdr bronze 1762	15,0	—	94	—	343,3	14,39	6,54	20,9/8,5	0,9	4
7 Pdr bronze 1790										
Horse Artillery	14,9	—	94	83	504	14,39	6,08	22,2/7,4	0,7	6
10 Pdr long 1744	17,26	—	135	—	—	16,63	12,6	33,7/9,7	1,87-2,1	10
10 Pdr normal 1762	17,26	—	107	—	—	16,63	12,6	32,7/9,8	1,4	10
10 Pdr heavy 1766	17,26	—	121	137-138	1726-1738	16,63	12,6	31,9/11,1	2,3	10
25 Pdr 1767/1777	22,75	—	123	—	1105,6	22,18	29	31,4/13,1	2,8	12
30 Pdr 1762	24,19	—	130	—	1324	23,54	30,38	31,4/13,1	3,27	—
Schuwalow's Canister Piece	12,68	—	143	—	—	—	—	none	?	

Mortars

	Barrel					Projectile		Chamber	Charge	Team
	Calibre cm *	Length calibres	Length cm	Weight per pound of shot	Weight kg	Calibre cm	Weight kg	cm	kg	horses
Coehorn										
Wrought iron 1775	10,46	—	23,54	—	18,7	10,2	2,10	9,42/4,58	—	—
7 Pdr Pack model 1790 (T)	14,91	—	45,77	—	86,5	14,39	6,08	14,38/6,8	0,47	—
10 Pdr 1785 (T)										
Bronze	17,0	—	52,31	—	182,0	16,63	12,6	17,0/17,85	0,60	4
25 Pdr 1778/83										
Bronze (D)	22,9	—	68,79	—	397	22,19	29,0	19,62/9,15	1,16	—
50 Pdr 1752 (Bs)										
Bronze, light	28,5	—	71,94	—	715	27,93	85,55	24,8/10,46	—	8
50 Pdr 1769 (L)										
Bronze	28,5	—	74,82	—	807,35	27,93	85,55	29,0/13,08	—	6
50 Pdr 1774 (H)										
Bronze	28,5	—	93,1	—	803,5	27,93	85,55	28,77/11,77	2,8	—
50 Pdr 1782/96 (H)										

Bronze	28,5	—	93,1	—	803,5	27,93	85,55	28,77/11,77	2,8	—
50 Pdr 1785 (H)										
Bronze	28,5	—	93,1	—	803,5	27,93	85,55	28,77/11,77	2,8	—
50 Pdr 1777 (V)										
Iron	28,5	—	96,7	—	970,5	27,93	85,55	28,77/11,77	2,8	—
50 Pdr 1784 (H)										
Bronze	28,5	—	95,9	—	970,5	27,93	85,55	28,77/11,77	2,8	6
140 Pdr 1775 stone	39,75	—	70,63	—	874	—	—	20,9/11,8	1,87	—
140 Pdr 1784/1810										
stone	39,23	—	87,8	—	785	—	—	20,9/11,8	2,06	—

* Designers: Bs = Breslau; D = Dornstein; H = Holtzendorf; L = Linger; T = Tempelhoff; V - Vietzer.

16

The Engineers

In 1787 there were three brigades in this corps; this increased to four in 1800.

Chefs – 1778 von Pfau; 1788 von Regler; 1792 von Geusau.

Uniforms. Dark blue coat of line infantry cut, black facings, red turnbacks, white buttons, Swedish cuffs, no lace decoration, buff small clothes. The bicorn was decorated with a silver brooch, button and black and silver cockade. This was an all officer corps.

Engineers Post 1808. As for the artillery with black velvet facings, white buttons, black shoulder straps edged red.

17

The Miners

There were two companies of miners in the regiment of pioneers raised in 1741; when this unit was converted to the 49th Fusiliers in 1758, the miners became a separate corps. In the early days of their independent existence they wore very low, skull cap like headdress with a silver front plate, orange headband and cap, white tape and pompon. To each side was a silver flame, at the back was the three-flamed grenade in silver.

From 1779 there were two companies of these specialists in Glatz fortress, one each in Neisse and Graudenz; in 1791 there was one company each in the fortresses of Glatz, Graudenz and Schweidnitz.

Uniform of infantry cut, dark blue coat, dark blue lapels, red collar, Prussian cuffs, shoulder strap and turnbacks, yellow buttons, white belts and small clothes. The bicorn was decorated with a button, black loop and light blue over red over yellow pompon; the kasket bore the royal cipher 'FWR' in brass.

In 1798 the Miners received the same uniform as the artillery.

18

The Uniforms of the Kuerassiers

From early times (1688) the coat worn by these regiments of heavy horse was usually a buff leather 'Kollet', which had excellent protective properties against sword cuts and was relatively light and flexible. Facings were worn on the collar, the Swedish cuffs and on the narrow shoulder straps. The short turnbacks were in the coat colour. Apart from the colour of the facings and buttons, each regiment was distinguished by a braid edging to the front of the coat, the cuffs, turnbacks, waistcoats and carbine bandoliers. The skirts of officers' tunics were to the knee; those of the men were cut short so that they just covered the buttocks.

Breeches were in buff leather; white gauntlets and knee cuffs were worn as were stiff, high jacked boots.

Headgear was a large bicorn with national cockade, black loop and regimental button. Under the bicorn an iron skull cap was worn when in action. White plumes were worn for parades from 1762 onwards. For NCOs the plume tip was black, for officers the base was black. From about 1735 onwards, the leather tunics were replaced with white cloth items, only Regiment Nr 2 retained the yellow colour and were then known as 'the yellow riders'. Waistcoats were originally in the coat colour, but in 1735 they were changed to be in the regimental facing colour. Only the Regiment Gensdarmes Nr 10 retained them in blue as before. After the 1st Silesian War (1741-1742) the facings of some regiments changed; they then remained the same until 1806. From about 1720 a two-piece metal cuirass was worn, but from 1735 the back plate was discarded and the front plate held in place by straps. The cuirass was polished for the Garde du Corps, blackened for the other regiments. These breastplates were abandoned under King Friedrich Wilhelm II. The cuirasses of officers were edged in gilt metal and their shoulder straps were covered in gilt plates. On the top centre of the breastplate they wore the crowned Prussian crest in trophies of arms, all gilt. A waist sash in the facing colour was worn over the tunic. Leatherwork was white. Weapons were a brace of pistols and a heavy, straight-bladed sword. A pouch (very much like the sabretasche of the hussars) was carried from the waist belt and sat high up on the left hip. It was in the facing colour, edged with the regimental lace and bore various designs. Harness was black with steel fittings. Holster covers and the square shabraques were in either the facing colour or the colour of the coat and were edged in the regimental lace. The designs on the holster covers and in the rear corners of the shabraque varied from regiment to regiment.

Apart from the daily uniform, officers also had a white state uniformwith lapels, Swedish cuffs and collar in the colour of their waistcoats. On each lapel were six embroidered lace loops, two under them, two on the cuff, two on each pocket flap and four in the small of the back. The small clothes were buff.

The lids of officers‘ cartouches were in the facing colour, edged in the regimental braid, usually enclosing trophies of arms and the crowned royal cipher. Officers‘ cartouche lids of the 7th, 8th, 9th, 11th and 12th Regiments bore the crowned black Prussian eagle on trophies of arms. Officers of each regiment wore gold or silver lace decorations to their buttonholes that were unique to their regimentThose of NCOs were in the facing colour, edged in braid in the button colour, bearing a crowned round plate showing the royal cipher. Troopers had black pouch lids woth a round brass plate bearing the Prussian eagle. NCOs cuffs were decorated with gold or silver lace, the pattern varying from regiment to regiment. Standard bearers had standard bandoliers u-in the facing colour, edged and fringed in the button colour. The standard pike was in the form of a lance.

19

The Kuerassier Regimental Histories

Only the regiments von Wagenfeld Nr 4 and the Garde du Corps Nr 13 survived the catastrophe of 1806.

1. Kuerassier-Regiment.
Raised on 10 Jun 1666.
Title in 1806: **Henkel von Donnersmark.**
Inspectorate: Lower Silesia. Garrison Breslau.

Uniforms. White tunic and breeches, poppy red collar, cuffs, sash and waistcoat. The regimental lace was white with three red stripes; officers' lace was silver. Holster covers and the shabraque bore the crowned royal cipher within braid edging.

Standards. The five standards were lost at the capitulation of Pasewalk on 29 Oct 1806. The regiment was not re-raised.

Campaign history. Raised in 1666 by Johann Georg Fuerst von Anhalt-Dessau (died as GFM)at six companies; 1689 three companies added, the regiment now had three squadrons each of three companies. 1691 a cadre was given to the new Regiment Nr 9; 1718 now in five squadrons, each of two companies; the extra men came from the disbanded Regiment von Heyden. **1672** the regiment served in Holland; **1674** in Alsace, then recalled to Brandenburg to defend Prussia against the Swedes **1675** 28 Jun at the victory of Fehrbellin. That same year in Pommerania, capture of

Greifenhagen. **1677** S Stettin; **1686** in Hungary against the Turks; S Ofen. **1689** S Bonn; **1691** B Steenkerken; **1693-1695** in Holland, 1693 Chef GM Carl Friedrich Graf von Schlippenbach (died as GdC); 19 Jul B Neerwinden (very heavy losses). **War of the Spanish Succession 1706** 23 May B Ramillies; 22 Aug capture of Menin. **1708** 5 Jul capture of Ghent; 11 Jul B Oudenard; 28 Sep C Winendael; capture of Lille (23 Oct). **1709** 27 Jun-28 Jul S and capture of Dornik. 11 Sep B Malplaquet; **1715** in Pommerania and on Ruegen Island. 1723 Chef GM Cuno Ernst von Bredow (resigned in 1724); 1724 Oberst Wilhelm Dietrich von Buddenbrock (died as GFM). **1742** Against Austria, 16 May B Chotusitz, drove off 50 squadrons of Austrian cavalry and took a battery of artillery, thus sealing the victory; **1745** 4 Jun B Hohenfriedberg, the regiment took four colours and four sets of kettledrums. 30 Sep B Soor; B Graner Koppe. 15 Dec B Kesselsdorf. **Seven Years War 1756-1763. 1757** Chef GM Hans Kaspar von Krockow (died of wounds received at Hochkirch); **1757** 6 May B Prag; 18 Jun; Kolin; 22 Nov B Breslau; 5 Dec B Leuthen. **1758** 14 Oct B Hochkirch. **1759** Chef GM Gustav Albrecht von Schlaberndorff (died in 1765; the regiment remained vacant until 1768 when Oberst Friedrich Wilhelm von Roeder became Chef; he died in 1781). **1759** 23 Jul B Kay; 12 Aug B Kunersdorf. **1760** 3 Nov B Torgau. **1762** B Freiberg. **1778-1779** no major actions. 1781 Chef GM Levin Friedrich Gideon von Apenburg (retired); 1784 GM Philipp Christian von Bohlen (retired as GL); 1787 Chef GM Georg Dietrich von der Groeben (was appointed Chef of the 5th Dept in the Higher War College); 1788 Chef Oberst Diedrich Goswin von Dolffs (died as GL); **1794** In Poland, at 15 Jun capture of Krakau and Sk Posen. 1805 Chef Oberst Elias Maximilian Graf Henckel von Donnersmark. **1806** 14 Oct Jena; the regiment surrendered at Pasewalk and was not re-raised. The depot was used to supply men and horses to new formations.

2. Kuerassier-Regiment.
Raised in 1655.
Title in 1806: **von Beeren.**
Inspectorate: Mark Brandenburg. Garrisons Gransee, Kyritz, Perleberg, Wittstock,Wusterhausen.

Uniforms. Lemon yellow tunics, white breeches, crimson facings and braid (the latter was white on the waistcoat). Officers' lace was silver. Holster covers and the shabraque bore the crowned royal cipher within braid edging.

Standards. The five standards were lost at the capitulation of Ratkau on 7 Nov 1806.

Campaign history. Raised on 13 Dec 1666 by Oberst Graf von Russow; 1670 Chef Oberst Christoph von Strauss (killed as GM in Hungary). 1672 Oberst Churprinz Friedrich von Brandenburg; received IR Nr 15. In 1674 it received the title 'Churprinz', later 'Kronprinz'. **1674** Against France; in Alsace, capture of Wesselsheim castle. **1675** 28 Jun B Ferhbellin. **1677** Against Sweden in Pommerania, S Stettin; invasion of Ruegen Island. **1686** Against the Turks in Hungary 14 Jun-3 Sep S Ofen. **1689** Against France, 24 Jun-12 Oct S Bonn. **1692/1693** Against France on the upper Rhine. **1694** at Luxemburg; Tournay; Tirlemont. **War of the Spanish Succession 1708** 11 Jul B Oudenarde. **1709 11** Sep B Malplaquet. **1710** 22 Apr-27 Jun S Douay;15 Jul-29 Aug S Bethune; 6-29 Sep Aire. **1715** Against Sweden in Pommerania. **1731** Chef Oberst Prinz August Wilhelm von Preussen, the elder brother of the king; died in office. **1742** Against Austria 17 May B Chotusitz; broke through two enemy lines and scattered two Hungarian regiments. **1745** 4 Jun B Hohenfriedberg; broke a Saxon regiment; 30 Sep B Soor; together with the 1st and 10th Kuerassiers it broke the enemy line to capture the Graner Koppe heights and 22 guns. **Seven Years War 1756** 1 Oct B Lobositz; **1757** 9 May-20 Jun S Prague, 18 Jun B Kolin, the regiment led the charge of Krosigk's brigade, which broke three enemy infantry regiments. 22 Nov B Breslau, a Prussian defeat. 5 Dec B Leuthen. **1758** Oberst Prinz Heinrich von Preussen, second son of Prinz August Wilhelm. S Schweidnitz; raid into Franconia. B Zorndorf; played a key part in the Prussian victory. Relief of Dresden. **1759** Raid into Franconia to destroy Austrian magazines. 12 Aug B Kunersdorf, a Prussian defeat. **1760** 19 Jul Bombardment of Dresden; 15

Aug B Liegnitz, took a battery. 3 Nov B Torgau. **1761** 15 Feb C Langensalza; invaded Poland to destroy the Russian magazines there. **1762** 7 Aug-9 Oct S Schweidnitz; 27 Oct C Reichenbach; then into Saxony, Sk Spechtshausen, Sk Tharanter Forest; together with the 2nd Hussars they broke aa Austrian corps and took 400 prisoners and a gun. 1767 Chef Oberst Georg Ludwig von Wirsbitzki (died as GM). 1778 Chef GM Christian Rudolf von Weyher; died in office. **1778/1779** in the War of the Bavarian Succession they were in Prinz Heinrich's army. 1782 Chef GM Friedrich August von Saher; died in office. 1783 Chef GM Karl von Backhoff; became Director of the 2nd Department of the War Ministry. 1789 Chef GL Gustav Ludwig von der Marwitz; died in office. 1797 Chef Oberst Peter Ewald von Malschitzki (died as GM). 1802 Chef Oberst Andreas Dietrich von Schleinitz; retired as GM. 1805 Chef Oberst Carl Friedrich Hermann von Beeren. **1806** 14 Oct Auerstaedt, in Kalckreuth's Reserve. The regiment withdrew with Bluecher's corps, surrendered at Ratkau and was not re-raised. 70 men and horses returne to East Prussia; they went into the new 4th Kuerassiers.

3. Kuerassier-Regiment.
Raised in 1672.
Title in 1806: **Leibregiment.** Inspectorate: Magdeburg. Garrisons. Calbe, Egeln, Frose, Salze, Schoenebeck, Wanzleben.

Uniforms. White tunic and breeches, dark blue facings, dark blue velvet braid with a wide central white stripe; officers‘ lace gold. Holster covers and the shabraque bore the crowned royal cipher within braid edging.

Standards.

Campaign history. Raised in 1672 by Oberst Nikolaus von Below and entitled the 'Leibregiment'. Below was later posted from the regiment. 1673 Chef GM Ulrich Graf von Promnitz; transferred into Saxon service. **1675** The regiment first came into action at Fehrbellin on 28 Jun. **1677** 24 Jul-22 Dec S Stettin. 1679 Chef Oberst Hans von Sydow; diedin office. 1680 Oberst Joachim Balthasar von Dewitz; later GL and governor of Colberg. **1689** S Bonn. **1695** Chef GL von Wangenheim; died in office. 2 Jul-3 Aug S Namur. **1706** S Ath in Brabant. **1708** 11 Jul B Oudenarde, Oct S Lille. **1709** Chef GM Wolf Christoph von Hackeborn; died as GL.

S Tournay; S Mons; B Malplaquet. **1710** 22 Apr-27 Jun S Douai, 6 Sep-9 Nov S Aire, S Landrecy, S Quesnoy. 1719 Chef GL Gottfried Albrecht von Bredow; he became governor of Peitz. 1725 Chef GL Friedrich Wilhelm von Dewitz; died in office. 1736 Chef Oberst Adam Friedrich von Wreech; retired as GL and died that same day. **1745** Campaign in Pommerania; 15 Dec B Kesselsdorf. 1746 Chef GM Nikolaus Andreas von Katzler; became commandant of the Gensdarmes. 1747 Chef GM Hans Friedrich von Katte; retired as GL. **1756** 1 Oct B Lobositz, distinguished. **1757** Distinguished at 6 May B Prague, 18 Jun B Kollin, 5 Nov B Rossbach. **1758** Chef GM Robert Scipio Freiherr von Lentius; retired as GL. Took part in the raid on Olmuetz; Aug C Gross-Sedlitz; C Dresden in Sep; Nov C Eilenburg. **1759** B St Sebastiansberg; together with two grenadier companies, two squadrons of the regiment captured the opposing commander at Passberg, together with three guns, eight colours, three standards and 800 men. 12 Aug B Kunersdorf; heavy losses, also lost a standard. **1760** 10-22 Jul Bombardment of Dresden; 15 Aug B Liegnitz; took seven colours and five guns from the Austrian IRs Nrs 1, 24 and 35. 3 Nov B Torgau. **1761/1762** operations in Saxony. 1778 GM Rudolf von Merian; retired. **1778/1779** In Prinz Heinrich's army. 1782 Chef GM Ernst Christian von Kospoth; retired as GL. 1795 Chef MG Heinrich Leopold Graf von der Goltz; retired. **1793** On the Rhine. **1794** 23 May B Kaiserslautern. 1797 Chef Oberst August Friedrich von der Droessel; retired as GM. 1800 Chef GM Ulrich Karl von Froreich; died in office. 1801 Chef Oberst Hermann von Koelichen; died in 1805 as GM. 1805 Chef Oberst Friedrich August Carl Leopold Graf von Schwerin. **1806** In Hohenlohe's corps at Jena; most of the regiment surrendered at Prenzlau on 28 October. The regiment was not re-raised. A detachment escaped to East Prussia and was used to form the new 4th Kuerassiers; see the newly raised regiments below.

4. Kuerassier-Regiment.
Raised in 1674.
Title in 1806: **von Wagenfeld.** Inspectorate: Warsaw. Garrison Warsaw.

Uniforms. White tunic and breeches, black facings, white braid with three lines of dark blue squares; officers' lace gold. Holster covers and the shabraque bore a crowned shield bearing the crowned Prussian eagle within braid edging.

Standards. In 1808 they carried the following standards; king's – white ground, purple centre, silver corner fields, gold embroidery and fringes; regimental – purple ground, whit centre and corner fields, gold embroidery.

Campaign history. Raised in mid 1674 from the Stabs-Dragoner, titled the Lieb-Dragoner-Regiment, and given to Oberst Joachim Ernst von Grumbkow; later War Commissar. 1675 28 Jun B Fehrbellin. **1677** fought in Pommerania against the Swedes; 24 Jul-22Dec S Stettin. **1678** 23 Sep capture of Ruegen Island. 1684 Chef Oberst Dietrich, Graf zu Dohna, died in 1686. 1686 Oberst Joachim Friedrich von Wreech; later GL. **1688** Against the Turks in Hungary; 14 Jun-3 Sep S Ofen. **1689** 24 Jun-12 Oct S Bonn. **1703** B Noerdlingen; 24 Apr-15 May S Bonn; Dec S Geldern. **1704** 13 Aug B Hochstaedt, heavy losses and lost three standards. 1713 the regiment lost its title of Leib-Regiment. 1713 Chef GM Peter von Blanckensee. **1715** in Pommerania. 1718 the regiment was increased to five squadrons and became a Kuerassier regiment. 1733 Chef Oberst Friedrich Leopold von Gessler; later Graf and FM. **1742 Against Austria.** 17 May B Chotusitz. **1745** 22 May C Neustadt, 4 Jun B Hohenfriedberg, 30 Sep B Soor. **1756-1763 Against Austria in the Seven Years War. 1757** Chef MG Johann Ernst von Schmettau. 6 May B Prague; secured the left wing and ensured victory. Almost all the officers were killed or wounded. 9 May-20 Jun S Prague. Covered the retreat of the army after the defeat of Kollin. 22 Nov B Breslau; 5 Dec B Leuthen; took a standard. **1758** C Olmuetz, 30 Jun C Domstadl, C Bautsch, 14 Oct B Hochkirch. 1 Nov relief of Neisse; 16 Nov advance to Dresden. **1759** Schmottseifen, then in Fouque's corps at 17 Apr C Troppau. 25 Sep C Hoyerswerda-took a general, 28 officers and over 1,700 men. **1760** the regiment was ambushed near Torgau, but rallied and drove the enemy off with loss. C Bublitz; 3 Nov B Torgau. **1762** 12 May C Doebeln. 21 May C Chemnitz; 29 Oct B Freiberg; took 10 guns, 4 howitzers and 8 colours. Seven officers of the regiment received the *Pour-le-Merite* decoration for this action. 1764 Chef MG Hans Georg Woldeck von Arneburg. 1769 Chef Oberst Georg Christoph von Arnim; later GL. 1785 MG Carl Friedrich Freiherr von Mengden, ex-Garde du Corps; later LG. **1794** In Poland; no actions of note. 1796 Chef MG Carl Friedrich Ernst Graf Truchses zu Waldburg. 1800 Chef Oberst Ernst Philipp von Wagenfeld; GM in 1806. **1806** In the reserve corps in East Prussia. **1807** 7/8 Feb B Preussisch-Eylau. 26 Feb C Braunsberg. 10 Jun B Heilsberg. 14 Jun C

Koenigsberg. It became the new 1st Kuerassiers and took in a squadron of the Prittwitz Dragoons as its 6th Squadron.

5. Kuerassier-Regiment.
Raised in 1683.
Title in 1806: **von Balliodz.** Inspectorate: Pommerania. Garrisons Coerlin; Dramburg, Greiffenberg; Treptow on the Rega and Wollin.

Uniforms. White tunic and breeches, sky blue facings, white and sky blue diced braid, officers' lace gold. Holster covers and the shabraque bore a crowned shield bearing the crowned Prussian eagle within braid edging.

Standards. One standard was lost in combat at Steckenitz on 6 Nov; the other four were destroyed by their bearers in Erfurt prior to the capitulation. The remnants were never found.

Campaign history. Raised in May 1683 at six companies for GM Heinrich Graf de Briquemault Sieur de St-Loup. The basis was one of the *Freicompagnien* raised by Brigadier von Iselstein. **1686** Against the Turks in Hungary; 3 Sep storm of Ofen. In 1687 it was increased to ten companies; next year it gave four companies to raise Kuerassier R Nr 6; in 1691 another company went to raise Kuerassier R Nr 9. **1689** 24 Jun-12 Oct S Bonn. 1693 Chef MG Markgraf Philipp Wilhelm von Brandenburg; died as GL. **1702** 15 Apr-15 Jun S Kaiserswerth and 29 Au-Sep Venlo. **1704** 13 Aug B Hochstaedt; took a French standard. 1711 Chef Oberst Markgraf Friedrich von Brandenburg-Schwedt, son of the previous Chef; until his death in Mar 1771 the regiment had the following commanders: 1711 GM von Bredow; 1718 Oberst von Rochow (became Chef of Kuer R Nr 8; the regiment was increased to five squadrons. 1742 GM von Podewils; retired; 1753 GM von Luederitz (killed at Kollin); 1756 Oberst von Krosigk (killed at Kollin); 1757 Oberst von Zieten (killed at Zorndorf); 1758 GM von Aschersleben (died); 1761 Chef Oberst Friedrich Wilhelm Loellhoefel von Loewensprung; in 1771 he became GM and Chef of the regiment, died. **1715** Against the Swedes in Pommerania. **1741** Against Austria; 10 Apr B Mollwitz. **1742** 17 May B Chotusitz; heavy losses. **1745** B Hohenfriedberg; B Soor; B Kesselsdorf; very distinguished in all three actions. **1757 Against Austria and others in the Seven Years War.** **1757** 6 May B Prague; 9 May-20 Jun S Prague; 22 Nov B Breslau; 5 Dec

B Leuthen. **1758** 31 Mar-18 Apr S Schweidnitz; 25 Aug B Zorndorf. **1759** 23 Jul B Kay; 12 Aug B Kunersdorf; 21 Sep C Meissen. **1760** 19 Jul bombardment of Dresden; 15 Aug B Liegnitz, inflicted heavy losses on the enemy; 3 Nov B Torgau, took 14 guns and 13 colours. **1762** 29 Oct B Brand (Freiberg). **1778** in the army of Prinz Heinrich. 1780 Chef GM Maximilian von Mauschwitz; died. 1782 Chef GM Prinz Ludwig Alexander von Wuerttemberg; retired as FM. **1794** in Poland; defence of Posen. 1800 Chef Oberst Abraham von Balliodz, GM in 1806. **1806** 14 Oct Jena; most of the regiment surrendered at Erfurt; one squadron was destroyed at Steckenitz; the regiment was not re-raised. Parts of the regiment under Oberst von Stulpnagel fought their way out to East Prussia; they went to form the new 5th Dragoons.

6. Kuerassier-Regiment.

Raised in 1688.

Title in 1806: **von Quitzow.** Inspectorate: Magdeburg. Garrisons: Aschersleben, Kroppenstaedt, Oschersleben.

Uniforms. White tunic and breeches, light brick red facings, white braid with a light brick red pattern; officers' lace gold. Holster covers and the shabraque bore the crowned royal cipher within braid edging.

Standards. Three standards were lost at the capitulation of Anklam on 1 Nov 1806; two were saved.

Campaign history. Raised in 1688 with four companies of Kuer R Nr 5 and two companies of new recruits. Title GM Marquis Franz du Hamel; retired. The regiment went into Dutch service. **1689** S Bonn. . In 1691 it gave a company to help raise Kuer R Nr 9; completed itself with new recruits. Its first garrison was in Cleve. In 1718 it was increased to five squadrons recruited in the Dessau area. **1695** S Namur. **1702 War of the Spanish Succession.** Chef MG Charles Comte de l'Ostange; died in office. S Kaiserswerth. **1704** November Chef Oberst Benjamin Hieronimus Courold du Portail; resigned his post and received Kuer R Nr 12. 13 Aug B Hoechstaedt; lost a standard, but took two colours. **1705** 16 Aug B Cassano. **1709** 11 Sep B Malplaquet; took two kettle drums. **1715** In the Pommeranian campaign against Sweden. 1716 Chef GL Wilhelm Gustav Erbprinz von Anhalt-Dessau. 1737 GM Prinz Eugen von Anhalt-Dessau;

left Prussian service. 1742 Chef GM Christoph Ludwig von Stille; died in office. **1744** 10-16 Sep S Prague. **1745** 15 Dec B Kesselsdorf; took a colour.. **1756-1763 Against Austria and others in the Seven Years War. 1753** Chef GM Georg Philipp Gottlob Freiherr von Schoeneich; retired. **1757** 6 May B Prague; 18 Jun B Kolin, 22 Nov B Breslau; 5 Dec B Leuthen; defeated the enemy grenadiers. **1758** 27 May-2 Jul S Olmuetz; 14 Oct B Hochkirch; overthrew the Austrian infantry, took 500 prisoners and a colour and covered the retreat of the army. **1759** Chef GM Heinrich Rudolf Wilhelm von Vasold; retired. The regiment was captured in Maxen. By **1760** the regiment had been re-raised by use of taking five companies from Kuer R Nr 7 and recruiting five new. It fought against the Russians in Pommerania. **1761** raids into Poland. **1762** 7 Aug-9 Oct S Schweidnitz; 27 Oct B Reichenbach. 1769 Chef Oberst Just Rudolf von Seelhorst; died as GM. **1778/1779** In the army of Prinz Heinrich. 1779 Chef GM Theophil Ernst Freiherr von Overbeck; died in office. **1781** Chef Oberst Hans Ludwig von Rohr; took Drag R Nr 6 as GM. **1787** On campaign in Holland. Chef Oberst Carl August, reigning Duke of Sachsen-Weimar; retired as GL. **1792** On the middle Rhine; capture of Longwy and Verdun. 20 Sep Valmy. **1793** 1-7 Mar blockade of Koenigsstein Sk Stromberg; 19 Jun-23 Jul S Mainz. Sk Marienborn. Nov-Dec Blockade of Landau. **1794** In Prinz Hohenlohe's corps; Sk Meckenheim, Sk Kirweiler, Sk Edinghofen, 20 Sep B Kaiserslautern. 1800 Chef Oberst Christian Heinrich von Quitzow; GM in 1806. **1806** 14 Jun in the Centre at Auerstaedt; 3 squadrons destroyed. Many of these surrendered at Anklam. The regiment was not re-raised. 12 officers and 250 men fought their way out to East Prussia; they went to form part of the new 4th Kuerassiers.

7. Kuerassier-Regiment.
Raised in 1688.
Title in 1806: **von Reitzenstein.** Inspectorate: Magdeburg. Garrisons Osterburg, Salzwedel, Seehausen, Tangermuende

Uniforms. White tunic and breeches, lemon yellow facings; white braid with three lemon yellow stripes, officers' lace silver. Holster covers and the shabraque bore the crowned royal cipher within braid edging.

Standards. The five standards were lost at the capitulation of Magdeburg on 11 Nov 1806.

Campaign history. Raised in 1689 by Captain Friedrich Wilhelm Graf von Wittenhorst zu Sonsfeld (died later as GL) in Prussia as a dragoon regiment including some Free Companies. In 1717 it was formed into a Kuerassier regiment. **1689** 24 Jun-12 Oct S Bonn. **1704** 13 Aug B Hoechstaedt; took two kettledrums. 1711 Chef GM Georg Friedrich von der Albe; died later as GL. **1715** in Pommerania against the Swedes. New Chef Oberst Ludwig Graf von Wylich und Lottum; died as GL. 1729 Chef Oberst Carl Friedrich von Papstein; died in office. 1733 Oberst Friedrich Sigismund von Bredow; retired as GdC. **1742 Against Austria.** 17 May B Chotusitz; broke through both enemy lines. **1745** 4 Jun B Hohenfriedberg; drove off the Saxon cavalry and destroyed the Saxon IR Schoenberg. 15 Dec B Kesselsdorf. 1755 Chef GM Goerg Wilhelm von Driesen; died as GL. **1756** 1 Oct B Lobositz. **1757** 6 May B Prague; 18 Jun B Kolin; 5 Nov B Rossbach; 5 Dec B Leuthen. **1758** Chef GM Christian Sigismund von Horn; retired. 31 Mar-18 Apr Blockade of Schweidnitz; 27 May-2 Jul S Olmuetz. **1759** 23 Jul B Kay; scattered the Russian infantry and took 18 guns, but was unable to bring them off. 12 Aug B Kunersdorf, stormed a redoubt. 20 Nov C Maxen; the much-weakened regiment was captured. **1762** Chef Oberst Leopold Sebastian von Manstein; died as MG. 7 Aug-9 Oct S Schweidnitz. 1777 Chef GM Gustav Ludwig von der Marwitz; retired as GL and Chef of Kuer R Nr 2. **1778/1779** in the army of Prinz Heinrich. 1784 Chef Oberst Friedrich Adolf Graf von Kalckreuth; received Drg R Nr 5 as GM. **1787** on campaign in Holland; 17 Sep captured the Oude Meer redoubt with 80 prisoners and 18 guns. Together with a company of infantry, the regiment cut the fortress of Nieuversluis off from Amsterdam and forced it to surrender with 800 prisoners and 95 guns. Two of these were taken into service, crewed by the troopers, and used to bombard Amsterdam. These guns were then presented to the regiment. In this campaign the regiment took a total of 127 guns. 1788 Chef GM Otto Friedrich von Ilow; died in office. **1792** Against France. Chef GM Hans Friedrich Heinrich von Borstel; died in 1804 as GL. Covered the siege of Longwy; blockade of Verdun, cannonade of Valmy. **1793** 3 May storm of Kostheim; 19 Jun-23 Jul S Mainz; 14 Sep B Pirmasens; took over 1,000 prisoners, 2 howitzers and 11 guns. All field officers of the regiment received the *Pour-le-Merite*. The king ordered that the 13 captured guns should be included in the regimental seal. 28-30 Nov B Kaiserslautern. **1794** 23 May B Kaiserslautern; captured an entire enemy brigade. 1804 Chef MG Heinrich August von Reitzenstein. **1806** 14 Oct Auerstaedt; very heavy losses. The remainder served with

Bluecher, surrendered at Magdeburg. The regiment was not re-raised. The depot became part of the new 4th Kuerassiers.

8. Kuerassier-Regiment.
Raised in 1690.
Title in 1806: **von Heising.** Inspectorate: Lower Silesia. Garrisons Grottkau, Loewen, Olau, Strehlen.

Uniforms. White tunic and breeches, dark blue facings, white braid with two wide, dark blue stripes and narrow dark blue edge piping, officers lace silver. Holster covers and the shabraque bore a crowned shield bearing the crowned Prussian eagle within braid edging.

Standards. The five standards were lost at the capitulation of Pasewalk on 29 Oct 1806.

Campaign history. Raised in 1690 for Oberst Markgraf Christian Ernst of Brandenburg-Bayreuth from two companies of cavalry raised in 1689 by Oberst Freiherr von Lethmate and two of Mecklenburg cavalry. **1689** the two companies fought against France; 24 Jun-12 Oct S Bonn. **1690** Against the Turks in Hungary. **1691** 19 Aug C Salankement. **1697** 11 Sep B Zenta. **1702** Against France; April S Kaiserswerth. **1704** 13 Aug B Hoechstaedt. 1718 Chef GM Stephan von Dewitz; died as GL. Set at five squadrons including some companies of the disbanded regiment von Heyden. 1712 GM von Letmathe; died in office. 1714 Chef GM Stephan von Dewitz; died as GL. 1723 Chef Oberst Friedrich von Egeln; retired as GL. 1734 Chef Oberst Friedrich Sigismund von Waldow; died in office. **1742** Against Austria. Chef Oberst Friedrich Wilhelm von Rochow; retired. 17 May B Chotusitz. **1744** 10-16 Sep S Prague. **1745** 22 May C Neustadt in Upper Silesia. 4 Jun B Hohenfriedberg; 30 Sep B Soor; 23 Nov C Katholisch-Hennersdorf, took a standard, three colours and four guns. B Kesselsdorf. **1756** B Lobositz. **1757** Chef GL Friedrich Wilhelm Graf von Seidlitz; died as GdC in 1773. 6 May B Prague; 18 Jun B Kolin; 5 Nov B Rossbach; 5 Dec B Leuthen; 8-21 Dec S Breslau. **1758** 31 Mar-18 Apr S Schweidnitz; 27 May-2 Jul S Olmuetz; 25 Aug B Zorndorf, took a battery of heavy guns and overthrew the enemy right wing. 14 Oct B Hochkirch. **1759** 25 Sep C Hoyerswerda. **1760** Jul S Dresden; 15 Aug B Liegnitz-took 12 guns, four colours and five battalions. 3 Nov B Torgau. **1761** 15 Feb C Langensalza; 14 Aug C Kloster Wahlstadt. **1762** 21 Jul C

Leutmannsdorf; 16 Aug C Reichenbach; 7 Aug-9 Oct S Schweidnitz. 1774 Chef GM Maximilian Sigismund von Pannewitz; retired as GL. **1778** In the army of the king. 1787 Chef GL Graf von Schlitz, known as Goertz; died as GdC. **1794** In Poland; many minor actions around Krakau and Sendomir; C Polinowka near Casimir. 1797 Chef Oberst Ludwig Ferdinand Friedrich von Heising; GL in 1806. **1806** 14 Oct in the Left Wing at Auerstaedt; heavy lossess. The rest of the regiment surrendered at Pasewalk. The regiment was not re-raised. Remnants went into the new 1st Kuerassiers.

9. Kuerassier-Regiment.
Raised in 1691.
Title in 1806: **von Holzendorf.** Inspectorate: Upper Silesia. Garrisons Falkenberg, Krappitz, Neustadt, Oppeln.

Uniforms. White tunic and breeches, crimson facings, white braid with three crimson stripes, officers' lace gold. Holster covers and the shabraque bore a crowned shield bearing the crowned Prussian eagle within braid edging.

Standards. The five standards were lost at the capitulation of Pasewalk on 29 Oct 1806.

Campaign history. Raised in 1691 from one company each of the regiments Nrs 1, 2, 3, 5 and 6 for Oberst Hans Ehrentreich von Schoening. **1691** in Brabant and on the Rhine. 1703 Schoening sold the regiment for 8,000 Thaler to Oberst Philipp Ludwig Freiherr von Canstein. In 1705 von Canstein sold it on for the same price to Oberst and General Adjutant Hans Heinrich von Katte; later FM. **1708** 11 Jul B Oudenarde 1718 increased by one company to five squadrons. 1741 Oberst Herman Graf von Wartensleben; resigned this same year. **1742** 17 May B Chotusitz. 1743 GM Bernhard Heinrich von Bornstedt; retired as GL. **1744** 10-16 Sep S Prague. **1745** 4 Jun B Hohenfriedberg; took three standards; 30 Sep B Soor took ten colours and captured three enemy regiments. 23 Nov C Katholisch-Hennersdorf, took two standards. 1751 GM Johann Carl Friedrich Erbprinz von Schoenaich-Carolath; retired as GL. **1757** 6 May B Prague, took two standards. Sk Moys; 22 Nov B Breslau; 5 Dec B Leuthen. **1758** GM Jakob Friedrich von Bredow; retired. 31 Mar-18 Apr S Schweidnitz; 27 May-2 Jul S Olmuetz; 14 Oct B Hochkirch. **1759** 20

Nov C Maxen; the regiment was captured. **1762** 7 Aug-9 Oct S Schweidnitz. 1769 Chef GM Friedrich Wilhelm von Podewils; died. **1778/1779** in the army of the king. 1784 Chef Oberst Christian Friedrich von Braunschweig; died as GL. 1787 Chef GM Johann Wilhelm von Manstein; retired as GL. **1794** parts of the regiment were in Poland. 1797 Chef Oberst Friedrich Jakob von Holtzendorf; GL in 1806. **1806** 14 Oct; on the right wing at Jena; heavy losses. The regiment surrendered and was not re-raised. Remnants were used in the new regiments.

10. Kuerassier-Regiment.
Raised in 1691.
Title in 1806: **Gens d'armes.** Inspectorate: Mark Brandenburg. Garrison Berlin.

Uniforms. White tunic and breeches, red facings, red braid with a golden stripe, officers' lace gold. Officers' holster covers and the shabraque bore a crowned, eight-pointed star, as did their cartouche lids within braid edging. Prior to 1803, troopers' cartouche lids were white, edged in regimental braid and in the centre the round brass plate bearing the crowned royal cipher. After this date they became plain black with the same plate.

Standards. Four standards were lost in combat at Wichmannsdorf on 27 Oct; the fifth at the capitulation of Anklam on 1 Nov 1806.

Campaign history. Raised at one company in 1691 by Oberst Dubislav Gneomar von Natzmer in the area of Halberstadt from new recruits. Von Natzmer became the Chef. A second squadron was raised in 1713, then a third. In 1714 the White Squadron of the Garde du Corps was brought into the regiment; a fifth squadron was raised in 1718. The sergeants of the regiment had the rank of Cornets in the army.

1702 On the Rhine; S Rheinbergen. **1703** End Mar blockade of Geldern; 24 Apr-15 May S Bonn; Sk Lauingen. **1704** 13 Aug B Hoechstaedt. **1706** 23 Jul-22 Aug S Menin. **1708** 11 Jul B Oudenarde. **1709** 11 Sep B Malplaquet. **1710** 22 Apr-27 Jun S Douay; 15 Jul-28 Aug S Bethune; 6 Sep-9 Nov S Aire. **1711** 7 Aug-10 Sep S Bouchaine. **1715** Four squadrons at S Stralsund (17 Jul-24 Sep). 1739 Chef Oberst Wolf Adolf von Pannewitz; retired as GM. **1741** 10 Apr B Mollwitz. **1742** Ambushed at night in its village quarters in Senitz, south of Wsetin in March, the regiment fought off its attackers. 1743 Chef Oberst Georg

Konrad Freiherr von der Goltz; died as GM. 1745 4 Jun B Hohenfriedberg; 30 Sep B Soor; together with Kuer R Nr 1 they drove 50 Austrian squadrons from a hill and captured the main enemy battery. 1747 Chef GM Nikolaus Andreas von Katzler, ex-Chef of Kuer R Nr 3; died as GL. **1756** 1 Oct B Lobositz; took two standards. **1757** 9 May-20 Jun S Prague; 5 Nov B Rossbach; 5 Dec B Leuthen; took a cannon. Together with the Garde du Corps, they led the pursuit and brought in 15 more, as well as colours, standards and several hundred prisoners. **1758** 27 May-2 Jul S Olmuetz; 25 Aug B Zorndorf; together with the Garde du Corps and Drag R Nr 4 they created havoc in the Russian grenadiers; took seven colours and 14 cannon. 14 Oct B Hochkirch. **1759** 25 Sep C Hoyerswerda. **1760** July S Dresden; 17 Sep C Hoch-Giersdorf; 15 Aug B Liegnitz; 3 Nov B Torgau. **1761** Chef Oberst Friedrich Albert von Schwerin; retired as GM. Sk Adelsbach; C Reichenbach (200 of the regiment) took a standard. 1768 Chef GM Hans Friedrich von Krusemark; died. 1775 Chef GM Joachim Bernhardt von Prittwitz; died as GdC. **1778** 25 Oct C Jaegerndorf. 1793 Chef GM Carl Friedrich von Elsner; GL in 1806. **1794** In Poland; minor actions. **1806** 14 Oct in von Kalckreuth's Reserve at Auerstaedt; heavy losses. Joined Hohenlohe's corps and surrendered at Anklam. The regiment was not re-raised. The depot went into the new 4th Kuerassiers.

11. Kuerassier-Regiment.

Raised in 1692.

Title in 1806: Leib-Carabiniers. Inspectorate: Magdeburg. Garrisons Genthin, Havelberg, Neu-Haldensleben, Rathenau, Sandau.

Uniforms. White tunic and breeches, light blue facings, white braid with a light blue central, open diamond pattern between two light blue stripes; officers' lace silver. Holster covers and the shabraque bore the crowned royal cipher within braid edging.

Standards. The five standards were lost at the capitulation of Pasewalk on 29 Oct 1806.

The regiment was not re-raised.

Campaign history. Raised in 1692 by Oberst Paul von Brandt as a regiment of dragoons by addition of companies from existing regiments. In 1718 it was converted to a Kuerassier regiment of five squadrons. King Friedrich II dubbed it the Leib-Carabiniers. **1675** 28 Jun B Fehrbellin. **1677** in Pommerania against the Swedes. **1679** helped drive the Swedes out of Prussia. **1689** On the Rhine; 11 Mar C Nuess; Jun S Kaiserswerth; 24 Jun-12 Oct S Bonn. 1692 Chef Albrecht Markgraf von Brandenburg; died. **War of the Spanish Succession** In Brabant and Flanders; present at many sieges. 1731 Markgraf Carl von Brandenburg, son of the deceased Chef. That same year, he left to become Chef of IR Nr 19. 1731 Chef MG Carl Ludwig Graf von Truchses zu Waldburg; died. 1738 Chef Oberst Hermann Graf von Wartensleben; became Chef of Kuer R Nr 9. **1741**Chef Oberst Kaspar Ludwig von Bredow; retired as GL. 10 Apr B Mollwitz; heavy losses. **1744** 10-16 Sep S Prague. **1745** 4 Jun B Hohenfriedberg; 15 Dec B Kesselsdorf. 1751 Chef GM Peter Ernst von Pennevaire; died. **1756** B Lobositz. **1757** 6 May B Prague; very distinguished; 18 Jun B Kolin, took a standard. 22 Nov B Breslau; 5 Dec B Leuthen, took a cannon. **1758** 27 May-2 Jul S Olmuetz; Sk Wischau; 25 Aug B Zorndorf, very effective charges against the Russians; all field officers of the regiment were awarded the *Pour-le-Merite*. 14 Oct B Hochkirch. 1759 Chef GM Joachim Christian von Bandemer; died in 1764, the regiment remained vacant until 1768. **1760** 15 Aug B Liegnitz; 3 Nov B Torgau. **1761** 15 Feb C Langensalza; took two cannon. **1762** 29 Oct B Freiberg. 1768 Chef Oberst Reinhold Friedrich Baron von Hoverbeck; died as GM. 1771 Chef Oberst Reimar von Kleist; retired as GM. 1775 Chef Oberst Philipp Christian von Bohlen; became Chef of Kuer R Nr 1. **1778/1779** in Prinz Heinrich's army. 1784 Chef Oberst Heinrich Sebastian von Reppert; retired as GL. **1792** On the Rhine; no actions of note. **1793** Took part in the successful expedition to Schwalm, 3 Mar Roermond and Venlo; 8 May C Raismes; 1-2 May C Vicogne and St Amand; it charged at St Imbert (27 Sep), and Saarbruecken; 28-30 Nov B Kaiserslautern, cut down a regiment of French infantry in square and took a colour. In the subsequent pursuit, it cut down and captured hundreds. **1794** 23 May C Schifferstadt; Sk Meckenheim; Sk Fischlingen. 1795 Chef MG August von Bismark; GL in 1806. **1806** 14 Oct Auerstaedt; in the Prince of Orange's Right Wing; heavy losses. The regiment surrendered at Pasewalk and was not re-raised.

The depot went into the new 4th Kuerassiers.

12. Kuerassier-Regiment.
Raised in 1695.
Title in 1806: **von Bunting.**
Inspectorate: Upper Silesia. Garrisons Katscher, Leobschutz, Ober-Glogau, Ratibor.

Uniforms. White tunic and breeches, dark orange facings, white braid with two wide orange stripes; officers' lace gold. Holster covers and the shabraque bore a crowned shield bearing the crowned Prussian eagle within braid edging.

Standards. The five standards were lost at the capitulation of Pasewalk on 29 Oct 1806.

Campaign history. Raised in 1695 as a regiment of dragoons from two companies each from the regiments Taschen and Kuechen; increased to eight companies. It saw much good service in Brabant, Flanders and on the Rhine. 1705 Chef August Graf von Sayn und Wittgenstein. 1710 Chef GM Ludolf von Pannewitz; retired as GL. **1715** In Pommerania; capture of Usedom. As a reward for this distinguished service it was converted to a Kuerassier regiment that same year and increased to five squadrons. 1716 Chef GM Erbprinz Gustav von Anhalt-Dessau for only three days, he then became Chef of KuerR Nr 6. In November Oberst Benjamin Hieronimus Courold du Portail; resigned his post and received Kuer R Nr 12. 1717 Chef Oberst Georg Levin von Winterfeld; died as GL. 1728 Oberst Arnold Christoph von Waldow; died as GL. **1742** 17 May B Chotusitz; broke through both enemy lines and cut down two Hungarian infantry regiments. **1743** Chef Oberst Friedrich Wilhelm Freiherr von Kyau; died as GL. **1744** 10-16 Sep S Prague. **1745** 4 Jun B Hohenfriedberg; 30 Sep B Soor; 15 Dec B Kesselsdorf. **1757** 6 May B Prague; 18 Jun B Kollin; 22 Nov B Breslau; 5 Dec B Leuthen. **1758** C Domstaedl; saved the greater part of a large convoy of bullion and ammunition supplies destined for Olmuetz. **1759** Chef GM Johann Heinricg Friedrich Freiherr von Spaen; died. 23 Jul B Kay; 12 Aug B Kunersdorf. **1760** 3 Nov B Torgau; scattered two enemy infantry regiments, captured a corps of grenadiers and six cannon. Over half the regiment were casualties. Each squadron commander was rewarded with the *Pour-le-Merite* and 500 Thalers. **1762** Chef Oberst Georg Ludwig von Dallwig; died as GdC. **1778** in the army of the king. 11 Sep C Burkersdorf; Sk Leutmannsdorf; Sk

Reichenbach. **1794** In Poland. 1796 Chef GM Jakob Friedrich von Berg; died 1797. 1798 Chef Oberst Georg Christian von Werther; retired as GM. 1804 Chef GM Carl Wilhelm von Buenting. **1806** 14 Oct Auerstaedt; the regiment was in the Left Wing; heavy losses. It surrendered at Pasewalk and was not re-raised.

13. Kuerassier-Regiment.
Raised in 1740.
Title in 1806: **'Gardes du Corps'.**
Inspectorate: Mark Brandenburg. Garrisons Berlin, Charlottenburg, Potsdam.

Uniforms. White tunic and breeches, red facings, red braid with silver stripes; blue waistcoats; officers' lace silver. Officers' holster covers and the shabraque bore a crowned, eight-pointed star, as did their cartouche lids within braid edging. From 1803 this star was also worn on troopers' cartouche lids, which had been white up to that point.
In 1808 facings were red, buttons white.

Standards. In 1741 they were presented with a standard having a silver eagle instead of the usual spear point; the cloth hung from silver chains. On 29 Feb 1757 two new standards were presented. On 7 Sep 1798 five new standards, with conventional spear points, were presented. One was damaged at Auerstaedt. In 1808 they carried a Leib-Standarte and three regimental standards. The Leib-Standarte was of silvered cloth with silver and gold embroidery and silver fringes. The regimental standards were also of silvered cloth, but the centres and corner fields were orange; embroidery was silver and gold, the fringes silver. On 17 Mar 1813 three of these were handed in to the arsenal; taken out again in December 1814. Two of these were given to the Garde-Ulanen (later to become the Garde-Kuerassiers) and the Garde-Husaren. On 5 Dec 1814 another one was given to the Light Guards Cavalry Regiment. The standard retained by the Garde-du-Corps was fitted with a silver eagle on 16 Nov 1840, in the same style as the original.

Campaign history. Raised in 1740 in Potsdam at one squadron of 178 men. Candidates for all ranks were selected from existing cavalry regiments. The king was Chef of this regiment; the commanders are listed as they changed. 1740 Rittmeister Otto Friedrich von Blumenthal; died as

Major. 1741 Rittmeister Georg Christoph von Jaschinsky; became Chef of Kuer R Nr 9. **1745** 4 Jun B Hohenfriedberg; 30 Sep B Soor. 1747 Oberst Hans August von Blumenthal; retired due to his wounds. **1756** raised to three squadrons, mostly using men of the Saxon Trabanten-Leibgarde, who had been captured at Pirna. 1 Oct B Lobositz; together with the Gensd'armes and two squadrons of the Kuer R Prinz von Preussen Nr 2, they overthrew two Austrian cavalry regiments, took three standards and captured General Fuerst Lobkowitz prisoner. **1757** 6 May B Prague; the 2nd and 3rd Sqns also stayed here for the siege. The 1st Sqn fought at Kolin on 18 Jun. 5 Nov B Rossbach; all three sqns distinguished; took two pairs of kettle drums and four standards from the beaten Austrian cavalry. In the pursuit next day, at Spielberg, they took two howitzers and five cannon. 5 Dec B Leuthen again distinguished; together with the Gens d'armes they took a number of colours and standards and 15 cannon. **1758** Rittmeister Wilhelm Dietrich von Wackenitz; promoted Oberst after his actions in the battle of Zorndorf. 27 May-2 Jul S Olmuetz; 25 Aug B Zorndorf; cut up the Russian infantry, took a number of colours and cannon. B Hochkirch. **1760** July S Dresden; 15 Aug B Liegnitz; 3 Nov B Torgau; heavy losses. **1762** C Burkersdorf; 16 Aug B Reichenbach; 7 Aug-9 Oct S Schweidnitz. 1770 Major Carl Heinrich von Schaetzel; retired as Oberst. 1774 Oberst Carl Friedrich Freiherr von Mengden; went as GM to be Chef of Kuer R Nr 4. **1778** in the army of the king. 1785 Chef Major Carl Wilhelm von Byern; went as GM to be Chef of Kuer R Nr 6. **1794** in Poland. 1795 Chef Oberst Friedrich Wilhelm von Zollikofer; retired in 1798. 1798 Chef Oberstlt Christian Friedrich von Rabenau; died as Oberst. The regiment was raised to five squadrons; sergeants of the regiment ranked as Premierleutnants in the army. 1802 Major Ernst August von Wintzingerode; Oberst in 1806. **1806** 14 Aug Auerstaedt; on the Right Wing. Following the battle, it escaped to eastern Prussia and joined l'Estocq's corps. **1807** 5 Jan Sk Heilsberg (a detachment); 9 Jan C Schippenbeil (a detachment); 30 Jan Sk Bialochowo (a detachment); 13 Jun Sk Gollau; 14 Jun C Koenigsberg. It became the new 3rd Kuerassiers. In **1813** it lost the number '3'. 25 Apr Sk Neu-Flemmingen (a detachment); 2 May B Gross-Goerschen; 20/21 May B Bautzen; 26 May C Haynau; 16-18 Oct B Leipzig. **1814** 20 Mar B Arcis-sur-Aube. **1815** No actions.

20

Kuerassier Regiments Raised After 1806

Garde-Kuerassier-Regiment.
Raised on 21 Feb 1815 as the 1Garde-Ulanen-Regiment' from the Garde-Ulanen-Eskadron (1st Sqn), and the Garde-Kosacken-Eskadron (4th Sqn) of the Guards Light Cavalry Regiment, and two squadrons of the Schlesischen NKR (2nd and 3rd Sqns) 3 Aug 1821 – present title awarded. The regiment was mobilised in 1815 but saw no action.

Standards. On 24 Feb 1815 they were presented with the King's standard of the Gardes du Corps; a new cloth was presented in April 1891.

Leichtes Garde-Kavallerie-Regiment. (Guards Light Cavalry Regiment) Raised on 23 Feb 1813 from the Normal-Dragoner-Eskadron (*27 Mar 1811), the Garde-Ulanen-Eskadron (* 29 Mar 1809), the Garde-Kosacken-Eskadron (* 13 Feb 1813) and the Normal-Husaren-Eskadron (* 27 Mar 1811).

Campaign history. 1813 22 Apr C Umpferstaedt (a detachment); 22 Apr C Saalfeld (a detachment); 28 Apr Sk Schoeps (a detachment); 2 May B Gross-Goerschen (Hus. Sqn); 4 May C Borna; 21/22 May Bautzen; 24 May Sk Taubentraenke; 26 May C Haynau; 30 Aug B Kulm (a detachment); 16-18 Oct B Leipzig; 20 Oct C Weissenfels (Kosacken-Esk); 31 Oct C Hoechst (a detachment). **1814** 20 Mar B Arcis-sur-Aube (Drag-Esk); 30 Mar B Paris.

2. Schlesisches National-Kavallerie-Regiment. Raised by Order of 17 Feb 1813 from volunteers in Silesia at two squadrons; the 3rd, 4th and depot were only raised after the end of the war in 1814. The regiment was disbanded on 21 Feb 1815; two squadrons went into the Garde-Ulanen-Regiment; the rest went to form the 7th Hussars.

Campaign history. 1813 26/27 Aug B Dresden; 28 Aug Sk Passendorf; 30 Aug B Kulm; 10 Sep C Weissenfels; 12 Sep C Naumburg; 18 Sep Sk Merseburg; 28 Sep C Altenburg; 16 Oct B Leipzig; 21 Oct C Freiburg / Unstrutt; 31 Oct C Hoechst. **1814** 2 Jan Sk Alzey; 11 Jan C St-Avold; 12 Jan Sk Nancy; 23 Jan Sk Ligny; 27 Jan St-Dizier; 30 Jan B Brienne; 1 Feb B Brienne and La-Rothiere; 4 / 5 Feb C Chalons; 12 Feb B Mont-Mirail; 12 Feb B Chateau-Thierry; 9 Mar B Laon; 27 Mar C Meaux; 30 Mar B Paris.

1st Kuerassier-Regiment, Schlesisches Kuerassier-Regiment. Facings black, buttons yellow. Raised from the depots and survivors of the following regiments by Order of 16 Oct 1807: 4th, 8th, 9th, 12th Kuerassiers and 13th Dragoons. By Order of 7 Sep 1808 this regiment became the Kuerassier-R Nr 1, Schlesisches Kuer-R. On 27 Jan 1889 it was awarded the title borne in 1918: **Leib-Kuerassier-Regiment Grosser Kurfuerst (Schlesisches) Nr 1.**

Standards. In 1808 they were presented with a Leibstandarte; white cloth, purple centre, silver corner fields, golden embroidery and fringes. The three regimental standards had purple cloth, white centre, silver corner fields, gold embroidery and fringes.

2nd Kuerassier-Regiment, Ostpreussisches Kuerassier-Regiment. Facings light blue, white buttons. Raised from the depots and survivors of the following regiments: 6th, 9th and 10th Dragoons. By Order of 16 Oct 1807, the old 6th Dragoons was split into two; the 1st Bn became the Zieten Kuerassiers; the 2nd Bn became the Zieten Dragoons, which then became the 2nd Dragoons in 1808. The men of the 9th and 10th Dragoons became the Kuerassier-Regiment Nr 2, Ostpreussisches Kuer-R by Order of 7 Sep 1808.

Standards. In 1808 they were presented with four old regimental standards of the 1st Bn, Auer Dragoons Nr 6. These were of dark blue cloth, sivler centre and corner fields, gold embroidery and fringes.

4th Kuerassier-Regiment, Brandenburgisches Kuerassier-Regiment. Facings red, buttons yellow. Raised from the depots and men of the following regiments: 2nd, 3rd, 6th, 7th, 10th and 11th Kuerassiers as the Maerkischen Kuerassier-Brigade by Order of 16 Oct 1807. By Order of 7 Sep 1808 it became the 4th Kuerassiers. In May 1810 it was ordered that the facings would be changed to dark blue, but the king then said that cornflower blue might be worn and many regiments adopted this.

Standards. In 1808 they were presented with four old regimental standards. Two were from the old 6th Kuerassiers (von Quitzow): dark blue cloth, white centre and corner fields, gold embroidery and fringes. The other two were from the 6th Dragoons (von Auer): dark blue cloth, silver centre centre and corner fields, gold embroidery and fringes.

On 25 March 1815 a fourth Kuerassier regiment was raised; the Regiment Garde-du-Corps lost its number and the Brandenburg-Kuerassier R became Nr 3. The 7th and 8th Dragoon regiments were raised, as were the 7th – 12th Hussars and the 4th – 8th Ulans

Kuerassier-Regiment Nr 4. Raised on 25 March 1815 from levies from the existing Kuerassier regiments and from the newly absorbed Saxon Kuerassiers. **Standards.** *New Pattern.*

Kuerassier-Regiment Nr 5. Raised on 25 March 1815 from levies from the existing Kuerassier regiments and from the newly absorbed Saxon Kuerassiers. **Standards.** *New Pattern.*

Kuerassier-Regiment Nr 6. Raised on 25 March 1815 from levies from the existing Kuerassier regiments and from the newly absorbed Saxon Kuerassiers. **Standards.** *New Pattern.*

Kuerassier-Regiment Nr 7. Raised on 25 March 1815 from levies from the existing Kuerassier regiments and from the newly absorbed Saxon Kuerassiers. **Standards.** *New Pattern.*

21

Kuerassier Uniforms Post 1808

With the reorganisation of 1808, the following uniform, very similar to that of the Russian cuirassiers, was adopted. A black leather helmet with leather combe and stiff, black horsehair crest (red for trumpeters), with brass chinscales, front to combe and edging to the front peak. A large brass front plate bore the Prussian eagle, except for the Garde du Corps, who wore an eight-pointed star. Underneath the lefthand chinstrap boss was the black-within-white national cockade.

The double breasted white tunic had two rows of eight buttons on the front, Swedish cuffs and a standing collar in the facing colour; the white turnbacks and the bottom of the lapels were edged in the facing colour. Initially the shoulder straps were in the facing colour, but this rapidly became white with an piping in the facing colour. The white knee breeches were now covered in grey overalls, with side piping and regimental buttons down the outside of the leg. The inside of the leg and the bottoms of the overalls were trimmed with black leather. The sword fist strap for rank and file was of red leather with a tassel in the squadron colour: 1st –

Karabinier of Dragoon Regiment von Gilsa Nr 5 1792; note the distinctive plume. From Kling's great work; author's collection.

NCOs 1806. On the left a musketier NCO, IR Courbiere Nr 58; on the right a grenadier NCO, IR von Kropff Nr 31. From Kling's great work; author's collection.

Trumpeters of Kuerassier Regiments Markgraf Friedrich (Nr 5, left) 1737 and von Quitzow Nr 6 1806. From Kling's great work; author's collection.

Dragoon troopers 1806. Left Regiment von Esebeck Nr 8; right Koenig von Bayern Nr 1. From Kling's great work; author's collection.

Königl. Preuß. General der Infanterie 1806.
Allgemeine Uniform der Generalität seit 1803.
(Vergleiche Seite 187, unter 5).

Zu Seite 187.

General of Infantry in 1806. From Kling's great work; author's collection.

Contemporary colour plate of four great characters of the Prussian Napoleonic era: General L'Estocq, General von Kalkreuth, Major von Schill and von Bluecher in his regimental uniform. Author's collection.

Contemporary plate of the Death's Head Hussars (Nr 5) during the Seven Years War. Author's collection.

The 1st Leib-Husaren Nr 5 in action in 1813. To the left is a Freiwilliger-Jaeger in his green dolman, to the right rear an officer. It was not yet the custom in the Prussian army that only trumpeters rode greys. A Knoetel plate, author's collection.

Fuesilier, IR Nr 41 of the 1760 era; a reconstruction. Wesel Military Museum.

An officer of the Towarczys (9th Hussars) 1806. A plate by Goddard and Booth; author's collection.

An officer and private of the Jaegers of the Guard in 1813. A plate by Goddard and Booth; author's collection.

Kuerassier sabretasches 1735 – 1806. Kling. Top row, left to right: 4th Regiment, 1757; 6th Regiment, 1757; 8th Regiment 1757; 11th Regiment, 1757. 2nd row, left to right: 4th Regiment, 1743; 2nd Regiment, 1743; 12th Regiment, 1743; 7th Regiment, 1743. 3rd row, left to right: 5th Regiment 1806; 3rd Regiment, 1806; 1st Regiment, 1806; 9th Regiment, 1806. bottom row, left to right: Gendarmes Nr 10, 1759 and 1806; Garde du Corps Nr 13, 1740 and 1806.

Saddle furniture: top 1786, Leib-Regiment, Kuerassier Regiment Nr 3; bottom, 1806 Kuerassier Regiment von Quitzow Nr 6. Kling.

Feuerwerker (NCO) and gunner of the artillery of the Guard of the 1813 – 1815 period. Author's collection.

Private and NCO, Fuesiliers of the Guard of the 1813 – 1815 period. Author's collection.

Officer of the Normal Infantry Battalion of the 1813 – 1815 period. Author's collection.

Officer, Normal Dragoon Squadron of the 1813 – 1815 period. Author's collection.

Officer of the foot artillery of the Guard of the 1813 – 1815 period. Author's collection.

Trooper, Brandenburg Hussars of the 1813 – 1815 period. Author's collection.

Gunner, horse artillery of the Guard, 1813 – 1815. Author's collection.

Officer, Garde du Corps, 1813 – 1815. Author's collection.

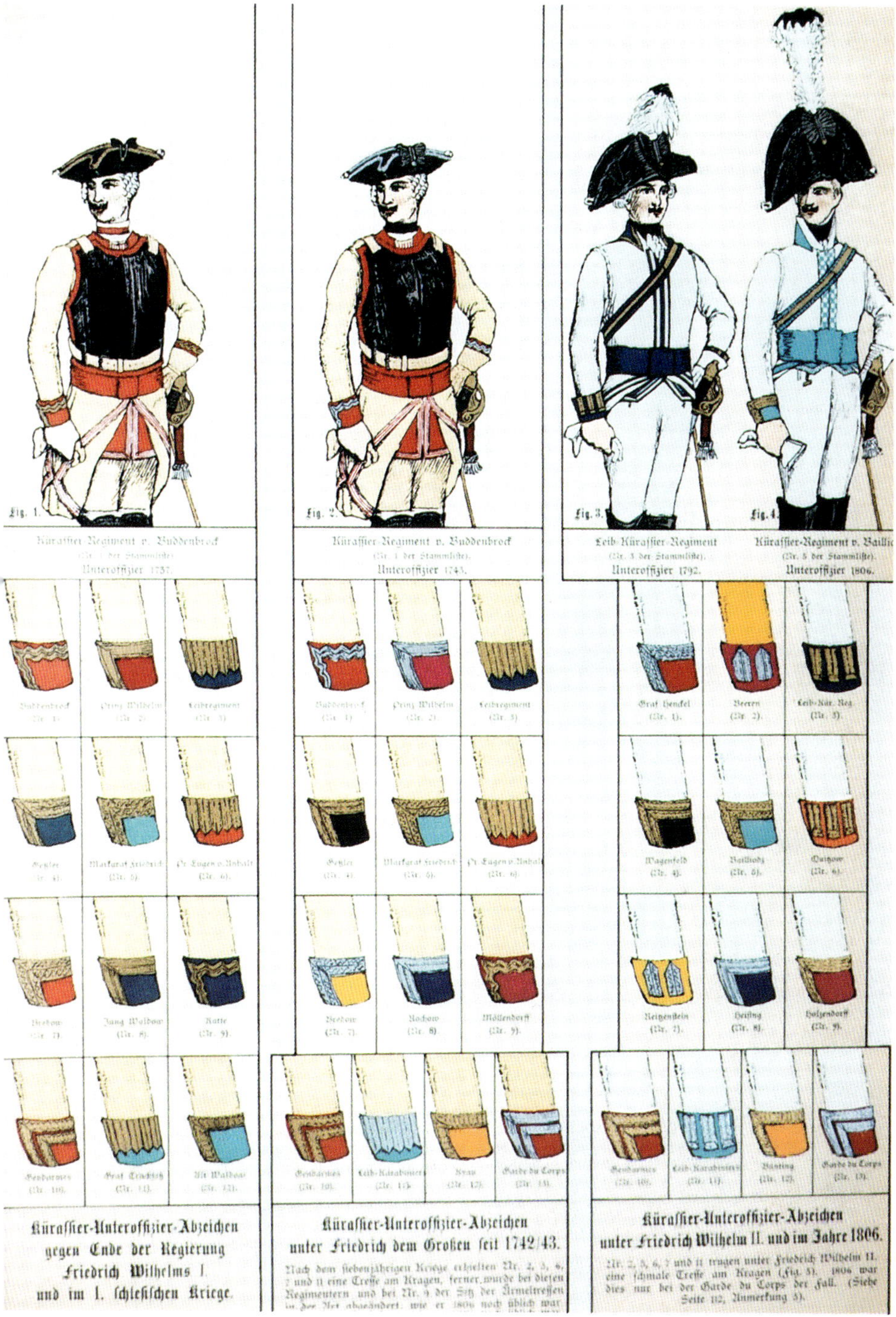

Various Kuerassier regiments' rank badges for NCOs, worn on the cuffs. Left hand column: at the end of the reign of Frederick I, at the time of the 1st Silesian War. Centre column: the time of Frederick the Great, after 1742 / 1743. Right hand column: the period 1792 – 1806. Kling.

Officer, Brandenburg Hussars, 1813 – 1815. Author's collection.

Trooper, Normal Dragoons, 1813 – 1815. Author's collection.

Grenadier private Footguards, 1813 – 1815. Author's collection.

The 1st Leib-Husaren Nr 5 in action in 1807. Knoetel; Author's collection.

Musketiers, IR Prinz von Oranien Nr 19 in 1806. The regiment surrendered at Magdeburg and was not re-raised. Kling; author's collection.

Officer and trooper, Hussar Regiment Prinz Eugen von Wuerttemberg Nr 4 in 1806. From Ramm; author's collection.

Officer and trooper, Hussar Regiment von Bluecher Nr 8 in 1806. From Ramm; author's collection.

Private, 2nd Battalion, 4th Reserve Infantry Regiment in British-supplied uniform in 1813. Author's collection.

Trooper, 4th Squadron, Ostpreussischen NKR in 1813. He wears a Russian Kiwer shako with the typical Prussian Litewka tunic. Author's collection.

Private, 2nd Battalion, 9th Reserve Infantry Regiment in 1813. As with many Prussian troops at this time, he wears a uniform supplied by Britain, with the typical hussar-style coat worn by the rifle regiments. Author's collection.

Men from the Pommeranian Landwehr infantry (left) and the East Prussian Landwehr cavalry in 1813. Landwehr uniforms were very often much less luxurious than these; many men went barefoot. Author's collection.

white; 2nd – yellow; 3rd – light blue; 4th – red. The tassel of honour for good conduct in the war of 1806/1807 was as for the infantry. NCOs had black and white tassels as for the infantry. Since 1789 the Wachtmeisters had worn officer's portepees (sabre fits straps) in black leather and silver thread. Scabbards were supposed to be of steel, but in the period 1813 1815 many made of brass were used. In 1807 NCO's lace went along the front and bottom of the open collar and along the top and back of the cuff; in 1814 the collar was shorter, worn closed and their lace now ran along top and front of the collar.

Until August 1812 officers wore shoulder straps showing their ranks with silver and black lace of infantry pattern. These were then replaced by metal scale epaulettes in the button colour. In 1813 these were replaced by a common model. The strap was white and the lining edging of the strap and the silver cross piece were red for all regiments until 1834. The metal halfmoons at the shoulders were of the button colour; majors and above wore fringes to the halfmoons also in the button colour. Trumpeters wore swallows nests in the facing colour, laced in the button colour. Standard bandoliers were in the facing colour in velvet, edged in gold or silver according to the button colour. Those of the guards regiments and of the 1st, 4th, 7th and 8th Kuerassiers were also fringed. Kuerasses had been withdrawn by King Friedrich Wilhelm II; they were re-introduced in April 1814, when the four regiments then in existence were equipped with items taken from the arsenal in Versailles. The Garde-du-Corps and the Brandeburg regiments added brass sheeting to theirs. The shoulder scales and the rivets on all Kuerasses were brass. They were edged in red. In 1814 Czar Alexander presented the Garde-du-Corps with black, Russian-style Kuerasses edged in red; these were worn on special occasions. In April 1814 the Prussians took French-pattern cuirassier swords from Versailles to replace their own. Officers of the guards wore a small guards star on the hilt of their swords. Their sashes were worn under the Kuerass, except by adjutants who wore then bandolier fashion.

22

Dragoon Uniforms Up to 1808

Originally the dragoons wore the same bicorns, breeches, boots, gauntlets, waistcoats and buff leather tunics as the Kuerassiers; their cuffs were usually light blue. Under Friedrich Wilhelm I red facings were also to be seen, and light blue woollen coats of infantry style were introduced; all cuffs were of Swedish style, and there was an aiguillette on the right shoulder strap in the button colour. Facings were worn on the collar, cuffs, lapels and turnbacks. Officers' parade uniform coats were decorated with embroidered lace loops in the button colour, six on and two under each lapel, two on the cuff, two on each pocketflap and four in the small of the back. Leatherwork was white and the cartridge pouch was worn on the carbine bandolier. The brass-hilted 'Pallasch' (the long, straight sword) was carried in a brown leather sheath. It had a basket pattern guard and an eagle's head for the pommel. The dragoons bore swallow-tailed guidons instead of the square cavalry standards of the Kuerassiers. Harness was black, heavy-cavalry style with steel fittings. The holster covers and shabraques were usually in the facing colour and bore lace edging and ciphers or eagles as with the Kuerrassiers. Some regiments had square ended shabraques, others had rounded ends.

At the time of the death of Frederick the Great (Friedrich II 1740-1786) the uniforms of the dragoons were as follows.

Regiment	Collar & Cuffs	Lapels	Turnbacks	Buttons
1 Graf Lottum	black	white	black	yellow
2 von Mahlen	white	pink	white	yellow
3 von Thun	pink	buff	pink	white
4 von Goetzen	buff	dark red	buff	white
5 Anspach-Bayreuth	dark red	white	dark red	white
6 von Rohr	white	none	white	white
7 von Borcke	scarlet	scarlet	scarlet	yellow
8 von Brausen	scarlet	none*	scarlet	white
9 von Zitzewitz	light blue	none*	blue	white
10 von Rosenbruch	orange	none	orange	white
11 von Bosse	yellow	yellow	yellow	white
12 von Kalckreuth	black	black	black	white

*white laces to the buttonholes on the chest.

In the following period, the turnbacks came to be in the coat colour, edged in the facing colour; the waistcoat was abandoned and in 1797 the cartouche received its own, narrow, bandolier, which was worn over the right shoulder. The facings (worn on collar, cuffs, lapelsand turnback edgings) and buttons worn in 1806 are shown in the regimental sections. All badges of rank; plumes etc were as for the Kuerassiers.

23

Dragoon Uniforms Post 1808

Only the regiments Koenigin Nr 5, Auer Nr 6, vacant Rhein Nr 7, von Esebeck Nr 8 and von Roquette Nr 13 survived the catastrophe of 1806 intact. The 1st Dragoons did provide the basis for the new 2nd Dragoons and their history was incorporated into that regimen in the Prussian Army List of 1905 and it has been incuded here as well.

All uniform details as for musketiers except as noted below. The shako was reinforced with leather at the top and bottom and to the sides. The cap badge was a brass Prussian eagle and white plumes and cords were worn for parades. NCOs plumes had a black tip, those of officers had a black base. Trumpeters wore red plumes.

Brass chinscales were worn on campaign. The light blue, double-breasted tunic had two rows of eight buttons down the front; collar, shoulder straps, Swedish cuffs and edging to the light blue turnbacks in the regimental colour. Grey overalls with buttoned sides. White belts. The old, straight swords of the 1806 era were carried at this point. The facings and buttons were as follows.

Lithuanian Dragoons Nr 3 – red, yellow. 2nd West Prussian Dragoons Nr 4 – red, white.

Swords had red leather fist straps with tassels in the squadron colour: 1st – white; 2nd – yellow; 3rd – light blue; 4th – red. On 27 Aug 1813 officers were ordered to wear the same scale epaulettes as the Kuerassiers; the field was in the facing colour, the lining, edging and cross strap were red.

The Normal-Dragoner-Eskadron had red facings and yellow buttons and two lace loops to each side of the collar and on he cuffs. Their shako badge was the guard's, eight-pointed star

Saddle furniture was a light blue shabraque with a piping and a wide braid in the facing colour. Over

this was laid a black sheepskin saddle cover. Harness was black with steel fittings. Rank and file had a round brass plate bearing the Prussian eagle on trophies of arms on their pouch lids and the Regiment Koenigin also had small brass grenades flanking the plate. The Garde and the Normal Dragoons wore the star. Officers did not wear cartouches until 1808; those of the line bore the crowned royal cipher; the Garde and Normal-Eskadron wore silver and enamelled stars.

In the Wars of Liberation the Brandenburg Dragoons captured many French cuirassiers' swords, which they used instead of the issued items. Officers of the 1st Dragoons had small silver eagles on their sword guards, in commemoration for their deeds in the Seven Years War.

24

The Dragoon Regimental Histories

1. Dragoner-Regiment.
Raised in 1690. Title in 1806: **Koenig von Bayern.**
Inspectorate: The Mark. Garrisons Greifenhagen, Lippehne, Schoenfliess, Schwedt, Wriezen.

Uniforms. Facings black, lining and buttons yellow, white waistcoat. Saddle furniture was light blue, edged in wide white and light blue barred lace with four narrow red stripes along it. Inside this ran an intricate squirl of thin red lace edged in white and light blue. The crest was a black eagle on a white, crowned cartouche

Guidons. One guidon was lost on 6 Nov at the capitulation of Hansfeld, the other four at that of Boitzenburg on 12 Nov 1806. The king's guidon was plain white, the regimental models were bright yellow; embroidery and fringes were gold. The central black eagle was flying against the sun on a white disc; the scroll bore 'NON SOLI CEDIT' in gold. The corner ciphers were the crowned 'FWR' within olive wreaths. The finial bore 'FWR' interlocked.

Campaign history. In 1690 the Markgraf Georg Friedrich of Anspach gave two companies of dragoons to Prussia; new recruits increased the unit to eight companies and the regiment was given to the Markgraf as Chef until his death from wounds on 29 March 1703. The regiment remained 'vacant Anspach' until 1713. **1689** 24 Jul-12 Oct S Bonn. **1695** 2

Jul-3Aug S Namur. **1706** in Brabant; 23 Jul-22 Aug S Menin. **1708** 11 Jul B Oudenarde. **1709** 11 Sep B Malplaquet. **1710** 22 Apr-27 Jun S Douay; 7 Aug-10 Sep S Bouchain, 1-8 Nov S Aire. **1714** Chef Oberst Andreas Reveillas du Voyne (who had previously commanded the regiment. He retired as LG. **1715** in Pommerania against the Swedes. 1719 Oberst Georg Joachim von der Wensen; died as GM. 1725 Chef Oberst Hans Friedrich von Platen; later LG. The regiment was split into two, five companies going to form the 2nd Dragoons. The remaining five companies were increased to five squadrons. 1727 two squadrons of light dragoons were raised; these were increased to five squadrons in 1734. In 1740 five more squadrons of light dragoons were added; the regiment thus had 15 squadrons. **1741** Chef Oberst Carl Friedrich von Posadowsky; died. Blockade of Glogau. 10 Apr B Mollwitz. 1743 the regiment was split; the light dragoons went to form the 9th and 10th Dragoons. **1744** 10-16 Sep S Prague. **1745** 4 Jun B Hohenfriedberg, distinguished. 1747 MG Berend Christian von Katte; retired. 1751 Chef MG Johann Ernst von Ahlemann; retired. 1755 Chef MG Carl Ludwig von Normann; retired. **1757** 21 Apr C Reichenberg; 6 May B Prague, very distinguished. 18 Jun B Kolin; took four enemy colours and a standard of the Saxon Carabiniers. 7 Sep C Moys or Goerlitz. 22 Nov B Breslau; 5 Dec B Leuthen, overthrew two enemy cuirassier regiments and took two Bavarian infantry battalions prisoner. After the battle, they took part in the pursuit of the enemy under General Zieten; took many prisoners and two cannon. **1758** 25 Aug B Zorndorf; broke the enemy cavalry and pushed part of them into a swamp. 14 Oct B Hochkirch; after the battle, on the march to Neisse, they attacked an enemy corps at Ebersbach near Goerlitz on 26 Oct and took many prisoners. **1759** 25 Sep C Hoyerswerda; 29 Oct C Pretsch. **1760** S Dresden, 15 Aug B Liegnitz, 3 Nov B Torgau. 1761 Chef Oberst Johann Wenceslaus Freiherr von Zastrow; died as MG. 1774 Chef Oberst Friedrich Albrecht Carl Hermann, Graf von Wylich and Lottum; retired as GdC. **1762** 21 Jul C Burkersdorf. **1778/1779** In the army of Prinz Heinrich. 5 Feb 1779 C Brix; took two cannon. **1787** in Holland. **1792** Against France on the upper Rhine; 20 Sep Valmy; storm of Frankfurt / Main. **1793** 27 Mar C Waldalgesheim; 17 and 20 Aug C Kettricher Hof, took three cannon and two howitzers. 20 Aug B Weissenburg Lines (mounted a diversionary raid through the Vosges mountains). 14 Sep B Pirmasens, took a cannon. 28-30 Nov B Kaiserslautern. **1794** 23 May C Kaiserslautern; Sk Koeshofen; Sk Hermersberg. 1795 Chef MG Ludwig Prinz von Preussen; died in 1796. 1797 Chef LG Ludwig, reigning Herzog von Pfalz-

Zweibruecken; 1799 Kurfuerst (Elector) of Pfalz-Bayern, 1806 King of Bavaria. Under him, the regiment had two commanding officers: 1797 Oberst von Gloeden (died in 1802 as GM) and GM von Pelet. **1806** 14 Oct Auerstaedt; then in Bluecher's corps. Some of the regiment surrendered at Boitzenburg; the remnants went into the new **5th Dragoons.** Raised from the depots and men of the 5th Kuerassiers and 1st Dragoons. On 16 Oct 1807 its title became the Dragoner-Brigade Prinz Wilhelm. By Order of 7 Sep 1808 it changed to Dragoner Nr 5, Brandenburgisches Drag-R (Prinz Wilhelm). On 7 Nov 1867 the title became **1. Brandenburgisches Dragoner-Regiment Nr 2.**

Standards from 1808. They received no King's standard, but carried four regimental standards of the old 5th Dragoons (Koenigin): black cloth, silver centre, gold embroidery and fringes.

Campaign history. 1812 Two squadrons were mobilised as part of the 2nd Combined Dragoons. 19 Jul C Eckau; 7 Aug C Wolgund; 22 Aug Sk St Olay / Aa; 1 Oct C Garossenkrug. 1813 The squadrons of the regiment that had not been mobilised in 1812 were at the siege of Stettin from 1 Oct-5 Dec 1813.

The 1812 squadrons. 5 Apr C Dannigkow; 2 May B Grossgoerschen; 4 May Sk Borna. The combined regiment fought at: 20/21 May B Bautzen; 26 May C Haynau; 23 Aug B Gross-Beeren (2 sqns); 5 Sep Sk Zahna (the same two sqns); 6 Sep B Dennewitz; 7 Sep Sk Herzberg; 18 Oct B Leipzig; 30 Nov capture of Arnhem. **1814** 1 Mar C Soissons; 9 / 10 Mar B Laon. **1815** 16 Jun B Ligny; 18 Jun B Waterloo; 27 Jun C Villers-Cotterets; 2 Jul C Issy.

2. Dragoner-Regiment.

Raised in 1690. Title in 1806: **von Prittwitz.** Inspectorate: Lower Silesian. Garrisons; Beuthen, Hainau, Lueben, Polckwitz, Rauden.

Uniforms. Facings, waistcoats and lining white, buttons yellow. Saddle furniture was white with a rounded rear corner. The edging was three light blue stripes, the crowned cipher was 'FR'.

Guidons. The king's guidon was white with red corner piles, the regimental models were yellow with red corner piles; embroidery and fringes were gold. The central black eagle was flying against the sun on a white disc; the scroll bore 'NON SOLI CEDIT' in gold. The corner ciphers were the crowned 'FWR' within olive wreaths. The finial bore 'FWR' interlocked.

The five guidons were lost at the capitulation of Prenzlau on 28 Oct 1806.

Campaign history. Raised in 1690 as Nr 1. 1725 the regiment was halved and the new 2nd Regiment was made up to strength again with new recruits. In the early campaigns the regiment was brigaded together with the 1st Dragoons; it was very distinguished at Oudenarde and Malplaquet. 1725 Chef Oberst Friedrich Otto Freherr von Sonsfeld; later GL. **1734/1735** on the Rhine. 1742 Chef Oberst Prinz Ludwig von Wuerttemberg; transferred into French service. **1744** 10-16 Sep S Prague. **1745** 31 Jul C Neustadt; after the action all field received the *Pour-le-Merite*. 4 Jun B Hohenfriedberg; 20 Aug-5 Sep S Cosel. 1749 Chef GM Reimar Julius von Schwerin; died as GL. 1754 Chef GM Christian Friedrich von Blankensee; killed at Prague on 6 May 1757. **1757** Oberst Anton von Krockow; died as GL. 6 May B Prague; 18 Jul B Kolin; 22 Nov B Breslau; 5 Dec B Leuthen. **1758** 14 Oct B Hochkirch. **1759** 12 Aug B Kunersdorf; lost half its strength in the heavy fighting. In the subsequent retreat to Saxony, it clashed with an enemy corps at Spremberg; captured an entire regiment with two colours and three guns. 14 Nov C Gorbitz. **1760** July S Dresden; 15 Aug B Liegnitz, charged the enemy in flank and rear; pushed the first two enemy lines into the third; took over 1,000 prisoners, 12 guns, 3 colours and 2 standards, but lost a standard of their own. The king rewarded the regiment with 1,000 Thalers. 17 Sep C Hohen-Giersdorf. **1762** 30 Aug C Friedberg (three squadrons). **1778** Chef Oberst Prinz Friedrich von Wuerttemberg; retired as GL. Sk Trautenau. 1781 Chef Oberst Johann Christoph von Mahlen; died in 1789 as GM. 1790 Chef GM Carl Wilhelm Friedrich von Schmettau; retired as GL. **1792** On the Rhine. Valmy, **1793** 6 Jan C Hochheim, 28 Mar Bingen, Sk Gundersheim, 13 Aug C Limbach; Sk Eschweiler, Sk Bubenhausen, Sk Wachenheim, Sk Kirrweiler, **1794** 20 Sep B Kaiserslautern. Took a colour and 600 prisoners. 1797 LG Wolf Moritz von Prittwitz. **1806** 14 Oct Jena, in Hohenlohe's Reserve; heavy losses. The regiment surrendered at Prenzlau on 28 Oct 1806 and was not re-raised. The depot and some remnants went into the new 2nd Dragoons.

3. Dragoner-Regiment.

Raised in 1705. Title in 1806: **von Irwing.** Inspectorate: Pommeranian. Garrisons: Arenswalde, Berlinchen, Driesen, Friedeberg.

Uniforms. Facings pink, buttons and waistcoat white. The saddle furniture was pink with a rounded rear corner. The edging was a narrow white lace outside a wide one.

Guidons. The king's guidon was plain white, the design of the regimental models is not known; embroidery and fringes were gold. The central black eagle was flying against the sun on a white disc; the scroll bore 'NON SOLI CEDIT' in gold. The corner ciphers were the crowned 'FWR' within olive wreaths. The finial bore 'FWR' interlocked. Four guidons were lost at the capitulation of Erfurt on 16 Oct; the Leib-Guidon was saved.

Campaign history. Raised in 1705 with 8 companies by GM Friedrich Freiherr von Derfflinger, who had retired in 1691 as GL. **1709** 27 Jun-28 Jul S Dornik; 11 Sep B Malplaquet. **1710** 5-26 Jun S Douay; 15 Jul-28 Aug S Bethune; 12 Sep-9 Nov S Aire. **1711** 11 Aug-12 Sep S Bouchain. **1712** 8 Jun-30 Jul S Le Quesnoy; S Fagel; Jul S Landrecies. In 1713 it was converted into a regiment of Grenadiere zu Pferde; in 1718 it was raised to five squadrons and then to ten in 1725; in 1740 it was converted to dragoons. **1715** Against Sweden. 17 Jul-24 Sep S Stralsund; 16 Oct capture of Ruegen Island. 1724 Chef Oberst Adolf Friedrich von der Schulenburg; killed as GL at Mollwitz. Early in the XVIII Century it was in Brabant and present at many sieges and actions. **1741** 9 Jan C Glatz; 27 Mar Sk Baumgarten; 10 Apr B Mollwitz. Chef Oberst Friedrich Rudolf Graf von Rothenburg; died as GL. **1742** 17 May B Chotusitz. **1744** 10-16 Sep S Prague. **1745** 4 Jun B Hohenfriedberg; 30 Sep B Soor. 1752 Chef GM Georg Philipp Gottlob Freiherr von Schoenaich; received Kuer R Nr 6. 1752 Chef GM Graf von Truchses; retired. **1756** 1 Oct C Lobositz; heavy losses. **1757** Chef GM Peter von Meinicke; retired. 6 May very distinguished in B Prague on 6 May; 18 Jun B Kolin and 5 Nov B Rossbach; C Gotha, where General von Seidlitz and 15 squadrons of cavalry drove the whole French and imperial Austrian army out of the town. **1758** 1 Feb C Horneberg, overthrew a French force and took 400 prisoners. 23/24 Feb C Schladen; 4 Mar C Hildesheim; 20 Jul Marienberg. **1759** In the raid into Franconia. 11 May C Himmelskron, where they charged the Saxon corps of General von Riedesel, scattered it and took over 800 prisoners, 4 colours 3 standards 2 guns and the baggage train. 16 May C Bamberg; 30 May C Moenchsberg. **1760** 7 Jul C Goedau; 15 Aug B Liegnitz. **1761** 30 Jul C Kunzendorf; Gross-Wandries; 8 Aug C Bunzelwitz. Chef GM Kurt Friedrich von Flans; died in office. **1762** 31 May C Teschen; 18 Aug C Reichenbach, where, practically alone, they charged 14 Austrian cavalry regiments and held them in check until the king arrived from Peterswalde with support. In this action they lost almost 200 men. As a reward, the regiment received a large cash sum. 1763 Chef Oberst Achaz Heinrich von Alvensleben; died. 1777 Chef GM Otto Balthasar von Thun; retired as GL. **1778/1779** Against Austria; in the king's army; 22 Sep C Weisskirch; drove the Austrian Croats out of the village. Ambushed at night in Moeckern by 4,000 Pandurs and 1,000 Austrian hussars, they drove the enemy off with their carbine fire. As a reward, all field officers received the *Pour-le-Merite*. 25 Oct C Moeckern. **1779** 24 Feb Sk Paulwitz. 1788 Chef GM Georg Ludwig von Gilsa; retired. 1792 Chef GM Wolf Moritz von Prittwitz; received Drag R Nr 2. **1794** In Poland; 27 Jul-6 Sep S Warsaw; 19 Aug C Powonske and other actions. 1797 Chef GM Hans Carl Ludolf von Stranz. 1800 Chef Oberst Friedrich Wilhelm von Irwing; GM in 1806. **1806** 14 Oct Auerstaedt; parts of the regiment surrendered at Erfurt. 5 Nov C Gadebusch; 6 Nov storm of Luebeck. Most of the regiment surrendered at Ratkau; it was not re-raised. The depot went into the new 6th Dragoons; see in the 4th Dragoons von Katte below.

4. Dragoner-Regiment.

Raised in 1741. Title in 1806: **von Katte.** Inspectorate: Pommeranian. Garrisons: Baerwalde, Landsberg on the River Warte, Woldenberg.

Uniforms. Facings buff, buttons, lining and waistcoat white. The saddle furniture was red with a rounded rear corner. The edging was a broad white lace with three royal blue stripes. Officers had eight embroidered silver lace loops on each lapel, plus the usual others to cuffs, pocket flaps and rear of the coat.

Guidons. The king's guidon was plain white, the design of the regimental models was not known; embroidery and fringes were gold. The central black eagle was flying against the sun on a white disc; the scroll bore 'NON SOLI CEDIT' in gold. The corner ciphers were the crowned 'FWR' within olive wreaths. The finial bore 'FWR' interlocked. Two guidons were lost at the capitulation of Ratkau on 7 Nov; the Leib-Guidon and two others were saved.

Campaign history. Raised in 1741 from the 2nd and 3rd Regiments of dragoons. Chef Oberst Wilhelm Ludwig von Bissing; retired as GL. **1741** 28 Apr-3 May S Brieg; 27-31 Oct S Neisse. **1742** Chef Oberst Friedrich Wilhelm Freiherr von Kannenberg; retired. Chef GL von Spiegel; died that same year and was succeeded by Oberst Casimir Wedig von Bonin. Many actions in Moravia including Fulnek on 18 Apr. **1744** 10-16 Sep S Prag. **1745** 4 Jun B Hohenfriedberg; 3 Sep C Soor; 15 Dec B Kesselsdorf. 1752 Chef GM Henning Ernst von Oertzen; killed at Lobositz. **1756** 1 Oct B Lobositz. **1757** Chef GM Ernst Heinrich von Czettritz; retired as GL. 21 Apr C Reichenberg; 6 May B Prague, took over 100 prisoners; S Prague, then 18 Jun B Kolin; 5 Nov B Rossbach; 5 Dec B Leuthen. **1758** 26 May-2 Jun S Olmuetz; 25 Aug B Zorndorf, charged with the Gens d'armes and the Garde du Corps into the Russian grenadiers; overthrew them and took 7 colours and 14 guns. 10 Sep C Zenkwitz; 14 Oct B Hochkirch; after this battle, the regiment attacked a large body of enemy cavalry at Ebersbach near Goerlitz, scattered it and took almost 500 prisoners. **1759** 12 Aug B Kunersdorf. **1760** S Dresden, incl. Sks Kosdorf and 7 Jul Goedau; 15 Aug B Liegnitz; 3 Nov B Torgau. **1761** C Schnellendorf; Sk Gross-Wandries; 15 Aug C Kloster Wahlstadt two squadrons held off ten times their number. **1762** 6 Jun C Wernersdorf; 27 Jun C Konradswaldau; 21 Jul C Burkersdorf; 16 Aug C Reichenbach; 7 Aug-9 Oct S Schweidnitz. **1772** Chef Oberst Georg Ludolf von Wulfen; dismissed as GM. **1778/1779** in the army of the king; 14 Sep 1778 C Jungbuchau. **1779** 9 Jan Sk Doebersdorf. 1782 Chef GM Carl Ludwig von Knobelsdorf; died in office. 1786 Chef Oberst Carl Ludwig von Goetzen; died in office as GM. 1789 Chef GM Georg Balthasar von Normann; retired. **1792** Chef GM Friedrich Heinrich von Katte; GL in 1806. On the Rhine; 14-14 Aug blockade of Longwy; 30 Aug capture of Verdun; 20 Sep Valmy; 2 Dec storm of Frankfurt / Main. **1793** 28 Mar C Waldalgesheim; 10 Aug-28 Dec S Landau; 13 Aug C Limbach; Altstadt took a cannon. Blieskastel; Hornbach; 14 Sep B Pirmasens; 13 Oct Sk Herzogenhand; 16 Nov failed raid on Bitsch; 28-30 Nov B Kaiserslautern. 26 Dec Weissenburg lines. **1794** 29 May C Kirrweiler; 20 Sep C Kaiserslautern; Sk Hochspier. **1806** 14 Oct Jena; scattered; 5 Nov C Gadebusch; 6 Nov storm of Luebeck; three squadrons surrendered at Ratkau on 7 Nov 1806; two squadrons should have surrendered at

Pasewalk, but escaped, joined up with their depot and reached East Prussia. The regiment was not re-raised; the two squadrons and the depot went into the new 6th Dragoons. The Prussian Army List of 1905 credits the history of the old Katte-Dragoons to the new 6th; for the sake of conformity, we will follow suit here. **6th Dragoons.** Raised from the depots and men of the 3rd and 4th Dragoons. On 16 Oct 1807 it became the Dragoner-Brigade von Wedell; by Order of 7 Sep 1808 this changed to **Dragoner Nr 6, Neumaerkisches Drag-R.**

Standards after 1806. King's standard: the old king's standard of the 3rd Dragoons (von Irwing): white cloth, rose red centre, silver embroidery and fringes. Regimental standards: three of the old 4th Dragoons (von Katte): white cloth, blue centre, gold embroidery and fringes.

Campaign history. 1813 2 May B Gross-Goerschen; 4 May Sk Borna; 20/21 May B Bautzen; 24 May Sk Rothkretscham; 26 May C Hainau; 22 Aug Sk Saida; 26/27 Aug B Dresden; 26 Aug B Katzbach; 28 Aug C Lobkowitz; 30 Aug C Kulm and Arbesau; 8 Sep Sk Dohna; 17 Sep Sk Liebertwolkwitz; 28 Sep Sk Altenburg; 18 Oct B Leipzig; 30 Oct-20 Dec S Erfurt. **1814** 13 Jan-24 Apr blockade of Thionville; 28 Feb C Gue-a-Tresmes; 2 Mar Sk May; 3 Mar Sk Neuilly; 9 Mar B Laon; 21 Mar Sk Oulchy-le-Chateau; 27 Mar Sk Trilport; 28 Mar C Clay-Souilly; 30 Mar B Paris. **1815** 16 Jun B Ligny; 18 Jun B Waterloo; 20 Jun storm of Namur; 24 Jun-21 Jul S Landrecy; 30 Jun-19 Aug S Givet; 20 Jun-12 Jul S Maubeuge; 8 Jul Sk Philippville; 16 Aug C Rocroy.

5. Dragoner-Regiment.

Raised in 1717. Title in 1806: **Koenigin.** Inspectorate: Pommeranian. Garrisons: Bahn, Garz, Gollnow, Massow, Naugard, Pasewalk, Treptow on the River Tollense, Uckermuende.

Uniforms. Facings and lining dark crimson, buttons and waistcoat white. Saddle furniture in the facing colour with a square rear corner and no crests. The edging was a wide white lace with three narrow dark crimson stripes. On the daily uniform officers have silver embroidered loops, two under each lapel and two in the small of the back.

Guidons. The king's guidon was white, the regimental models were black, the central disc was white; embroidery and fringes were gold. The central black eagle was spread under a scroll bearing 'PRO GLORIA ET PATRIA' in gold. The corner ciphers were the crowned 'FR' within olive wreaths. The finial bore 'FR'. One guidon was lost on 26 Oct in the clash at Zehdenick, another at the capitulation of Ratkau; eight guidons were saved.

Campaign history. Raised in 1717 at four squadrons in Halberstadt from contingents from all Kuerassier and dragoon regiments. Increased to five squadrons in 1718 and to ten in 1724. Chef in 1717 Oberst Achaz von der Schulenburg; died as GL. 1731 Chef Oberst Friedrich Erbprinz (later Markgraf) von Brandenburg-Baireuth; died as GL. **1741** 10 Apr B Mollwitz. **1742** 17 May B Chotusitz. **1744** 10-16 Sep S Prague. **1745** 4 Jun B Hohenfriedberg; the regiment charged a force of twenty battalions, overthrew them and took 2,500 prisoners, 67 colours and 4 guns. In commemoration of this deed, the regiment received a cash reward and a new regimental seal bearing details of the triumph. B Kesselsdorf. **1756** 1 Oct B Lobositz. **1757** 9 May-20 Jun S Prague; 7 Sep C Moys; 22 Nov B Breslau;5 Dec B Leuthen, crushed the enemy infantry on their left wing. **1758** 27 May-2 Jul S Olmuetz; ambushed at Holitsch, Moravia; lost the regimental kettle drums and 300 prisoners. 14 Oct B Hochkirch; heavy losses. In the subsequent retreat, they overthrew an enemy Kuerassier regiment and took 64 prisoners. **1760** 3 Nov B Torgau; charged the enemy right wing and captured four regiments. **1762** 21 Jul C Leutmannsdorf. 1763 Chef GL Friedrich Wilhelm Markgraf von Brandenburg-Baireuth; died in office. 1769 Chef GL Carl Friedrich Alexander von Brandenburg-Baireuth; died in 1806. **1778/1779** with the king's army. **1792** On the Rhine. 14 Oct Verdun; 20 Sep Valmy; 2 Dec C Bockenheim. 1793 6 Jan C Hochheim. 30 Mar C Alsheim, Bacharach and Rheintuerkheim; captured a French column of 43 officers, 1,411 men, 6 colours, three guns, several ammunition wagons, the war chest and much baggage. Blockade of Landau; Sk Frankenthal. **1794** The 2nd Bn was in the expedition to Creuznach. The whole regiment was at B Kaiserslautern on 20 Sep, in the Hunsrueck mountains and on the Moselle. **1795** In Westphalia, defending the Bishopric of Muenster against French attack. **1806** Title 'Regiment Koenigin'. 14 Oct at Aeurstaedt. In the Left Wing; fought their way through Nordhausen (17 Oct), Zehdenick (26 Oct), except for a detachment which capitulated at Nossentin on 1 Nov 1806. **1807** 27 Jan C Dirschau to Danzig. 15 May C Danziger Nehrung. The regiment survived and became the 1st Dragoons, Regiment Koenigin on 7 Sep 1808. 1810 the title 'Koenigin' to be retained permanently. **1813** 28 May C

Hoyerswerda; 4 Jun C Luckau; 22 Aug C Wittstock; 23 Aug B Grossbeeren; 5 Sep C Zahna; 6 Sep B Dennewitz; 16 Oct B Moeckern; 18 Oct B Leipzig; 19 Oct C Halle; 29 Nov C Arnhem. **1814** 1 / 2 Feb C Merxem; 10 Mar B Laon; 28 Mar-24 Apr occupation of Soissons; 1 Apr C Compeigne. **1815** 16 Jun B Ligny; 18 Jun B Waterloo. 27 May 1819 it became the 2nd Kuerassiers (Pommersches) and in 1846 the title was **Kuerassier-Regiment Koenigin (Pommersches) Nr 2.**

6. Dragoner-Regiment.

Raised in 1717. Title in 1806: **von Auer.** Inspectorate: Prussian. Garrison: Allenburg, Darkehmen, Gerdauen, Koenigsberg, Labiau, Welau.

Uniforms. Facings, buttons, waistcoats white. Saddle furniture was light blue, square ended, edged in wide white lace with two narrow red stripes. On the daily uniform officers wear two embroidered silver loops under each lapel, two on the pocket flap and four in the small of the back.

Guidons. Five guidons presented in 1727; in 1741 five more were presented; 1807 two guidons handed in to the arsenal at Koenigsberg, four given to the 1st Westpr. Drag R. 18 Mar 1813 three guidons were handed in to the arsenal in Glatz. The king's guidon was white with dark blue, wavy corner flames and light blue central disc; the regimental guidons were dark blue with buff corner flames, white central disc. Embroidery and fringes were gold. The central eagle was black, the scroll bore 'PRO GLORIA ET PATRIA in gold. The corner ciphers were crowned 'FR' within laurel wreaths. The finial bore 'FR'.

Campaign history. In 1717 King August II of Poland (August the Strong) presented King Friedrich Wilhelm I of Prussia 600 Kuerassiers and dragoons in exchange for the gift of a very expensive set of porcelain and amber items. These men were used to found the regiment of four squadrons, which bore the nickname of the 'Porcelain Dragoons'. First Chef GM Heinrich Jordan von Wuthenow; died as GL. In 1718 a fifth squadron was added and in 1725 it was increased to ten. In 1727 five squadrons were detached to form the 7th Dragoons. Chef now Oberst Hans Caspar von Cosel; retired as GL. In 1739 regimental strength was back at ten squadrons. **1734/1735** On the Rhine; no actions. Chef Oberst Friedrich Christian von Moellendorf; died as GL. 1741 Chef GM Ludwig Wilhelm von Schorlemmer; retired as GL. **1745** 22 Mar C Landshut; 4 Jun B

Hohenfriedberg; 15 Dec B Kesselsdorf. **1757** 30 Aug B Gross-Jaegersdorf, took a large enemy battery. **1758** 25 Aug B Zorndorf. **1759** 23 Jul B Kay; 12 Aug B Kunersdorf, heavy losses. **1760** Chef GMCarl Friedrich von Meyer; died in 1775 as GL. July S Dresden; 20 Aug C Strehla, the 2nd Bn saved Hulsen's corps from surrender. Three Austrian regiments (including IR Esterhazy) aimed to capture the Prussian battery at Oschatz, which had run out of amunition; the battalion charged and wrecked all three enemy regiments. Scarcely had they done this, when the Austrian Ch L R Loewenstein came up in their rear; they too were rapidly scattered. They were then charged in flank by the Hungarian cavalry regiment Baronay, but beat off this attack as well. 1777 Chef GM Christoph Wilhelm Sigmund Freiherr von Posadowski; retired as GL. **1778/1779** In the army of Prinz Heinrich; **1779** 5 Feb C Brix. 1787 GM Hans Ludwig von Rohr; retired as GL. 1790 Chef GM Philipp August Wilhelm von Werther; died as GL. **1794** In Poland; in action at Piontki; 10 Jun Raigrod, 8 Oct Czarnowo, Lublinitza, 31 Oct Wyczna and Kowalewen. 1803 Chef Oberst Johann Casimir von Auer; GM in 1806. **1806** 23 Dec Sk Gurszno; 25 Dec C Soldau. **1807** 5 Feb Sk Liebstadt; 8 Feb Preussisch-Eylau (Wackern); 4 / 5 Jun Spanden; 10 Jun Heilsberg; 13 Jun Sk Gollau; 14 Jun Koenigsberg. **1813** 2 May Gross-Goerschen; 20/21 May Bautzen; 26 May Hainau; 26/ 27 Aug Dresden; 30 Aug Kulm; 14 Oct Liebertwolkwitz; 16-18 Oct Leipzig. **1814** 19-30 Jan blockade of Luxemburg; 12 Feb Chateau-Thierry; 14 Feb Etoges; 28 Feb Gue-a-Tremes; 9 Mar La Ferte-Milon; 9 / 10 Mar Laon; 21 Mar Oulchy-le-Chateau; 26 Mar Sezanne; 28 Mar Claye-Soully; 31 Mar Paris. **1815** No actions. The regiment later became the 2nd Kuerassiers (**Kuerassier-Regiment Graf Wrangel (Ostpreussisches) Nr 3** on 27 Jan 1889) and the 2nd Dragoons.

7. Dragoner-Regiment.

Raised in 1725. Title in 1806: **vacant von Rhein.** Inspectorate: Prussian. Garrison: Tilsit.

Uniforms. Facings scarlet, buttons yellow, white waistcoat. Saddle furniture was dusty pink, square ended. The edging was a wide white lace with a geometrical pattern in light orange.

Guidons. The king's guidon was white with red corner piles, the regimental models were black with red corner piles; embroidery and fringes were gold. The central black eagle was flying against the sun on a white disc; the scroll bore 'NON SOLI CEDIT' in gold. The corner ciphers were the crowned 'FWR' within olive wreaths. The finial bore 'FWR' interlocked. Five guidons were presented in 1717 (to the squadrons in the 6th Dragoons); in 1741 five new ones were presented. In 1812 those of the 1st, 2nd and 4th Sqns were handed in; one of these was given to Drag R Nr 8 (raised on 7 Mar 1815) on 2 Apr 1815

Campaign history. Raised in 1725 from the 6th Dragoons; Chef Oberst Martin Arend von Dockum; killed in a duel as GM. By 1730 it had 10 squadrons. 1732 Chef Prinz Eugen von Anhalt-Dessau; he became Chef of Kuer R Nr 6. 1734 on the Rhine. 1737 Chef Oberst Christoph Friedrich von Thuemen; went to a Grsn R as GM. 1741 Oberst Ernst Friedrich von Werdeck; killed at Chotusitz. In 1744 five of these squadrons were used to form the 8th Dragoons. **1742** 17 May B Chotusitz; heavy losses. **1742** Oberst von Roehl; killed near Meissen in 1745. 1744 Five squadrons were detached to form Drag R Nr 8. **1745** Fell into an ambush in a sunken road near Meissen and lost its General, von Roell, its kettle drums and a standard as well as many casualties. 15 Dec B Kesselsdorf; took a colour and a standard. The regiment was given the silver kettle drums taken from the Saxon Karabinier-Garde in this battle. 1745 Chef GM Erdmann Ernst von Ruets; retired as GL. 1756 Chef GM Christoph Friedrich Stephan von Plettenberg; retired as GL. **1757** 30 Aug B Gross-Jaegersdorf; took an enemy battery and rode down two infantry regiments. **1758** 25 Aug B Zorndorf. At Linum near Fehrbellin it ruined a Swedish cavalry regiment, capturing those not killed. **1759** Present at the capture of Demmin (17 Jan) and Anclam (21 Jan); 26 Aug-5 Sep S Dresden; 8 Sep C Torgau, took eight cannon; 21 Sep C Gorbitz (Meissen) scattered the Austrian Serbelloni Kuerassier Regiment. **1760** A detachment of 60 men took 160 Swedes prisoner on 3 Sep at Taschenberg. **1761** In Pommerania, 12 Sep C Treptow; C Corlin; 27 Sep C Spie. **1762** 12 May C Doebeln; 14 Oct C Brand or Freiberg. 1763 Chef Oberst Levin Friedrich Gideon von Apenburg; received Kuer R Nr 1 as GL. **1778/1779** In the army of the king. 1781 Chef Oberst Gottlieb Matthaeus Siegfried von Borcke; retired as GM. 1790 Chef GM Johann Ernst von Zabeltitz; died. 1792 Chef GM Wilhelm Friedrich von Schenck; retired as GL. **1794** Parts of the regiment were in Poland. 1803 Chef GM Friedrich Ludwig von Pastau; died in 1805. Chef Oberst Adrian Gottlieb von Rhein. **1806** The silver kettle-drums were handed into the depot in Memel at the start of the campaign and fell into French hands. 15 Nov Sk Schulitz; 6 Dec C Thorn; 23 Dec C

Biezun; 25 Dec C Soldau. 1807 In l'Estocq's corps. 11 Jan C Schippenbeil; 8 Feb B Preussisch-Eylau; 4 / 5 Jun Spanden; 9 Jun Sk Dietrichsdorf; 10 Jun B Heilsberg. At the end of the campaign they took in the depot squadrons of the disbanded regiments Herzberg and Heyking Dragoons. On 7 Feb 1808 they were retitled the 'Ostpreussisches Drag R'; on 14 Sep of that year they became the Drag R Nr 3 Litthauisches Drag R. **1812** Two squadrons were mobilised as part of the 1st Combined Drag R. 19 Jul Eckau; 26 Sep Dahlenkirchen and Tomossna; 27 Sep Bauske; 29 Sep Graefenthal; 16/17 Nov Dahlenkirchen; 26 Dec Piktupoehnen; 28 Dec Ragnit. **1813** 5 Apr Dannigkow; on 7 Apr the two squadrons mobilised in 1812 returned to the regiment. 29 Apr Mersebuerg; 2 May Gross-Goerschen; 5 May Colditz; 19 May Koenigswartha-Weissig; 21 May Bautzen; 24 May Naumburg/Saale; 21 Aug Loewenberg; 26 Aug Katzbach; 3 Oct Wartenburg; 16 Oct Moeckern. **1814** 11 Jan St Avold; 19 Jan Sk Manheules (1 sqn); 3 Feb La Chausee; 4 Feb Chalons-sur-Marne; 11 Feb Montmirail; 12 Feb Chateau-Thierry; 9 / 10 Mar Laon; 26 Mar Sezanne; 30 Mar Paris. **1815** no actions. In 1819 the regiment's title became the Litthauisches Drag R Nr 1. **Dragoner-Regiment Prinz Albrecht von Preussen (Litthauisches) Nr 1** from 1872.

8. Dragoner-Regiment.
Raised in 1744. Title in 1806: von Esebeck. **Inspectorate: Prussian. Garrison Insterburg.**

Uniforms. Facings scarlet, buttons and waistcoat white. Saddle furniture was red and square ended. The edging was a wide white lace with two narrow royal blue stripes close to the outer edge.

Guidons. The king's guidon was white with red corner piles, the regimental models were black with red corner piles; embroidery and fringes were gold. The central black eagle was spread under a scroll bearing 'PRO GLORIA ET PATRIA' in gold. The corner ciphers were the crowned 'FR' within olive wreaths. The finial bore 'FR'.

Campaign history. Raised in 1744 from half the men of Regiment Nr 7 at five squadrons; Chef GM Friedich von Stoch; retired. **1745** 15 Dec B Kesselsdorf. 1752 GM Adolf Friedrich von Langemann; died. **1757** Chef GM Dubislav Friedrich von Platen; died as GdC and Governor in Prussia. 30 Aug B Gross-Jaegerndorf; caused heavy losses to the Russians; **1758** 20 Jun C Ratzeburg; 25 Aug B Zorndorf, distinguished; 27 Oct C Greifenberg and Radam; 18 Nov C Lustow; 25 Nov Sk Pasewalk and Wilschow. **1759** 4 Jan Sk Schlatkow; 15/16 Jan S Demmin; 23 Jul B Kay; 12 Aug B Kunersdorf, very distinguished, but heavy losses. 20 Sep C Sagan. **1760** 17 Jun C Gruessau; 23 Jun B Landshut (4 sqns only); 4 Aug C Neumarkt. **1761** 3 Jul Sk Kriewen; 5 Sep C Reichenbach. **1762** 6 Jun C Wernersdorf; 14 Jul C Klentsch; 16 Jul C Peilau. **1778/1779** In the army of Prinz Heinrich. 1787 Chef GM Carl Wilhelm von Brausen; retired. 1790 GM Georg Friedrich von Bardeleben; died as GL. **1794** In Poland; split up into detachments. 9/10 Jul C Raygrod; 19 Nov Bialla. 1801 Chef GL Carl Gottlieb von Busch; died. 1803 GM Christian Carl von Esebeck. **1806** In l'Estocq's reserve corps. 14 Nov C Schulitz; 7 Dec C Gollub; C Biezun, Zibor, Gurschno; 25 Dec C Soldau; 30 Dec C Schimanen. **1807** 8 Feb B Preussisch-Eylau; 26 Feb C Braunsberg; 13 Jun Sks Jesau and Ludwigswalde; 14 Jun C Koenigsberg. 1808 The regiment became Drag R Nr 4 (2. Westpreussisches). **1812** 1st and 2nd Sqns were mobilised as part of the 1st Combined Drag R. 17 Jul C Dalety; 19 Jul C Eckau; 26 Jul C Dahlenkirchen; 5 Sep Sk Bersamuende; 27 Sep C Eckau; 28 Sep C Mesoten; 28/29 Sep C Graefenthal; 29 Sep C Bauske; 16 Oct C Dahlenkirchen; 22 Dec C Chelm; 26 Dec Piktupoenen. **1813** 5 Apr C Moeckern, Zehdenick and Dannikow; 13 Apr C Herrenkrug; 2 May C Halle; 4 Jun C Luckau; 22 Aug C Wittstock; 23 Aug B Grossbeeren; 5 Sep C Zahna; 6 Sep B Dennewitz; 17 Oct Taucha; 16/17 Nov Sk Wesel; 23 Nov Sk Duisdurg; 24 Nov capture of Zuetphen; 25 Nov Sk Middagten and Velp; 29/30 Nov S Arnhem. **1814** 11 Jan C Westmale and blockade of Antwerp; 1 Mar C Soissons; 9/10 Mar B Laon; 20 Mar C Soissons; 31 Mar Sk Crespy and Grondeville. **1815** No actions. On 7 Mar 1815 the 3rd Sqn was given to help form the 8th Dragoons. On 5 Nov 1816 the title changed to 4th Drag R (Ostpreussisches); on 27 May 1819 the regiment was converted to the 5th Kuerassier R (2. Ostpreussisches). In 1889 the title became: **Kuerassier-Regiment Herzog Friedrich Eugen von Wuerttemberg (Westpreussisches) Nr 5.**

9. Dragoner-Regiment.
Raised in 1743. Title in 1806: **von Herzberg.** Inspectorate: Prussian. Garrisons: Bischofswerder, Christburg, Deutsch-Eylau, Riesenburg, Salfeld.

Uniforms. Facings scarlet, buttons, lining and waistcoat white. Saddle furniture was light blue and square ended. The edging was a wide white lace with two narrow red stripes near to the outer edges.

Guidons. The king's guidon was white with red corner piles, the regimental models were yellow with red corner piles; embroidery and fringes were silver. The central black eagle was spred under a scroll bearing 'PRO GLORIA ET PATRIA' in silver. The corner ciphers were the crowned 'FR' within olive wreaths. The finial bore 'FR'. The five guidons were lost at the capitulation of Ratkau.

Campaign history. Raised in 1743 in Silesia from five of the ten squadrons of light dragoons of the 1st Dragoons and completed with new recruits. The old Chef had been LG Hans Friedrich von Platen; on 27 Oct 1743 Oberst Georg Ludwig Prinz von Holstein-Gottorp took over; died as GL. **1745** 15 Dec B Kesselsdorf. **1757** 30 Aug B Gross-Jaegersdorf. At the start of the action they charged and broke through the three enemy lines and took eight cannon. Due to lack of teams, they could not bring these trophies off. **1758** 23 Jun B Krefeld. Overthrew a French Cuirassier regiment and took a standard and a pair of kettledrums. **1759** 13 Apr B Bergen; 1 Aug B Minden, took four enemy battalions prisoner as well as ten cannon. For this victory, Herzog Ferdinand of Braunschweig presented the regiment with several thousand Thalers. **1760** July S Dresden; 15 Aug C Liegnitz. **1761** Chef Oberst Nikolaus Alexander von Pomeiske; died as GL. In Pommerania; S Colberg; 12 Dec C Spie. **1778/1779** In the army of Prinz Heinrich. 1785 Chef GM Johann Bogislav von Zitzewitz; retired. 1788 GM Hans Friedrich von Borstell; received Kuer R Nr 7. 1792 Chef Oberst Hyronimus von Brueckner; retired as GM. **1793** In the occupation of South Prussia. **1794** In Poland. Storm of Warsaw; saved the Russian troops from defeat. 23 Oct saved Suchaczew from being stormed by the insurgents. 10 Nov C Kapinos. 1798 Chef Oberst Friedrich Wilhelm Graf von Herzberg; GL in 1806. **1806** In the Herzog von Wuerttemberg's Reserve Corps. 17 Oct C Halle; then with Bluecher's corps. 6 Nov storm of Luebeck; the regiment surrendered next day at Solmsdorf and Ratkau and was not re-raised. The depot went to the new 2nd Kuerassiers, 2nd and 4th Dragoons and 1st Ulans.

10. Dragoner-Regiment.

Raised in 1743. Title in early 1806: **vacant Manstein.** Inspectorate: Prussian. Garrisons: Locbau, Neidenburg, Ortelsburg, Osterode, Strasburg.

Uniforms. Facings orange, buttons and waistcoats white. Saddle furniture was orange and square ended. The edging was a plain, wide white lace.

Guidons. The king's guidon was white with red corner piles, the regimental models were yellow with red corner piles; embroidery and fringes were silver. The central black eagle was spread on a white disc under a scroll bearing 'PRO GLORIA ET PATRIA' in silver. The corner ciphers were the crowned 'FR' within olive wreaths. The finial bore 'FR'. The five guidons were lost at the capitulation of Ratkau.

Campaign history. Raised in 1743 in Silesia from five of the ten squadrons of light dragoons of the 1st Dragoons and completed with new recruits. Chef: until November GL Hans Friedrich von Platen; died; then GM Johann Adolf von Moellendorf; died as GL. **1745** 15 Dec B Kesselsdorf. 1754 Chef GM Friedrich Ludwig Graf Fink von Finkenstein; died as GL. **1757** 30 Aug B Gross-Jaegersdorf, crushed a large body of Russian infantry. **1758** Jan-15 Jun S Stralsund; 21 Feb C Rothenburg, broke a Franco-Swiss infantry battalion and overthrew three French cavalry regiments, taking a standard and over 300 prisoners, before capturing an entire Swiss battalion. At Emmerich on the Rhine they defeated the French cavalry regiment Bellefond and took their kettledrums. 23 Jun B Krefeld, took the French infantry in rear and drove them out of their fortifications; also overthrew the French Royal Cravattes cavalry regiment and took their kettledrums. 18 Oct C Soest, charged two battalions of French grenadiers in a redoubt and captured them. **1759** 13 Apr B Bergen; 1 Aug B Minden. **1760** July S Dresden; 15 Aug B Liegnitz. **1761** 15 Aug C Kloster Wahlstadt; overthrew two regiments of Austrian Kuerassiers and a regiment of horse grenadiers. For their actions here, the king gave the officers silver-hilted swords with the black eagle badge (as carried in 1806) and awarded the regiment 3,000 Thaler. The regiment also went to Poland and was involved in the destruction of the great wagon train at Gostin on 15 Sep, capturing a battalion of Russians and two cannon. Sk Regenwalde; thence in the campaign in Pommerania. **1762** 6 Jul C Adelsbach; 7 Jul C Friedland – overthrew three regiments of cuirassiers; 21 Jul C Leutmannsdorf; 7 Aug-9 Oct S Schweidnitz. **1778/1779** 25 Oct C Jaegerndorf in Upper Silesia. 1785 Chef GM Wilhelm Leopold von Rosenbruch; died. 1790 Chef GM Carl Wolfgang von Franckenberg; died in office. **1794** In Poland. Bombardment of Blonien; Sk Raczin; 26 Jul C Opalin; 27 Jul-6 Sep S Warsaw; Sk Wawrcize against Poniatowski's corps; 19 Aug C Powonski; Sk Karniowko (1 sqn); Sk Lubienitza (1 sqn); 26 Oct Sk Gross-Magnisczewo (1 sqn). 1795 Chef GM Carl Gottlieb von Busch; received Drag R Nr 8. 1801 Chef Oberst Christian Heinrich von Manstein; retired. **1806** Chef from 19 Aug GM Ulrich Leberecht von Heyking. The regiment was in Bluecher's corps, surrendered at Ratkau and was not re-raised. The men were distributed among the new 2nd Kuerassiers, the 2nd and 4th Dragoons and the 1st Ulans.

11. Dragoner-Regiment.
Raised in 1743. Title in early 1806: **von Voss.** Inspectorate: Lower Silesian. Garrisons: Freistadt, Gruenberg, Sagan, Sprottau.

Uniforms. Facings, lining and saddle furniture were lemon yellow, buttons and waistcoat white. The saddle furniture was square ended and edged with a plain wide white lace.

Guidons. The king's guidon was white with yellow corner piles, the regimental models were yellow with white corner flames; embroidery and fringes were silver. The central black eagle was spread on a white disc under a scroll bearing 'PRO GLORIA ET PATRIA' in silver. The corner ciphers were the crowned 'FR' within olive wreaths. The finial bore 'FR'. The five guidons were lost at the capitulation of Prenzlau.

Campaign history. Raised in Silesia in 1741 from new recruits at five squadrons; the officers came from the Saxon cavalry regiment von Nassau. Chef GM Christoph Ernst Graf von Nassau; died as GL. **1741** 31 Oct capture of Neisse. **1742** capture of Olmuetz. **1744** 10-16 Sep S Prague. **1745** 4 Jun B Hohenfriedberg; distinguished by cutting down a corps of Saxon grenadiers. 5 Sep Capture of Cosel; 27 Sep C Leobschuetz; scattered a battalion of Hungarian infantry and took 170 prisoners. Chef GM Christoph Ludwig von Stechow; retired. **1757** 6 May B Prague; overthrew three enemy Kuerassier regiments singlehanded. 18 Jun B Kolin; 22 Nov B Breslau; 5 Dec B Leuthen, took many prisoners and a convoy of 90 flour wagons. **1758** Chef Oberst Leopold Johann von Platen; retired as GL. 17 Oct Ambushed at Troppau losing some hundreds of men captured. C Domstaedtel (1 sqn); 14 Oct B Hochkirch, in Rentzow's corps at Weissenburg which covered the retreat of the army. **1759** C Spremberg captured the enemy Regiment Wuerzburg of 1,300 men. 12 Aug B Kunersdorf, at the Judenberg; heavy losses. 21 Sep C Gorbitz; 29 Oct C Pretsch; overthrew two battalions of enemy grenadiers, took many prisoners and two cannon. As a reward, all captains and field officers received the *Pour-le-Merite*. 20 Nov C Maxen; most of the regiment was captured. It was at once re-raised. **1760** 3 Nov B Torgau; took a standard. **1762** 14 Oct C Freiberg; one squadron of the regiment took an enemy battalion prisoner and overthrew a whole regiment of imperial dragoons. 1770 Chef Oberst Franz Gustav von Mitzlaff; retired as GL. **1778/1779** In the army of the king. Chef now GM Friedrich Leopold von Bosse; retired. 1789 GM Carl Wilhelm von Tschirschki; retired. **1792** On the Rhine and in Champagne; 20 Sep Valmy. **1793** Chef now GM Ludwig Ernst von Voss; retired as GL. 17 and 20 Aug C Kettricher Hof; 14 Sep B Pirmasens; 28-30 Nov B Kaiserslautern. **1794** 23 May C Kaiserslautern; 3 Jul C Trippstadt. **1806** Chef from 19 Sep GM August Friedrich Erdmann von Krafft. 14 Oct Jena, in the right wing. The regiment surrendered at Prenzlau on 28 October and was not re-raised. The depot and the few survivors went into the new 2nd Dragoons and the new regiments raised in Silesia.

12. Dragoner-Regiment.

Raised in 1734. Title in 1806: **von Bruesewitz.** Inspectorate: Prussian. Garrisons: Kosten, Krochostin, Meseritz, Peysern, Schmiegel.

Uniforms. Facings black, buttons, lining and waistcoat white. Officers' coat lining was buff. Saddle furniture had rounded rear corner and plain, wide white lace edging.

Guidons. The king's guidon was white with red central disc, the regimental models were red with white central disc; embroidery and fringes were silver. The central black eagle was spread on a white disc under a scroll bearing 'PRO GLORIA ET PATRIA' in silver. The corner ciphers were the crowned 'FR' within olive wreaths. The finial bore 'FR'.

The five guidons were lost, partly in combat on 6 Nov at Schwartau, partly next day, at the capitulation of Ratkau.

Campaign history. Raised in 1734 by the Carl Alexander Herzog von Wuerttemberg-Stuttgard and placed into Austrian service; in 1742 it came into Prussian service at five squadrons. It was titled Alt-Wuerttemberg to differentiate it from other regiments of the same name.

From 1734-1749 the Chef was Marie-Auguste Herzoegin von Wuerttemberg, the Herzog's widow. **1734/1735** On the Rhine as a Wuerttemberg regiment. **1744** 10-16 Sep S Prague. **1745** 4 Jun B Hohenfriedberg; 30 Sep B Soor. 1749 Friedrich Eugen Herzog von Wuerttemberg-Stuttgard; later FM. **1757** 21 Apr C Reichenberg; together with the 1st Dragoons, they overthrew the Austrian cavalry, taking three standards and some cannon. 6 May B Prague; 18 Jun B Kolin; 7 Sep C Moys; 22 Nov B Breslau; 5 Dec B Leuthen. **1758** 27 May-2 Jul S Olmuetz; 14 Oct B Hochkirch. **1759** 25 Sep C Hoyerswerda; 29 Oct C Pretsch; 20 Nov C Maxen (4 sqns only). **1760** 3 Nov B Torgau. **1761** In Pommerania 9-20 Sep S Colberg; 12 Dec storm of Spie. **1762** C Burkersdorf; 21 Jul C Leutmannsdorf; 19 Aug C Reichenbach; 7 Aug-9 Oct S Schweidnitz. 1769 Oberst Carl Erdmann von Reitzenstein; retired as GM. **1778/1779** In the army of Prinz Heinrich. 1780 GM Johann Nikolaus von Kalckreuth; retired as GL. **1794** GM Carl Gustav von Biberstein; retired. **1794** In Poland 6 Jun B Scelze; charged the enemy led by Prinz Ludwig von Preussen, brother of the later King Friedrich Wilhelm III, and scattered them. For this all squadron commanders received the *Pour-le-Merite*. S Warsaw. 1797 GM Carl Friedrich von Bruesewitz; GL in 1806. 24 Sep **1806** Oberst Friedrich von der Osten. The regiment was part of Bluecher's corps, fought at Schwartau, surrendered next day and was not re-raised. The depot was incorporated into the new regiments raised in Silesia.

13. Dragoner-Regiment.
Raised in February 1802. Title in 1806: **von Roquette.** Inspectorate: Prussian. Garrisons: Kollno, Mysziniec, Prasniz, Sczuczin.

Uniforms. Facings and lining crimson, buttons yellow; white waistcoats.

Guidons. Not known

Campaign history. Raised in February 1802 from drafts from all other Kucrassier and dragoon regiments in the area of Bialystock. Chef Oberst von Roquette; GM in 1806. **1806.** The regiment was in the defence of Danzig; in 1807 the survivors were incorporated into the new 1st Kuerassiers.

14. Dragoner-Regiment.
Raised in March 1803. Title in 1806: **von Wobeser.** Inspectorate: West Prussian. Garrisons: Duderstadt, Hildesheim, Muenster, Warburg, Warendorf.

Uniforms. Facings buff, buttons yellow. Officers had white lining to the coat.

Guidons. Four guidons were lost at the capitulation of Prenzlau on 28 Oct 1806, one at that of Ratkau.

Campaign history. Raised in March 1803 from the old Cavalry Regiment von Nagel of Muenster service; Chef Oberst von Wobeser; GM in **1806.** 14 Oct Jena, in Ruechel's corps. The depot surrendered at Wolgast on 2 Nov 1806; regiment was not re-raised.

25

Dragoon Regiments Raised After 1808

1. Garde-Dragoner-Regiment Koenigin Viktoria von Grossbritannien und Irland.
Raised on 21 Feb 1815 as Garde-Dragoner-Regiment from the Garde-Drag-Esk of the Guard Light Cavalry Regiment, two squadrons of the Pommersches National-Kavallerie-Regiment and the Leib-Esk of the Drag-R Koenigin Nr 5. The title shown above was adopted on 23 Jan 1901. The regiment saw no action in 1815.

Standards. On 24 Feb 1815 it was presented with a standard of the old Gardes du Corps, presented itself on 7 Sep 1798, which had been transferred to the Guards Light Cavalry Regiment on 5 Dec 1814.

Campaigns of the 1st Garde-Drag-Esk. Raised on 27 Mar 1813 from volunteers in Pommerania at three squadrons and a depot. Disbanded on 6 Feb 1815; the best men and horses went to the Garde-Drag-R, the rest went to form the 4th Ulanen-R.

1813 22 Aug Sk Wittstock; 23 Aug B Grossbeeren; 31 Aug Sk Linde (2nd and 3rd Sqns); 6 Sep B Dennewitz; 14 Sep-4 Oct S Wittenberg; 30 Nov storm of Arnhem; 16/17 Dec Sk St-Andre (1st Sqn); 19 Dec Sk Geffen (2nd and 3rd Sqns); 26 Dec Sk Nulans (2nd and 3rd Sqns). **1814** 7 Jan Sk Poppel (3rd Sqn); 11 Jan C Hoogstraten; 2, 3 and 7 Mar Sks Courtray; 7 Mar Sk Sweweghem.

2nd Dragoon-Regiment.
Raised from the depots and survivors of the following regiments: 1st Kuerassiers, 2nd and 11th Dragoons. By Order of 7 Sep 1808 this unit became the **Dragoner Nr 2, 1st Westpreussisches Drag-R.** On 21 May 1819 it was converted to the 4th Kuerassier-Regiment (Westfaelisches); on 27 Jan 1889 it became the **Kuerassier-Regiment von Driesen (Westfaelisches) Nr 4.**

Standards. In 1807 they were presented with four guidonss of the old Zieten Hussars Nr 2, originally presented in 1737; details not known. Two of these were handed in to the arsenal at Glatz in 1813, and a third was given to the new 4th Dragoons.

Campaign history. 1812 The 1st and 3rd Sqns were mobilised against Russia in the Combined Dragoon Regiment. 19 Jul C Eckau; 7 Jul C Wolgund; 27 Sep C Eckau and the Aa River. **1813** The 1st and 3rd Sqns were initially in von Yorck's corps, the 2nd abd 4th were in von Buecher's corps; fron 2 May the regiment was united again. 5 Apr C Dannigkow; 2 May B Gross-Goerschen; 19 May C Koenigswartha and Weissig; 20/21 May B Bautzen; 26 Aug Katzbach; 15 Sep C Reichenbach; 16 Oct B Moeckern; 17 Oct C Markroehlitz. **1814** Sk Manheulles; 3 Feb C La Chaussee; 5 Feb Sk Marieul; 11 Feb B Montmirail; 12 Feb B Chateau-Thierry; 10 Mar B Laon; 26 Mar C Sezanne; 30 Mar B Paris. **1815** 15 Jun C Gilly; 16 Jun B Ligny; 18 Jun B Waterloo; 28 Jun C Nantuil.

3rd Dragoons.
Raised from the depots and men of the old 7th Dragoons. By Order of 7 Sep 1808 it became the Ostpreussisches Drag-R. The Order of 14 Sep 1808 retitled the unit the **Dragoner Nr 3, Litthauisches Drag-R.** On 15 Oct 1872 the title became **Dragoner-Regiment Prinz Albrecht von Preussen (Litthauisches) Nr 1.**

Standards. King's standard: their own old one. White cloth with red corner flames and balck centre; gold embroidery and fringes. Regimental standards: three of their old ones. Black cloth, red corner flames, silver centre, gold embroidery and fringes.

Campaign history. See the old 7th Dragoons above.

4th Dragoons.

Raised on 7 Mar 1815 from the 2nd Sqn, 1st Westpreussisches Drag-R (4th Kuer-R in 1905), 3rd Sqn of the 5th Dragoons (2nd Kuer-R in 1905) and 2nd Sqn, 6th Drag-R (3rd Kuer-R in 1905). On 5 Nov 1816 it was re-titled the 7th Drag-R (Rheinisches), and the title in 1905 was **Dragoner-Regiment von Bredow (1. Schlesisches) Nr 4.**

Standards. On 2 Apr 1815 they were presented with the old standard of the 2nd Sqn, 1st West Prussian Dragoons (see new Kuer-R Nr 4 above). This had originally been presented in 1737 to the old 6th Drag-R von Auer. For the description of the standard see there.

Campaigns 1815 16 Jun B Ligny; 18 Jun C Wavre; 20 Jun storm of Namur.

In 1815 the regimental distinctions were as shown below:

Title	**Facings**	**Buttons**
Garde-Dragoner-Regiment	red	yellow
Koenigin Nr 1	crimson	white
1st Westpreussisches Nr 2	white	white
Litthauisches Nr 3	poppy red	yellow
2nd Westpreussisches Nr 4	poppy red	white

26

Hussar Uniforms

They wore the traditional Hungarian costume of a richly laced dolman and pelisse, coloured barrel sash, leather breeches and short boots with the top front cut out in a 'V' shape with a tassel. The pelisse was lined with fur in the regimental colour and this was worn slung over the left shoulder. NCO s and trumpeters often wore fox fur edging. The original headgear was the fur trimmed cap, in 1756 this had given way to the mirliton and by 1786 the brown fur busby was being worn. By 1806 Knoetel shows them with black felt shakos with a leather peak. They were armed with a carbine, curved sabre and a pair of pistols. Harness was black, Hungarian-style with steel fittings. They used the Hungarian-style 'Bock' saddle made on a birch wood frame.

In 1976 the *Deutsche Gesellschaft fuer Heereskunde* published a hitherto unknown collection of watercolours of the hussar regiments of the Prussian army in 1791 (1792 in the case of the 11th Regiment), which threw much new light on the uniforms of this arm. This booklet was called *'Preussische Husarenbilder um 1791.'* 1,100 copies were printed. They were obviously painted by an artist with much detailed knowledge of his subject; where the details of these plates conflict with other sources, these are mentioned separately. Interestingly enough, regiments 1 – 4, 9 and 10 of this water colour series are shown wearing brown fur busbies, with bags in the colour of the dolman. Number 9, the Bosniaken, wear black Astrakhan caps in one plate and brown fur colpacks in the other, Regiments 5 – 8 wear mirlitons or winged caps. Officers' (parade?) harness is shown as being elaborately decorated with white cowrie shells.

Initially, the hussars carried swallow-tailed guidons; these were withdrawn in 1743. No standards were then carried until the Wars of Liberation.

27

Regimental Histories of the Hussars

Nr 1.

Raised in 1721; often referred to as the Green Hussars. Title in 1806: **Von Gettkandt.** Inspectorate: Lower Silesian. Garrisons: Guhrau, Herrnstadt, Koeben, Militsch, Prausnitz, Steinau, Sulau, Trachenberg, Wintzig, Wohlau

Light green Dolman, dark green Pelisse, red collar and cuffs, white buttons and lace, red and white barrel sash. Light green sabretasche with white crowned royal crest and edging. Dark green round portmanteau edged light green, dark green shabraque with light green wolfs tooth edging and white piping. The regiment was destroyed in 1806.

Campaign history. Raised in 1721 at a strength of 30 men by GL von Wuthenow (died in office). He recruited mainly Ulans and in 1722 they were increased by two companies; the commander was a Major Schmidt. Chef in 1727 Oberst von Dockum; died in office. In 1730 they were increased to three squadrons and command passed to Major Johann von Brunikowski. Chef in 1732 was Prinz Eugen von Anhalt-Dessau. In 1737 von Brunikowsi became Chef; retired as GM. Three more squadrons were added, but in 1740, three of them were sent to Silesia to reinforce the Leibhusaren (Nr 2) and two more were used in 1740 to found the 3rd Hussars. The remaining squadron, under Major von Mackerodt, was again reinforced by two squadrons, which later went to form the 5th and 7th Hussars in 1741. Brunikowski, now a colonel, had kept back ten men from each squadron that he had been ordered to send away. With these and new recruits, he raised a new regiment of five squadrons in Prussia in 1741. These he took to Silesia and increased to ten squadrons.

1741 31 Oct Capture of Neisse. **1742** 17 May B Chotusitz. **1744** 10-16 Sep S Prague, Sk Neustadt. **1745** 1 Mar C Hirschberg, took over 300 prisoners; 22 Mar C Landshut; 4 Jun B Hohenfriedberg. 1747 Chef Oberst Otto von Dewitz; retired. 1750 Oberst Michael von Szekeli; retired as GM. **1756** 1 Oct B Lobositz (8 sqns). **1757** 6 May B Prague (3 sqns); 18 Jun B Kolin (5 sqns); 5 Nov B Rossbach; took four cannon and pursued the beaten enemy through Erfurt. 5 Dec B Leuthen (5 sqns). 1758 Chef Oberst Friedrich Wilhelm Gottfried Arend von Kleist; died in 1767 as GM. **1759** In Franconia; B Kunersdorf. **1760** July S Dresden; 20 Aug C Strehla, very successful against the enemy cavalry. 3 Nov B Torgau. **1761** 2 Apr C Saalfeld. **1762** 14 Oct C Frieberg. 1770 Chef Oberst Georg Oswald Graf von Czetteritz; died in 1796 as GdC. **1778/1779** In the army of Prinz Heinrich. **1793** One squadron in Poland; Sk Szurawe. **1794** Sk Pilica River. The 2nd Bn and a squadron of the 1st were in C Skala on 18 May; the whole regiment in B Scelze, 6 Jun; Sk Nowamiesta / Nidda (2 sqns); C Radzin 14 Jul (5 sqns); cannonade at Wola 28 Jul (8 sqns); S Warsaw 27 Jul-6 Sep (8 sqns); Sk Collo (1 sqn); C Konin (2 sqns); Sk Lekno and Exin (2 sqns); Sk Bromberg (1 sqn). 1797 Oberst von L'Estocq; in 1806 he was GM and Chef of the Towarczys. 1803 Chef GM Ernst Philipp von Gettkandt. **1806** 14 Oct Jena, in the Right Wing; heavy losses. The regiment surrendered at Anklam on 1 November; a detachment surrendered at Ratkau; the regiment was not re-raised. The depot went into the new 3rd Hussar Brigade, raised in Silesia. See the old 6th Hussars below

Nr 2.

Raised in 1730; often referred to as the Red Hussars. Title in 1806: **Leibhusaren-Regiment von Rudorff.** Inspectorate: Mark Brandenburg. Garrisons: Beeskow, Berlin, Fuerstenwalde, Muellrose.

Dark blue Pelisse, scarlet Dolman with dark blue collar and cuffs, white buttons and lace, dark blue and white barrel sash. Scarlet Sabretasche with white crowned cipher and white edging. Dark blue shabraque with scarlet wolfs tooth edging and white piping, dark blue, round portmanteau edged scarlet.

Campaign history. Raised in 1730 from a draft from the 1st Hussars and new recruits at one company with Obstlt von Beneckendorf as commander. A 2nd company was added in 1731, a 3rd in 1732 and in 1733 they were increased to three squadrons. 1735 Oberst von Wurmb took over command; posted as Oberst to a Grsn R. In 1739 King Friedrich Wilhelm I declared this unit to be his **Leibhusarencorps** and in 1740 three more squadrons were added from the 1st Hussars. In 1741 in Silesia the regiment was increased to ten squadrons in two battalions. **1741** Oberst Hans Joachim von Zieten; died as GdC. 18 Mar C Frankenstein; 10 Apr B Molwitz; 17 May C Rothschloss; 10 Jun C Grottkau; Neisse. On 24 Jul 1741 the regiment absorbed the Prussian Husarenkorps of Oberst Bronokowski and had ten squadrons in two battalions. From now until 1806 it was known as the Leibhusaren-Regiment. **1742** Sk Goeding; Sk Skalitz; Ungarisch-Brodt; 17 May B Chotusitz; Pardubitz. **1744** Sk Smetschau-Munczisay; 23 Sep C Tabor; 1 Oct Frauenberg; 4 Oct C Moldauthein; together with the 5th Hussars they put the enemy to flight. 19 Nov Teinitz. **1745** 22 May C Neustadt in Upper Silesia under Markgraf Carl; 20 May C Jaegerndorf; again 22 May Huellberge; 4 Jun B Hohenfriedberg; 5 Jun Faulbrueck; 23 Nov C Katholisch-Hennersdorf; with the 5th Hussars they took the kettledrums that they were still using in 1806. **1756** 10 Sep-15 Oct blockade of the Saxon army in the camp at Pirna; 22 Sep Lang-Hennersdorf. **1757** 27 Apr Budin; 6 May B Prague; 12 Jun Moleschau; 18 Jun B Kolin; took thousands of prisoners from the fleeing enemy army. 26 Jun Sk Opatzkau; 3 Jul Sk Wellmina; 7 Sep C Moys; 13 Sep Sk Birkenbrueck; 10 Nov Wultischau; 21 Nov Sk Klettendorf; 22 Nov B Breslau; 5 Dec B Leuthen. **1758** 2 May Sk Littau; 22 May Sk Konitz; 28 Jun Sk Domstaedtl; 12 Jul Sk Hohenmauth; 29 Jul Sk Czernilow; 25 Aug B Zorndorf; overthrew the enemy cavalry. 14 Oct B Hochkirch; 29 Oct Sk Pfaffendorf; 1 Nov River Queis. **1759** 26 Mar Sk Greifenberg; 23 Jul C Kay (500 men); 12 Aug B Kunersdorf (500 men); 1 Sep Sk Luebben; 29 Oct Sk Sadowitz. **1760** 27 May Sk Kalkowitz; 2 Jun Sk Torgau; 15 Jun Sk Ober-Roedern; 7 Jul Sk Nieder-Gurka; 14-31 Jul S Dresden; 31 Aug Sk Bunzlau; 15 Aug B Liegnitz; 13 Sep Sk Hohenfriedberg; 17 Sep C Hohengiersdorf; 31 Oct Sk Zobtelberg; 3 Nov B Torgau; in the advanced guard, they clashed with the Austrian St Ignon Dragoons, captured the enemy general, 20 officers and nearly 400 men. **1761** 14 Feb Merxleben; 15 Feb C Langensalza; the 1st Bn captured the Saxon Guard, a battalion of grenadiers and six cannon. 9 Mar Sk Greiz. On 2 April, the 1st Bn clashed with the Austrians again at Saalfeld; two squadrons charged a force of two squadrons of Kuerassiers and six companies of grenadiers; captured the latter together with two colours and four cannon. Both battalions then united in the chase of the beaten enemy at Hoheneiche, where they took almost 900 prisoners, two colours and six cannon. 5 Apr Sk Plauen; 9 Jun Sk Nimptsch; **1762** The regiment was in Silesia; cleared the enemy from the heights of Burkersdorf (21 Jul) and Leutmannsdorf; at Spechtshausen in the Taranter forest they attacked the Austrian advanced guard, taking several hundred prisoners and six cannon. Jul-Aug S Schweidnitz; 7 Nov Sk Landsberg. **1778/1779** In the army of the king. **1778** 7 Jul Sk Skalitz; 11 Sep Sk Trautenau. **1787** In the expedition to Holland; captured a frigate. 14 Sep Vianen; 16 Sep the locks at Arkel; 17 Sep Nieuwport; 19 Sep Altphen; 18 Sep charged the Wiering redoubts. 1786 Oberst Carl August Baron von Eben und Brunnen; retired as GL. **1792** 11 Sk Aug Sierck; 19 Aug Sk Fontoy; 2 Sep Verdun; 18 Sep St-Menehoud. The regiment took over 100 men of the French Light Bns Nrs 1 and 20, 61 horses and the military chest of 57,000 Livres.; 20 Sep Valmy. **1793** On the Rhine. 16 Jan Capture of Hochheim (7 sqns); 30 Mar Sk Ober-Flersheim; C Alzei; 13 Aug C Limbach (1st Bn took 2 cannon); cannonade of Russweiler (2nd Bn); 26 Sep C Hornbach; Sk Bubenhaeuser Heights; 28/29 Nov B Kaiserslautern (1st Bn); 6 Dec Sk Zweibruecken; **1794** 8 Jan Sk Kreuznach; Sk Kirchheim-Poland; Sk Moesheim; 23 May C Kaiserslautern (the regiment took two colours, two howitzers, 12 guns, 29 ammunition wagons, two field smithies, 60 baggage wagons; two limbers, 1,500 prisonersand 450 horses); 13 Jul Schaenzel; 3 Jul Sk Trippstadt; 13 Jul Sk Hermersberg; Sk Koeshofen; Sk Mainz. 1795 Chef GL Friedrich Eberhard Siegmund Guenther von Goecking; retired in 1805 as GdC. He was succeeded by Oberst Wilhelm Heinrich von Rudorff; GM in 1806. **1806** 11 Oct Sk Coburg; 26 Oct Sandau; withdrew with Bluecher's corps. 3 Nov C Criwitz; 5 Nov Waren; 6 Nov storming of Luebeck. The main body of the regiment surrendered at Ratkau next day and was not re-raised. A detachment of 40 men surrendered at Hameln on 22 Nov 1806. **1807** 10 Jun B Heilsberg (1 sqn). Three weak squadrons were formed in East Prussia; they then absorbed the Hus-Bn von Bila Nr 11 and some men of the Berlin Husaren-Kommando. They formed the Husaren-Brigade von Rudorff; on 7 Sep 1808 the title changed to Hus-R Nr 3 (1. Brandenburgisches). 31 May 1809 this changed to Brandenburgisches Hus-R. **1812** Two squadrons mobilised as pert of the 2nd Combined Hus-R for the Russian campaign. 28 Jun Sk Wilna; 5 Jul C Kosziany; 25 Jul Ostrowo; 27 Jul Witepsk; 30 Jul Sk Paniszani; 16 Aug Sk Smolensk; 21 Aug Dniepr River; 25 Aug C Dorogobusch; 27 Aug Rupki; 28 Aug C Wiasma; 5 and 7 Sep B Borodino; 4 Oct C Winkowo; 28 Oct B Malo-Jaroslawetz. The regiment in Russia ceased to exist. **1813** 2 May B Gross-

Goerschen; 20/21 May B Bautzen; 29 May Sk Zwickau; 31 May Sk Neukirchen; 21 Aug C Loewenberg; 20 Aug Sk Graeditzberg; 26 Aug B Katzbach; 27 Aug Sk Kroitsch; 30 Aug C Bunzlau; 5 Sep C Reichenbach; 6 Sep Sk Camenz; 22 Sep Bischofswerda; 15 Sep Sk Dresden; 20 Sep Sk Goldbach; 23 Sep Sk Roth-Nauselitz; 3 Oct Wartenburg; 10 Oct C Naumburg; 14 Oct Sk Schkeuditz; 16 Oct B Moeckern; 18 Oct B Leipzig; 21 Oct C Freiburg; 30/31 Oct B Hanau; 16-25 Nov blockade of Kastel. **1814** 12-14 Jan blockade of Thionville; 19-21 Jan blockade of Luxemburg; 24-27 Jan blockade of Metz; 4 Feb Sk Chalons; 5 Feb C La Chaussee; 7 Feb Sk Epernay; 11 Feb B Montmirail; 12 Feb C Chateau-Thierry; 23 Feb Sk Mery; 25 Feb C Sezanne; 26 Feb C La Ferte; 28 Feb C Meaux; 9 Mar B Laon; 14 Mar C Berry-au-Bac; 21 Mar Sk Coincy; 23 Mar C Clay-Souilly; 30 Mar B Paris. **1815** 16 Jun B Ligny; 18 Jun B Waterloo; 1 Jul Ambushed at Versailles by Excelmanns‘ cavalry division; heavy losses. On 7 Mar 1815 they gave a squadron to form the new 1st Westfalian Hus-R Nr 8. On 27 Jan 1889 the title became **Husaren-Regiment von Zieten (Brandenburgisches) Nr 3**, which was still maintained in 1905.

Nr 3.

Raised in 1740. Title in 1806: **Von Pletz.** Inspectorate: Upper Silesian. Garrisons: Bernstadt, Constadt, Festenberg; Juliusburg, Medzibor, Oels, Pitschen, Reichthal, Trebnitz, Wartenburg.

In 1791 this regiment wore white dolman and breeches, dark blue pellise. By 1806 that had changed to dark blue Dolman and Pelisse, yellow collar, cuffs, buttons and lace, yellow and white barrel sash. The fur trim was white, fox fur for NCOs. Yellow sabretasche with white, crowned royal cipher and edging. Dark blue shabracque withwhite wolfs tooth edging and white piping; round, dark blue portmanteau with white piping.

Campaign history. Raised in **1740** from two squadrons of Regiment Nr 1 and new recruits at five squadrons; Chef Oberst Friedrich Asmas von Bandemer; retired. It went into action, was ambushed at Kloster Leubus and suffered heavy casualties. In 1742 it was completed up to 10 squadrons. 1741 Chef Oberst Hyazinth Malachow von Malachowski; killed at Gross-Strehlitz in 1745. **1744** In action against the rebels in Upper and Lower Silesia. 27 Nov Sk Plesse Attacked a force of Austrian cavalry and took over 200 prisoners. **1745** Sk Gross-Strehlitz; defeated a superior

Austrian cavalry force; took almost 300 prisoners. Oberst von Malachowski was killed. New Chef Oberst Hartwig Carl von Wartenberg; killed at Alt-Bunzlau in 1757. 28 Sep Sk Oderberg; overthrew an enemy dragoon regiment and took a standard and over 100 prisoners. **1756** Formed the advanced guard in the invasion of Bohemia; attacked at Aujest but beat the enemy of causing over 100 casualties. **1757** 3 May Sk Alt-Bunzlau; scattered over 1,500 Croats, but lost their Chef, GM von Wartenburg. New Chef Oberst Carl Emanuel von Warnery; retired and went into Polish service. 6 May B Prague; fought against the superior Austrian cavalry on the left wing and repelled them three times. 18 Jun B Kolin; again on the left wing. When the Prussian right wing collapsed, they covered the army's retreat. 5 Dec B Leuthen (3 sqns). **1758** Oberst Christian von Moehring; died as GL. S Olmuetz; Sk Landskron; overthrew an enemy cavalry force. The campaign in Pommerania against the Swedes; 25 Sep Sk Fehrbellin. **1759** 12 Aug B Kunersdorf (5 sqns); 29 Oct Sk Pretsch. **1760** July S Dresden; 15 Aug B Liegnitz; 17 Sep C Hohengiersdorf; 3 Nov B Torgau; 4 Nov Sk Zobten. **1761** In Saxony, various small actions. **1762** 24 Jul Sk Rattibor in Upper Silesia; 16 Aug C Reichenbach, distinguished. C Burkersdorf and Leutmannsdorf. 1773 Oberst Stephan von Samogy; died in office. 1777 Chef Oberst Hans Christoph von Rosenbusch; died as GM. **1778/1779** 2nd Bn in the invasion of Bohemia with the king's army. Sk Teutsch-Prausnitz, clashed with the Austrian Kaiser Hussars, defeated them and took some 80 prisoners. 1785 Chef Oberst Carl Franz von Keoszegy; retired as GM. 1788 Oberst Georg Ludwig Egidius von Koehler; received the 7th Hussars, GL in 1806. **1792** On the Rhine: 2 Sep Verdun; 14 Sep Sk Grandpre; 20 Sep Valmy. **1793** Sk Gundersheim; several actions in the Zweibruecken area, especially distinguished at Altstadt and Limbach on 13 Aug. 26 Sep Sks Blieskastel and Hornbach, drove out the French with the help of a company of Jaegers. Sk Remlingen (Lorraine). **1794** Minor actions around Trier, Zweibruecken and Saarbruecken. **1795** In the Muenster area against the French. 1796 Chef GM Friedrich Ludwig von der Trenck; died. 1797 Chef Oberst von Schultz; died in 1803 as GM. 1803 Chef Oberst August Wilhelm von Pletz; GM in 1806. **1806** 14 Oct Jena; the 1st Bn was in Ruechel's corps, the 2nd Bn was in Weimar's corps. They then joined Bluecher's corps and surrendered at Ratkow on 7 November. A detachment surrendered in Hamelin on 22 Nov 1806. The regiment was not reraised. Some survivors of the battle in Silesia went to form part of the 2nd Silesian Hussars and the 6th Hussars in 1808.

Nr 4.

Raised in 1741. Title in 1806: **Prinz Eugen von Wuerttemberg.** Inspectorate: Upper Silesian. Garrisons: Boleslawice, Dzialoczyn, Kempen, Namslau, Ostrowe, Radomsk, Rosenburg, Wielun, Wieruschau.
Light blue Dolman and Pelisse, red collar and cuffs, white buttons, silver lace for officers, blue and white lace for other ranks. Yellow and white barrel sash. The pelisse fur was white for officers and men, red fox for NCOs. Light blue shabracque and round portmanteau with white and blue piping; white sabretasche with light blue crowned royal cipher and edging.

Campaign history. Raised in 1740 and 1741 in Prussia by Oberst Georg Christoph von Natzmer (died in 1751 as GM) as a regiment of Ulans from Polish and Lithuanian recruits at five squadrons. In 1742 it was converted to a hussar regiment and increased to ten squadrons. **1741/ 1742** on campaign, no actions of note. **1745** 14 Feb C Planitz; 1 Mar C Hirschberg; 4 Jun B Hohenfriedberg; captured the kettledrums of the Saxon Karabinier-Garde. 30 Sep B Soor; Sk Gross-Strehlitz; charged an enemy column and took 112 prisoners. 15 Dec B Kesselsdorf. 1751 Oberst Heinrich Sigismund von Vippach; died in office. 1755 Oberst Georg Ludwig von Puttkamer; killed as GM at Kunersdorf. **1757** 21 Apr C Reichenberg; distinguished. 6 May B Prague, took many prisoners after the battle; 18 Jun B Kolin; 7 Sep C Kay; 22 Nov B Breslau; 5 Dec B Leuthen. **1758** Sk Harta, attacked Laudon's rearguard and caused them a loss of 500 casualties. 14 Oct B Hochkirch; 26 Oct C Goerlitz, it attacked the enemy cavalry and took 450 prisoners. **1759** The raid to destroy the great Russian magazines in Posen. 23 Jul B Kay; 12 Aug B Kunersdorf. **1760** Oberst August Lavin von Dingelstaedt; retired. 21 Jun C Landshut; 18 Sep relief of Colberg; 3 Nov B Torgau (200 men only). **1761** Pursued the Austrian rearguard; the 2nd Bn took many prisoners and three guns at Plauen. 12 Sep Sk Treptow. **1762** Chef Oberst Balthasar Ernst von Bohlen; retired. 16 Aug C Reichenbach. 1770 Oberst Carl von Podjurski; died in 1781 as GM. **1778/1779** In the army of Prinz Heinrich. **1778** 11 Aug Sk Glomnitz. 1782 Chef Oberst Eugen Friedrich Heinrich von Wuerttemberg; appointed Governor of Glogau in 1795, but the regiment continued to bear his name. **1794** In Poland; Sk Skala; 6 Jun B Scelze (8 sqns); 15 Jun capture of Krakow (5 sqns); 8 sqns were involved in the skirmishes at Tarzin, Raczin, 26 Jun Opalin and 28 Jul Wola near Warsaw. 26 Aug capture of redoubts at Warsaw (6 sqns); Wawrize (2 sqns); Sk Konin (2 sqns) took two guns. Sk Bromberg (1 sqn). Relief of Thorn. Sk Suchasczew (2 sqns). 1795 Chef Oberst von Lediwari; received the 10th Hus R. 1797 Chef LG Eugen Friedrich Heinrich von Wuerttemberg, as in 1781; GdC in 1806. **1806** 14 Oct Auerstaedt; 16 Oct C Buttstedt (6 sqns); 17 Oct Sk Nordhausen (3 sqns). Some 407 reached East Prussia with their horses. **1807** 14 Jun C Koenigsberg (Hafelstrom). On 9 Nov 1807 they became the Hussar Brigade von Dziengel. On 7 Sep 1808 the title changed to Oberschlesisches or 1st Schlesisches Hus-R. The depot went to form what later became the 6th Hussars. On 5 Dec 1808 they were absorbed the Niederschlesisches or 2nd Schlesisches Hus-R and the rumps of the 6th and 10th Hussars to form the Husaren Nr 4, 1st Schlesisches Hus-R. **1812** Two squadrons mobilised as part of the 3rd Combined Hus R. 5 Aug C St Annen; 7 Aug C Wollgund; 22 Aug C Dahlenkirchen; 29 Sep C Kosackenkrug; 30 Sep C Schlockhoff; 1 Oct Garossenkrug. **1813** 4-20 Apr blockade of Wittenberg (1st and 3rd Sqns); 18 Apr C Weimar (2nd and 4th Sqns); 28/29 Apr C Halle (1st and 3rd Sqns); 2 May B Gross-Goerschen; 19 May Koenigswartha; 20/21 May Bautzen; 26 May ambush at Hainau (2nd and 4th Sqns); 21 Aug C Neuhausen; 26/27 Aug B Dresden; 29 Aug C Glashuette; 30 Aug Culm; 16 Sep C Peterswalde; 17 Sep C Tellnitz; 16 and 18 Oct B Leipzig; 20 Oct Sk Kitscher; 26 Oct-24 Dec S Erfurt. **1814** 13 Feb B Etoges; 14 Feb B Champeaubert; 28 Feb C Gue-a-Tresmes; 2 Mar C May; 9 Mar B Laon; 26 Mar B La-Fere-Champenoise; 28 Mar C Claye-Souilly; 30 Mar B Paris. **1815** 7 Mar 1 sqn went to form the 9th Hussars. 15 Jun C Gosselies; 16 Jun B Ligny; 18 Jun B Waterloo; 22 Jun C Avesnes; 23/24 Jun Guise; 27 Jun C Compeigne; 28 Jun C Nanteuil. On 3 Dec 1816 title changed to 4th Hus R (1st Schlesisches); on 27 Jan 1889 the title became **Husaren-Regiment von Schill (1st Schlesisches) Nr 4.**

Nr 5.

Raised in 1741; often referred to as the Black Hussars or 'Der ganze Todt' (The Whole Death). Title in 1806: **Von Prittwitz.** Inspectorate: Prussian. Garrisons: Kalwary, Marienpoll, Prenn, Przeroslen, Schirwind, Serry, Suwalken, Wilkowischken, Wirballen, Wystiten

Black Dolman, Pelisse and sabretasche. White fur for officers, black for the men, scarlet collar and cuffs, white buttons and lace, red and white barrel sash, white breeches. The Mirliton, or winged cap, was originally decorated on the front with a white skeleton; by 1791 this had been reduced to a skull and crossed bones. Black shabraque with red wolfs tooth edging and white piping; round portmanteau piped red.

Standards King Friedich II granted the regiment the privilege of carrying the standards and guidons that they had captured in 1758 from the French as their own. These were later handed in to the arsenal in Berlin. On 28 Sep 1814 and 15 Jun 1815 two new-pattern standards were presented.

Campaign history. Raised in 1741 with one squadron of Regiment Nr 1 and new recruits in Brandenburg, at five squadrons. Chef Oberst von Mackeroth; died in 1743. In 1742 it was increased to ten squadrons. 1744 Chef Oberst Joseph Theodor von Ruesch; retired as GM. **1744** Against Austria. 29 Aug Sk Smetschna-Muncisay; 10-16 Sep S Prague; 9 Oct Sk Tecyn; together with the 2nd Hussars they contributed much to the success of the day. 21 Dec Sk Patschau. **1745** The Bosniakencorps (1 'Fahne' or squadron) was attached to the regiment; in 1760 it was increased to ten Fahnen, but reduced to one again in 1763 (see 9th Hus R). 1 Mar C Hirschberg; 7 Mar Sk Guttentag; 17 Mar Sk Neudorf; 22 May C Landshut; took 300 prisoners. 4 Jun B Hohenfriedberg, completely overthrew the opposing cavalry. 31 Aug Sk Lewin; 23 Sep Sk Trautenbach; 8 Nov Schlesisch-Friedland; 8/9 Nov Sk Oppau; 23 Nov C Katholisch-Hennersdorf; captured the kettledrums of a Saxon Kuerassier regiment and still used them in 1806. 25 Nov Sk Goerlitz; 28 Nov Sk Zittau. **1756-1763 The Seven Years War against Austria and Others. 1756** 1 Aug Sk Kummeln and Kattenau; 8 Aug Sk Poetschkamen; 9 Aug Sk Stannaitschen; 26 Aug Sk Plibischken; 30 Oct B Gross-Jaegerndorf; distinguished. 1758 Chef Oberst von Beust; died. Three squadrons went to the allied army, the other seven fought in Pommerania against the Swedes and Russians. Towards the end of 1761 the three re-joined the Prussian army. **1758** Chef Oberst von Beust. The squadrons in Pommerania. Jan-Jun Blockade of Stralsund; 14 Apr Sk Buetow; 20 Jun Sk Ratzebuhr. 1758 (The main body of the regiment) 5 Aug Sk Botschow; 13 Aug Sk Dexel; 25 Aug B Zorndorf; 19 Sep Sk Reppen; 22 Oct C Doelitz; 27 Oct C Greifenberg; 18 Nov Sk Guestow; 25 Nov Sk Werbelow. **1759** The main body. 23 Jul B Kay; 12 Aug B Kunersdorf, 21 Sep Sk Sagan; 24 Sep Sk Bauna; 22 Oct defence Herrenstadt (7 sqns); 14 Nov Sk Meissen (7 sqns); 17 Nov C Zoellmen (7 sqns). **1760** The regiment was still divided, part in Pommerania, part in Poland and Silesia. The main body: 24 Jun Sk Regenwalde; 25 Jun Sk Schievelbein (7 sqns); 13 Jul Sk Meseritz (5 sqns) 19 Jul Sk Tirschtiegel (5 sqns); 6 Aug Sk Neumarkt (5 sqns); 23 Aug Sk Medzibor (4 sqns); 30 Aug Sk Grabow (4 sqns); 31 Aug Sk Wartenberg (4 sqns); 13 Oct Sk Weissig (7 sqns); 18 Nov Sk Tannhausen (5 sqns); 18 Nov Sk Charlottenbrunn (5 sqns); 19 Nov Sk Tannhausen (5 sqns) 22 Nov Wuestegiershausen (5 sqns).

Apart from these actions, three squadrons of the regiment, together with two of the Regiment Malachowski Nr 7 were with the army of Prinz Ferdinand in Germany fighting against the French in Hanover, Hessen-Kassel and Westphalia. Their actions are shown here. **1758** 23 Feb C Stoecken-Drebber; the regiment overthrew the French Polleretzky Hussars and captured 300 horses, their standards and kettledrums. 3 Mar Sk Lauenau; 26 Mar Sk Lippstadt; 28 Mar Sk Soest; 1 Jun C Emmerich; 2 Jun Sk Duesselward; 12 Jun Sk Kloster Camp; 15 Jun C Toenisberg; 23 Jun B Krefeld. At the Rhine crossing the squadrons attacked a French cuirassier regiment (Bellefonds?), overthrew them and took four standards and two kettledrums. 29 Sep Sk Koppenberg; 18-22 Oct blockade of Soest. **1759** Chef Oberst Daniel Friedrich von Lossow. 1 Mar Sk Fridewald (2 sqns); 31 Mar Sk Mellrichstadt (3 sqns); 1 Apr Sk Meiningen (3 sqns)20 Apr Sk Bingart and Vacha; 9 Apr Sk Lauterbach (3 sqns); 13 Apr B Bergen (3 sqns); 18 Apr Sk Reichelsheim; 19 Apr Sk Gruenberg; 13 May Sk Ulrichstedt and Wasungen; 4 Jun Sk Elberfeld; 1 Jul Sk Gutersloh; 12 Jul Sk Diepenau; 19 Jul Sk Hille; 1 Aug Sk Gohfeld (1 sqn); 1 Aug B Minden (2 sqns) cut down two regiments of French infantry. 17 Aug C Naumburg; 2 Sep Sk Weimar; 3 Sep Sk Marburg; 29 Nov Sk Lauterbach (1 sqn); 30 Nov Sk Fulda (1 sqn). **1760** 19 Mar C Flieden; 29 May Sk Fulda (1 sqn); 10 Jul Korbach-Emsdorff and Siebertzmuehle; 31 Jul Sk Warburg; 10 Aug capture of Sababurg (2 sqns); 6 Sep Sk Zierenberg; 16 Oct Sk Kloster Camp; 18 Oct Sk Ginderich. **1761** The squadrons with the Duke of Brunswick: 15 Feb capture of Fritzlar; 21 Mar Sk Gruenberg; 25 Mar Sk Dittershausen (2 sqns); 9 Jul Sk Oestinghausen and Untrop; 16 Jul B Vellinghausen; 20 Jul Sk Ruhne; 27/28 Jul Sk Soest and Gesicke; 18 Aug Sk Bredenborn; 19 Sep Sk Hoof near Kassel; 21 Sep Sk Fritzlar; 5 Nov Einbecker Hube. 1762 In the army of the Crown Prince. 5 Aug Sk Laaspe; 9 Aug Sk Berggemuende; 17 Aug Sk Battenberg; 22 Aug Sk Gladenbach; 30 Aug C Friedberg; 11 Sep Sk Atzenhein; 16 Sep Sk Alsfeld; 17 Sep Sk Manberg; 18 Sep Sk Treysa; 27 Sep Asterode; 28 Oct Sk Boedefeldt.

1761 The regiment was re-united at the end of the year and took part in the expedition to Gostin, where the Russians were put to flight. 5 Jul Sk Dolzig (4 sqns); 6 Jul Sk Gostyn (4 sqns); 8 Jul Sk Lubin (7 sqns); 20 Jul Sk Peuke (7 sqns); 14 Aug Sk Klein-Wandris; 19 Aug Sk Kuhnern; 28 Aug Sk Pitchenberg; 28 Aug Sk Hohen-Postritz; 14 Sep Sk Kobylin; 15 Sep Sk Gostyn; 18 Sep Sk Posen; 30 Sep Sk Koerlin; 1 Oct Sk Rambow; 2 Oct Sk Prettmin; 20 Oct Sk Zargelow; 21 Oct Sk Gollnow; 1 Nov Sk Stargard; 2 Nov Passberg near Stargard; 9 Nov Sk Arnswald 11 Nov C Reichenbach; 13/14 Nov Sk Naugard; 15 Nov Sk Koldewang; 18 Nov Sk Regenwalde; 23/24 Nov Sk Falkenberg; 25 Nov Sk Moitzelwitz; 11 Dec Sk Neumuehle near Kolberg; 12 Dec Sk Spie; 17 Dec C Gueltzow.

1762 2 May Sk Gross-Mohnau; 3 Jul Sk Nonnenwalde near Zedlitz; 6 Jul Sk Adelsbach; 7 Jul Sk Dietmannsdorf (5 sqns); 21 Jul Sk Burkersdorf.

1778 Against Austria. 7 Jul Sk Kowalkowitz; 11 Jul Sk Schmirsitz; 22 Jul Sk Dolan; 23 Jul Sk Kralowalhota; 5 Aug Sk Zwol; 19 Aug Sk Schlaney and Nachod; 22 Aug and 6 Sep Sk Tscherna; 25 Aug Sk Trautenau; 10 Sep Sk Geishuebel (1 sqn); 14 Sep C Dreihaeuser (5 sqns); 22 Oct C Weisskirch (5sqns); 21 Nov, 6 and 11 Dec Sks Gross-Herlitz. **1779** 2 Jan Sk Schnellenwalde (part); 11 Jan Sk Olbersdorf; 12 Jan Sk Braunsdorf, Schoenewiese and Frei-Hennersdorf; 10 Feb Sk Leitersdorf; 24 Feb Sk Ottendorf (part); 28 Feb defence of Neustadt. 1783 Chef Oberst Carl August von Hohenstock (ex-Chef of Hus R Nr 8); died as GM. 1788 Chef Oberst Friedrich Eberhard Siegmar von Goeckung; received Hus R Nr 2. **1794** In Poland. Sk Lyk; 10 Jun Sk Raygrod (3 sqns); 23 Jul and 5 Sep Sk Wilkowiszki; 30 Oct Sk Gawrichi (1 sqn); 31 Oct ambush at Wyczna; 1 Nov Sk Zboyna. 1795 GM Friedrich Wilhelm von Suter; retired in 1804 as GL. 1804 Chef Oberst Moritz von Prittwitz; GM in 1806. **1806** In l'Estocq's reserve corps in East Prussia. 23 Dec C Biezun; 25 Dec Soldau. **1807** 5 Feb C Waltersdorf; 7/8 Feb B Preussisch-Eylau; 26 Feb C Braunsberg (2nd Bn); 1 Apr Sk Danzig; 4 / 5 Jun C Spanden; 10 Jun B Heilsberg. 13 Jun C Gollau; 14 Jun C Koenigsberg. On 7 Sep 1808 it became the Leib-Hus-R. 20 Dec 1808 the regiment was split into two; the 1st Bn became the 1st Leib-Hus-R; the **2nd Bn became the 2nd Leib-Hus-R (see the newly-raised regiments below). 1812** two squadrons mobilised with the 1st Combined Hus R in von Yorck's corps in Russia. 5 Jul C Poniewiesz; 18 Jul C Borusk; 18 Aug Wyszki; 3 Sep Czeszawen; 1 Oct Garossenkrug; 7 Nov Neugut; 15 Nov Schoeppingsmuehle and Baldohnen; 16 Nov Neumuehle; 18 Nov Friedrichstadt. **1813** 5 Apr C Dannikow (1st and 2nd Sqns); 8-23 Apr blockade of Magdeburg; 28 Apr Sk Bernburg; 2 May C Halle; 9 May Sk Friedeburg (1st Sqn); 19 May C Luckau (Fr Jg Sqn); 3 Jun Sk Kalau; 28 May Hoyerswerda; 4 Jun Sk Luckau; 23 Aug B Grossbeeren; 28 Aug B Jueterbog; 29 Aug Sk Klausdorf; 6 Sep Dennewitz; 19 Sep Sk Burxdorf; 24/25 Sep Sk Streumen (2nd Sqn); 26 Sep Sk Colmnitz; 18 and 19 Oct B Leipzig; 23/24 Nov Sk Zuetphen; 4 Dec C Gorkum; 14 Dec Sk Heusden; 19 Dec Sk 's Hertogenbosch; 21 Dec Sk Breda. **1814** 11 Jan C Hoogstraten and Turnhout; 13 Jan C Antwerp; 31 Jan Sk Lier; 11 Feb C Conde; 2 Mar C Soissons; 4 Mar Sk Fontenay; 12 Mar B La Fere; 15 Mar C Joinville; 31 Mar/1 Apr C Compeigne. **1815** Not mobilised. 7 Mar 1815 the 4th Sqn was detached to help form the 7th Hussars. 5 Nov 1816 the title became 1st Hus-R (1st Leib-Hus R); 7 May 1861 **1. Leib-Husaren-Regiment Nr 1.**

Nr 6.

Raised in 1741; often referred to as the Brown Hussars. Title in 1806: **Schimmelpfennig von der Oeye.** Inspectorate: Upper Silesian. Garrisons: Beythen, Gleiwitz, Gross-Strehlitz, Losslau, Lubinitz, Nicolai, Peisskretscham, Pless, Riebnick, Ujest.

Dark brown Dolman and Pelisse, yellow collar cuffs and buttons and lace, yellow and white barrel sash. Privates had white fur trim to the pelisse, NCOs had black, officers' pelisses were edged with the white fur of foxes' throats. Originally white breeches, by 1806 these were light blue. Shabraque and sabretasche were brown with yellow edging and decoration. The regiment was destroyed in 1806.

Campaign history. Raised in 1741 from new recruits in Breslau and Ohlau as the 'Brown Hussars' at ten squadrons. 1742 Chef Oberst Isidor Graf von Hoditz; retired. 1743 Chef now Oberst Carl Gustav von Soldau; died in office. 1743 two squadrons went to form Hus R Nr 7. **1744** C Rattibor; the regiment was attacked by superios numbers, but thoroughly defeated them. **1745** 14 Feb Sk Planitz; 1 Mar ambush at Hirschberg; 4 Jun B Hohenfriedberg; 15 Dec B Kesselsdorf. 1746 Chef Oberst Ludwig Anton Graf von Wechmar; retired. **1756** The regiment was the advanced guard of the invasion of Bohemia. Sk Koeniggraetz; they were attacked by 2,000 enemy cavalry, but drove them off and took almost 800 prisoners. The then Chef, Oberst von Wechmar was detached by FM Graf von Schwerin with 250 of the regiment to conduct a raid behind enemy lines. They were detected by a larger enemy force near Reichenau on 10 Oct and attacked, but were able to overthrow the enemy completely for very little loss and to rejoin the army. **1757** Chef Oberst Johann Paul von Werner; died as GL in office. In the second invasion of Bohemia they took the major enemy magazines at Jung-Bunzlau. Then it was in action, and distinguished, on 6 May, at B Prague and on 18 Jun at B Kolin. At Prague they were sent to support the left wing; on the way they were attacked by six enemy cavalry regiments, defeated them and took 1,200 prisoners; 10 standards and the military chest. On the retreat from Bohemia five squadrons of the regiment were attacked at Gabel by a much superior force, but were able to disengage with only light losses. 7 Sep Sk Kay; Sk Moys; 22 Nov B Breslau; 5 Dec B Leuthen. **1758** 14 Oct B Hochkirch; counter-attacked to save some infantry regiments which were in disorder. A few days after this, on 26 Oct, they were hurrying to the relief of Neisse, when, near Goerlitz they charged a superior enemy cavalry force and defeated them, taking over 500 prisoners. **1760** 23 Jun C Landshut (5 sqns), saved a gun. Sk Neumark in Silesia; attacked two Austrian dragoon regiments and scattered them. Then sent to Pommerania to the relief of Colberg (12 Sep) from the Russians and Swedes. Took four cannon and many prisoners. The king awarded seven medals to the regiment for gallantry. 3 Nov B Torgau; together with Hus R Nr 2 they took most of the Austrian Drag R St Ignon Nr 31* prisoner. **1762** 16 Aug C Reichenbach. **1778/1779** In the king's army. **1778** 11 Aug Sk Glomnitz. When the Austrian General Knebel was ambushed in the village of Teschen, the regiment cut down many Croats and took 480 prisoners. 1785 Chef Oberst Johann Benedikt von Groeling; retired as GL. **1792** Chef Oberst Erich Magnus von Wolfradt; died as GL in office. On the Rhine. 6 Aug Capture of Sierck; 16 Aug Sk Volmerange; 19 Aug Sk Aumetz; 19 Aug Sk Fontoi; 14 Sep Sk Grandpre; 20 Sep Valmy; Sk Esch. **1793** Sk Ober- and Nieder-Ursel; 6 Jan Sk Hochheim; Sk Guntersblum, Sk Homburg; 18 Jul Sk Zweibruecken; 17 and 20 Aug Sks Kettricher Hof; 14 Sep B Pirmasens. From 15 Dec 1793 to 5 Jan 1794 the regiment was in skirmishes daily on the retreat from Landau to Worms. **1794** 3 Jan Sk Frankenthal; Sk Diedesheim; Sk Kirweiler; Sk Fenningen; 20 Sep B Kaiserslautern. 1800 Chef Oberst Ludwig Schimmelpfennig von der Oye; GM in 1806. **1806** 10 Oct C Saalfeld; 14 Oct Jena, in the advanced guard; heavy losses. 26 Oct Sk Zehdenick. **1807** Chef GM Friedrich Ferdinand Prinz von Anhalt-Koethen-Pless. Thirty officers and 660 men made their way to East Prussia; together with the remnants of Hus-R von Gettkandt Nr 1 and the depot of

Hus-R von Koehler Nr 7, they formed the 3rd Hussar Brigade. In L'Estocq's corps; 22 Feb B Heilsberg; four squadrons were captured at Koenigsberg (Hafelstrom) on 14 June. Chef Fuerst von Anhalt-Pless. The regiment was re-raised from the remaining squadron and one from the 1st Hussar Brigade. On 9 Nov 1807 this unit became the Hussar Brigade von Dziengel. On 7 Sep 1808 this became the Oberschlesisches or 1st Schlesisches Hus-R. On 5 Dec 1808 it was combined with the Niederschlesisches or 2nd Schlesisches Hus-R (* 9 Nov 1807 as the Hussar Brigade von Zieten from the old 4th Hussars. See there for future history.

* This was the 11th Dragoons in 1798, and the 4th Chevaulegers in 1802.

Nr 7.
Raised in 1743; uniform red dolman and pelisse, white buttons and lace. Chef in 1743 Hallasz; 1747 Major Alexander Gottlieb von Seydlitz; 1759 von Gersdorff.

Campaign history. 1757 5 Dec B Rossbach; distinguished; captured in von Fink's corps at Maxen on 20 Nov 1759 and not re-raised.

Nr 7. **Originally Nr 8; often referred to as the Yellow Hussars.**
Raised in 1743. Title in 1806: **Von Koehler.** Inspectorate: Prussian. Garrisons: Klodawa, Kollo, Konin, Kowall, Kuttno, Pionteck, Slupce, Stawiszin, Szadeck, Uniewo.

Lemon yellow Dolman, light blue collar and cuffs, light blue Pelisse with black fur, white buttons and lace, light blue and white barrel sash. White breeches, light blue shabraque and sabretasche with lemon yellow edging and decoration. The regiment was destroyed in 1806.

Campaign history. Raised in 1743 in Koepenik, Pommerania, from two squadrons of Regiment Nr 5 and new recruits(some of them prisoners of war); by 1744 it had ten squadrons. It was originally Nr 8 in the hussars, but was renumbered 7 in 1763. First Chef was Oberst Peter von Dieury; retired as GM. When West Prussia was taken over, it was garrisoned in Bromberg and Schneidemuehle. **1744** 10-16 Sep S Prague. **1745** 15 Dec B Kesselsdorf, very distinguished. 1746 Chef Oberst Heinrich Wilhelm von Billerbeck; retired. 1753 Oberst Paul Joseph Malachow von Malachowski; died as GL. **1757** 30 Aug B Gross-Jaegerndorf.

Apart from these actions, two squadrons of the regiment, together with three of the Regiment Nr 7 were with the army of Prinz Ferdinand in Germany fighting against the French. Their actions are shown here. **1758** 22 Feb ambush at Stoecken-Drebber; 3 Mar Sk Lauenau; 26 Mar C Lippstadt; 28 Mar Sk Soest; 1 Jun Sk Emmerich; 2 Jun Sk Duesselward; 12 Jun Sk Kloster Camp; 15 Jun C Toenisberg; 23 Jun B Krefeld; 29 Sep Sk Koppenberg; 18-22 Oct blockade of Soest. **1759** 1 Mar Sk Friedewald (2 sqns); 31 Mar Sk Mellrichstadt (3 sqns); 1 Apr Sk Meiningen (3 sqns); 20 Apr Sk Bingart and Vacha; 13 May Ulrichstedt and Wasungen; 9 Apr Sk Lauterbach (3 sqns); 13 Apr Sk Bergen (3 sqns); 18 Apr Sk Reichelsheim; 19 Apr Sk Gruenberg; 4 Jun Sk Elberfeld; 1 Jul C Gutersloh; 12 Jul Sk Diepenau; 19 Jul Sk Hille; 1 Aug Sk Gohfeld (1 sqn); 1 Aug B Minden (2 sqns); 17 Aug Sk Naumburg; 2 Sep C Weimar; 3 Sep Sk Marburg; 29 Nov Sk Lauterbach (1 sqn); 30 Nov ambush at Fulda (1 sqn). **1760** 19 Mar Sk Flieden; 29 May C Fulda (1 sqn); 16 Jul ambush at Korbach-Emsdorff and Siebertzmuehle; 31 Jul Sk Warburg; 10 Aug capture of Sababurg (2 sqns); 6 Sep Sk Zierenberg; 16 Oct Sk Kloster Camp; 18 Oct Sk Ginderich. **1761** 15 Feb capture of Fritzlar; 21 Mar Sk Gruenberg; 25 Mar Sk Dittershausen (2 sqns); 10 Jul Sk Oestinghausen and Untrop; 16 Jul B Vellinghausen; 20 Jul Sk Ruhne; 27/28 Jul C Soest and Gesicke; 18 Aug C Bredenborn; 19 Sep Sk Hoof near Kassel; 21 Sep Sk Fritzlar; 5 Nov Einbecker Hube. **1762** In the army of the Crown Prince. 5 Aug C Laaspe; 9 Aug Berggemuende; 17 Aug Sk Battenberg; 22 Aug Sk Gladenbach; 30 Aug C Friedberg; 11 Sep Sk Atzenhein; 15 Sep Sk Alsfeld; 17 Sep Sk Momberg; 18 Sep Sk Treysa; 27 Sep Asterode; 28 Oct Sk Boedefeldt.

The other 8 squadrons were in the campaign in Pommerania against the Swedes. **1758** 25 Aug B Zorndorf, distinguished. **1759** 23 Jul B Kay; 12 Aug B Kunersdorf. **1760** 23 Jun C Landshut (6 sqns). **1761** 15 Sep Sk Gostyn, then back to Pommerania. **1762** The whole regiment was at the siege of Schweidnitz, 7 Aug-9 Oct. 1775 Chef Oberst Adolf Detlof von Usedom; died in office as GL. **1778/1779** In the army of Prinz Heinrich; 5 Feb 1779 Sk Brix. 1792 Chef Oberst Friedrich Ludwig von der Trenck; received Hus R Nr 3. **1794** In Poland. 6 Jun B Seelze/Rawka (two sqns); capture of Petrikau; C Blonien and Utrata (6 sqns); 26 Jul Sk Opalin; 28 Jul capture of Wola; 27 Jul-6 Sep blockade of Warsaw; 23 Oct Sk Suchaczew. 1796 Chef GL Georg Ludwig Egidius von Koehler; GdC in 1806. **1806** 14 Oct Jena, in Ruechel's corps; then with Bluecher's corps. Four squadrons of the 2nd Bn surrendered at Krempelsdorf on 7 November; one squadron surrendered at Boitzenburg on 12 November; one squadron at Lueneburg on 11 November. The regiment was not re-raised. The depot and some remnants crossed the Vistula and were used to form the new 3rd Hussar Brigade, see the old 6th Hussars above.

Nr 8. Originally Nr 9;
Raised in 1758. Title in 1806: **Von Bluecher.** Inspectorate: Pommeranian. Garrisons: Belgard, Buetow, Lauenburg, Neu-Stettin, Rummelsburg, Schlawe, Stolpe, Zahnow.

In 1758 (according to Menzel) the regiment was clothed completely in black with green lace and yellow buttons. On the Mirliton was a complete reclining skeleton in white.

Other services give crimson Dolman and Pelisse, black collar and cuffs, white buttons and lace, black fur, red and white barrel sash. Black sabretasche.

Standards. The first standards were presented to the regiment on 28 Sep 1814 and on 15 Jun 1815.

Campaign history. Raised in 1758 in Halberstadt from new recruits by Prinz Heinrich von Preussen; commander was Oberstleutnant von Belling. Drafts from other cavalry regiments completed the regiment at five squadrons. It was originally Nr 9 in the hussars, but was renumbered 8 in 1763. In 1761 it was increased to ten squadrons and in 1762 it was raised again to 15 squadrons or three battalions, but the 3rd was disbanded at the peace of 1763. **1758** In Bohemia and then Franconia. 15 Jun C Rehau; 15 Oct raid on Freiberg; cut down many of the enemy and took over 200 prisoners. 15 Oct Sk Conradsdorf. Sk Passberg in Bohemia, attacked an enemy corps of 2,000 men and took it prisoner, complete with the commanding general. Then in Franconia and the Upper Pfalz, before returning to the Oberlausitz. **1759** 14 Oct B Hochkirch; 20 May Sk Liebau. 12 Aug B Kunersdorf, distinguished. Transferred to Pommerania; together with some light infantry they held the whole Swedish army in check. In the clash in the Uekermark they captured over 200 Swedes. **1760** 28 Jan C Anklam; 29 Aug Sk Kabelpass (Bluecher was captured here); 26 Aug C Neu-Brandenburg; 29 Aug Sk Treptow; 3 Oct Sk Brenster Pass; 20 and 24 Dec C Demmin. 1761 In Pommerania against the Russians and Swedes. Sk Treptow. 1762 With Prinz Heinrich's army in Saxony; made a raid into Bohemia, almost to Prague. During this raid they took over 1,000 prisoners and much booty. 26 Jun Sk Reichenbach; 13 Jul Sk Oelnitz; 21 Jul Sk Auerbach; 25-29 Aug occupation of Eger; 29 Oct B Freiberg. **1778/1779** 2 Aug 1778 C Gabel, captured two Austrian battalions. 4 Dec: Chef was Oberst Carl August von Hohenstock. 1783 Chef Oberst August Ferdinand von der Schulenberg; died in office in 1787. 1787 Chef Oberst Johann Wilhelm Graf von der Goltz. **1793** On the Rhine; the 1st Bn went to Flanders; 3 Mar Roermond; 1 May Sk Vicogne; 9 May Sk St Amand; capture of Hasnon and Marchennes. The 2nd Bn was on the upper Rhine and at the blockade of Landau (Nov and Dec). **1794** Chef Oberst Gerhard Leberecht von Bluecher. From September both battalions were on the upper Rhine. 20 Sep B Kaiserslautern; 27 Sep Sk St Ingbert; 29 Sep Sk Saarbruecken. In the advance on C Kaiserslautern (23 May) in 1794; C Weidenthal (1st Bn); 8 Jan Sk Kreuznach; 12 Jan Sk Offenheim; 16 Jan C Tuerkheim; 23 May C Diedesheim (2nd Bn); 28 May C Kirrweiler, took almost 400 prisoners. 2 and 13 Jul C Edesheim; 20 Sep Sk Eselsfurth; 20 Sep B Kaiserslautern; threw the French cavalry back on their infantry in Edesheim and masked the enemy artillery with the mass of fugitives. The commanders of the French cavalry, the artillery and about 1,200 men were taken. They also took the teams from 12 cannon, but could only bring off three guns. In the three actions of Weidenthal, Kirrweiler and Edesheim, the regiment had taken 2 howitzers, 9 cannon, 3 colours and hundreds of prisoners. 16 Oct Sk Klein-Bockenheim. **1806** 14 Oct Auerstaedt;28 Oct Lychen and Criewitz; 6 Nov storm of Luebeck. The regiment surrendered at Ratkau next day. Over the next few days all members of the regiment managed to escape and make their way to join up with the remaining Prussian forces. None of the regiment was registered as crossing the Rhine into captivity in France and in 1807 there were 900 men in the regiment in east Prussia. **1812** The 1st and 3rd Sqns were mobilised as part of the 2nd Combined Hus-R together with two squadrons of the Brandenburg Hussars. 5 Jul Sk Kosziany; 25 Jul Sk Ostrowo; 27 Jul C Witepsk; 30 Jul Sk Paniszawi; 17/18 Aug B Smolensk; 20/21 Aug B Valutina-Gora; 27 Aug S k Rupki; 7 Sep B Borodino; 13 Sep C Moscow; 4 Oct C Winkowo; 18 Oct B Malo-Jaroslawetz; after this action the regiment had only 20 horses and ceased to exist. **1813** 16 Aug Sk Buchholz; 23 Aug B Gross-Beeren (1st and 3rd Sqns); 6 Sep B Dennewitz; 18/19 Oct B Leipzig; 2 Dec ambush at Neuss; 24 Dec storm of Wesel. 1814 11 Jan C Hoogstraten; 12 Jan Sk Lier; 25 Feb S Conde; 16/17 Feb C Tournay. **1815** 16 Jun B Ligny; 18 Jun B Waterloo; 19 Jun C Wavre; 20 Jun storm of Namur; 20 Jun – 12 Jul S Maubourg; 1 Jul ambushed at

Versailles by Gen Exelmans' 9th Cav. Div. Suffered very heavy casualties. On 27 Jan 1889 this regiment's title became: **Husaren-Regiment Fuerst Bluecher von Wahlstatt (Pommersches) Nr 5.**

Nr 9. Corps Towarczys.
Raised in 1745. Title in 1806: **Von L'Estocq.** Inspectorate: Prussian.

Garrisons of the regiment: Boczky, Bransk, Drohyczin, Knyszyn, Lomza, Ostrolenka, Ostrow, Tykoczyn, Wyzkowo, Zabludow.

Garrisons of the battalion: Augustowa, Janow, Lipsk, Sololka, Suchawolla.

Uniform. Initially they received no uniforms, but clothed themselves in red tunics and wide, red Turkish trousers, black 'Katanken' (frock coats), red fez and white turban. In 1787 they were given a new uniform: The Katanka became dark blue and the Turkish trousers gave way to white leather breeches. The sabretasche was of plain black leather, belts were black and the shabracque was red with white lace piping. The turban was replaced by black Astrakhan caps. The short red tunic had dark blue collar and cuffs. In 1799 the Bosniaken were disbanded and a corps of Towarczys (Comrades) was raised in the new provinces taken from Poland. There were 15 squadrons; split into a regiment of ten and a battalion of five. The regiment had yellow buttons, the battalion white. The 5th Squadron of the battalion was composed of native Tartars. The Towarczys were armed with a lance, a sabre and a brace of pistols. Each squadron had its own colour code worn as a cockade on the hat, in the cap lines and on the lance pennants. The first and third colours were the central part of the cockade and the upper half of the lance pennant; the second was the central ring of the cockade and the lower half of the pennant. In the regiment these colours were: 1st Sqn – white; 2nd Sqn – black; 3rd Sqn – red; 4th Sqn – vermillion; 5th Sqn green; 6th Sqn – green-white-green; 7th – green and red; 8th Sqn – white and red; 9th Sqn – black, yellow, black; 10th Sqn – orange. In the battalion the squadron colours were repeats of the first five squadrons of the regiment. The lance pennants of NCOs bore the sun in the black upper half and the eagle and 'NON SOLI CEDIT' in the white lower half.

Harness was black with steel fittings. In summer, officers wore red frock coats with dark blue collar and cuffs, the winter frock coat was black with red collar and cuffs; by 1791 this had changed to dark blue. They were decorated with hussar-style lace and buttons and long aiguillettes. By 1806 these uniforms had been replaced by dark blue, Polish-style jacket with poppy red collar, lapels, turnbacks and cuffs, white buttons and lace. On the shoulders were white cords. The Astrakhan caps were replaced with black felt shakos with white plume, cords and national cockade. The ten squadrons of the regiment had yellow buttons, the battalion wore white buttons. They all wore dark blue breeches, with twin red stripes down the side and hussar boots. NCOs cuffs were edged in the button colour.

The 5th Squadron of the battalion consisted of native Tartars; officers of this squadron wore silver epaulettes NCOs and troopers wore a red aiguillette.

Standards. In the early part of their career, the regimental standard was red and white; the 1st Battalion had a white standard; the 2nd Battalion bore a sky blue one. All emblems were embroidered and included the black Prussian eagle, flying against the sun and the motto 'NON SOLI CEDIT'. Each squadron carried a standard; the decorations on them were painted. On 15 Jun 1815 standards were presented to the regiment.

Campaign history. In 1745 one company of 'Bosniaken', from Albania, was taken into Prussian service from that of Saxony and attached to Regiment Nr 5, 'The Black Hussars'. In 1760 it was increased to ten squadrons, but reduced to one again in 1763. They took part in the Seven Years War together with the 5th . In 1771 the Bosniaken were again attached to the 5th Hussars and increased to ten squadrons on 13 April 1773. By Order of 14 Oct 1799 the Bosniaks were disbanded; the men distributed to the other hussar regiments. A corps of 'Towarczys' was raised instead from the lesser nobility in the newly acquired provinces of the old Poland; there was a regiment of ten squadrons and a battalion of five.

1745 Chef of the 5th - Oberst von Ruesch. 1758 Chef of the 5th - Oberst von Beust. 1757 30 Aug B Gross-Jaegerndorf. **1758** 25 Aug C Zorndorf. 1759 Chef of the 5th - Oberst Daniel Friedrich von Lossow; died as GL. The regiment was in Pommerania and Silesia. 23 Jul Sk Kay. **1761** in Silesia. 8 Jul C Lubien; C Canth. **1762** 3 Jul C Nonnenbusch; 7/8 Jul C Dittmannsdorf; 16 Aug C Reichenbach; C Adelsbach; B Hohenfriedberg; C Langenbeil; C Burkersdorf; distinguished in all these actions. Part of the regiment now went to Saxony; the rest were at the siege of Schweidnitz. **1778/1779** In the army of the king. 23 Jul Sk Rohemicz; 25 Aug C Burkersdorf. 1783 Chef of the 5th - GM Carl August von Hohenstock; resigned. 1788 Chef of the Bosniaken - Oberst Heinrich Johann von Guenther; died in office as GL. **1794** In Poland. 25 Jun D Piontnica, together with the 6th Fuesilier Bn. 10 Jul C Kollno (4 sqns); 18 Jul B Demnicki (3rd Sqn). C Kordowa; 26 Oct C Magniszewo (4th Sqn); scattered a Polish corps of 6,000 men. Took a general, 11 officers, over 400 men and six guns. On 14 Oct 1799 the title was changed to the 'Korps Towarczys'. **1803** Chef GM Anton Wilhelm von L'Estocq; GL in 1806. **1806** In L'Estocq's Reserve Corps. 23 Dec C Biezun. **1807** 5 Feb C Waltersdorf; 7/8 Feb B Preussisch-Eylau; 10 Jun B Heilsberg. 6 Oct 1807 the regiment and battalion were combined into a regiment of Ulanen of eight squadrons. 16 Nov 1808 split into two regiments; the 1st Bn became the 1st Ulan-R (Schlesisches); on 26 Dec 1808 it was renumbered the 2nd Ulan-R; it was brought up to strength by men of the old Hus-R von Gettkandt Nr 1. 15 May 1809, one squadron given to the newly-formed 3rd Ulan-R. 31 May 1809 title changed to Schlesisches Ulan-R. **1812** The 3rd and 4th Sqns were mobilised as part of the Combined Ulan-R. 8 Aug C Inkowo; 7 Sep B Borodino; 14 Sep C Moscow; 3 Oct C Woronowo; 29 Nov B Beresina. **1813** 2 May Gross-Goerschen; 20/21 May B Bautzen; 26 May C Haynau; 26/27 Aug B Dresden; 28 Aug Sk Possendorf, Peterswalde, Floessberg; Kitscker. 14 Oct B Liebertwolkwitz. 18 Oct B Leipzig; 30 Oct-23 Dec S Erfurt. 1814 30 Jan-12 Feb blockade of Luxemburg; 28 Feb C Gue-a-Tresmes; 2 Mar Sk May; 9 Mar B Laon; 30 Mar B Paris. **1815** 7 Mar: 3rd Sqn given to the newly-formed 5th Ulan-R. 16 Jun B Ligny; 18 Jun B Waterloo. On 27 Jan **1889** the regiment's title became: **Ulanen-Regiment von Katzler (Schlesisches) Nr 2.**

The regiment also gave men to found the **Ulanen-Regiment Alexander III von Russland (Westpreussisches) Nr 7.**

Nr 10.

Raised in 1773. Title in 1806: **Usedom.** Inspectorate: Prussian. Garrisons: Biezun, Blonien, Lipno, Mszczanow, Neuhof, Raczions, Rawa, Ryppin, Skierniewice, Warsaw.

Uniform. In 1791 the regiment wore a dark blue pelisse with white fur edging, red lace and white buttons for the men, gold lace and buttons for officers. Yellow dolman with dark blue collar and cuffs edged red, red lace and yellow buttons for the men, gold for officers. Red and dark blue sashes for the men, dark blue and gold for officers. White breeches, dark blue sabretasches with yellow decorations. Dark blue shabraque with yellow wolfs tooth edging and red piping, round, dark blue portmanteau eged in red.

By 1806 this had changed to: dark blue dolman and pelisse, edged black for NCOs and men; officers had a white fur edging of the foxes' throats. Sulphur yellow collar and cuffs, white buttons and lace, crimson sash with dark blue barrels. Dark blue breeches. Black, polished sabretasche for the men, dark blue, decorated in silver for officers.

Campaign history. Raised in 1773 from drafts from the other hussar regiments at ten squadrons in the area between the Vistula and Drage rivers. Chef Oberst Carl Christoph von Owstein; retired. **1778/1779** In the army of Prinz Heinrich. 1780 Chef Oberst Friedrich Wilhelm von Wuthenow; retired as GM. 1787 Chef Oberst Friedrich Gideon von Wolkiretired as GL in 1797. **1794** In Poland; S Warsaw, together with the Russians. Sk Schielitz; 26 Oct B Magniszewo; took six cannon. The on security duties along right bank of the River Narew from Zakroczim to past Roczan for six months. 1797 Chef Oberst Carl Ludwig von Lediwary; retired as GL. 1799 Chef Oberst Friedrich Daniel von Glaser; died in office in 1804. 1804 Chef Oberst Friedrich von Usedom; GM in 1806. **1806** In Wuerttemberg's corps; 17 Oct C Halle; then with Bluecher's corps; 6 Nov storming of Luebeck. The main body of the regiment surrendered at Ratkau next day. Two squadrons and the depot got away to East Prussia. On 9 Nov 1807 they were combined with the rump of the 4th and 6th Hussars to form the **Husaren-Brigade von Zieten**. For future actions see Hussar-R Nr 4 above.

Nr 11.

Raised in 1792. Title in 1806: **von Bila.** Inspectorate: Franconian.

Garrisons: Neustadt on the River Aisch and Troppach in Bayreuth. Uniform: From 1802-1804 the battalion wore dark green pelisses with black fur for NCOs, white for the men, yellow dolman with dark green collar and cuffs, white lace and buttons; green sash with white knots, white breeches. The shabraque was dark green with yellow wolfs tooth edging and white piping, the sabretasche was black with the royal cipher in brass. By 1806 the uniform had changed: dark green pelisse, white fur trim for the privates, black for NCOs; dark green dolman with scarlet collar and cuffs, yellow lace and buttons, red sash with white knots, light blue breeches. Red sabretasche with yellow decoration, dark green shabraque with red wolfs tooth edging and yellow piping.

Campaign history. Raised in 1792 as a battalion by taking ten men and horses from evey other hussar regiment at a strength of five squadrons. Chef Oberst Christian Alexander von Frankenberg; died in office as GM. 1795 Chef Oberst Johann Anton Gustav von Sass; retired. In this year the unit's title became the Ansbachsches Husarenbataillon. **1806** 1 May-Chef GM Ernst Christian von Bila. 14 Oct B Jena, in Tauentzien's Left Flank Corps. Most surrendered at Anklam on 1 November. The regiment was not re-raised.

28

Hussar Regiments Raised After 1806

Leib-Garde-Husaren-Regiment.
Raised on 21 Feb 1815 as the Garde-Husaren-Regiment from the Garde-Huasen-Eskadron of the Guads Light Cavalry Regiment and three squadrons of the Ostpreussischen National-Kavallerie-Regiment. The title carried in 1918 was adopted on 19 Jun 1888. The regiment saw no action in 1815.

Standards. On 24 Feb 1815 the regiment was presented with a standard of the 2nd Squadron of the Gardes du Corps, which they had received on 7 Sep 1798.

The Ostpreussisches National-Kavallerie-Regiment was raised in response to an appeal by the king to the men in East Prussia on 27 Mar 1813 at four squadrons, a depot and later a Freiwillige-Jaeger Squadron. The regiment was disbanded on 21 Feb 1815; the 1st Sqn went to form the 4. Ulanen-Regiment; the rest of the regiment became the Garde-Hus-R.

Campaign history. 1813 21 Aug C Bunzlau; 26 Aug Katzbach; 13 Sep C Bischofswerda; 20 Sep C Goldbach; 22 Sep Sk Bischofswerda; 23 Sep Sk Rot-Nauslitz (3rd, 4th and Fr-J Sqns); 8 Oct Sk Breitenfelde (4th Sqn); 16 Oct B Moeckern; 21 Oct C Freiburg / Unstrutt; 16-25 Nov blockade of Mainz. **1814** 15-21 Jan blockade of Thionville; 1 Feb Sk Vitry; 11 Feb B Montmirail; 12 Feb B Chateau-Thierry; 22 Feb C Mery; 9 Mar B Laon;

20 Mar Sk Fismes (2nd Sqn); 26 Mar Sk La Ferte-Gaucher and Coulommiers; 27 Mat Sk Triport; 30 Mar B Paris.

2nd Leib-Husaren-Regiment. On 20 Sep 1808 the Leib-Regiment (see the old Husaren-R Nr 5 above) was split into two; the 1st Bn became the **Husaren Nr 1, 1st Leib-Hus-R; the 2nd Bn became the Husaren Nr 2, 2nd Leib-Hus-R.**

Standards. Two new-pattern standards were presented on 28 Sep 1814 and 15 Jun 1815.

Campaign history. 1812 The 2nd and 3rd Sqns were mobilised in the 1st Combined Hus-R together with two squadrons of the 1st Leib-Husaren-R; see there for details. They fought in Curland. **1813** 2 May B Gross-Goerschen; 21 May B Bautzen; 5 Jun Sk Colditz; 19 Aug Sk Groeditzberg; 23 Aug C Goldberg; 26 Aug B Katzbach; 27 Aug Sk Rochlitz; 2 Sep Sk Reichenbach; 3 Oct C Wartenburg; 16 Oct B Moeckern; 18 Oct B Leipzig; 22 Oct C Freiburg; 26 Oct C Eisenach; 17-28 Nov blockade of Mainz. **1814** 12 Jan Sk Noisseville; 13-26 Jan blockade of Metz; 30 Jan C St Dizier; 3 Feb Sk Chaloms; 21 Feb Sk Mery; 27 Feb Sk Lizy; 28 Feb C Gue-a-Tresmes; 9 Mar B Laon; 14 Mar C Bery-au-Bac; 21 Mar Sk Coincy; 25 Mar B Etoges; 28 Mar C Claye-Souilly; 30 Mar B Paris. **1815** no actions.

2nd Schlesisches Husaren-Regiment.
Raised on 21 Nov 1808. **1815** 2 Mar the 4th Sqn went to form part of the new 8th Hussars. On 10 Mar 1823 the 2nd Schlesisches Hus-R became the 6th Hussars. 27 Jan 1889 the title became the **Husaren-Regiment Graf Goetzen (2. Schlesisches) Nr 6.** This was retained to 1905.

Standards. Presented on 15 Jun and 16 Nov 1815; new pattern.

Campaign history. 1812 1st and 2nd Sqns mobilised as part of the 3rd Combined Hus-R. 19 Jul C Eckau; 23 Aug C Dahlenkirchen; 26 Sep Sk Tomoczna; 27 Sep C Eckau; 28 Sep Sk Bauske; 29 Sep Sk Ruhenthal; 30 Sep C Graefenthal; 14 Oct C St Olay; 18 Nov Sk Friedrichstadt. **1813** 4-20 Apr blockade of Wittenberg; 2 May Sk Koenigsbrueck; 19 May C koenigswartha; 20/21 May B Bautzen; 26/27 Aug B Dresden; 28/29 Aug

C Halle; 12 Sep C Weissenfels; 18 Sep Sk Merseburg; 19 Sep Sk Koesen; 25 Sep C Altenburg; 28 Sep Sk Meuselwitz; 10 Oct Sk Stoessen; 16-18 Oct B Leipzig; 27 Oct Sk Weimar; 30/31 Oct Sk Gelnhausen. **1814** 11 Jan Sk St Avold; 22/23 Jan Sk St Aubin; 29 Jan B Brienne; 1 Feb B La Rothiere; 11 Feb B Montmirail; 22 Feb C Mery; 3 Mar C Neuilly; 9 Mar B Laon; 20 Mar Sk Fismes.

1813 The 3rd and 4th Sqns in the Combined Schlesisches Hus-R: 13 Apr C Langensalza; 17 Apr C Wanfried; 22 Apr C Apolda; 28 Apr Sk Skoelen; 2 May B Gross-Goerschen; 23 Aug B Gross-Beeren; 24 Aug Sk Speremberg; 27 Aug C Jueterbogk; 24 Sep Wartenburg; 8 Oct Sk Leipzig; 16 Oct Sk Dippach. **1814** 7 Jan Sk Gross-Zaudert; 11 Jan C Hoogstraten; 28 Jan Sk Loewen; 25 Feb C Courtray; 2 Mar Sk Sweweghem; 31 Mar C Courtray.

1815 18 Jun B Waterloo; 27 Jun C Senlis.

7th Husaren-Regiment.

Raised on 7 Mar 1815 from the 4th Sqn, 1st Leib-Hus-R and two squadrons of the Schlesisches NKR (see the Garde-Kueassier-Regiment). On 25 Mar 1815 the title became the Westpreusssches Hus-R and on 5 Nov 1816 it changed to 7th Husaren-Regiment (Westpreussisches). On 21 Jun 1888 it became the **Husaren-Regiment Koenig Wilhelm I (1 Rheinisches) Nr 7.** This name was retained into the First World War.

Standards. On 15 Jun 1815 a new-pattern standard was presented.

Campaign history. See the parent regiments for pre-1815 actions. The regiment did not come into action in 1815.

8th Husaren-Regiment.

Raised on 7 Mar 1815 from squadrons of the 2nd Leib-Hus-R, 3rd and 6th Hus-R. On 25 Mar 1815 it became the 8th Hussars and on 5 Nov 1816 it was the 8th Hus-R (1. Westfaelisches). On 17 Nov 1894 the title became **Husaren-Regiment Kaiser Nicolaus II von Russland (1. Westfaelisches) Nr 8.** This was retained until 1914.

Standards. Presented on 15 June; new pattern.

Campaign history. 1815 18 Jun B Waterloo; 21 Jun C La Fere; 27 Jun C Creil; 29 Jun C St-Germain.

9th Husaren-Regiment. Raised on 25 Apr 1815 in Rochefort, Belgium from one squadron each of the 4th and 5th Hussars, and one from Luetzow's Freikorps cavalry regiment as the 9th Hussars. On 5 Nov 1816 the title changed to 9th Husaren-Regiment (Rheinisches); on 7 May 1860 it became **2. Rheinisches Husaren-Regiment Nr 9;** this was retained into the First World War.

Standards. Two new models were presented on 15 Jun 1815.

Campaign history. 1815 16 Jun B Ligny; 19 Jun C Wavre; 20 Jun storm of Namur.

10th Husaren-Regiment.

Raised on 19 Nov 1813 as the Elb-NKR from volunteers. 25 Mar 1813 the title changed to 10th Husaren-Regiment (Magdeburgisches); on 7 May 1861 the title became **Magdeburgisches Husaren-Regiment Nr 10.** This was retained into the First World War.

Standards. One each, new pattern, presented on 3 Oct and 12 Dec 1815.

Campaign history. 1814 1 Apr Sk Klein-Ottersleben; S Magdeburg. **1815** 18 Jun C Wavre; 28 Jun C Le Menil; 29 Jun Sk Livry.

11th Husaren-Regiment.

Raised on 7 Mar 1813 from the Berg Grand Ducal Hussars (*5 Dec 1813) when that state was annexed by Prussia. The title was then 11th Hussars. One squadron was given to help form the 5th Ulans. On 5 Nov 1816 the title became the 11th Hussars (2nd Westfalian); on 7 May 1861 it changed again to **2. Westfaelisches Husaren-Regiment Nr 11.**

Standards. Two new-pattern standards were presented, one on 3 Oct 1815, the other on 12 Dec 1815.

Campaign history. In Prussian service 1814 Blockade of Mainz. **1815** 15 Jun C Charleroi; 16 Jun B Ligny; 18 Jun C Wavre; 20 Jun storm of Namur; blockade of Maubeuge; blockade of Laon; blockade of Givet.

12th Husaren-Regiment.

Raised on 7 Mar 1815 as the 12th Husaren-R from the men of the Saxon Husaren-R (* 30 Jul 1791) living in those areas awarded to Prussia at the Congress of Vienna. On 7 May 1861 the title became **Thueringisches Husaren-Regiment Nr 12;** this was retained into the First World War.

Standards. One each of the new models presented on 3 Oct and 12 Dec 1815.

Campaign history.

In Prussian service. **1815** For the 1815 campaign they wore their old uniforms, light blue dolman and pelisse, black collar and cuffs, white buttons and lace. They joined Thielmann's III Prussian Corps on 19 June 1815 at Wavre and went into von der Marwitz's Reserve Cavalry Brigade. They entered Paris on 8 July 1815. 20 Jun storm of Namur; 27 Jun C Dammartin.

29

Hussar Uniforms Post 1807

Only the regiments von Wuerttemberg Nr 4, von Prittwitz Nr 5, Bluecher Nr 8 and the Towarczys Corps survived the catastrophe of 1806 intact. Although the regiment von Bluecher Nr 8 capitulated at Ratkau, every officer and man escaped and made their way to East Prussia. At the end of the war in 1807 the regiment mustered 900 horses and was reinstated in the Army List.

Shakos as for the dragoons, with cockade, loop and button, white cords and plumes as before. The cords of NCOs were white with black and white flounders and tassels; officers' cords were silver and black. The two Leib-Hussar regiments wore white/silver skull and cross bones badges to their shakos.

The dolman had fifteen rows of lace on the chest and three rows of buttons; the collar and cuffs were in the regimental colour as were the lace and buttons. The pelisse was in the same colours as the dolman.

Privates of all line regiments wore white fur trim to the pelisse; NCOs and officers had black fur trim.

Trumpeters had swallows nests in the collar colour.

Officers sashes had silver and black cords with silver barrels, the men had cords in the dolman collar with barrels and other fittings in the lace colour.

Weapons were almost identical to the British M 1794 pattern sabre in a steel sheath, with black fist strap. The sabretasche was red with crowned royal cipher and edging in the button colour, except for the two

Leib-Hussar regiments and the Pommeranians, which had plain black leather lids. Carbines were carried on black bandoliers; the best shots had rifled weapons. The mens' pouch lids were plain, except in the Leib-Hussar regiments where silver stars were worn. Officers of the line regiments wore the crowned royal cipher.

The shabraque was in the dolman colour, with wolf's tooth edging in the collar colour and lace in the regimental colour. It had a long rear corner. Harness was black with fittings in the regimental button colour and of Hungarian style.

Regimental distinctions were as shown below; the 1st Leib-Hussars had white shoulder straps, the 2nd had red ones.

Regiment	Dolman and Pelisse	Collar and cuffs	Buttons and lace
1st Leib	black	red	white
2nd Leib	black	black	white
1st Brandenburg	dark blue	red	white
2nd Brandenburg*	dark blue	red	yellow
Pommeranian (Bluecher's)	light blue	black	yellow
Oberschlesisches	brown	yellow	yellow
Niederschlesisches	green	red	white

* In 1809 revolts against Napoleonic domination flared up throughout central Europe and the Iberian peninsular. On 24 April of that year, Major Ferdinand Baptista von Schill led part of this regiment through north Germany, fighting Westphalian and other troops in the hopes of stirring up a rebellion. He was finally cornered in Stralsund by Dutch troops and killed on 31 March. His head was cut off, pickled and taken to Holland where it remained until 1837. His fellow officers who survived the action were later shot in Wesel fortress. On 16 May 1809 his old regiment was disbanded; a new 3rd Ulans was raised in its place. Later, von Schill's name was given to the 4th Hussar-R (1st Schlesisches) as a title of honour.

30

Ulans (Lancers) Uniforms Post 1806

All details as for the dragoons except as noted below. They wore traditional lancer costume in dark blue, with red facings and yellow buttons. The square-topped Czapka had a black base and a dark blue top and yellow piping. Belts were black. The shabraque was dark blue, edged red and had rounded rear corners. Weapons were hussar sabres and lances for the front rank; the rear rank carried carbines instead of lances. The lance pennants were dark blue over the shoulder strap colour.

As light cavalry, the Ulans carried no standards.

31

Ulan Regiments Raised After 1806

During the war of 1806, ten squadrons of cavalry were raised in Silesia by Graf Goetzen; these and the men of the 3rd Hussars in their depot area went into the 2nd Silesian Hussars in November 1808. By Order of 29 Mar 1809 selected men of this regiment went to form the **Leib-Ulanen Eskadron.** This later became the **Garde-Ulanen Eskadron.**

Ulanen-Regiment Nr 1, Westpreussisches Ulanen-R.
Raised from the depots and men of the following regiments: 8th, 9th and 10th Dragoons. By Order of 18 Nov 1808 these men became the foundations of Dragoner Nr 4, 2. Westpreussisches Drag-R and the Ulanen-Regiment Nr 1. White shoulder straps, yellow buttons. On 15 May 1809 one squadron was sent to form the 3rd Ulanen-R; these men were replaced by taking in 16 officers and 176 men of the disbanded 2nd Brandenburgisches Hus-R. On 7 Mar 1815 two squadrons were detached to the 4th Ulanen-R. On 5 Nov 1816 the title became the 1st Ulanen-Regiment (Westpreussisches). On 16 Sep 1884 the title of **Ulanen-Regiment Kaiser Alexander III von Russland (Westpreussisches) Nr 1;** this was retained into the First World War.

Standards. New models were presented on 5 Oct 1814 and 15 Jun 1815.

Campaign history. Prior to 1813 see the old regiments. **1813** 2 May B Gross-Goerschen; 4 May C Borna; 20/21 May B Bautzen; 26 May C

Haynau; 23 Aug B Gross-Beeren; 6 Sep B Dennewitz; 18/19 Oct B Leipzig. **1814** 7 Mar C Courtray. **1815** 18 Jun B Waterloo.

Ulanen-Regiment Nr 2, Schlesisches Ulanen-R.

Raised on 23 November 1808 from men of the old Hus-R von Gettkandt Nr 1 and two squadrons of the 1st Ulans. Red shoulder straps; yellow buttons. 15 May 1809 a squadron was detached to form the 3rd Ulanen-R. On 31 May 1809 the regimental title became Schlesisches Ulanen-R. 7 Mar 1815 a squadron was detached to form the 5th Ulanen-R; 5 Nov 1816 the title became the 2nd Ulanen-R (Schlesisches). On 27 Jan 1889 the title became **Ulanen-Regiment von Katzler (Schlesisches) Nr 2.** This was retained into the First World War.

Standards. New models were presented on 15 Jun 1815.

Campaign history. For actions prior to 1808 see the parent regiments.**1812** 3rd and 4th Sqns mobilised as part of the Combined Ulanen-Regiment. 8 Aug C Inkowo; 7 Sep B Borodino; 14 Sep C Moscow; 3 Oct C Woronowo; 29 Nov B Beresina. **1813** 2 May B Gross-Goerschen; 20/21 May B Bautzen; 26 May C Hainau; 26/27 Aug B Dresden; 28 Aug Sk Possendorf, Peterswalde, Floessberg, Kitscher; 14 Oct C Liebertwolkwitz; 18 Oct B Leipzig; 30 Oct-23 Dec S Erfurt. **1814** 30 Jan-12 Feb blockade of Luxemburg; 28 Feb C Gue-a-Tresmes; 2 Mar Sk May; 9 Mar B Laon; 30 Mar B Paris. **1815** 16 Jun B Ligny; 18 Jun B Waterloo.

Ulanen-Regiment Nr 3, Brandenburgisches Ulanen-R.

Raised on 16 May 1809 from the remnants of the new 2nd Brandenburg Hussars, who had not accompanied Schill on his raid into north Germany and from one squadron each of the 1st and 2nd Ulans. Yellow shoulder straps, yellow buttons.

On 27 Jan 1889 the title became **Ulanen-Regiment Kaiser Alexander II von Russland (1. Brandenburgisches) Nr 3.** This was retained into the First World War.

Standards. New models were presented on 15 Jun 1815.

Campaign history. 1812 Two squadrons were mobilised as part of the Combined Ulanen-Regiment. 5 Jul Sk Staroi-Dangelisi; 15 Jul Sk

Stranowczyn; 8 Aug C Rudnia; 7 Sep B Borodino; 3 Oct Sk Marwinna; 18 Oct B Tarutino; 24/25 Oct B Malojaroslawetz; 28 Nov B Beresina. **1813** 24 Mar-24 May blockade of Glogau; 18 Aug Sk Lauterseiffen; 21 Aug C Loewenberg; 22 Aug C Goldberg; 26 Aug B Katzbach; 30 Aug Sk Bunzlau; 4 Sep Sk Hochkirch; 23 Sep Sk Rothnauslitz; 16 Oct B Moeckern; 18 Oct B Leipzig. **1814** 31 Jan Sk Montier; 3 Feb C La Chaussee; 4 Feb C Chalons; 11 Feb B Montmirail; 12 Feb B Chateau-Thierry; 22 Feb C Mery-sur-Seine; 28 Feb C Meaux; 9 Mar B Laon; 21 Mar C Oulchy-le-Chateau; 26 Mar C Sezanne; 30 Mar B Paris. **1815** 13 Jun Sk Fleurus; 16 Jnu B Ligny; 18 Jun B Waterloo; 26 Jun C La Fere; 27 Jun Sk Crespy; 28 Jun C Nantuil; 2 Jul C Issy; 25 Jul-10 Aug blockade of Soissons and La Fere.

Normal-Ulanen-Eskadron.
Raised on 27 Mar 1811 from men from all existing Ulan regiments.

Standards.

Campaign history. 1815 This was retained into the First World War.

Ulanen-Regiment Nr 4.
Raised in 1815. The members of this new regiment took the field in 1815 still in the uniforms of their old regiments. 1st Squadron ex-1st Ulans. 2nd Squadron ex-1st Ostpreussisches NKR; 3rd Squadron ex-Pommersches NKR.

Standards.

Campaign history. 1815 This was retained into the First World War.

Ulanen-Regiment Nr 5.
Raised on 7 Mar 1815. The members of this new regiment took the field in 1815 still in the uniforms of their old regiments. 1st Squadron ex-Schlesisches Ulanen-R; 2nd Squadron ex-Brandenburgisches Ulanen-R; 3rd Squadron ex-Hussars of Berg. The title was the 5th Ulanen-R; on 5 Nov 1816 it became the 5th Ulanen-R (Westfaelisches) and on 7 May 1861 it was **Westfaelisches Ulanen-Regiment Nr 5.** This was retained into the First World War.

Standards. New models were presented on 15 Jun 1815.

Campaign history. 1815 16 Jun B Ligny; 18 Jun C Wavre; 20 Jun storm of Namur; 26 Jun capture of Ham.

Ulanen-Regiment Nr 6.

Raised as the Royal Prussian Freikorps (Luetzow's) on 18 Feb 1813. It consisted of three battalions of infantry, five squadrons of cavalry and a battery of four foot artillery and five horse artillery guns. In April 1814 the cavalry had been reduced to four squadrons. On 23 March 1815 the 6th Ulanen-Regiment was formed from these men in Dinant. On 7 march one squadron had been detached to form the 9th Hussars. The members of this new regiment took the field in 1815 still in the uniforms of their old regiments. On 5 Nov 1816 the title became 6th Ulanen-Regiment (2. Westpreussisches) and on 7 May 1861 the title changed to **Thueringisches Ulanen-Regiment Nr 6.** This was retained into the First World War.

Standards. New models were presented on 15 June and 12 December 1815.

Campaign history. 1813 12 May B Goehrde; 17 Jun ambushed at Kitzen; 17-19 Aug Sks Lauenburg; 26 Aug Sk Gadebusch; 5 Sep Sk Moelln; 18 Sep Sk Zarentin; 13/14 Sep C Bremen; 8 Nov Sk Steknitz River; 21-15 Dec blockade of Glueckstadt. **1814** Feb-4 May S Juelich (two sqns); 3 Mar Sk Chateau-Thierry; 5 Mar Sk Fidmes; 15 Mar Sk Vouziers (Chatres); 16 Mar Sk Les Aleux. **1815** 15 Jun C Gosselies; 16 Jun B Ligny; 18 Jun B Waterloo (1st, 2nd and Fr-J Sqns); 18 Jun Sk Limale (3rd and 4th Sqns); 21/22 Jun capture of Avesnes-sur-Helpe; 28 Jun Sk Nanteuil (Paris).

Ulanen-Regiment Nr 7.

Raised in 1815. The members of this new regiment took the field in 1815 still in the uniforms of their old regiments. 1st and 2nd Squadrons ex-Hellwig's Streifkorps; 3rd Squadron ex -Schill's Streifkorps.

On 23 Apr 1891 the title became **Ulanen-Regiment Grossherzog Friedrich von Baden (Rheinisches) Nr 7.** This was retained into the First World War.

Standards. Two new pattern standards were presented on 3 Oct and 12 Dec 1815.

Campaign history. 1815 18 Jun C Wavre.

Ulanen-Regiment Nr 8.

Raised on 6 Sep 1812 in Reval from 30 prisoners of war from the 2nd Dutch Lancers of Napoleon's Imperial Guard and 40 captured Prussian hussars, as the 1st Hussars of the Russo-German Legion. In February 1813 the 2nd Hussars of the Russo-German Legion was raised in Mitau. They were taken into British pay. On 2 June 1814 they were assigned to the III Prussian Corps, and on 1 September 1814 they entered Prussian service. On 19 March 1815 both regiments were combined into an Ulanen regiment, which, on 25 March of that year became the 8th Ulanen-Regiment. On 3 Nov 1816 this title changed to 8th Ulanen-Regiment (2. Rheinisches) and on 4 Jul 1860 it changed to Ostpreussisches Ulanen-Regiment (Nr 8). On 27 Jan 1889 it became **Ulanen-Regiment Graf zu Dohna (Ostpreussisches) Nr 8.** This was retained into the First World War.

Standards. New models were presented on 3 Oct and 12 Dec 1815.

Campaign history. 1813 1st Hussars: 16 Sep B Goehrde; 7 Oct C Ratzeburg; 18 Oct Sk Kegeler Muehle; 14 Nov Sk Moelln; 3 Dec Sk Oldesloe; 10 Dec C Sehestadt. **1814** 26 Jan-20 Feb blockade of Harburg; Maastricht; Kortryck.

1813 2nd Hussars: 20 Aug C Boitzenburg; 21 Aug Sk Vellahn; 6 Oct Sk Buchen; 7 Oct Sk Kehrsen; 12, 14 and 18 Oct Sks Rosengarten; 2 Nov Sk Roennenberg; 3 Dec Sk Steinhorst; 9 Dec Sk Cluvensieck and Eckernfoehrde. **1814** 17 Feb Sk Wilddorf and Venlo; Antwerp.

1815 As the 8th Ulanen-R. 16 Jun B Ligny; 18 Jun C Wavre; 19 Jun Sk Limale; 20 Jun storm of Namur; 24 Jun Sk Rodemachern; 28 Jun C Nantuil; 10 Aug S Longwy.

Appendix A: Prussian Battles and Clashes 1700-1815

Abbreviations used in this text. **B** – battle; **C** – clash; **S** – siege; **Sk** – skirmish; Battles are in bold print; Prussian victories are in italics.

War of the Spanish Succession 1701-1714. France, Bavaria, Portugal and Savoy took the field against Austria, England, Prussia, the Netherlands and most of the states of the Holy Roman Empire of German Nations over who should sit on the Spanish throne. Prussia and her allies were eventually victorious.

1702 15 Apr – 15 Jun S and capitulation of Kaiserswerth. 29 Aug – 25 Sep S and evacuation of Venlo. 29 Sep – 5 Oct S and surrender of Stephanswerth. 29 Sep – 6 Oct S and capitulation of Roermonde. 14 Oct B Friedlingen.

1703 6 Feb capture of Rheinbergen. 28 Mar Sk Einhoefen. End of Mar – 17 Dec blockade and surrender of Geldern. 19 Apr 1st Sk Stallhofen. 24 Apr – 15 May S and capitulation of Bonn. 8 May storm of Fort Burgundy. 12 May 2nd Sk Stallhofen. 17 May – 14 Oct S and capitulation of Thorn. 26 Jun Sks of Stecken, Stenbrock and storm of St Anthony. 30 Jun Sk Eckeren. 24 Jul Sk St Job. 17 – 25 Aug S and capitulation of Huy. 20 Sep C Hoechstaedt. 25 Sep bombardment of Limburg.

1704 2 Jul C Schellenberg. ***13 Aug B Hoechstaedt versus the Imperials.*** 12 Sep – 22 Nov S and capitulation of Landau. **1705** 27 Jun Sk Urago d'Oglio. 28 Jun C Bergamo. 11 Jul capture of Huy. 18 Jul storm of the lines of Tirlemont. 15 Aug Sk Il Paradiso. 16 Aug C Cassano. 17 Sep S Hagenau. 11 Oct Sk Montodino.

1706 19 Apr 1st C Castiglione or Calcinato. 24 May capture of Loewen. 18 Jun storm of Plassendael. 17 Jul capture of Polisello on the River Po. 18 Jul C Polisello. 23 Jul – 22 Aug S and capitulation of Menin. 11 – 14 Aug blockade and surrender of Reggio. 20 Aug capture of Stradello. 5 Sep surrender of Dendermonde. 5 Sep storm of Pianezza. ***7 Sep B Turin versus the French.*** 9 Sep 2nd B Castiglione. 15 Sep surrender of Chivas. 30 Sep capture of Lodi. 6 Oct storm of Fort Ghera near Pizzighetone. 16 Oct surrender of Tortona. 17 Oct surrender of Alexandria. 23 Oct surrender of Pizzighetone. 7 Dec surrender of Casale.
1707 5 Jul storm of the pass at the Col di Tenda or Col di Cornio. 26 Jul blockade of Toulon. 20 Sep capture of the camp at Savigliano. 28 Sep storm of Fort Catinat near Susa. 3 Oct capitulation of Susa. **1708** 5 Jul raid on Ghent. 8 Jul capture of Lessines; 13 Jul Sk Ypres. 14 Jul capture of Arras. 16 Jul destruction of the French lines at Ypres. 9 Aug S Exilles. 10 Aug storm of Fort Louis near Fenestrelles. 11 Aug bombardment of Exilles. 11 Aug capitulation of Perusa. 14 Aug – 23 Oct S Lille town or Ryssel. 8 Dec capitulation of Lille citadel. 11 Sep cannonade of Seclin. 12 Sep sortie at Lille. 28 Sep Sks of Ichteghem and Wynendael. 27 Nov Sks at Gavre and Oudenarde.
1709 2 Jan capture of Bruges, Plassendael, Lessinghem and Oudenburgh. 2 Jan surrender of Ghent. 27 Jun – 28 Jul S and capitulation of Tournay or Dornick. 3 Sep capitulation of the citadel of Tournay. 5 Sep Sk Mons. 25 Sep – 20 Oct S and capitulation of Mons.
1710 20 Apr C Scarpe (Dyle). 22 Apr – 27 Jun S and capitulation of Douay. 27 Jun surrender of Scarpe. 15 Jul – 28 Aug S Bethune. 22 Aug storm of Arche castle. 28 Aug storm of Fort Fogliosa. 31 Aug storm of Fort Castelet. 6 Sep – 9 Nov S and capitulation of Aire. 6 – 29 Sep S and capitulation of St Venant.
1711 5 Aug Sk Aubanchoeuil-au-Bac. 7 Aug – 10 Sep S and capture of Bouchain. 10 Aug Sk Marquette. 31 Aug ambush at Hordaing. 16 Sep Sk Chaumont.
1712 20 Mar sortie at Porto Hercole. 4 May surrender of Porto Hercole. 25 May opposed crossing of the Sheldte at Bouchain. 24 Jul C Denain. 10 Sep Sks at Kievrain and Mormal. 6 Nov storm of Moers.

The Great Northern War 1700-1721. Prussia against Sweden
1715 22 Apr capture of Wollin. 17 Jul – 24 Sep S and capitulation of Stralsund. 16 Nov C Stresow. **1716** 8 Apr surrender of Wismar.

The War of the Polish Succession 1733-1735. A very low-key dispute. **1734** At Philippsburg.

1st Silesian War 1741-1742. Prussia invaded Silesia and then offered to support Maria Theresia's claim to the Hapsburg throne in return for ownership of the province. Prussia was allied with France, Bavaria, Spain and Saxony; Austria was supported by England. By the Treaty of Breslau of 11 June 1742, Maria Theresia ceded most of Silesia to Prussia, keeping only the districts of Jaegerndorf, Teschen and Troppau.
1741 9 Jan storm of Glatz; capture of Ohlau. 16 Jan capture of the Jablunka pass. 30 Jan surrender of Namslau. 7 – 8 Mar storm of Glogau. ***10 Apr B Mollwitz versus Austria.*** 28 Apr – 3 May S Brieg. 17 May Sk Rothschloss. 30 Jul Sk Zobten. 27 – 31 Oct S Neisse. **1742** 1 Mar Sk Wissewitz. 14 Mar C Busch. 25 Apr surrender of the citadel of Glatz. ***17 May B Chotusitz versus Austria.***

2nd Silesian War 1744-1745. Frederick the Great made a preemptive invasion of Bohemia, fearing the rise of Austrian power; he was soon expelled by Austro-Saxon forces. France supported Spain and Prussia; Austria was allied to England, the Netherlands and Sardinia. Fighting spread to North America, northern Italy, the Netherlands and the East Indies. The Austrians were decisively beaten at the battles of Hohenfriedberg and Kesselsdorf in December 1745, and the war was concluded by the Treaty of Dresden on 25 Dec 1745, which confirmed Frederick the Great's ownership of Silesia
1744 6 Sep C Beraun. 10 – 16 Sep S and capitulation of Prague. 18 Sep Ambush at Ziska mountain. 23 Sep capture of Tabor. 30 Sep capture of Budweiss. 1 Oct capture of Frauenberg castle. 9 Oct Sk Tecyn. 22 Oct surrender of Budweis. 23 Oct surrender of Tabor and Frauenberg castle. 26 Oct C Kammerburg. 27 Oct Sk Schurz-Kosteletz on the Sazawa River. 9 Nov Sk Pardubitz. 14 Nov C Kollin. 19 Nov Sk at Telczitz or Selmitz. 23 Nov Sk Nechanitz. 27 Nov Sks at Pless and Jaromier. 30 Nov Sk Lewin. 2 Dec Sk Mittelwalde. 3 Dec Sks at Johannisberg and Leitmeritz. 7 Dec C Reichstaedt. 14 Dec Sk Alt-Waltersdorf. 22 Dec Sk and occupation of Leobschuetz.
1745 19 Jan C Troppau. 14 Feb C Habelschwerdt. 1 Mar ambush at Hirschberg. 22 Mat C Landshut. 11 Apr Sk Slawentzitz. 22 May C Neustadt. ***4 Jun B Hohenfriedberg versus Austria.*** 20 Jul C Loewen near Neisse. 28 Jul Sk Ziegenhals. 31 Jul Sk Neustadt. 7 Aug Sks

Falkenberg and Neustadt in Silesia. 20 Aug – 5 Sep S and capitulation of Cosel. 6 Sep Sk Metau River. 7 Sep defence of Neustadt in Bohemia. 12 Sep Sk and relief of Neustadt in Bohemia. 27 Sep C Leobschuetz. 28 Sep Sk Oderberg. 30 Sep B Soor. 14 Oct Sk Troppau. 20 Oct Sks at Beneschau, Holtschin and Grulich; and capture of Ratibor. 7 Nov ambusch at Bennisch. 23 Nov C Katholisch-Hennersdorf. 27 Nov Sk and capture of Zittau. 28 Nov Sk Zittau. 29 Nov Sk Partha. 29 Nov Sk Schkeuditz. 30 Nov capture of Leipzig. 5 Dec capture of Torgau. 6 Dec ambush at Schwarzwalde. 12 Dec capture of Meissen. 13 Dec Sk Zehre. 14 Dec capture of Wittenberg. ***15 Dec B Kesselsdorf versus Austria.*** 18 Dec surrender of Dresden.

Seven Years War or the Third Silesian War 1756-1763. In this world-wide struggle, Austria invaded Silesia in an attempt to retake the province with its vital natural resources. Prussia was allied with England and Hanover; Austria was supported by France, Russia, Saxony and Sweden. Prussian victoriy at Prague was cancelled out by that of the Austrians at Kolin. The French defeated the Duke of Cumberland in September 1757 and Sweden and Russia invaded Pommerania and East Prussia. Frederick's victories at Rossbach and Leuthen at the end of 1757 and the British victory over the French at Krefeld in June 1758 restored the balance. In August 1758 the Prussians defeated the Russians at Zorndorf and gained some breathing space. In August 1759 the Austrian FM Daun heavily defeated the Prussians at Kunersdorf, but this victory was balanced out by the British win at Minden that same month. Daun invaded Saxony in December 1761. The Prussians were cornered, but the death of Empress Elisabeth of Russia made way for Peter III (an ardent admirer of Frederick's) to the Russian throne. He made peace with Prussia and entered the lists on her side against Austria. He was soon assassinated, but his successor, Catherine II (The Great) chose neutrality.

1756 18 Apr Sk Gueldene-Else. 7 Sep ambush at Stolpen. 17 Sep Sk Nachod. 22 Sep Sk Jassena and Aujezd. 23 Sep C Nollendorf. 24 Sep capture of Teschen. ***1 Oct B Lobositz versus Austria.*** 10 Oct Sk Reichenau. 11 Oct Sks at Lilienstein and Pirna. 15 Oct surrender of the Saxon army at Lilienstein. 15 Oct surrender of Sonnenstein castle. 22 Oct Sk Skalitz. 20 Dec Sk Herwigsdorf.

1757 20 Feb Sk Hirschfeld. 12 Mar Sk Busch-Ullersdorf. 18 Apr Sk Schatzlar. 20 Apr Sk Kommotau. 20 Apr C Reichenberg. 21 Apr Sk Linay. 3 May Sk Alt-Bunzlau. ***6 May B Prague versus Austria.*** 8 May Sk Arnsdorf. 9 May C Boehmisch-Brod. 9 May storming of Mount Ziska. 9

May – 20 Jun S Prague. 16 May encirclement of Nuernberg. 5 Jun Sk Kuttenberg. **18 Jun B Kolin versus Austria.** 3 Jul Sk Welmina. 22 Jul defence of Zittau. **26 Jul B Hastenbeck versus France.** 8 Aug Sks Gottleube and Gieshuebel. 12 Aug Sk Landshut. 16 Aug Sk Burkersdorf. 30 Aug ambush at Lomnitz and Nieder-Rudolsdorf. **30 Aug B Gross-Jaegersdorf versus Russia.** 7 Sep C Moys or Goerlitz. 9 Sep Sk Naumburg. 19 Sep Sk Gotha. 26 Sep cannonade of Barsendorf. 12 Oct artillery battle at Pirna. 16 Oct the burning of Berlin. 31 Oct Sk and capture of Weissenfels; and the bombardment of Schweidnitz. ***5 Nov B Rossbach versus the French and Imperials;*** a Prussian victory. 6 Nov Sk Freiburg. 21 Nov C Kleinburg, and Sk at Klettendorf. **22 Nov B Breslau; defeat by the Austrians.** 28 Nov C Parchwitz. 30 Nov bombardment of Harburg. ***5 Dec B Leuthen* versus Austria** (a Prussian victory) and C Borne. 7 Dec Sk Olatschin. 8 Dec Sk Rohrau. 8 – 21 Dec S and capitulation of Breslau. 15 Dec Sk Landshut. 28 Dec surrender of Liegnitz. 29 Dec surrender of Harburg and the capture of Anclam.

1758 In Prinz Heinrich's corps. 5 Jan capture of the Anclam ferry redoubt. Jan – 15 Jun S Stralsund. 12 Jan capture of Regenstein castle. 21 Feb capture of Ottersberg, Rothenberg and Werden. 22 Feb ambushes at Hoya, Nord-Drebber and on the River Ocker. 24 Feb Sk Schladen. 25 Feb capture of Bremen. 4 Mar Sk at Hildesheim and Eldagshausen. 8 Mar encirclement of Minden. 9-13 Mar S and capture of the Peenemuende redoubt. 20 Mar Sk Habelschwerdt. 24 Mar C Bentheim. 31 Mar – 18 Apr S and capture of Schweidnitz by the Prussians. 9 Apr Sk Wesel. 3 May Sk Marienburg. 13 May capture of Predlitz. 14 May surrender of Minden. 19 May Sk Sternberg. 22 May Sk Konitz. 27 May capture of Duisburg; Sk at Aue. 27 May – 2 Jul S Olmuetz. 28 May Sk Repschin. 30 May ambush at Homburg. 1 Jun Sk Nieder-Elten. 3 Jun capture of Zschoppau. 7 Jun C Pasberg. 12 Jun Sk Kloster Kampen. 15 Jun Sk Rehan. 16 Jun 1st Sk Holitz. 17 and 19 Jun Sks Asch. 20 Jun Sk Ratzebur. ***23 Jun B Krefeld versus France.*** 25 Jun 1st Sk Koeniggraetz. 26 Jun bombardment of Duesseldorf. 27 Jun capture of Roermond. 28 Jun Sks at Bautsch, Domstaedl and Gundersdorf. 29 Jun capitulation of Duesseldorf. 30 Jun 2nd Sk Domstaedl. 30 Jun Sks Gibau, Heiligenberg and Pohorz. 7 Jul Sk Krenau. 10 Jul Sk Driesen. 11 Jul 2nd Sk Holitz . 15 Jul C Friedeberg. 25 Jul 2nd Sk Koeniggraetz. 29 Jul crossing of the River Metau. 29 Jul Sks at Eipel, Jassewitz and Sebastianberg. 3 Aug ambush at Wachtendonk. 4 Aug Sk Nachod. 4 Aug – 1 Nov S Neisse. 5 Aug defence of the Rhine bridges at Rees and Meer. 7 Aug Sk Kloster Gruessau. 22 Aug destruction of the Schaumburg redoubts. ***25 Aug B Zorndorf versus Russia.*** 26 Aug Sk and capitulation of Peitz. 29 Aug Sk Guben. 16 Sep C Fischbach. 17 Sep Sk Gross-Roehrsdorf. 25 Sep Sks Riechstadt and Linum (Fehrbellin). 28 Sep C Fehrbellin. 29 Sep Sk at Bork im Hessischen. 10 Oct C Lutternberg and Sk Jenkwitz. 11 Oct Sks at Meschwitz, Schecknitz and Steindoerfel. **14 Oct B Hochkirch versus Austria;** a Prussian defeat. 15 Oct Sk Freiberg and ambush at Boitzenburg. 17 Oct Sk Troppau. 18 Oct Sk Soest. 26 Oct Sk Ebersbach. 27 Oct Sk Greifenberg in Pommerania. 29 Oct Sk Pfaffendorf. 3 Nov Sk Loecknitz. 7 Nov relief of Neisse. 9 Nov defence of the Great Garden at Dresden. 11 Nov Sk Pirna. 15 Nov Sk Eilenburg. 16 Nov relief of Dresden. 1 Dec ambush at Rheinfels. **1759** 2 Jan ambush at Frankfurt / Main. 6 Jan C Ebsdorf. 7 Jan Sk Dillenburg. 8 – 17 Jan S and capitulation of Demmin. 8 – 21 Jan S and capitulation of Anklam. 28 Feb Sk Posen and the capture of Erfurt. 26 Mar Sks at Saalfeld and Greifenberg in Silesia. 1 Apr Sk at Meinungen and ambush at Wasungen. 7 Apr capitulation of Ulrichstein. 10 Apr capture of the Peenemuende redoubts. 11 Apr Sk Birstein. **13 Apr B Bergen versus France.** 15 Apr Sks at Peterswalde and Toeplitz. 16 Apr Sks at Buchin, Leitmeritz, Libochowitz, Lowositz and Welmina. 16 Apr storm of Kommotau. 17 Apr Sk Troppau. 19 Apr Sk Brunersdorf. 30 Apr Sk Arnoldsdorf. 1 May Sk Zuckmantel. 8 May Sk Asch. 9 May Sk Kronach. 11 May Sks at Himmelskorn, Holfeld, Kemnach and Nagel. 13 May Sk Reichenbach. 15 May ambush at Gasbach. 16 May Sk and capture of Bamberg. 20 May Sks at Bayersdorf, Landshut and Liebau. 21 May Sk Reichenberg. 30 May Sk Moenchsberg. 4 Jun ambush at Elberfeld. 11 Jun capture of Cassel. 29 Jun Sk Trautenau and capture of Schatzlar. 1 Jul Sk Guetersloh or Rheda. 3 Jul Sk Owinsk. 4 Jul capture of Halle. 8 Jul Sk Samter. 12 Jul cannonade of Wilzyn. 13 Jul Sk Diepenau. 18 Jul C Reichhennersdorf. 20 Jul 1st Sk Zuellichau. 21 Jul defence of Freiburg near Schweidnitz. 21 Jul relief of Vechte. 22 Jul 2nd Sk Zuellichau. **23 Jul B Kay versus Russia;** a Prussian defeat, lost over 1,000 men. 23 Jul Sk Luebbecke. 23 Jul storm of Osnabrueck. 24 Jul Sk Hochkirch. 24 Jul capture of Muenster. 25 Jul Sk Crossen. 27 Jul Sk Conradswalde. 28 Jul Sk Friedland. ***1 Aug B Minden*** and Sk Gohfeld. 2 Aug capture of Minden. 5 Aug capture of Detmold. 9 Aug Sk Lemgo. 11 Aug Sk Reichenbach. **12 Aug B Kunersdorf versus Russia and Austria** (a Prussian defeat) and 1st storm of Torgau. 13 Aug 2nd storm of Torgau. 17 Aug Sk Sachsenhausen and Naumburg. 19 Aug capture of Cassel. 20 Aug Sk Sorau. 26 Aug – 5 Sep S and surrender of Dresden to the imperials. 27 Aug capture of

Wittenberg and Sk Wrietzen. 27/28 Aug Sk Wetter. 28 Aug Sk Sprottau. 2 Sep Sk Einhausen. 3 Sep Sk Marburg. 5 Sep Sks Dresden, Gorgast and Reichenberg. 6 Sep Sk Trebatsch. 8 Sep Sk and relief of Torgau. 9 Sep 1st Sk Friedland. 12 Sep 2nd Sk Friedland and capture of Leipzig. 18 Sep Sk Neukirch. 21 Sep Sk Meissen or Corbitz. 23 Sep Sk Bandelow. 25 Sep ambush at Hoyerswerda and Sk Zawelwitz near Bautzen. 26 Sep capture of Werbelow. 3 Oct Sk Buetzow and capture of Anklam. 22 Oct defence of Herrenstadt. 29 Oct Sk Pretzsch. 7 Nov Sk Homburg and capture of Demmin. 14 Nov Sk Korbitz. **20 Nov C Maxen**; Finck's Prussian corps was captured by the Austrians. 29 Nov Sk Lauterbach. 30 Nov ambush at Fulda. 26 Dec ambush at Siegen or Dillenburg. 31 Dec Sks at Kaiserswerth and Stauffen.

1760 1 Jan Rhine crossing and clash at Uerdingen. 20 Jan Sk Greifswald. 23 Jan Sk Priemen, Stolp. 28 Jan ambush at Anklam. 22 Feb Sk Pyritz. 15 Mar rearguard action at Neustadt in Upper Silesia. 17 Mar ambush at Zeitz. 28 Apr ambush at Vach. 25 May Sk Butzbach near Giessen. 12 Jun ambush at Duisburg and Sk at Schievelbein. 14 Jun Sk Zadel. **23 Jun C Landshut** (Fouquet's Prussian corps captured by the Austrians) and Sk Steinau. 1 Jul Sk Fritzlar. 7 Jul Sks at Goedau and Salzfoerstgen. 8 Jul Sks at Bautzen and Bischofswerda. 10 Jul C Corbach. 16 Jul ambush at Emsdorf and Erxdorf. 19 Jul bombardment of Dresden. 21 Jul Sk Priesnitz. 22 Jul ambush at Gruena. 23 Jul Sk Oppeln. 24 Jul Sk Anraff. 26 Jul Glatz falls to the Austrians with its 5,000 garrison. 31 Jul Sk Warburg. 2 Aug Sk at Polish Friedland. 5 Aug Sk Protsch. 10 Aug capture of Nordheim and Sababurg. 15 Aug B Liegnitz. 17 Aug Sk Demmin. 20 Aug Sk Strehla. 25 Aug Sks at Brest and Spantekow. 27 Aug Sk Waserow. 29 Aug Sk at the Kabel Pass near Friedland. 31 Aug Sks at Langen-Seifersdorf and Koetschen near Scweidnitz, and Buenzlau and Loewenberg. 3 Sep Sk on the Taschenberg near Pasewalk; at Jagow and at Striegau. 6 Sep Sk at Zierenberg and the storm of Prenzlau. 9 Sep Sk Marburg and ambush at Zierenberg. 10 Sep ambush at Butzbach. 12 Sep Sk Striegau. 13 Sep Sk Hohenfriedberg and Sk at Radern. 17 Sep C Hohengiersdorf / Kunzendorf and Sks at Roepersdorf and Wilhelmsburg. 18 Sep cannonade at Neu-Reussendorf. 22 Sep Sk Fredersdorf. 1 Oct raid on Cleve. 2 Oct Sk at Wittenberg or Eupen. 3 Oct Sk Loecknitz near Pasewalk. 3 – 14 Oct S Wittenberg. 4 Oct defence of Berlin and Koepenick. 5 Oct relief of Berlin. 7 Oct Sks at Britz, Nixdorf and Zehlendorf; cannonade at Lichtenberg. 9 Oct surrender of Berlin. 12 Oct evacuation of Berlin. 14 Oct surrender of Wittenberg, Sk Taschenberg. 15 Oct Sk Belzig. 16 Oct Sks at Kloster Kamp and Rheinbergen and the evacuation of Wittenberg. 23 Oct occupation of Wittenberg. 25 Oct ambush at Koethen. 27 Oct Sk Rabis. 1 Nov Sk Greifenhagen. 2 Nov Sks at Langen-Reichebach, Schoene and Staupitz. ***3 Nov B Torgau versus Austria.*** 4 Nov Sk Zobten. 6 Nov capture of Schwedt. 7 Nov Sk Zarnewanz. 15 Nov Sk of Lager near Kammin. 18 Nov Sks at Augustenburg, Kloster-Zelle, Siebenlehn, Tannhausen and Wuestergiersdorf. 26 Nov the encirclement of Goettingen. 28 Nov ambush at Hedemuenden. 29 Nov Sk at Coeslin and Belgard.

1761 1 Jan ambush at Duderstadt. 3 Jan Sk Duderstadt. 14 Jan Sks at Belgard and Coeslin. 21 Jan attacks on Greiffenberg, Plate and Regenswalde. 26 Jan Sk Ebeleben. 9 Feb Sk Rosenthal. 12 Feb Sk Doerna. 13 Feb ambush at Marburg. 14 Feb Sk Merxleben. 15 Feb C Langensalza and capture of Fritzlar. 19 Feb capture of Hersfeld. 24 Feb Sk Gelnhausen. 26 Feb capture of Marburg. 6 Mar sortie from Cassel. 7 Mar Sk Wilsdorf. 12 Mar Sk Lich. 15 Mar capture of Neustadt. 16 Mar Sk Stangerode. 17 Mar cannonade at Schleitz. 19 Mar Sks at Laubach and Wetter. 21 Mar Sks at Atzenhain, Gruenberg and Stangenrode. 26 Mar Sk Leimsfeld. 2 Apr Sks at Saalfeld and Schwarza. 5 Apr Sks at Strasburg in Saxony and Plauen. 24 Apr Sk Uslar. 4 May Sk Hohenfriedberg. 5 May ambush at Nordheim. 8 May Sk Loesnitz. 10 May Sk Goerlitz. 17 May Sk Waldkirchen. 21 May Sks at Alzenhain and Chemnitz. 12 Jun Sk Naundorf. 20 Jun C Landshut. 8 Jul Sk Kriewen. 9 Jul Sks Oestinghausen and Vellinghausen. 11 Jul Sk Kostenblut. 13 Jul Sk Sende. ***16 Jul B Vellinghausen versus France.*** 19 Jul Sk Peucke. 20 Jul Sks at Ruhme and Sommershausen. 21 Jul Sk Zinkwitz. 22 Jul Sk Gross-Nossen. 23 Jul Sk Klodebach. 25 Jul Sk Bartow. 30 Jul Sks at Kunzendorf and Maidelberg. 31 Jul Sks at the Kabel Pass and the Klempenower Pass. 2 Aug Sk Peterwitz. 5 Aug ambush at Bartow and Spantekow. 8 Aug Sk Kenzlin. 14 Aug Sks at Kloster Wahlstadt, Ludhorst, Ruehlow, Wandris and Woldeck. 15 Aug Sk Uslar. 17 Aug Sk Jauer. 21 Aug Sk Neu-Brandenburg. 22 Aug Sks at Neu-Brandenburg and Klein-Waltersdorf. 23 Aug ambush at Nordhausen. 28 Aug Sk Bunzelwitz and the storm of Treptow. 31 Aug Sk at the Klempenower Pass. 3 Sep Sk Breester Pass. 4 Sep Sk Klempenow. 7 Sep Sk Bunzelwitz. 9 Sep Sk Jatzke. 12 Sep Sk Treptow. 14 Sep Sk Kobylin. 15 Sep Sk Gostyn. 19/20 Sep loss and recapture of the Green Redoubt at Colberg. 21 Sep Sk Fritzlar. 27 Sep Sk Spie. 28 Sep ambush at Duerr-Brockutt. 29 Sep Sks at Friedland and Neuensund. 30 Sep Sk Coerlin. 1 Oct Sks at Torgelow and Uckermuende. 2 Oct Schweidnitz falls to the Austrians. 8 Oct Sks at Eschershausen and Halle. 10 Oct Sk Anklam. 13 Oct Sks at Braunschweig and Gollnow, and the capture of Woelfenbuettel. 20 Oct Sk Zargelow. 1 Nov Sk Stargard. 7 Nov Sk Seifersdorf. 8 Nov Sk Oldendorf. 13 Nov C Berka, Duehme and Kattenburg. 8 Dec Sk Treptow. 10 Dec Sk Volksdorf. 11 Dec Sks at Drenow, Glansee and Zarben. 12 Dec Sk Spie. 23 Dec Sk Neu-Kahlden; evacuation of Malchin. 24 Dec defence of Basedow castle. **1762** 3 Jan Sk Neu-Kahlden and the relief of Malchin. 12 Jan Sk Altenburg. 9 Mar Sk Gretha. 14 Apr Sk Goettingen. 16 – 19 Apr blockade and capture of Arnsberg. 12 May Sk Doebeln. 13 May Sk Nossen. 15 May Sks at Frauenstadt and Lampersdorf. 16 May Sks at Klingenburg and Wilsdruf. 21 May Sk Chemnitz. 24 May Sks at Goettingen and Weisdorf. 26 May Sks at Auerswalde, Ebersdorf and Zschoenengrund. 29 May Sk Clausnitz. 31 May Sk Gepueltzig and ambush at Teschen. 1 Jun Sk Reichstaedt. 6 Jun Sk Wernersdorf. 14 Jun Sk Lampersdorf. 21 Jun Sks at Grebenstein, Heidersdorf and Ottmachau. 22 Jun Sk Grumbach. 24 Jun Sks at Glauchau, Grebenstein and Wilhelmsthal. 25 Jun capture of Jaegerndorf; Sks at Schoenfels and Zwickau. 27 Jun Sks at Constapel, Hartha and Wilsdruf. 28 Jun Sks at Kanth and Striegau. 2 Jul Sk Einsiedel. 3 Jul Sk Hohenfriedberg. 6 Jul Sks at Adelsbach and Freiburg. 7 Jul Sks at Friedland and Taschendorf. 9 Jul Sk Goesfelden. 11 Jul capture of Waldeck. 18 Jul Sks at Boehmisch-Einsiedel and Holzhausen. 20 Jul Sk Kirchberg. 21 Jul C Leutmannsdorf and Sk at Auerbach. 23 Jul B Lutternberg and the capture of Homburg. 24 Jul Sks at Hersfeld, Ober-Glogau and Ratibor. 25 Jul Sks at Gensungen, Neu-Morschen and Ratibor. 26 Jul Sk Fulda. 28 Jul Sk Meckbach. 29 Jul ambush at Amoeneburg. 31 Jul Sk Kommotau. 1 Aug Sks at Aujezd, Dux and Osseg. 2 Aug Sks at Neu-Morschen and Teplitz. 6 Aug capture of Eschwege, Muehlhausen and Wanfried. 7 Aug – 9 Oct S and capitulation of Schweidnitz. 8 Aug Sk Fulda. 9 Aug Sk Buechenwerder, Kaltenbach and Moershausen. 11 Aug Sk Lembach. 14 Aug Sk Kleutsch. 16 Aug C Reichenbach. 17 Aug Sk Neu-Morschen. 20 Aug C Schauenstein. 21 Aug C Baireuth and Sk Rupertenrode. 22 Aug Sks at Alzenhain, Gladenbach and Kupferberg. 23 Aug capture of Hersfeld. 30 Aug C Friedberg. 11 Sep Sk Waldau. 16 Sep Sks at Alsfeld, Momberg and Neustadt. 18 Sep Sk Treisa. 21 Sep Sks at Amoeneburg, Bruecker-Muehle and Schwarzenborn. 27 Sep Sk Altenburg. 27 Sep Sks at Hattendorf, Ober-Grenzbach and Steina. 29 Sep Sks at Cunnersdorf, Gruelenburg, Grumbach, Hartha, Spechthausen, Tharandt and Wilsdorf. 30 Sep Sk Burggemuenden. 1 Oct Sk Helbigsdorf. 4 Oct Sks at Schotten and Stumpertenrode. 10 Oct ambush at Schnallenburg. 13 Oct Sk Moenchefrei. 14 Oct Sks at Brand, Erbisdorf, Freiberg and Muedisburg. 15 Oct Sk Brand. 22 Oct Sk Gross-Schirma. 27 Oct Sk Reichenbach. **29 Oct B Freiberg.** 3 Nov Sk Dittersbach. 4 Nov storm of Saatz and the burning of Brix. 6 Nov Sk Limburg. 7 Nov C Limburg. 19 Nov Sk Kronach. 20 Nov storm of Windheim. 23 Nov capture of Rothenburg an der Tauber. 29 Nov storm of Nuernberg. 30 Nov Sk in Bamberg. 2 Dec Sk Heman.

The War of the Bavarian Succession 1778-1779. Prussia and Saxony against Austria. Due to the low level of hostilities, this is also known as the Potato War, from a skirmish in which the opposing sides threw potatoes at each other.

1778 7 Jul Sk Kowalkowitz. 20 Jul Sk Trautenau. 23 Jul Sk Slavieten. 30 Jul Sks at Schoenborn and Zollenstein. 2 Aug Sk Gabel. 11 Aug Sk Glomnitz (Teschen village). 23 Aug Sk Molsche. 25 Aug Sk Trautenau. 26 Aug Sk Nieder-Kruppau. 4 Sep Sk at Huenewasser, Kloster-Poesig and Neuendorf. 8 Sep Sks at Lauterwasser, Niemes and Pulkinau. 11 Sep Sk Burkersdorf. 14 Sep Sk Hartmannsdorf. 19 Sep Sk Martinowes. 21 Sep Sk Forstberge in Bohemia. 28 Sep Sk Nickelsberg. 17 Oct Sk Troppau. 25 Oct ambush at Moeckern near Jaegerndorf. 27 Oct Sk Maehrisch-Ostrau. 8 Nov Sk Dittersbach. 22 Nov Sk Weisskirch.

1779 11 Jan Sk Moenick. 14 Jan Sk Zuckmantel. 18 Jan C Habelschwerdt and the defence of the blockhouse at Schweddeldorf. 5 Feb Sk Brix. 28 Feb defence of Neustadt in Upper Silesia.

The War against the Dutch Patriots 1787. Prussia sent forces to stifle the nascent republican movement in the country.

17 Sep Sk Arkel, surrender of Gorkum and the storm of the Uitermeer redoubt. 18 Sep raid and capture of the Groeversvell locks and the Wiering redoubt. 23 Sep storm of Wesep. 26 Sep surrender of Wesep and Naarden. 1 Oct Sks at Amstelveen, Halfwey and Oudekerk. 7 Oct surrender of Amsterdam.

The Revolutionary Wars Against France 1792-1795.

1792 19 Aug Sk Fontoi. 30 Aug capture of Verdun. 9 Sep Sks at Monthomme and St-Juvin. 12 Sep defence of the St-Croix defile. 14 Sep Sk Grandpre. 15 Sep Sk Montcheutin. 20 Sep cannonade of Valmy. 6

Nov C Limburg / Lahn. 9 Nov Sk Limburg / Lahn. 2 Dec Sk Bockenheim and storm of Frankfurt / Main.

1793 6 Jan Sk Hochheim. 7 Jan Sk Kreuznach. 20 Jan Sk Kirchheim. 6 Feb – 3 Mar defence of Mastricht. 1 Mar Sks at Aldenhoven and Dueren. 1 – 7 Mar blockade and capture of Koenigstein castle. 3 Mar Sk Roermonde. 4 Mar capture of Geertruidenburg. 14 Mar relief of Willemstadt castle. 20 Mar Sk Rheinbellen. 25 Mar Sk Guelzen. 26 Mar capture of Antwerp. 28 Mar Sks Bingen and Bockenheim. 28 Mar Sk Waldelgesheim or Weiler. 30 Mar Capitulation of Breda; Sks at Alsheim, Floersheim and Rheintuerkheim. 12 Apr Sk at Hemm and ambushes at Croix, Mouveaux, Roncq and Lincelles. 15 Apr Sk Weissenau. 27 Apr Sk Kostheim. 1 / 2 May Sks St-Amand and Vigogne. 3 May storm of Kostheim. 7 May Ske at St-Amand and Rossuth. 8 May Sk Hasnon. 9 May Sk St-Amand. 10 May Sk Hasnon. 22 May storm of Hasnon castle. 23 May capture of Marchiennes. 27 May capture of the abbey of the Quatre-Cloches d'Aniche. 30 May Sks at Bechingen and Wahlsheim. 31 May Sk Marienborn. 4 Jun Sk Bouvignie. 6 Jun Sk Gruson / Scarpe. 9 Jun Sk Lannoy. 19 Jun – 23 Jul S and capitulation of Mainz. 28 Jun Sk Weissenau. 12 Jul Sk Hemm. 18 Jul Sk Zweibruecken. 22 Jul Sk Edenkoben. 23 Jul Sk St Annen-Kapelle near Weyher. 13 Aug Sks at Altstadt, Bouvines, Limbach and Neukirchen. 14 Aug Sk Peronne. 17 Aug Sks at Kettricher Hoff and Rohrbach. 20 Aug Sk Kettricher Hoff. 27 Aug defence of Launoi. 12 Sep Sk Rohrbach and Spiesen. ***14 Sep B Pirmasens,*** Sks at Limbach, Spiesen and Zweibruecken. 26 Sep Sks at Blieskastel and Hornbach. 27 Sep Sks at Ensdorf and St Imbert. 29 Sep Sk Bischmischheim. 13 Oct Sks at Bondenthal, Erchingen, Herzogenhand and Lembach. 16 Nov failed raid on Bitsch fortress. 17 Nov C Biesingen. 26 Nov Sk Waldmoor. ***28 – 30 Nov B Kaiserslautern***. 29 Nov Sk Limbach. 2 Dec Sk Jaegersburg. 6 Dec Sk Zweibruecken. 22 Dec Sks at Engelsburg, Froeschweiler and Limbach. 24 Dec defence of Scheerhoehle. 26 Dec Sk Weissenburg. 27 Dec Sk Anweiler valley.

1794 3 Jan Sk Frankenthal. 8 Jan Sk Kreuznach. 4 Apr Sk Hirxheim. 1 May Sk Gruenstadt. ***23 May C Kaiserslautern;*** Sks at Diedesheim, Forst, Hardenkopf an der Schmelze, Neidenfels, Neustadt, Vogelweh, Wackenheim, and Weidenthal. 25 May Sks at Edenkoben and Neustadt. 13 Jun Sk Edesheim. 15 Jun Sk Lautern. 16 Jun C Fleurus. 2 Jul Sks Edenkoben, Edesheim and Fischlingen. 3 Jul Sks at Koesshofen and Trippstadt. 12 Jul defence of the Schaenzel. 13 Jul Sks at Edenkoben,

Edesheim, Hermersberg and the storm of the Schaenzel. 15 Jul Sks at Hardt and Neustadt. 1 Aug Sks at Neu-Leiningen and Ungstein. 12 Aug ambush at Klein-Carlebach and Weissen am Sand. 20 Aug Sk Tiefenthal. 28 Aug Sk Wachenheim. 17 Sep Sks at Malzberge, Plattenberg and Schorleberg. 20 Sep B Kaiserslautern. 20 Sep Sk Moorlautern. 16 Oct Sk Klein-Bockenheim. 2 Dec Sk Mainz.

The War in Poland 1793-1795. In cooperation with Russian forces, the Prussians crushed the Polish revolt.
1794 18 May Sk Skala. ***6 Jun B Rawka*** (aka Scelze or Szczekocin). 10 Jun Sk Raygrod. 15 Jun surrender of Krakow. 10 Jul Sk Kolno. 14 Jul Sk Radzin. 18 Jul Sk Demnicki or Sboyne. 26 Jul Sk Opalin. 27 Jul – 6 Sep S Warsaw. 28 Jul Sk Wola. 19 Aug Sk Powonsk. 13 Sep C Kamion. 20 Sep Sk Lublinitza. 28 Sep Sks at Barschin and Labischin. 8 Oct Sk Czarnowo. 8 Oct Sk Lasch. 13 Oct defence of Thorn. 19 Oct Sk Kamion. 23 Oct Sk Suchaczew. 24 Oct Sk Roczan. 26 Oct Sk Gross-Magnisczewo. 31 Oct ambush at Wyczna. 1 Nov Sk Piontki. 10 Nov Sk Kapinos. 11 Nov Sk Plonien.

Napoleonic Wars

1806 The Jena Campaign.
8 Oct Sk Saalburg (Reuss). 9 Oct Sk Schleitz. 10 Oct C Saalfeld. 12 Oct Sk Winzerle. **14 Oct B Auerstaedt. B Jena.** 15 Oct Sk Langensalza. 16 Oct Sk Creussen. 17 Oct Sks at Eichrode, Halle and Nordhausen. 23 Oct Sk Neuhaldensleben. 24 Oct Occupation of Berlin and Potsdam. 25 Oct Sk Sandau. 26 Oct Sks Altenzaun and Zehdenick. 27 Oct Storm of Boitzenburg. 28 Oct Sks at Bandelow, Lychen and Prenzlau. 29 Oct Sks at Krangen and Schulzendorf. 1 Nov Sk Wahren. 2 Nov Sk Frauenmark. 3 Nov Sks at Faehre, Klein-Welzin and Kriewitz. 5 Nov Sks at Gadebusch and Ratzeburg. 6 Nov C Luebeck. 7 Nov Capitulation of Bluecher's corps at Ratkau. 8 Nov Surrender of Magdeburg. 14 Nov Sks at Bellnow and Bromberg. 15 Nov Sks at Gniekowo and Schulitz. 18 Nov Sks at Hameln and bombardment of Thorn. 18 Nov – 3 Dec S and surrender of Thorn. 20 Nov Capitulation of Hameln. 25 Nov surrender of Nienburg and the Plassenburg fortress. 7 Dec Sk Guelzow. 13 Dec Sks at Biezun and Swienemuende. 23 Dec Sks at Biezun and Gurzno. 24 Dec Sk Strehlen. 25 Dec Sks at Mlawa and Soldau. 26 C Pultusk. 29 Dec C Ohlau. 30 Dec Sks at Breslau and Schimanen. **1807** 6 Jan Sk Wollin. 11 Jan Sk

Schippenbeil. 15 Jan Sks at Altwasser, Arnswald and Leinenburg. 18 Jan Sk Stepnitz. 22 Jan – to the end of the war: defence of Graudenz. 23 Jan – 24 Jun blockade of Cosel. 25 Jan C Mohrungen. 26 Jan Sk Preussisch-Stargard. 27 Jan Sk Dirschau. 28 Jan Sk Bialochowo. 29 Jan Sk Schoeneck. 31 Jan Sks at Altwasser and Vilgendorf, Silesia. 1 Feb Sk Alt-Cuestrinschen. 2 Feb Sk Zachau, Pommerania. 4 Feb Sk Allenstein. 5 Feb Sks at Crossen, Liebstadt and Waltersdorf. 6 Feb Sks at Dirschau, Glanden and Neu-Stettin. **8 Feb B Preussisch-Eylau,** Sks at Creuzburg or Wackern, Hussehnen, Pomnicken, Schlautinen and Wartha. 15 Feb Sks at Koenigswalde, Lichtenhagen, Mahnsfeld and Wernsdorf. 17 Feb Sks at Naugard and Stargard. 21 Feb Sk Bischoffstein. 22 Feb C Heilsberg. 23 Feb Sk Seeburg. 24 Feb 1st Sk Braunsberg. 25 Feb Sk Koenigswalde. 26 Feb 2nd Sk

Braunsberg. 26 Feb – July blockade of Glatz. 27 Feb – June blockade of Silberberg. 6 Mar Sk Dirschau and Mewe. 8 Mar Sk Schilde. 10 Mar – 27 May S and surrender of Danzig. 10 Mar Sks at Zigaukendorf near Danzig and Frankenstein. 12 Mar Sk Bischofsberg. 13 Mar Sk Neu-Fahrwasser. 18 Mar Sk Cosel. 19 Mar – 2 Jul defence of Colberg. 20 Mar storm of the Sellnow trenches and ambush at Torfmor at Colberg. 1Apr Sk Danzig. 17 Apr Sk Haffitz, Glatz. 10 May Sk Steegen. 14 May Sk Kanth. 15 May sortie at Weichselmuende, or Stutthof, Danzig. 16 May Sk Adelsbach. 17 May 1st storm of Wolfsberg at Colberg. 20 May Sk Bunzlau. 4 Jun Sk Rothwaltersdorf and cannonade at Wuhsen. 5 Jun Sk Spanden. 9 Jun Sks at Dietrichsdorf, Nuebrueck and Zanow. **10 Jun B Heilsberg.** 14 Jun B Friedland, defence of Koenigsberg, Sk Gollau or Ludwigswalde and the 2nd storm of the Wolfsberg at Colberg. 15 Jun Sks at Katschau and the Weydeuer forest at Koenigsberg. 16 Jun Sk Sellnow. 17 Jun Sk Baumwald and bombardment of Pillau. 18 Jun Sk Tilsit and surrender of Cosel. 19 Jun 3rd storm of the Wolfsberg and storm of Sellnow at Colberg. 19 – 20 Jun Sks at Haffitz at Glatz. 29 Jun storm of Silberberg. 6 – 20 Aug blockade of Stralsund.

1809 Schill's Raid. 5 May Sk Dodendorf. 24 May Sk Dammgarten. 31 May Sk Stralsund.

1812 The Russian Campaign. 4 Jul Sks at Szawle and Telsze. 5 Jul Sks at Kosziany ans Widszy. 7 J ul Sk Poniewitsch. 18 Jul Sk Drizna. 19 Jul 1st Sk Eckau. 23 Jul C Saltanowka. 24 – 26 Jul Sks at Ostrowno. 27 Jul Sk Witepsk. 28 Jul Sk Janowicze. 29 Jul Sk Paniszawi. 1 Aug C Drissa River and the capture of Duenaburg. 5 Aug Sks at Schlock and St Annen. 7 Aug Sls at Wolgund and Clivenhofen. 8 Aug Sk Malewa-Boloto or Lukowa. 14 Aug C Krasnoi. 16 Aug Sk Korytnia. 17 Aug B Smolensk. 18 Aug 1st B Polotzk. 19 Aug Sk Duenaburg. 20 Aug Sk Gedeonowo. 21 Aug Sk Solowiewo. 22 Aug Sks at 1st Dahlenkirchen, St Annen and St Olay. 23 Aug Sk Michalewka. 27 Aug 1st Sk Dorogobusch or Ribki. 28 Aug Sk Wiasma. 29 Aug Sk Zarewo-Saimitschch. 5 Sep C Kolotzkoi. 7 Sep B Borodino. 10 Sep Sk Moschaisk. 13 Sep Sk Moscow. 14 Sep entry into Moscow. 16 Sep Sk Wiasowska. 26 Sep 2nd Sk Dahlenkirchen and Sks at Plakhanen and Tomoszna. 27 Sep 2nd Sk St Olay and 2nd Sk Eckau. 29 Sep Sks at Graefenthal or Kiopen and Kósackenkrug or Mestoen. 30 Sep Sks at Rauenthal and Schlockhoff. 1 Oct Sk Garossenkrug. 3 Oct Sk Woronowo. 17 Oct Sk Wolgund. 18 – 19 Oct 2nd B Polotzk and Sk at Winkowo or on the Tschernitschna River. 24 Oct B Malo-Jaroslawetz. 29 Oct C Slonim. 1 Nov C Wiasma. 7 Nov 2nd Sk Dorogobusch and Sk Tomossna. 17 Nov Sk Tomsdorf. 18 Nov Sk Friedrichstadt. 23 Nov Sk Borissow. 26 Nov B Beresina River. 27 – 29 Nov B Beresina crossing. 26 Dec Sk Piktupoehnen.

1813-1815 The Wars of Liberation. End Jan – 4 Apr S and surrender of Thorn. 16 Jan – 30 Nov blockade, S and capitulation of Danzig. Mid-Feb – 21 Nov blockade, S and surrender of Stettin and Damm. Mid-Feb – 27 May blockade and S of Glogau. Feb – 7 Mar 1814 blockade, S and capitulation of Cuestrin. 24 Mar Sk Werben. 31 Mar capture of Leipzig. 31 Mar – 13 May 1st blockade of Wittenberg. 1 – 24 Apr S and capitulation of Spandau fortress. 2 Apr storm of Lueneburg. 5 Apr Sks at Dannikow, Moeckernand Vehlitz. 12 Apr occupation of Weimar. 12 / 13 Apr Sk Langensalza. 13 Apr Sk Ruhla and the storm of the Wittenberg suburbs on the left bank of the Elbe. 17 Apr Sk Wanfried. 18 Apr Sk Weimar. 27 Apr Sk Wettin. 28 Apr Sk Halle. 29 Apr Sk Merseburg and 1st Sk Zerbau near Glogau. **2 May B Gross-Goerschen or Luetzen** and Sks at Halle, Leipzig, Lindenau and Starsiedel. 3 May 2nd Sk at Zerbau. 4 Mat Sks at Bornau und Lausig. 5 May Sk Colditz. 7 May Sks at Schoenebeck and Zerrau. 11 May Sk Bischofswerda. 15 May Sk Goedau. 17 May Sk Dobrilugk. 19 May Sks at Dahme, Koenigswartha and Weissig. **20/21 May B Bautzen.** 22 May Sk Reichenbach. 23 May Sks at Goerlitz and Zuellnitz. 24 May Sk Naumburg / Saale. 25 May Sks at Gera and Walbau. 26 May ambush at Haynau. 28 May Sk Hoyerswerda. 29 May Sks at Zollenspiecker near Hamburg and at Zwickau. 30 May Sk at the

Nettelburger locks. 31 May Sk Neukirck. 1 Jun ambush at Rohda. 3 Jun Sk Kalau. 4 Jun Sk Luckau. 6 Jun Sk Ortrand. 7 Jun Sk Leipzig. 17 Jun ambush at Kitzen. 17 Aug Sks at Baruth, Buchholz, Dornenwalde, Lauenburg, Mueckendorf, Rochlitz and Zech. 18 Aug Sk Lauenburg. 19 Aug Sks at Siebeneichen, and at Hirseberg near Loewenberg or Lautcrseifern. 21 Aug C Loewenberg and Seifersdorf and Sks at Alt-Koenigsborn near Magdeburg, Moellen and Trebbin. 22 Aug Sk Wietstock. ***23 Aug B Gross-Beeren,*** C Goldberg, Sks at Blankenfelde, Strehlen and Ziesar. 24 Aug Sks at Sperenberg and Zitz. 25 Aug Sks at Dannenberg and Doemitz. ***26 Aug B Katzbach*** and Sk at Cotta. 26/27 **Aug B Dresden** and Sks at Crayn, Goldberg, Hagelsberg, Kroitsch, Luebnitz and Pilgramsdorf. 28 Aug Sks at Haeselich, Luebnitz and Lueckau. 29 Aug Sks at Falkenhayn and at Dippoldiswalde or Possendorf. 29 Aug – 10 Apr 1814 blockade, S and capitulation of Glogau. ***30 Aug B Kulm (Nollendorf),*** ambush at Pfaffendorf and Sk at Bunzlau. 31 Aug Sk at Naumburg/Saale and at Ullersdorf. 1 Sep Sk Ganzer. 3 Sep Sk Thiesen. 4 Sep Sks at Hochkirch, Pitchenberge and Zarrentin. 5 Sep Sks at Markersdorf, Moellen, Reichenbach and Zahna. ***6 Sep B Dennewitz.*** 6/7 Sep ambush at Dahme. 7 Sep Sk Herzberg. 8 Sep Sks Dohna and Pirna. 10 Sep Sks at Geiersdorf, Holzdorf, Ober-Graupen, Peterswalde, Taschendorf and Waldenburg. 11 Sep Sk Zeitz. 12 Sep Sk Weissenfels. 14 Sep blockade of Magdeburg on the right bank of the Elbe; from 20 Oct – 24 May 1813 also blockade of the city of the left bank as well. 15 Sep Sks at Dresden-Neustadt, Hellendorf and Lang-Wolmsdorf. 15 Sep – 11 Oct 2nd blockade of Wittenberg. 16 Sep B Goehrde and Sks at Nollendorf and Peterswalde. 17 Sep Sks at Arbesau, Ebersdorf, Kulm and Tellnitz. 18 Sep Sks at Gross-Hartha, Merseburg and Zarrentin. 19 Sep Sk Muehlberg. 20 Sep Sk Koesen. 22 Sep Sk Goldbach. 23 Sep storm of the Wittenberg suburbs and Sks at Bischofswerda and Goedau. 24 Sep Sk Wartenburg. 25 Sep Sk Braunschweig. 28 Sep C Altenburg and Zeitz (Meuselwitz). ***3 Oct B Wartenburg.*** 4 Oct Sks at Aken and Rosslau. 5 Oct Sk Dueben. 8 Oct Sks at Lindenthal and Soemmerda. 10 Oct Sks at Atzendorf, Frohburg, Mieneweh and Stoessen. 12 Oct Sks at Coswig, Dessau and Rosslau. 13/14 Oct Sks Bremen. 14 Oct C Liebertwolkwitz. 15 Oct Sk Bremen. ***16 Oct B Moeckern,*** B Wachau. 17 Oct C Eustritzsch and Gohlis. ***18 Oct B Leipzig. 19 Oct storm of Leipzig.*** 21 Oct Cs at Freiburg, Gleina, Koesen and Schulpforte. 22 Oct Sk Weimar. 23 Oct Sk Buttelstaedt. 23 Oct – 13 Jan 1814 S and storm of Wittenberg. 25 Oct Sks at Elsterode and Geisa. 26 Oct C at the Hoerselberge, Sk at Vacha. 27 Oct

Sks at Fulda, Huenfeld and Vacha. 28 Oct Sk Saalmuenster. 29 Oct Sk Orb. 30 Oct – 20 Dec blockade and surrender of Erfurt. 31 Oct Sks at Gelnhausen and Mehrholtz. 1 Nov Sks at Bergen, Schidlitz (Danzig) and Zinna. 2 Nov Sks at Bergen and Frankfurt / Main. 2 Nov – 26 Nov blockade of Torgau. 3 Nov Sks at Hochheim and Wickert. 8 Nov Sk Schoenebeck. 10 Nov Sks at Frohburg and Meineweh. 19 – 29 Nov blockade of Hamburg and Harburg. 21 Nov surrender of Stettin. 21 Nov – 8 May 1814 blockade of Wesel. 22 Nov storm of the Tesch redoubt at Torgau. 23 Nov – 5 May 1814 blockade of Delf-Zijl. 23 Nov capture of Duisburg and Zuetphen. 25 Nov Sk at Middachten and the storm of the glacis at Arnhem. 30 Nov storm of Arnhem and Sk at Elden. 2 Dec Sks at Juelich and Segeberg. 2 / 3 Dec ambush and Sk at Neuss. 3 Dec Sk Duesseldorf. 6 Dec Sks at Bornhoeft and Oldesloe. 9 Dec Sk Eckernfoerde. 11 Dec capture of Fort Zinna at Torgau. 14 Dec occupation of Gorkum, Sk at Aalst and ambush at Heerewarden. 15 Dec Sks at Bommeler-Waard, Dodendorf and Weel. 15 Dec – 26 Jan 1814 blockade and surrender of 'sHerzogenbosch. 16 Dec Sks at Gaebs, Pechau, St Andre, Turnhout, Wollmirstadt and the bombardment of 'sHerzogenbosch. 19 Dec Sk at Westwesel and capture of Friedrichsort. 21 – 25 Dec blockade of Glueckstadt. 22 Dec Sks at Doorst and Upelaar. 26 Dec Sk Druitten. Capitulation of Torgau. 31 Dec / 1 Jan Sk Bacharach.

1814 1 Jan Rhine crossing by Bluecher at Kaub; surrender of Danzig; Sk at Tillburg. 2 Jan Sk Alzey. 3 Jan Sks at Algersheim and Simmern. 4 Jan Sk Ebendorf and Frohse. 6 Jan Surrender of Erfurt, capture of Trier. 6 Jan – 4 May blockade of Mainz on the left bank of the Rhine. 8 Jan Sk Charmes; 9 Jan Sk Meyel in the Netherlands. 9 Jan – 24 Apr blockade of Saarlouis. 11 Jan Sk Hoogstraten. 12 Jan Sk Lier. 13 Jan storm of Wittenberg; Sk Merxem or Wyneghem. 13 Jan – 24 Apr blockade of Thionville. 14 Jan Sk Nancy. 14 Jan – 26 Apr blockade of Metz. 15 Jan – 24 Apr blockade of Antwerp. 17 Jan Sk Noiseville. 19 Jan cavalry Sk at Manheul. 20 Jan failed storm of Harburg. 22 Jan – 26 Apr observation of Deventer. 23 Jan Sk Ligny. 24 Jan Sks at Bar-sur-Aube and Liege. 25 Jan Sk St-Dizier. 26 Jan storm and capture of 'sHerzogenbosch. **29 Jan B Brienne.** 30 Jan Sk St-Dizier. 31 Jan storm of Lier. ***1 Feb B La Rothiere,*** Sks at Doeren and Wyneghem, bombardment of Vitry-le-Francais. 2 Feb storm of Merxem. 3 Feb Sk La Chaussee. 4 Feb Sk Chalons-sur-Marne. 5 Feb occupation of Vitry-le-Francais. 6 Feb – 4 May blockade of Juelich. 10 Feb Sk Terville/Moselle. 11 Feb Sk Montmirail. 12 Feb Sk Viffort and

Chateau Thiery. 13 Feb Sk Etoges. 14 Feb C Champcaubert and Vauchamps. 16/17 Feb occupation of Tournay. 22 Feb C Mery-sur-Seine. 22 Feb – 24 Apr blockade of Maubeuge. 25 Feb Sk Conde. 26/27 Feb Sk La Ferte-sur-Jouarre. ***27 Feb B Bar-sur-Aube***, capture of La Fere. 28 Feb Sks at Gue-a-Tresmes or Lisy, Vieux-Maison. 28 Feb – 16 May blockade of Grave. 2 Mar Sks Courtray, May or Neufchelles and Mareuil. 3 Mar Sk at Neuilly-St-Front and capture of Soissons. 5 Mar Sk Oudenarde. 6 Mar Sks at Courtray and Soissons. 7 Mar Sks at Rheims, Belleghem and Sweweghem. 8 Mar Sk Courtray. ***9/10 Mar B Laon.*** 12/13 Mar storm and loss of Rheims. 14 Mar C Berry-au-Bac. 17 Mar Sks at Lire and Werlos. 18 Mar C Berry-au-Bac; Sks at Asfeld-la-Ville and Pontavert. 19 Mar Sk Fismes. ***20/21 Mar B Arcis-sur-Aube.*** 21 Mar Sk Oulchy-le-Chateau. 22 – 28 Mar S Soissons. 23 Mar Sk Pont-a-Bouvines. 23 Mar Sk Vitry-le-Francais. 24 Mar Sk Marchais. ***25 Mar B Fere-Champenoise;*** Sk Conantray. 26 Mar Sks at Coulommiers; La Ferte-Gaucher; La Ferte-sous-Jouarre; Montis; St-Martin; Sezanne. 27 Mar Marne crossing and Sk at Trilport. 27 Mar – 11 Apr blockade of Venlo. 28 Mar ambush at Claye-Souilly. 28 Mar – 24 Apr blockade of Soissons. ***30 Mar B Paris.*** 31 Mar Sks at Courtray, Sweweghem and Tournay. 1 Apr Sk Compeigne; sortie of the French at Ottersleben by Magdeburg. 4 Apr Sk Courtray. 10 Apr surrender of Glogau. 25 Apr surrender of Landau. 29 Apr capitulation of Hamburg. 10 May surrender of Wesel. 14 May surrender of Magdeburg. 16 May surrender of Grave and of the fortresses at Erfurt.

1815 15 Jun Sks at Charleroi, Gilly and Gosselies. **16 Jun B Ligny and St Amand.** 17 Jun Sk St-Guibert. ***18 Jun B Waterloo;*** C Wavre. 20 Jun storm of Namur; Sks at Gembloux, Temploux. 20 Jun – 12 Jul S and capitulation of Maubeuge. 21 Jun Sk St-Dinant. 22 Jun capture of Avesnes. 24 Jun capture of Guise; capture of Rodemachern town. 24 Jun – 21 Jul S Landrecy. 24 Jun – 28 Jul S and capitulation of Marienbourg. 25 Jun – end of the war blockade of La Fere. 26 Jun capture of Sedan. 27 Jun Sk Senlis. 27/28 Jun Sks at Compeigne and Villers-Cotterets. 27 Jun – 8 Aug S Philippeville. 28 Jun Sks at Crespy, Levignan and Nanteuil. 28 Jun – 10 Aug S and surrender of Mezieres. 29 Jun capture of Charleville; Sk Gonesse, Sk St-Germain. 29/30 Jun Sk at St-Denis abd Stains. 29 Jun = 14 Jul 1st blockade of Longwy. 30 Jun Sk Aubervilliers. 30 Jun – 19 Aug blockade of both Givet forts and Charlemont; the Givets surrendered an 11 Sep; Charlemont was only given up after peace was signed. 1 Jul ambush at Versailles and Marli. 2 Jul Sks at Chatillon, Clamart, Issy, Meudon and Sevres. 3 Jul surrender of Paris; Le Quesnoy and the capture of Chalons. 9 Jul – 27 Aug blockade of Vincennes. 11 Jul – 30 Aug blockade of Bitsch. 18 Jul blockade of Montmedy. 20 Jul surrender of Sedan castle. 21 Jul surrender of Landrecy. 27 Jul surrender of Vincennes. 27 Jul – 10 Aug blockade of Laon. 6 Aug occupation of Angers. 9 Aug surrender of Philippeville. 10 Aug – 15 Sep 2nd blockade and surrender of Longwy. 16 Aug capitulation of Rocroy. 13/14 Aug storm of the Bourgogne redoubt at Longwy. 14/15 Sep storm of Medybas; the place capitulated on 19 Sep.

Appendix B: Alphabetical Index of the Regimental Colonels-in-Chief (Chefs) up to 1807

Albe, GM Georg Friedrich von der
Ahlemann, MG Johann Ernst von
Alvensleben, Achaz Oberst Heinrich von
Alvensleben, Oberst Ludolph August Friedrich von
d'Amandruez, GM Benjamin
Amstell, GM Georg Friedrich von,
Anhalt, GM Albert Graf zu
Anhalt, GL Heinrich Wilhelm von,
Anhalt, Oberst Friedrich Reichsgraf zu
Anhalt, GM Karl Philipp von Anhalt.
Anhalt-Dessau, Oberst Leopold Ludwig Graf zu
Anhalt-Bernburg, GM Franz Adolph Prinz von.
Anhalt-Dessau, GM Dietrich Prinz von
Anhalt-Dessau, GM Eugen Prinz von
Anhalt-Dessau, GM Gustav Erbprinz von
Anhalt-Dessau, Johann Georg Fuerst von
Anhalt-Dessau, Oberst Leopold Fuerst von
Anhalt-Dessau, GM Leopold Friedrich Franz Erbprinz von
Anhalt-Dessau, FM Leopold (II) Maximilian Fuerst von
Anhalt-Dessau Oberst Moritz Prinz von
Anhalt-Dessau, GL Wilhelm Gustav Erbprinz von
Anhalt-Koethen-Pless, GM Ferdinand Prinz von
Anhalt-Zerbst, Oberst Anton-Guenther Fuerst von
Anhalt-Zerbst, Oberst Christian August Prinz von

Anspach, Georg Friedrich Markgraf von
Apenburg, GM Levin Friedrich Gideon von
d'Arbaud, Oberst Peter
Arneburg, MG Hans Georg Woldeck von
Arnim, GM Alexander Wilhelm von
Arnim, GL Georg Abraham von
Arnim, Oberst Georg Christoph von
Arnstaedt, Oberst Adam Friedrich von.
Asseburg, Wilhelm GM Moritz von der.
Auer, Johann Casimir von
Auer, Oberst Jonas Casimir von

Backhoff, GM Karl von
Baden, GM Ludwig Prinz von
Balliodz, Oberst Abraham von
Bandemer, Col Christoph Friedrich von
Bandemer, Oberst Friedrich Asmus von
Bandemer, GM Joachim Christian von
Bardeleben, Oberst Christoph Carl Friedrich von
Bardeleben, Georg Friedrich von
Bardeleben, Oberst Hans Christoph von
Bardeleben, Chef Oberst Philipp Ernst von
Barfuss, Oberst Johann Albrecht von
Bauvrye, Obstlt Philipp Lambert de
Beaufort, Oberst von. Bechefer, GL Jakob von
Beckwith, GM Karl Friedrich von
Beeren, Oberst Carl Friedrich Hermann von
Belling, Oberstleutnant von
Belling, Oberst Hans Georg von.
Below, GM Lorenz Ludwig von.
Below, GM Matthias Wilhelm von
Below, Oberst Nikolaus von
Beneckendorf, Oberst von
Berg, GM Jakob Friedrich von
Berner, Oberst Christian Friedrich von
Berrenhauer, Oberst Sigismund August
Besser, GM Gottlieb Ehrenreich von
Beville, GM Gottlieb Ludwig von
Biberstein, Oberst August Friedrich Marschall von

Biberstein, GM Carl Gustav von
Bila, GM Ernst Christian von
Billerbeck, Oberst Heinrich Wilhelm von
Billerbeck, GM Konstantin von
Birckensee, Oberst Philipp Wolfgang Teuffel von
Bismark, MG August von
Bissing, Oberst Wilhelm Ludwig von
Blanckensee, GM Berendt Sigismund von
Blanckensee, GM Christian Friedrich von
Blanckensee, GM Peter von
Blanckenstein, GM Alexander Ernst von
Block, Oberst Karl Friedrich von
Bluecher, Oberst Gerhard Leberecht von
Blumenthal, Oberst Georg Ewald von
Bohlen, Oberst Balthasar Ernst von
Bohlen, GM Philipp Christian von
Bonin, Oberst Anselm Christoph von
Bonin, Oberst Bernd Eckard von
Bonin, GM Bogislav Ernst von
Bonin, Oberst Casimir Wedig von
Boerstell, Oberst Johann Heinrich von.
Borch, GM Otto Heinrich Friedrich von
Borck, Oberst Gottlieb Matthaeus Siegfried von
Borck, GM Adrian Berndt von
Borck, GM Christian Ernst Wilhelm von
Borck, GM Franz Andreas von
Borcke, Oberst Friedrich Adam von
Borck, Oberst Georg Heinrich von
Borck, GM Philipp Ernst von
Bornstedt, GM August Gottlieb von
Borstel, GM Hans Friedrich Heinrich von
Bornstedt, GM Dietrich Eugen Philipp von
Bornstedt, GM Hans Erich von.
Borstell, GM Hans Friedrich von.
Bose, Oberst Karl Ernst von.
Bosse, GM Friedrich Leopold von.
Bosse, GL Heinrich Guenther Gottfried von
Botzheim, Oberst Friedrich Casimir von
Boys, Oberst von.

Brandeis, GM Johann Christoph von.
Brandenburg, Albrecht Markgraf von
Brandenburg, Carl Markgraf von
Brandenburg, Carl Philipp Markgraf von
Brandenburg, Christian Ludwig Markgraf von
Brandenburg, Dorothea Kurfuerstin von
Brandenburg, Friedrich Kurprinz von
Brandenburg, Karl Emil Kurprinz von
Brandenburg, MG Philipp Wilhelm Markgraf von
Brandenburg-Baireuth, GL Carl Friedrich Alexander von
Brandenburg-Baireuth, Oberst Christian Ernst Markgraf von
Brandenburg-Baireuth, Oberst Friedrich Erbprinz (later Markgraf) von
Brandenburg-Baireuth, GL Friedrich Wilhelm Markgraf von
Brandenburg-Schwedt, Albrecht Markgraf von
Brandenburg-Schwedt, Oberst Friedrich Markgraf von
Brandenburg-Schwedt, GM Heinrich Markgraf von
Brandenburg-Schwedt, Heinrich Friedrich Markgraf von
Brandenburg-Schwedt, Oberst Karl Markgraf von
Brandenburg-Schwedt, Philipp Markgraf von.
Brandt, Oberst Paul von
Brandt, GL Wilhelm von
Braun, GM August Wilhelm von
Braun, Oberst Gottlob von
Braunschweig, Albrecht Prinz von
Braunschweig, Oberst Christian Friedrich von.
Braunschweig, GL Ferdinand Herzog von
Braunschweig, GL Friedrich August Herzog von
Braunschweig Oberst Friedrich Franz Prinz von
Braunschweig, GdI Karl Wilhelm Ferdinand Erbprinz von
Braunschweig, Oberst Wilhelm Ferdinand Erbprinz von
Braunschweig, Oberst Leopold Prinz (Herzog from 1780)
Braunschweig-Bevern, Oberst August Wilhelm Herzog von
Braunschweig-Bevern, Oberst August Prinz von
Braunschweig-Oels, Oberst Friedrich Wilhelm Prinz von
Brausen, GM Carl Wilhelm von
Bredow, GM Asmus Ehrentreich von
Bredow, GM Cuno Ernst von
Bredow, Oberst Friedrich Sigismund von
Bredow, GL Gottfried Albrecht von

Derschau, Oberst Christian Reinhold von
Derschau, Oberst Karl Friedrich von
Dessau, Leopold Prinz von
Dewitz, GL Friedrich Wilhelm von
Dewitz, Joachim Balthasar von
Dewitz, Oberst Otto von
Dewtz, GM Stephan von
Diericke, GM Christian Friedrich von
Diericke, GL Otto Friedrich von
Dieringshofen, Oberst Bernhard Alexander von
Dieury, Oberst Peter von
Dingelstaedt, Oberst August Levin von
Dittmar,Oberst Johann Wilhelm von
Doenhoff, GM Alexander Graf von
Doenhoff , GM Ernst Graf von
Doehnhoff, Gen-Lt Friedrich Graf von
Doehnhoff, Gen-Lt Otto Magnus Graf von
Dohna, GM Graf zu
Dohna, Oberst Alexander Burggraf zu
Dohna, Oberst August Graf zu
Dohna, GdI Christian Albrecht Graf zu
Dohna, GM Christoph Graf zu
Dohna, Oberst Dietrich, Graf zu
Dohna, Oberst Friedrich Ludwig Graf zu
Dohna, Oberst Wilhelm Alexander Graf zu.
Dolffs, Oberst Diedrich Goswin von
Dockum, Oberst Martin Arend von
Dossow, GL von (FM in 1756)
Dossow, Oberst Friedrich Wilhelm von
Driesen, GM Goerg Wilhelm von
Droessel, Oberst August Friedrich von der.

Eben und Brunnen, Oberst Carl August Baron von
Eckartsberg, GM Johann Ludwig von
Egeln, Oberst Friedrich von
Egloffstein, GM Baron Albrecht Dietrich Gottfried Graf von und zu
Eichmann, Oberst Martin Ludwig von
Einsiedel, Gottfired Emanuel von
Eller, GM von

Erlach, GM Friedrich August Freiherr von
Elsner, GM Carl Friedrich von
Esebeck, GM Christian Carl von
L'Estocq, Oberst Anton Wilhelm von
Eulenburg, Oberst Casimir Freiherr zu.

Falkenheyn, Oberst Friedrich Gotthelf von
Faergel, Oberst Johann von
Favrat, GL Franz Andreas von
Finck, GM Friedrich August von
Finck von Finckenstein, GFM Albrecht Konrad Graf
Fink von Finkenstein, GM Friedrich Ludwig Graf
Flans, GM Kurt Friedrich von.
Flemming, GM Hans Ludwig von
Forcade, GM Friedrich Wilhelm Quirin von
Forcade, Oberst Johann von
Frankenberg, Oberst Christian Alexander von
Frankenberg, GM Karl Ludwig Sylvirus von
Franckenberg, GM Carl Wolfgang
Froreich, GM Ulrich Karl von.

Gaetringen, GM Johann Rudolph Hiller von
Gabelentz, GM Georg Karl Gottlob von der
Gaudi, Oberst Friedrich Wilhelm Ernst von
Gentzkow, Oberst August von
Gersdorff, GM David Gottlieb von
Gessler,Oberst Friedrich Leopold von
Gettkandt, GM Ernst Philipp von
Gillern, GM Ernst Wilhelm von
Gilsa, GM Georg Ludwig von
Glasenapp, GM Kaspar Otto von
Glaser, Oberst Friedrich Daniel von
Glaubitz, GL Georg Rudolph von
Goecking, GM Friedrich Eberhard Siegmund Guenther von
Goetz, Oberst Henning von.
Goetz, GM Karl Ludwig Bogislav von
Goetzen, GM Friedrich Wilhelm von
Goetzen, Oberst Carl Ludwig von
Gohr, Oberst Ludwig von

Goltz, Oberst Georg Konrad Freiherr von der
Goltz, GL Christoph Heinrich von der
Goltz, MG Heinrich Leopold Graf von der
Goltz, Oberst Johann Wilhelm Graf von der
Goltz, GM Wilhelm Heinrich Freiherr von der.
Gotter, Oberst Heinrich Ludwig von.
Grabow, GM Christoph Heinrich
Grant, GM Johann von
Grape, Oberst Jakob Heinrich von
Grawert, Oberst Julius August von
Grevenitz, Oberst David Juergen
Grevenitz, Oberst Friedrich August von
Groeben, GM Georg Dietrich von der
Groeling Oberst Johann Benedikt von
Grolmann, Oberst Georg Arnold von
Groscreuz, Oberst Franz Thoro von
Gruenberg, Oberst Johann Ludwig von
Grumbkow, Oberst Friedrich Wilhelm von
Grumbkow, Oberst Joachim Ernst von
Guenther, Oberst Heinrich Johann von.

Hacke, Oberst Levin Friedrich von
Hackeborn, GM Wolf Christoph von
Hagen, GM Karl Friedrich Baron von
Hager, GM Eberhard von
Hagken, Oberst Christian Alexander von
Hakke, Col Johann Christoph Friedrich Graf von
Hallmann, Oberst Friedrich Sylvius von
Hamberger, GM Karl Friedrich von
Hamel, GM Marquis Franz du
Hanenfeldt, Oberst Nikolaus Reinhold von
Hanstein, Oberst Ernst Friedrich Karl von
Hartmann, Oberst Gottfried Ludwig von
Hasslocher, Oberst Johann Benjamin von
Hausen, GM Friedrich Wilhelm Heinrich von
Hautcharmoy, Obst Heinrich Karl Ludwig Herault de
Heiden, GM Fiedrich Freiherr von
Heiden, GL Johan Sigismund Freiherr von

Heising, Oberst Ludwig Ferdinand Friedrich von
Hellermann Oberst Volrath von
Henckel von Donnersmark, GL Graf
Henckel von Donnersmark, Oberst Elias Maximilian Graf
Henckel von Donnersmark, Oberst Viktor Amadeus Graf. Hertzberg
GM Johann Karl Graf von
Hertzberg, Oberst Johann Kaspar.
Herzberg, Oberst Friedrich Wilhelm Graf von.
Hessen-Cassel, GL Erbprinz Friedrich von
Hessen-Cassel, GM Prinz Georg von
Hessen-Cassel, GFM Landgraf Wilhelm von
Hessen-Darmstadt, Georg Wilhelm Prinz von
Hessen-Darmstadt, Oberst Ludwig Erbprinz von
Hessen-Philippsthal, Oberst Prinz Adolph von
Heuking, Oberst Ulrich von
Hueser, Oberst Heinrich Ernst von
Heyden, Oberst Ernst Friedrich von der
Heyking, GM Ulrich Leberecht von
Hoditz, Oberst Isidor Graf von
Hoffmann, Oberst Rudolph August von
Hohenlohe-Ingelfingen, Friedrich Ludwig Fuerst zu

Hohenlohe-Ingelfingen, GM Georg Prinz zu
Hohenstock, Oberst Carl August von
Holstein-Beck, Friedrich Karl Ludwig Herzog von
Holstein-Beck, Herzog Friedrich Ludwig von.
Holstein-Beck, GM Friedrich Wilhelm von.
Holstein-Gottorp, Oberst Georg Ludwig Herzog von
Holstein-Ploen, Herzog August von
Holtzendorf, Oberst Friedrich Jakob von
Holwede, Oberst Karl Friedrich von
L'Homme de Courbiere, Oberst Guillaume Rene de.
Horn, GM von.
Horn, GM Christian Sigismund von
l'Hospital, Oberst Ludwig de
Hoverbeck, Oberst Reinhold Friedrich Baron von
Huelsen, GM Johann Dietrich von
Hutten, Oberst.

Ingersleben, Oberst Friedrich de
Ilow, GM Otto Friedrich von
Ingersleben, Oberst Karl Ludwig von
Irwing, Oberst Friedrich Wilhelm von
Itzenplitz, GM August Friedrich von
Itzenplitz, GM Joachim Christian Friedrich von.

Jeetze, Oberst Adam Friedrich von
Jeetze, Oberst Joachim Friedrich von
Jeetze, Oberst Hans Christoph von.
Jung von Rothkirch, GM.
Jungken, GM Martin Eberhard von, known as Muenster von Mohrenstamm.

Kahlden, GM Hennig Alexander von
Kalinowa Zaremba, GM Michael Konstantin
Kalckreuth, Oberst Ernst Georg von
Kalckreuth, Oberst Friedrich Adolf Graf von
Kalckreuth, GM Johann Nikolaus von
Kalckreuth, GM Ludwig Gottlieb von
Kalckreuth, GM Samuel Adolph von
Kalckreuth, Oberst Wilhelm Heinrich Adolph von
Kalckstein, Oberst Christoph Wilhelm von
Kalckstein, GL Ludwig Karl von
Kalnein, GM Karl Eduard von
Kalsow, Oberst Christian Ludwig von.
Kamecke, GM Anton.
Kanitz, GM Hans Wilhelm von.
Kannenberg, GM Christoph von
Kannenberg, Chef Oberst Friedrich Wilhelm Freiherr von
Katte, GM Hans Friedrich von
Katte, MG Berend Christian von
Katte, GM Friedrich Heinrich von
Katte, Oberst and General Adjutant Hans Heinrich von
Katzler, GM Nikolaus Andreas von
Kauffberg, GM Friedrich August von
Keller, Oberst Georg Wilhelm Freiherr von
Keller, GL Johann Georg Wilhelm Freiherr von
Kenitz, GM Heinrich Gottloeb von

Keoszegy, Oberst Carl Franz von
Kleist, Oberst Alexander von
Kleist, Oberst Andreas Joachim von
Kleist, GL Franz Casimir von
Kleist, GM Franz Ulrich von Dieringshofen
Kleist, GM Friedrich Ludwig von
Kleist, Oberst Friedrich Wilhelm Gottfried Arend von
Kleist, GM Georg Friedrich von
Kleist, Oberst Heinrich Werner von
Kleist, Oberst Henning Alexander von
Kleist, Oberst Reimar von
Klinkwstroem, GM Karl Friedrich von
Klitzing, Oberst Karl Kuno Ludwig von
Knobelsdorff,GM Alexander Friedrich von
Knobelsdorff, GM Carl Ludwig von
Knobelsdorff, Oberst Kaspar Friedrich von
Knobloch, GM Karl Gottfried von
Knobloch, Oberst Dietrich Erhard von
Koehler, Oberst Georg Ludwig Egidius von
Koelichen, Oberst Hermann von
Koenen, Friedrich Wilhelm Frieherr von
Koenitz, Oberst Christian Ludwig von
Koenitz, Oberst Heinrich Gottlieb von
Koethen, Oberstleutnant Georg Bogislav von
Koschenbahr, Col Erich Julius von
Koschenbahr, GM Melchior Sylvius von
Kospoth, GM Ernst Christian von
Kowalsky, Oberst Georg Lorenz von
Kracht, Oberst Hillebrand von
Krafft, GM August Friedrich Erdmann von
Kreytzen, GM Georg Christoph von
Kreytzen, Oberst Johann Friedrich von.
Kroecher, GM von.
Krockow, Oberst Anton von
Krockow, GM Doering Wilhelm von
Krockow, GM Hans Kaspar von
Kroecher, Oberst Georg Volrath von
Kropff, GM Heinrich von
Krusemark, GM Hans Friedrich von

Kunheim, Col Johann Ernst von
Kunitzky, GM Franz Georg von
Kurland, Oberst Prinz Alexander von
Kurland, Prinz Ferdinand von
Kurssell, GM Heinrich Adolph von
Kyau, Oberst Friedrich Wilhelm Freiherr von.

Lange, Oberst Christian Henning
Langen, Oberst Karl Friedrich von
Langemann, GM Adolf Friedrich von
Langfeld, GL Christian August von
Larisch, GM Karl Leopold von
Larisch, Oberst Wilhelm Christoph von
Lattorff, GM Christian Friedrich von
Lattorf, GM Johann Sigismund von
Lattorf, Oberst Philipp Friedrich Lebrecht von
Lauenstein, Oberst Streiff von
Laujardiere, Oberst Wilhelm de
Laurens, Oberst Friedrich Gottlieb von
Laxdehnen, GM Otto Heinrich von
Lediwary, Oberst Carl Ludwig von
Lehmann, GM Johann Georg von
Lehwaldt, GFM Johann von
Lehwaldt, GM Wenzelslaus Christoph von
Leipziger, Oberst Heinrich Ernst von
Lengefeld, GM Christian August von
Lentius, GM Robert Scipio Freiherr von
Lentken, Oberst Johann Wilhelm von
Leoben, Oberst Kurt Lebrecht Freiherr von
Leps, Oberst Otto Friedrich von
Lestwitz, Oberst Hans Sigismund von
Lestwitz, GL Johann Georg von
Letmathe, GM von
Lettow, Oberst Ewald Georg von
Lettow, Major Heinrich Wilhelm von
Lettow, GM Karl Ernst von
Lichnowski, GM Stephan von
Lilien, GL von.
Linden, GM Christian Bogislav von

Lindstaedt, GM Daniel Georg von
Lochau, Oberst Alexander August von der
Loeben, GM Kurt Hildebrandt Freiherr von
Loellhoefel von Loewensprung, Oberst Friedrich Wilhelm von
Lossau, Oberst Matthias Ludwig von
Lossow, Oberst Daniel Friedrich von
Luck, Oberst Christoph Georg von
Luck, Oberst Kaspar Fabian Gottlieb
Luettwitz, GM Kaspar Siegmund von

Mackeroth, Oberst von
Mahlen, Oberst Johann Christoph von
Malakowski, Oberst Hyazinth Malachow von
Malakowsky, Oberst Paul Joseph Malachow von
Malschitzky, Oberst Johann von
Malschitzki, Oberst Peter Ewald von
Manstein, Oberst Christian Heinrich von
Manstein, GM Ernst Johann von
Manstein, GM Johann Wilhelm von
Manstein, Oberst Leopold Sebastian von
Mannstein, Oberst Wilhelm Dietrich von
Manteuffel, Oberst Franz Christoph von
Manteuffel, GM Gerd Heinrich von
Marwitz, Oberst Christian August von der
Marwitz, GM Friedrich Wilhelm von der
Marwitz, Oberst Heinrich Karl von der
Marwitz, GL Gustav Ludwig von der
Mauschwitz, GM Maximilian von
Meerkatz, Oberst Johann Friedrich von
Meerscheidt, GM Otto Kasimir von, known as Huellessem
Mellin, Oberst Henning Christian von
Mengden, MG Carl Friedrich Freiherr von
Merian, GM Rudolf von
Meyer, GMCarl Friedrich von
Meyerinck, GM Dietrich Richard von
Mitzlaff, Oberst Franz Gustav von
Moehring, Oberst Christian
Moellendorf, GM Johann Adolf von
Moellendorf, GL Wichard Joachim Heinrich von

Mohrenstamm, Oberst Martin Eberhard von
Jungkenn Muentzer von
Moller, Oberst Christian Friedrich von
Mosch, GM Christian Friedrich von
Mosel, GM Friedrich Wilhelm von der
Mosel, GM Konrad Wilhelm von der
Motte, Oberst Ernst August de la Chevallerie Baron de la
Motte Fouque, Oberst Heinrich August Baron de la
Moulin, Oberst Peter Ludwig von
Mueffling, Oberst Friedrich Wilhelm von
Muelbe, Oberst Christoph von der
Muenchow, Oberst Gustav Bogislav von
Muenchow, GM Lorenz Ernst von
Muetzschefahl, Oberst Friedrich Julius von

Nassau, GM Christoph Ernst Graf von
Nassau-Usingen, GM Prinz Johann Adolph von
Natalis, Oberst Johann de
Natalis, Chef Oberst Paul von
Natzmer, Oberst Georg Christoph von
Natzmer, Oberst Dubislav Gneomar von
Natzmer, Oberst Hans Christoph von
Nettelhorst, Oberst Georg Ernst
Noble, Oberst Francois de le
Normann, MG Carl Ludwig von
Normann, GM Georg Balthasar von

Oldenburg, GM Juergen Friedrich von
Oldenburg, Oberst Vollrad von
Oertzen, Chef GM Henning Ernst von
Oranien, GM Prinz Friedrich Ludwig von
Oranien-Fulda, GL Fuerst Wilhelm Friedrich von
l'Ostange, MG Charles Comte de
Osten, Oberst Friedrich von der
Overbeck, GM Theophil Ernst Freiherr von
Owstein, GM Karl Philipp von

Pannewitz, GL Anton von
Pannewitz, GM Gottlob Ernst von

Pannewitz, GM Ludolf von
Pannewitz, GM Maximilian Sigismund von
Pannewitz, Oberst Wolf Adolf von
Papstein, Oberst Carl Friedrich von
Pastau, GM Friedrich Ludwig von
Pelkowsky, GM Friedrich von
Pelchrzim, Oberst Karl von
Pennevaire, GM Peter Ernst von
Persod, Oberst Andreas Johann de
Petersdorff, Oberst Eggert Christian von
Pfalz-Bayern, Ludwig Kurfuerst von (1806 King of Bavaria)
Pfalz-Zweibruecken, LG Ludwig, reigning Herzog von
Pfau, GM Theodor Philipp von
Pfuhl, Oberst Ernst Ludwig von
Pfuhl, GM Karl Ludwig von
Pirch, GM Franz Otto von
Pirch, Oberst Georg Lorenz von
Platen, GM Dubislav Friedrich von
Platen, Oberst Hans Friedrich von
Platen, Oberst Leopold Johann von
Plettenberg, GM Christoph Friedrich Stephan von
Pletz, Oberst August Wilhelm von
Ploetz, Oberst Christian Franz Heinrich von
Ploetz, GM Karl Christoph von
Plotho, Obstlt Friedrich Franz Ernst Freiherr von
Podjurski, Oberst Carl von
Podewils, GM Friedrich Wilhelm von
Podewils, Oberst Otto Wilhelm von
Poellnitz, GM Ernst von
Pomeiske, Oberst Nikolaus Alexander von
Portail, Oberst Benjamin Hieronimus Courold du
Posadowsky, Oberst Carl Friedrich
Posadowski, GM Christoph Wilhelm Sigmund Freiherr von
Potthausen, Oberst Kaspar von
Preussen, Oberst August Wihelm Prinz von
Preussen, Oberst Ferdinand Prinz
Preussen, Friedrich Wilhelm Kronprinz von
Preussen, Heinrich Prinz von
Preussen, GM Louis Ferdinand Prinz von

Preussen, MG Ludwig Prinz von
Prittwitz, GM Joachim Bernhardt von
Prittwitz, Oberst Moritz von
Prittwitz, GM Wolf Moritz von
Pritz, GM Hans Samuel von
Pritzelwitz, Oberst Adam Heinrich von
Promnitz, GM Ulrich Graf von
Puttkamer, Oberst Christian Ernst von
Puttkamer, Oberst Georg Ewald von
Puttkamer, Oberst Georg Henning von
Puttkamer, Oberst Georg Ludwig von
Puttkamer, Oberstleutnant Martin Anton Freiherr von
Puttkamer, GM Nikolaus Lorenz von
Puttkamer, Oberst Peter Georg von
Puttkamer, Oberst Werner Friedrich von

Quadt (or Quaad) von Wickeradt, Oberst Friedrich Freiherr
Quadt von Wickeradt, GM Johann Christ Rulemann Freiherr
Queiss, GM Julius Dietrich von
Quitzow, Chef Oberst Christian Heinrich von

Raders, GM von
Radziwill, Fuerst Bogislav Prussian Regent
Ramin, GM Friedrich Ehrentreich von
Rampusch, GM Maximilien von
Raumer, GL Karl Albrecht Friedrich von
Rautern, GM Karl Friedrich von
Rebentisch, GM Johann Karl Freiherr von
Recke, Oberst Karl Friedrich von der
Reinhart, GM Joachim von
Reitzenstein, GM Christian Ludwig Rudolph von
Reitzenstein,Oberst Carl Erdmann von
Reitzenstein, Chef MG Heinrich August von
Renouard, GM Johann Jeremias von
Rentzel, GM Christoph Friedrich von
Reppert, Oberst Heinrich Sebastian von
Rettberg, GM Arnold Goswin von
Retzow, Oberst Wolf Friedrich von
Rhedern, Oberst Adam Valtin von

Rhein, Oberst Adrian Gottlieb von
Riedesel, GM Johann Volprecht Freiherr
Rochow, Oberst Friedrich Wilhelm von
Roeder, GFM von
Roeder, Oberst Christoph Ernst von
Roeder, GM Erhardt Ernst von
Roeder, GM Friedrich Adam Dietrich von
Rohdich, GM Friedrich Wilhelm von
Rohr, Oberst Albrecht Ehrentreich von
Rohr, Oberst Hans Ludwig von
Rohr, Oberst Kaspar Friedrich von
Romberg, GM Gisbert Wilhelm Freiherr von
Romberg, GM Johann Wilhelm Friedrich Moritz Freiherr von
Roeseler, Oberst von
Rosenbusch, Oberst Hans Christoph von
Rosieres, GM Franz Ludwig von
Rosen, GM Karl Gustav Freiherr von
Rosenbruch, GM Wilhelm Leopold von
Rothenburg, Oberst Friedrich Rudolf Graf von
Rothkirch, Oberst Hans Christoph von
Rothkirch, Oberst Friedrich Alexander von
Rudorff, Oberst Wilhelm Heinrich von
Ruechel, GL Ernst Friedrich Wilhelm von
Ruechel, Oberst Valentin Friedrich von
Ruesch, Oberst Joseph Theodor von
Ruits, GM Erdmann Ernst von
Ruits, GM Friedrich Leopold von
Russow, Oberst Graf von
Rutowsky, GM Friedrich August Graf von

Sachsen-Weimar, Oberst Carl August Herzog von
Sack, Oberst Sigismund von
Saher, GM Friedrich August von
Salenmon, GM Konstantin Natanael von
Saldern, GM Friedrich Christoph von
Saldern, GL Friedrich Christoph von
Saldern, GM Heinrich von
Salmuth, Oberst Friedrich Wilhelm von
Sanitz, Oberst Karl Wilhelm von

Sass, Oberst Gerhard Alexander Freiherr von
Sass, Oberst Johann Anton Gustav von
Sayn und Wittgenstein, August Graf von
Schenck, Oberst Johann Friedrich von
Schenck, GM Wilhelm Friedrich von
Schenkendorff, GM Balthasar Rudolf von
Schenckendorff, GM Friedrich August von
Schimmelpfennig von der Oye, Oberst Ludwig
Schimonsky, Oberst Dietrich Leberecht von
Schlaberndorff, GM Gustav Albrecht von
Schlaberndorff, GL Otto Freiherr von
Schladen, Oberst Karl Friedrich von
Schleinitz, Oberst Andreas Dietrich von
Schlichting, GL Samuel von
Schlieben, GM Friedrich Karl von
Schierstedt, GM August Ludwig von
Schlieffen, GL Martin Ernst von
Schliewitz, Oberst Christoph Rudolf von
Schlippenbach, GM Carl Friedrich Graf von
Schlitz, GL Graf von, known as Goertz
Schmettau, GM Carl Wilhelm Friedrich von
Schmettau, MG Johann Ernst von
Schoenaich, GM Georg Philipp Gottlob Freiherr von
Schoenaich-Carolath, GM Johann Carl Friedrich Erbprinz von
Schoenberg, GM Karl Graf von
Schoenebeck, GM Berend Christian von
Schoeneich, Georg Philipp Gottlob Freiherr von
Schoenermark, Oberst Georg Friedrich Wilhelm von
Schoenfeldt, GM Friedrich Wilhelm von
Schoenfeldt, GM Georg August von
Schoening, Ernst Sigismund von
Schoening, Oberst Hans Adam von
Schoening, Oberst Hans Ehrentreich von
Scholten, GM Johann Andreas Anton von
Schott von Schottenstein, GM Karl Adolph
Schulenburg, Oberst Achaz von der
Schulenburg, Oberst Adolf Friedrich von der
Schulenberg, Oberst August Ferdinand von der
Schulenburg, Obstlt Georg Werner von der

Schultz, Oberst Dietrich Wilhelm von
Schultze, GM Hans Caspar Ernst von
Schwandes, Oberst von
Schwartz, Oberst Karl August von
Schwendy, GM Johann Sigismund Freherr von
Schwerin, General Wachtmeister Bogislav von
Schwerin, Oberst Friedrich Albert von
Schwerin, Oberst Friedrich August Carl Leopold Graf von
Schwerin, Chef Oberst Friedrich Julius von
Schwerin, Oberst Karl Magnus
Schwerin, GM Kurt Christoph von
Schwerin, GM Philipp Albrecht von
Schwerin, Philipp Bogislav von
Schwerin, GM Reimar Julius von
Schwerin, GM Wilhelm Friedrich Karl Graf von
Seelhorst, Oberst Just Rudolf von
Seidlitz, GL Friedrich Wilhelm Graf von (the famous cavalry commander)
Selchow, Oberst Wolf Balthasar von
Sers, Oberst Peter de
Sers, Oberst Philipp Loth de
Sobeck, Oberst Karl Franz Freiherr von
Soldau, Oberst Carl Gustav von
Somoggy, Oberst Stephan von
Sonsfeld, Oberst Friedrich Otto Freiherr von
Stechow, GM Christoph Ludwig von
Stechow, GM Johann Ferdinand von
Stechow, Oberst Kaspar Heinrich von
Steensen, GL Georg von
Steinkeller, Oberst Anton Abraham von
Steinwehr, GM Friedrich Wilhelm von
Steinwehr, GM Johann Christian Wilhelm von
Stille, Chef GM Christoph Ludwig von
Stille, GM Ulrich Christoff von
Stoch, GM Friedich von
Stockhausen, Oberst Johann Friedrich Gustav von
Stojentin, GM Peter Heinrich von
Strachwitz, 1803 GM Ludwig von
Stranz, GM Hans Carl Ludolf von
Strauss, Oberst Christoph von

Stutterheim, GM August Ludwig von
Stutterheim, GL Joachim Friedrich von
Stutterheim, GM Otto Ludwig
Stwolinsky, Oberst Sylvius Ferdinand von
Suter, GM Friedrich Wilhelm von
Syburg, GM Friedrich Wilhelm von
Sydow, Oberst Balthasar Friedrich von
Sydow, Oberst Egidius Ehrenreich von
Sydow, Oberst Gustav Adolf von
Sydow, Oberst Hans von
Sydow, Oberst Hans Sigismund
Szekeli, Oberst Michael von

Tauentzien, GL Bogislav Emanuel von
Tauentzien, GL Bogislav Friedrich von
Tempelhoff, Oberst Georg Friedrich von
Tettau, Oberst Daniel von
Tettenborn, GM Hans von
Thadden, GM Leopold von
Thadden, GM Georg Reinhold von
Thiele, GM Alexander Heinrich von
Thiele, GM Friedrich Wilhelm von
Thiele, Oberst Martin von
Thuemen, GM Christoph Friedrich von
Thuena, Oberst August Wilhelm von
Thun, GM Otto Balthasar von
Tiedemann, Oberst Karl Eduard von
Trenck, GM Friedrich Ludwig von der
Tresckow, GM Joachim Christian von
Treskow, GM Johann Otto von
Tresckow, GM Karl Peter von
Treuenfels, Oberst Karl von
Troschke, Oberst Karl Ludwig Gottlieb von
Trossel, GM Etienne du
Truchsess von Waldburg, GM Wolf Christoph Freiherr
Truchses zu Waldburg, Carl Ludwig Graf von
Truchses zu Waldburg, Oberst Friedrich Sebastian Wunibald Graf
Tschammer, Oberst Friedrich Wilhelm Alexander von
Tschepe, Oberst Karl von

Tschirschki, GM Carl Wilhelm von
Tuempling, Oberst Georg Wolf von

Uchtlander, GM Christian Gottfried von
Unruh, Oberst Karl Philipp von
Usedom, Oberst Adolf Detlof von
Usedom, Oberst Friedrich von

Varenne, Oberst Jacques L'Aumonier Marquis de
Varenne, Oberst Friedrich Wilhelm Marquis de
Vasold, GM Heinrich Rudolf Wilhelm von
Vietinghoff, GM August Wilhelm Freiherr von
Vietinghoff, Chef Oberst Detlof Freiherr von
Vippach, Oberst Heinrich Sigismund von
Voigt, Oberst Johann August von
Voss, GM Karl von
Voss, GM Ludwig Ernst von
Voyne, Oberst Andreas Reveillas du

Wachholtz, Oberst von
Wachholtz, Oberst Kaspar Jochen von
Wagenfeld, MG Ernst Philipp von
Waldburg, Oberst Friedrich Sebastian Wunibald Graf zu
Waldburg, MG Carl Friedrich Ernst Graf zu
Waldow, Oberst Arnold Christoph von
Waldow, Oberst Friedrich Sigismund von
Walrave, GM Gerhard Kornelius von
Wangenheim, GL von
Wangenheim, GM Friedrich Wilhelm von
Warnery, Oberst Carl Emanuel von
Wartenberg, Oberst Hartwig Carl von
Wartensleben, FM Alexander Hermann Graf von
Wartensleben, GM Alexander Leopold Graf von
Wartensleben, Oberst Herman Graf von
Wechmar, Oberst Ludwig Anton Graf von
Wedell, Oberst Johann Heinrich von
Wedell, Oberst Karl Alexander von
Wedell, GM Karl Heinrich von
Wegnern, Oberst Georg Friedrich von

Weitersheim, GM Leopold Friedrich Ludwig von
Wendessen, GM Balthasar Ludwig Christian von
Wensen, Oberst Georg Joachim von der
Werdeck, Oberst Ernst Friedrich von
Werner, Oberst Johann Paul von
Werther, GM Philipp August Wilhelm von
Werther, Oberst Georg Christian von
Weyher, GM Adam von
Weyher, GM Christian Rudolf von
Wied zu Neuwied, Oberst Franz Karl Ludwig Graf
Wiersbitzki, Oberst Georg Ludwig von
Wildau, GL Franz Ludolf Ferdinand von
Winning, Oberst Christian Ludwig von
Winterfeld, Oberst Georg Levin von
Winterfeldt, GL Hans Karl von
Winterfeld, GM Karl Ludwig von
Winterfeld, Oberst Kaspar Dietlof von
Wittenhorst zu Sonsfeld, Oberst Friedrich Wilhelm Graf von
Wobeser, Oberst Georg Bogislav von
Wobeser, Joachim Wocislaus von
Woldeck, GM Alexander Friedrich von
Woldeck, GM Hans Christoph von
Wolki, Oberst Friedrich Gideon von
Wolffersdorff, GM Karl Friedrich von
Wolframsdorf, Oberst Adam Heinrich von
Wolfrath, Oberst Erich Magnus von
Wreech, Oberst Adam Friedrich von
Wreech, Oberst Joachim Friedrich von
Wulffen, Oberst Georg Ludolf von
Wunsch, GM Johann von
Wuerttemberg, Oberst Eugen Friedrich Heinrich Prinz von
Wuerttemberg, Karl Eugen Erbprinz von

Wuerttemberg, Marie-Auguste Herzoegin von
Wuerttemberg, Oberst Friedrich Prinz von
Wuerttemberg, Oberst Ludwig Prinz von
Wuerttemberg, GM Prinz Ludwig Alexander von
Wuerttemberg-Stuttgard, Carl Alexander Herzog von
Wuerttemberg-Stuttgard, Friedrich Eugen Herzog von
Wutginau, Oberst Hans Leonhard von
Wuthenow, Oberst Friedrich Wilhelm von
Wylich and Lottum, Oberst Friedrich Albrecht Carl Hermann, Graf von
Wylich und Lottum, GM Friedrich Wilhelm Graf von
Wylich und Lottum, Oberst Johann Christoph Graf von
Wylich und Lottum, Chef Oberst Ludwig Graf von
Wythenow, GM Heinrich Jordan

Yorck, Major Johann David Ludwig von (later Feld Marshal, Graf von Wertenberg

Zabeltitz, GM Johann Ernst von
Zastrow, GM Berend Asmus von
Zastrow, Oberst Friedrich Wilhelm von
Zastrow, GM Karl Anton Leopold Freiherr von
Zastrow, Oberst Jakob Ruediger von
Zastrow, Oberst Johann Wenceslaus Freiherr von
Zenge, Oberst August Wilhelm von
Zeuner, GM Karl Christoph von
Zieten, GM Christian Wilhelm von
Zieten, Oberst Hans Joachim von
Zieten (the legendary Prussian cavalry commander), Oberst Johann von
Zimmernow, Oberst Joachim Ernst von
Zitzewitz, GM Johann Bogislav von
Zitzewitz, GM Peter Christoph von
Zweiffel, Oberst Johann Christian von

Appendix C: Prussian Orders of Battle 1792-1815

THE PRUSSIAN ARMY AT VALMY IN 1792.

Commander: Field Marshal the Duke of Brunswick

Avantgarde GL Erbprinz von Hohenlohe-Ingelfingen.

1st Brigade. GM von Wolffradt. IR Hohenlohe Nr 32 (3 bns).

2nd Brigade. GM Graf von Hertzberg. IR von Kleist Nr 12 (3 bns).

3rd Brigade. GM von Kleist. IR von Forcade Nr 10 (1 bn), Fus Bn von Renouard Nr 2 (1 bn), Feldjaeger (5 coys), DragR von Schmettau Nr 2 (5 sqns), HusR von Wolffradt Nr 6 (10 sqns), 1 FAB (8 guns), 1 HAB (8 guns), plus 14 x 3 pdr regimental guns.

The First Rank FM The Duke of Brunswick.

1st Division GL von Schoenefeld.

1st Brigade. GM von Romberg. IR von Braunschweig (3 bns), IR von Woldeck Nr 41 (3 bns).

2nd Brigade. GM von Borch. IR von Thadden Nr 3 (3 bns), IR von Romberg Nr 10 (3 bns).

Artillery: 24 x 3 pdr regimental guns.

2nd Division GL von Budberg.

1st Brigade. GM von Thadden. IR von Budberg Nr 9 (3 bns), IR von Schoenfeld Nr 30 (3 bns).

Artillery: 36 x 3 pdr regimental guns.

The Second Rank GL l'Homme de Courbiere.

1st Division GL von Koenitz.
1st Brigade. GM von Vietinghof. IR von Koenitz Nr 39 (3 bns), IR von Borch Nr 31 (3 bns).
2nd Brigade. GM Crown Prince Friedrich Wilhelm von Preussen. IR Graf Hertzberg Nr 47 (3 bns).
2nd Division GM von Wolframsdorf.
1st Brigade Colonel Ludwig, Prinz von Baden. IR von Wolframsdorf Nr 37 (3 bns), IR von Vietinghof Nr 38 (3 bns).
Artillery: 30 x 3 pdr regimental guns.
Cavalry GL Graf von Lottum .
1st Division GL Graf von Kalckreuth.
1st Brigade GM Herzog von Sachsen-Weimar. KuerR von Weimar Nr 6 (5 sqns).
2nd Brigade GM von Normann. DragR Ansbach-Bayreuth Nr 5 (10 sqns), DragR von Normann Nr 4 (5 sqns).
2nd Division GL Ludwig Prinz von Wuerttemberg.
1st Brigade GM von Tschirsky. DragR von Lottum Nr 11 (5 sqns).
2nd Brigade GM von Ilow. KuerR von Ilow Nr 7 (5 sqns).
The Corps of GM von Koehler
Fues Bn von Mueffling Nr 18, Fues Bn von Ernest Nr 19, HusR von Koehler Nr 3 (10 sqns), _ HAB (5 guns), and 2 x 3 pdr regimental guns.
The Corps of GM von Eben
Fues Bn von Schenke Nr 1, Fues Bn von Legat Nr 20, HusR von Eben Nr 2 (10 sqns), _ HAB (4 guns) and 2 x 3 pdr regimantal guns.
Reserve Artillery Colonel von Tempelhof. 2 x 10 pdr mortar batteries (8 pieces each), 2 x 12 pdr gun batteries (8 pieces each), 7 x 6 pdr gun batteries (8 pieces each).
Totals: 35,000 men, 104 field pieces, 108 regimental pieces.
Comment. The Duke of Brunswick was not confident that he could defeat the two French armies (*Armée du* Nord and *Armée du Centre*) opposing him here. They totalled some 50,000 infantry, 2,000 cavalry, with 60 field guns, plus the light regimental pieces. The Prussian troops had outrun their supply trains and morale was low due to the appalling weather. After a largely ineffective artillery bombardment, the Prussian commander decided to withdraw. Paris (and the Revolution) were saved and lived to fight another day. Losses in this non-action were some 300 French to 200 Prussians.

THE PRUSSIAN-SAXON ARMY AT KAISERSLAUTERN ON 28-30 NOVEMBER 1793.
Commander FM Herzog Karl Wilhelm Ferdinand von Braunschweig.

Avantgarde. Colonel von Szeckely. *Saxon HusR* (2 sqns), Combined Cavalry Regiment (5 sqns), IR von Vietingenhof Nr 38 (1 bn).
1st Division GL von Kalckstein.
IR Prinz Heinrich Nr 35, IR von Braunschweig Nr 21 (3 bns each).
GM von Roeder. IR Garde Nr 15 (2 bns), Gren Bn Rohdich.
Cavalry. GL von Schoenfeld. KuerR von Borstell Nr 7, DragR von Lottum Nr 1 (5 sqns each), Prussian Jaegers (1 coy), *Kur-Trier Jaegers* (1 coy), IR von Crousaz Nr 39 (3 bns), Fues Bn Legat Nr 20, HusR von Eben Nr 2 (2 sqns).
Artillery: 1 x HAB, 1 FAB (8 guns each).
2nd Division. GL Graf von Kalckreuth.
Fues Bn Wedel Nr 1, IR von Kalckstein Nr 5 (3 bns), IR Garde Nr 15 (2nd Bn), IR von Braunschweig Nr 21 (3 bns), IR von Knobelsdorff Nr 27 (3 bns), IR von Vietinghof Nr 38 (2 bns), Gren Bn von Rohdich.
Cavalry. DragR Voss Nr 11 (5 sqns).
Attached Saxon troops.
IRs Kurfuerst, Prinz Anton, Prinz Clemens, Prinz Gotha (1 bn each).
Cavalry. Karabiniers, Leib-Kuerassier-Regiment, Chevaulegers-Regiment Kurland (5 sqns each)
Artillery: 1 x 12 pdr FAB (8 guns), 4 x 6 pdr FAB (26 guns), 1 x howitzer battery (8 pieces).
Total About 26,000 men. Each battalion had two 3 pdr regimental guns.
Comment. The allies convincingly defeated a French attempted invasion of Germany here this day. They inflicted casualties of 2,400 killed and wounded, 700 captured, 2 guns and a colour on General Hoche's 35,000-strong *Armée de la Moselle* for a loss of 44 officers and 762 men on their own side. The Prussian king forbade any pursuit, thus no fruits of this victory were reaped.

THE PRUSSIAN-SAXON ARMY AT KAISERSLAUTERN ON 23 MAY 1794. FM GRAF MOELLENDORF COMMANDING

GL von Kalckreuth.
Prussian IR von Crousaz Nr 39 (1st and 2nd Bns), IR von Borch Nr 49 (3 bns), Gren Bn von Hertzberg.
Saxon IRs von Lindt (1st Bn), von Langenau (2nd Bn), Prinz Maximilian (1st Bn), Prinz Xavier (2nd Bn).
Prussian Cavalry. DragR Anspach-Bayreuth Nr 5 (3 sqns).
Saxon Cavalry. KuerR Kurfuerst, ChevaulegerR Prinz Albrecht (4 sqns each), HusR (2 sqns).
Artillery: 2 x FAB, _ HAB 1 Saxon howitzer battery.
GL von Romberg.
Prussian IRs von Braunschweig Nr 21, Prinz Heinrich Nr 35, von Koethen Nr 48 (3 bns each), Jaegers (1 coy), HusR von Eben Nr 2 (3 sqns).
Artillery: 1 x FAB.
GL von Knobelsdorff.
Prussian IRs von Thadden Nr 3, von Kalckstein Nr 5, von Kleist Nr 12 (3 bns each), Fues Bn von Thadden Nr 13, Jaegers (1 coy), HusR von Eben Nr 2 (2 sqns).
Artillery: 1 x FAB.
Cavalry GL Prinz Louis von Wuerttemberg.
Kuer R Nr 3, KuerR von Borstell Nr 7, Drag R von Lottum Nr 1 (5 sqns each).
Artillery: 1 x HAB.
GL von Ruechel.
IRs von Ruechel Nr 30, von Wolframsdorff Nr 37 (3 bns each), Fues Bn von Ernest Nr 19, Jaegers (3 coys), DragR von Voss Nr 11 (3 sqns), HusR von Eben Nr 2 (3 sqns).
Artillery: 2 x FAB, 1 x HAB.
GL von Kleist.
IRs von Knobelsdorff Nr 27, Prinz Ferdinand Nr 34 (3bns each); Drag R von Voss Nr 11 (3 sqns).
Artillery: 1 x FAB.
Total 46,000 men.

Comment. Once again, a French invasion attempt by General Ambert, with 5,000 men of the *Armée du Rhin* was frustrated. The French lost 1,000 men, 17 guns and 2 colours. The allies lost about 150 men. There was no follow up by the allies.

THE PRUSSIAN CONTINGENT AT KAISERSLAUTERN ON 17-20 SEPTEMBER 1794.
General der Infanterie Prinz von Hohenlohe commanding.

IRs von Manstein Nr 9, von Romberg Nr 10, von Kunintzky Nr 44 (3 bns each); Fus Bns von Martini Nr 10, von Bila Nr 2 and von Mueffling Nr 18 (1 bn each); Jaegers (5 coys); Drag R von Katting Nr 4 and von Schmettau Nr 2, Hus R von Wolffradt Nr 6 and von Bluecher Nr 8 (5 sqns each).
Artillery not known. Contingents of Austrian and Hessen-Darmstadt troops were also present.
Comment. French General Schaal, with 16,000 men of the *Armée du Rhin*, made another thrust at Germany; they were again overthrown with a loss of over 4,000 men, 4 guns and 3 colours. Allied losses were about 1,000 men. There was no pursuit of the beaten foe. At the end of the campaigning season, Prussia withdrew from the allied coalition on the Rhine to conolidate her new possessions in the east.

THE PRUSSIAN-SAXON ARMY CORPS AT JENA ON 14 OCTOBER 1806.
General der Infanterie Prinz von Hohenlohe commanding.

Regiments in italics are Saxon.
Avantgarde GM von Schimmelpfennig.
GM von Pelet. Fues Bns von Rabenau Nr 13, von Pelet Nr 14 and von Ruehle Nr 15, Jaegers (2 coys).
Hus R von Schimmelpfennig Nr 6 (10 sqns).
Artillery: 1 x HAB, 1 x FAB 6 guns each.
GM von Truetzschler. Hus R (8 sqns).
GM von Bevilaqua. IR von Mueffling Nr 49 (2 bns), *IR Prinz Clemens, IR Kurfuerst (2 bns each).*
Artillery: 1 x 6 pdr FAB (6 guns).
The Left Wing *General der Kavallerie von Zeschwitz.*
GL von Niesemeuschel.

GM von Burgsdorff. IR von Thuemmel, IR Prinz Xavier (2 bns each), IR Prinz Friedrich August (2 bns)
Artillery: 2 x 8 pdr FABs (16 guns).
GM von Dyherrn. IR von Low and von Niesemueschel (2 bns each), IR von Bevilaqua (2nd Bn).
Artillery: 1 x 12 pdr FAB (10 guns).
GL von Kochtitsky. Carabiniers, Kuer R von Kochtitsky (4 sqns each), Chevaulegers R Prinz Albert (4 sqns).
Artillery: 1 x HAB (10 guns).
GL von Polenz. Fues Bn Boguslawsky Nr 22, *Chevaulegers R von Polenz (4 sqns).*
Artillery: _ HAB (5 guns).
Left Wing total: 8,200 infantry, 1,250 cavalry, 41 guns, 300 gunners.

Left Flank Guard. GM von Tauentzien.
GM von Zweiffel. IR von Zweiffel Nr 45 (2 bns), Gren Bn von Herwarth, GM von Schoeneberg. Gren Bn a. d. Winkel, IR von Rechten (2 bns), IR Prinz Maximilian (2 bns).
Artillery: 1 x howitzer battery (6 guns).
GM von Bila. Jaegers (2 coys), Fues Bn Rosen Nr 7, *Chevaulegers R Prinz Clemens (4 sqns).* Hus R von Bila Nr 11 (5 sqns).
Left Flank Guard total: 4,500 infantry, 1,800 cavalry, 6 guns, 70 gunners.

The Right Wing. GL von Grawert.
GM von Mueffling. Gren Bn Hahn, IRs von Hohenlohe Nr 32 and von Sanitz Nr 50 (2 bns each)
Artillery: 1 x 12 pdr FAB (6 guns).
Colonel von Schimonsky. Gren Bn Sack, IRs von Zastrow Nr 39 and von Grawert Nr 47 (2 bns each).
Artillery: 1 x 12 pdr FAB (6 guns).
GL von Holtzendorff.
GM von Henkel. Kuer Rs von Henkel Nr 1 and von Holtzendorff Nr 9, Drag R von Krafft Nr 11 (5 sqns each).
Artillery: 1 x 6 pdr HAB (6 guns).
Colonel von Erichsen. Fues Bn von Erichsen Nr 10, Hus R von Gettkandt Nt 1 (10 sqns).
Artillery: _ x 4 pdr HAB (4 guns).
Right Wing total: 7,500 infantry, 1,900 cavalry, 22 guns, 230 gunners.

The Reserve. GL von Prittwitz.
GM von Sanitz. Gren Bns von Borcke, Dohna, Kollin and Losthin (3 _ bns),
Artillery: 1 x 12 pdr FAB (8 guns).
GM von Cerrini. Gren Bns Hundt, Le Coq, Lichtenhayn, Thiollaz, Metsch (5 bns).
Artillery: 1 x 6 pdr FAB (6 guns).
GM von Krafft. Drag R von Prittwitz Nr 2 (5 sqns), *Chevaulegers R Prinz Johann (4 sqns).*
Artillery: 1 x 6 pdr HAB (7 guns).
Reserve totals. 6,900 infantry, 1,000 cavalry, 23 guns, 190 gunners.

GL von Ruechel's Corps (arrived very late in the battle).
Avantgarde. GL von Winning.
Fues Bns von Kayserlink Nr 1, von Bila Nr 2, Jaegers (2 coys), Hus R von Pletz Nr 3 and von Koehler Nr 7 (5 sqns each)
Artillery: 1 x 6 pdr HAB (8 guns), 1 x 6 pdr FAB (8 guns).
GM von Wobeser. Fues Bn von Ernest Nr 19, Jaegers (1 coy), Drag R von Wobeser Nr 14 (5 sqns).
Artillery: _ x 4 pdr HAB (4 guns).
Corps de Bataille. Herzog von Sachsen-Weimar.
IR von Schenck Nr 9 and von Winning Nr 23 (2 bns each), Gren Bn von Borstell.
IRs von Treuenfels Nr 29 and von Strachwitz Nr 43 (2 bns each), Gren Bn von Hellmann.
IRs von Wedell Nr 10 and von Tschepe Nr 37 (2 bns each), Fues Bn von Sobbe Nr 18.
Kuer R von Bailliodz Nr 5, Drag Rs Koenig von Bayern Nr 1 and von Katte Nr 4 (5 sqns each).
Artillery: 2 x 6 pdr FABs, (16 guns), _ x 6 pdr HAB (4 guns).
Corps de Bataille totals 15,000 men, 20 guns.

Comment. Some 54,000 men out of the 96,000 under Napoleon's command of the *Grande Armée* actually came into action here, at the amazingly successful conclusion of his legendary march round the Prussian left flank.

In this crushing defeat, the allies lost the vast majority of their artillery as well as much of their manpower. 30 Prussian officers were killed or mortally wounded as were 19 Saxons. 168 Prussian officers were wounded as were 95 Saxons. Due to the scale of the devastation, it is not possible to be exact as to numbers of soldiers lost here this day, but 27,000 is usually accepted. 66 field artillery pieces were taken in addition to the regimental guns. Twelve colours and standards were also taken as was most of the baggage train.

The French lost under 7,000 men at Jena.

THE PRUSSIAN ARMY MAIN BODY AT AUERSTAEDT ON 14 OCTOBER 1806.
King Friedrich Wilhelm III commanding through FM Herzog Karl Wilhelm Ferdinand von Braunschweig.

Avantgarde. GM von Bluecher.
GM von Oswald. *Fues Bn von Sachsen-Weimar*, Fues Bn Greifenberg Nr 4, von Oswald Nr 16, von Kloch Nr 18.
Artillery: 1 x FAB (6 guns).
GM von Bluecher. Hus Rs Wuerttemberg Nr 4 and von Bluecher Nr 8 (10 sqns each), Drag R von Irwing Nr 3 (5 sqns).
Artillery: 1 x HAB (8 guns).
Avantgarde total: 2,350 infantry, 3,500 cavalry, 14 guns, 150 gunners.
The Left Wing. GL von Schmettau.
GM von Alvensleben. IRs Prinz Heinrich Nr 35, von Alvensleben Nr 33 (2 bns each), Gren Bn von Schack.
Artillery: 1 x FAB (6 guns).
GM von Schimonsky. IRs von Schimonsky Nr 40 and von Malschitsky Nr 28 (2 bns each), Gren Bn von Krafft.
Artillery: 1 x FAB (6 guns).
GM von Irwing. Drag R Koenigin Nr 5 (10 sqns).
GM von Bunting Kuer Rs von Bunting Nr 12 and von Heising Nr 8 (5 sqns each).
Artillery: 1 x HAB (6 guns).
Left Wing totals: 8,600 infantry, 2,700 cavalry, 18 guns, 200 gunners.

The Centre. GL von Wartensleben.
GM von Wedel. Gren Bn von Hanstein, IRs von Renouard Nr 3 and von Kleist Nr 5 (2 bns each).
Artillery: 1 x FAB (6 guns).
GM von Renouard. Gren Bn Alt-Braun, IRs Prinz Louis Nr 20 and von Braunschweig Nr 21 2 bns each).

Artillery: 1 x FAB (6 guns).
GM von Quitzow. Kuer Rs von Quitzow Nr 6 and von Reitzenstein Nr 7 (5 sqns each).
Artillery: 1 x HAB (6 guns).
The centre totals: 8,300 infantry, 1,800 cavalry, 18 guns, 200 gunners.

The Right Wing. GL Wilhelm Prinz von Oranien.
Colonel Heinrich Prinz von Preussen. Gren Bn Rheinbaben, IRs Prinz Ferdinand Nr 34 and von Puttkammer Nr 36 (2 bns each)
Artillery: 1 x FAB (6 guns).
Colonel von Luetzow. Gren Bn von Knebel, IRs von Moellendorf Nr 25 and von Wartensleben Nr 59 (2 bns each).
Artillery: 1 x FAB (6 guns).
Lieutenant-Colonel Wilhelm Prinz von Preussen. Kuer Rs Leibkarabiniers Nr 12 and Garde du Corps Nr 13 (5 sqns each).
Artillery: 1 x HAB (6 guns).
Right Wing totals: 7,300 infantry, 1,700 cavalry, 18 guns, 200 gunners.

The Reserve. GL Graf von Kalckreuth.
1st Division. GL von Kuehnheim.
Colonel von Pletz. Gren Bns Prinz August and Rabiel, IR Koenig Nr 18 (2 bns).
Artillery: 1 x howitzer battery (6 guns).
GM von Hirschfeld. Grenadier-Garde Nr 6 (1 bn), IR Garde Nr 15 (2 bns).
Artillery: _ x FAB (3 guns).
Kuer Rs von Beeren Nr 2 and Gens d'armes Nr 10 (5 sqns each).
Artillery: 1 x HAB (6 guns).
2nd Division. GL von Arnim.
GM von Malschitsky. Gren Bns von Schlieffen and von Hulsen, IR von Zenge Nr 24 (2 bns).
Artillery: 1 x FAB (8 guns).
Reserve totals: 13,000 infantry, 2,550 cavalry, 23 guns, 200 gunners.
Prussian-Saxon totals: 39,550 infantry, 12,250 cavalry, 163 guns, 950 gunners.

Comment.

Marshal Davout's III Corps single-handedly defeated the main body of the Prussian-Saxon army here this day. The Duke of Brunswick was fatally wounded early in the battle; the king's failure to appoint a successor for some hours undoubtedly contributed greatly to the catastrophe. In addition, GL Graf von Schmettau, GMs von Quitzow and von Greiffenberg, 7 field officers and 36 junior officers also lost their lives. FM von Moellendorf, 5 other generals, 34 field officers and 181 junior officers were wounded. Without counting regimental pieces, 57 field guns were lost as were two colours of the IR von Puttkammer Nr 36. Bodart quotes French figures of 10,000 allies killed and wounded and 3,000 captured. Davout gave the losses of his corps as 7,000 men. The Prussian army literally fell to pieces after these twin defeats.

THE PRUSSIAN CORPS AT LUEBECK ON 6 NOVEMBER 1806.
GL von Bluecher commanding.

Avantgarde. GM von Usedom. Hus R von Usedom Nr 10 (10 sqns), GM von Oswald. Hus R von Bluecher Nr 8 (10 sqns), the remnants of some fusilier battalions and of Gren Bns von Schmeling and Vieregg.
Artillery: _ HAB.
1st Division. GM von Natzmer. IRs von Tschammer Nr 27, Kauffberg Nr 51, Natzmer Nr 54 and von Manstein Nr 55 (2 bns each), Drag R von Hertzberg Nr 9 (5 sqns).
Artillery: 1 x FAB, _ HAB.
2nd Division. GM von Larisch. IRs von Kalckreuth Nr 4, von Austein Nr 7, von Braunschweig Nr 21, Jung-Larisch Nr 53 (2 bns each), Drag R von Heyking Nr 10 (5 sqns).
Artillery: 1 x FAB.
Totals: 17 battalions, 30 squadrons, 12 guns, ca 15,000 men.

Comment. Following the disaster of Auerstaedt, Bluecher was one of those very few commanders of steely resolve, who motivated their men and held them to the colours. Gathering up other units *en route*, he fought his way north to the Baltic coast, hoping to escape the clutches of the French I and IV Corps, some 30,000 men with 90 guns, under Bernadotte. Bluecher stormed into Luebeck (a neutral Hanseatic city) hoping to requisition food, clothing and transport. He was attacked by Bernadotte and beaten off, losing some 2,000 killed and wounded and 4,000 captured. The French lost about 2,000 men. Now cornered, and out of ammunition and food, von Bluecher was forced to capitulate at Ratkau with the remainder of his force on 7 November.

THE PRUSSO-RUSSIAN CORPS AT PREUSSISCH-EYLAU ON 7/8 FEBRUARY 1807. LG von l'Estocq commanding.

Avantgarde. Drag R von Auer Nr 6 (10 sqns), Towarczys (1/2 sqn).
Artillery: 1 x HAB (6 guns).
1st Division. GM von Diericke.
Gren Bn von Fabecky, IR von Ruechel Nr 2 (2 bns), Kuer R von Wagenfeld Nr 4 (4 sqns), Drag R von Baczko Nr 7 (5 sqns).
Artillery: _ x HAB (3 guns).
2nd Division. GM von Rembow. IR von Schoening Nr 11 (2 bns), Gren Bn von Schlieffen.
3rd Division. GM von Auer. *Russian IR Wyborg (3 bns), 21st Russian Jaegers (3 bns), _ HAB*, two Fues coys. Towarczys (9 _ sqns).
Prussian totals: 12 _ bns, 29 sqns, 9 guns. Ca 5,584 men.
Comment. The Prussian corps arrived late on the field, having had to march quite a distance to get there. They were just in time to stabilise the Russian left wing and thus prevent a disaster. They lost about 800 casualties in the battle; the Russians lost 18,000 killed and wounded, 3,000 captured, 24 guns and (according to French claims) up to 24 colours. Napoleon losses are shrouded in controversy. Bodart gives '23,000 men, 5 eagles, 7 colours'. This topic is still hotly debated today. The battered allies withdrew to the northeast; all the bloodshed achieved nothing.

THE PRUSSIAN CORPS AT HEILSBERG ON 10 JUNE 1807. LG von l'Estocq commanding.

Avantgarde. GM von Stutterheim. Hus R von Pletz Nr 3 (4 sqns), Hus R Wuerttemberg Nr 4 (4 sqns), Hus R von Prittwitz Nr 5 (8 sqns), Hus R von Koehler Nr 7 (1 sqn). Fues Bns von Wackenitz Nr 3, von Bergen Nr 11 and von Stutterheim Nr 21. Drag Brigade von Wedell (4 sqns).
Artillery: 1 x HAB.
Main Body.
1st (Prussian) Division GM von Diericke. Gren Bns von Fabecky and von Schlieffen, IRs von Ruechel Nr 2, Prinz Heinrich Nr 35 and von Ploetz Nr 42 (2 bns each).
Kuer Rs von Wagenfeld Nr 4 (6 sqns), von Stuelpnagel Nr 6 (4 sqns), Garde du Corps Nr 13 (4 sqns), Drag R von Esebeck Nr 8 (5 sqns).

Artillery: 2 x 12 pdr FABs, 1 x 6 pdr FAB, 1 x HAB

2nd (Russian) Division. GM von Rembow. *Musk Rs Sievsk, Perm and Kaluga (3 bns each).*

Artillery: 1 x 6 pdr FAB.

Comment. This was a successful rearguard action by General Bennigsen, who commanded the Russo-Prussian forces here; they totalled about 53,000 men, of whom some 6,000 were lost as were 2 guns. Napoleon opposed him with some 68,500 men. French losses were, as is so often the case, controversial. It seems that they were in the region of 12 – 13,000 men and three eagles. The war was finally ended by the battle of Friedland on 14 June.

THE PRUSSIAN CONTINGENT (28th Division) IN RUSSIA IN 1812.

The Prussians formed part of Marshal Macdonald's X Corps, together with the 7th Division.

All Prussian regiments which took part in this campaign were combined from two existing regiments, except for the Leib-Infanterie-Regiment, which was mobilised as a whole. The commander at the outset was General of Infantry von Grawert, but he fell sick and was replaced by Lieutenant-General von Yorck. The Prussians were employed against Riga on the northern flank and saw little serious fighting.

Chief of Staff Lieutenant-General von Massenbach.

Deputy Chief of Staff Major-General von Kleist.

Horses
Officers Men Officers' Train
& men's

	Officers	Men	Officers' & men's	Train
1st **Brigade.** Colonel von Below.				
1st Combined Infantry Regiment. Major von Soeholm I.	61	2006	49	55
2nd and Fusilier Bns, 1st East Prussian IR Nr 1;				
1st Bn, 2nd East Prussian IR Nr 3.				
2nd Combined Infantry Regiment. Major von Soeholm II.	65	2091	49	55
1st Bn 3rd East Prussian IR Nr 4; 1st and				
Fusilier Bns, 4th East Prussian IR Nr 5.				
2nd Brigade. Lieutenant-Colonel von Horn.				
3rd Combined Infantry Regiment. Major von Steinmetz.	58	1910	49	55
2nd and Fusilier Bns 1st Pommeranian IR Nr 2;				
1st Bn Colberg IR.				
4th Combined Infantry Regiment. Major von Zielinsky.	60	1971	49	55
1st, 2nd and Fusilier Bns, Leib-Infantererie-Regiment Nr 9.				
3rd Brigade. Colonel von Raumer.				
5th Combined Infantry Regiment. Major von Schmalensee.	60	2019	49	55
1st Bn 1st West Prussian IR Nr 6; 1st and				
Fusilier Bns, 2nd West Prussian IR Nr 7.				
6th Combined Infantry Regiment. Major von Carnall.	61	2047	49	55
2nd Bn, 1st Silesian IR Nr 11, 2nd and Fusilier Bns,				
2nd Silesian IR Nr 12.				
East Prussian Feld-Jaeger Bn. (1 bn).	19	464	19	17
Total infantry	384	12,508	313	347
1st Cavalry Brigade. Colonel von Huenerbein.				
1st Combined Dragoon Regiment. Major von Treskow.	23	574	659	23
2nd and 4th Squadrons, Lithuanian Dragoons Nr 3;				
1st and 2nd Squadrons, 2nd West Prussian Dragoons Nr 4.				
2nd Combined Dragoon Regiment.	24	586	668	24
Lieutenant-Colonel von Juergas.				
1st and 3rd Squadrons, 1st West Prussian Dragoons Nr 2;				

1st and 3rd Squadrons, Brandenburg Dragoons Nr 5.				
Combined Ulan Regiment. Major von St Paul. 3rd and 4th Squadrons, Silesian Ulans; 3rd and 4th Squadrons, Brandenburg Ulans. II Cavalry Corps;	This regiment was detached in June to the 2nd Light Cavalry Brigade, see there.			
2nd Cavalry Brigade. Colonel von Jeanneret.				
1st Combined Hussar Regiment. Major von Cosel.	26	614	697	23
(Attached to Grandjean's 7th Infantry Division of the X Corps.) 3rd and 4th Squadrons, 1st Leib-Hussars; 3rd and 4th Squadrons, 2nd Leib-Hussars.				
2nd Combined Hussar Regiment Colonel von Gzarnowski. 3rd and 4th Squadrons, Brandenburg Hussars; 1st and 2nd Squadrons, Pommeranian Hussars.	This regiment was detached in June to the 1st Light Cavalry Division, I Cavalry Corps; see there.			
3rd Combined Hussar Regiment. Major von Eicke. 1st and 3rd Squadrons, 1st Silesian Hussars; 1st and 2nd Squadrons, 2nd Silesian Hussars.	21	504	582	23
Total cavalry	94	2,278	2,605	93
Artillery.				
Horse Artillery				
1st, 2nd and 3rd East Prussian Horse Artillery Companies	13	439	377	282
1st – 3rd Artillery Train Companies. ? ? ? ?				
Foot Artillery				
3rd Silesian 12 pounder Foot Artillery Battery (half a battery),	3	106	7	8
1st, 2nd and 3rd East Prussian 6-pounder Batteries, 6th Brandenburg 6-pounder Foot Battery.	16	564	32	349
1st – 5th Artillery Park Companies.	6	452	30	563
1st and 2nd Bridging Companies -	2	6	2	48
East Prussian Pionier Company	6	248	-	-
Total Artillery etc.	44	1835	448	1310

Comment.

THE PRUSSIAN ARMY CORPS AT LUETZEN (GROSS-GOERSCHEN) ON 2 MAY 1813.

Russian General of Cavalry Count Wittgenstein commanding the allied army.
First Line. General of Cavalry von Bluecher commanding.
Right Wing. GM von Ziethen. Oberschlesischen Brigade. 1st Schles Hus R (2nd and 4th Sqns), 2nd Schles Hus R (3rd and 4th Sqns), Schles Ul R (1st and 2nd Sqns), 1 x HAB.
Colonel von Pirch I. Schles Gren Bn, 1st Schles IR (1st and 2nd Bns), 1 x Prussian 6 pdr FAB, 1 x Russian 12 pdr FAB.
Colonel von Losthin. 1st Schles IR (Fus Bn), 2nd Schles IR (2 bns), Schles Schuetzen (1 bn), 1 x 6 pdr FAB, x 12 pdr F AB.
Left Wing Colonel von Kluex. Niederschlesischen Brigade. Colonel von Mutius. Neumaerk Drag R (4 sqns), 1st Westpr Drag R (2nd and 4th Sqns), 1 x Prussian HAB.
Colonel von Jagow. Westpr Gren Bn, 1st Westpr IR (3 bns), 2nd Westpr IR (2nd and 3rd Bns), artillery 1 x Prussian 6 pdr FAB, 1 x Russian 12 pdr FAB.
Reserve of the First Line. GM von Roeder. Brandenburg Brigade. Westpr Ul R (4 sqns), Brand Hus R (1st and 2nd Sqns), artillery 1 x Prussian HAB.
Lt – Col von Tippelskirch. Garde-Jaeger-Bn, Garde-IR (1st and 2nd Bns), Normal I-Bn, 2 x 6 pdr FABs.
Major von Natzmer. Leib-Gren-Bn, 1st Ostpr Gren-Bn, Leib IR (3rd Bn), Garde Freiwilliger Jaeger-Bn.

Second Line. GL von Yorck.
Right Wing: Russians.
Left Wing. GM von Huenerbein. 1st Combined IR (3 bns), Lieb IR (Fues Bn), Lithuanian Drag R (4 sqns), artillery 1 x F ABs. Colonel von Horn. 5th Combined IR, 6th Combined IR (3 bns each), 2nd Combined Drag R (4 sqns), artillery 1 F ABs.
Reserve of the Left Wing. Lt-Col von Steinmetz. 2nd Leib HusR (4 sqns), Colberg IR (3 bns), artillery 1 x HAB, 1 x 3 pdr FAB.
With the Russian 1st Reserve Army was the Prussian Reserve Cavalry Brigades of Colonel von Dolffs (Garde-du-Corps (5 sqns), artillery 1 x HAB) and Colonel von Wahlen-Juergass (Ostpr Kuer R and Brandenburg Kuer R (4 sqns each), artillery 2 x HABs).

Comment. This was the first major battle of the 1813 campaign and the first major test of the new Prussian army in action with their Russian allies. The combined regiments are still serving as they were in Russia the previous year. Napoleon mustered some 144,000 men (of whom 78,000 came into action) against 93,000 allies, of whom 70,000 came into action. Due to allied mismanagement of their superior cavalry, Napoleon won the day, but the allied armies withdrew eastwards in good order. The French lost over 19,000 men; the allies some 12,000.

THE PRUSSIAN ARMY CORPS AT BAUTZEN ON 20/21 MAY 1813.

Russian General of Cavalry Count Wittgenstein commanding the allied army.
General of Cavalry von Bluecher commanding the Prussian army.
Prussian Element of the Allied Advanced Guard. GL von Kleist.
Lt-Col von Steinmetz. Colberg IR (3 bns), 1st and 2nd Schles Hus Rs (2 sqns each), artillery 1 x HAB.
Main Prussian Body.
GM von Ziethen. Oberschlesischen Brigade. 1st Schles Hus R (2nd and 4th Sqns), 2nd Schles Hus R (3rd and 4th Sqns), Schles Ul R (1st and 2nd Sqns), 1 x HAB.
Colonel von Pirch I. Schles Gren Bn, 1st Schles IR (3 bns), 1 x Prussian 6 pdr FAB.
Colonel von Losthin. Leib IR (2 bns), 2nd Schles IR (2 bns), Schles Schuetzen (1 bn), 1 x 6 pdr FAB, 2 x Russian 12 pdr FABs.
Colonel von Kluex. Niederschlesischen Brigade. Colonel von Mutius. Neumaerk Drag R (4 sqns), 1st Westpr Drag R (2nd and 4th Sqns), 1 x Prussian HAB.
Colonel von Jagow. Westpr Gren Bn, 1st Westpr IR (3 bns), 2nd Westpr IR (2nd Bn), 3rd Westpr IR (3rd Bn), Leib IR (2 bns), artillery 1 x 6 pdr FAB.
Corps of GL von Yorck.
Right Wing. GM von Corswandt. Lt-Col von Wutenow. 1st Westpr Drag R (2 sqns), artillery 1 x HAB.
Col Zielinsky. 2nd Ostpr IR (Fues Bn), 1st Ostpr IR (1st and 2nd Bns), Leib IR (1st and 2nd Bns), Combined Fues Bn (of the 1st Ostpr and Leib IRs), artillery 2 x FABs.

Colonel von Horn. 1st Westpr IR (1st Bn), 2nd Westpr IR (2nd Bn), 1st Schles IR (2nd Bn), 2nd Schles IR (2nd Bn), Combined Fues Bn (of the 5th and 6th Combined IRs), 2nd Westpr IR (Fues Bn), 2nd Schles IR (Fues Bn), Lithuanian Drag R (4 sqns), Brandenburg Drag R (2 sqns), artillery 1 x 3 pdr FAB, 1 x 6 pdr HAB.
Reserve. GM von Roeder.
Lt – Col von Tippelskirch. Westpr Ul R (4 sqns). Garde-Jaeger-Bn, Garde-IR (3 bns), Normal I-Bn, 1 x 6 pdr FAB.
Major von Natzmer. Leib-Gren-Bn, 1st Ostpr Gren-Bn, Leib IR (2 bns), Garde Freiwilliger Jaeger-Bn. Brand Hus R (1st and 2nd Sqns), 2nd Leib HusR (4 sqns).
Cavalry Brigades of Colonel von Dolffs (Garde-du-Corps (5 sqns), Garde Light Cavalry R (6 sqns), artillery 1 x HAB) and Colonel von Wahlen-Juergass (Brandenburg Kuer R, Ostpr Kuer R and Schles Kuer R (4 sqns each), artillery 1 x HAB.

Comment. Hampered by a chronic lack of well-trained cavalry, Napoleon was unable to mount his habitually ruthless pursuit of his beaten foe after Luetzen. In this second great battle of the 1813 campaign, he brought 167,000 men to the field to oppose 66,000 Russians and 31,000 Prussians. Again, it was a French victory, again it was limited. The Russians lost some 7,300 men, the Prussians 3,700. 9 guns were also lost. Napoleon's forces lost 22,000 men. The allies again withdrew to the east.

PRUSSIAN TROOPS AT HAYNAU ON 26 MAY 1813.

General der Kavallerie von Bluecher commanding.
Prussian Troops at Haynau.
GM von Corswant, commanding von Yorck's cavalry.
GM von Ziethen commanding the 11th Brigade, II Corps. Colonel von Dolffs, Garde du Corps (5 sqns), Guard Light Cavalry Regiment (6 sqns), Colonel von Juergass, Brandenburg Kuerassiars, East Prussian Kuerassiers, Silesian Kuerassiers (4 sqns each). Three horse artillery batteries.
Colonel von Mutius, Newmark Dragoons (3 sqns).
Colonel von Pirch I, commander 10th Brigade. 1st and 2nd Silesian Infantry Regiments (3 bns each), Silesian Grenadier Battalion, Silesian Sharpshooters (2 companies), two foot artillery batteries.
Total: 6,400 cavalry, 8,000 infantry, 40 guns.

Comment. This was a very successful ambush of French troops under General Count Maison and was planned by Bluecher himself. Maison's cavalry (Chastel's 3rd Light Cavalry Division) fled and abandoned their infantry to the Prussians. Of 3,700 men of his 16th Infantry Division, Maison lost 1,400 together with 11 of his 18 guns.
Prussian losses were 86 killed and wounded.

THE III PRUSSIAN ARMY CORPS AT GROSS-BEEREEN ON 23 AUGUST 1813.

GL von Buelow commanding.
3rd Brigade. GM Prinz Ludwig von Hessen-Homburg. 2nd Ostpr Gren Bn, 3rd Ostpr LIR (2 bns), 4th Res IR (3 bns) 3rd Ostpr LWIR (4 bns), 1st Leib-Hus R (4 sqns).
Artillery: 1 x FAB.
4th Brigade. GM von Thuemen. 4th Ostpr LIR , 5th Res IR (3 bns each), Elb-IR (2nd Bn), Ostpr Jaegers (2 coys), Pomm NKR (4 sqns).
Artillery: 1 x FAB.
5th Brigade. GM von Borstell. 1st Pomm LIR (Fues Bn), Pomm Hus R Nr 5 (4 sqns).
Artillery: 1 x FAB.
6th Brigade. Colonel von Krafft. Colberg IR, 9th Res IR (3 bns each), 1st Nuemaerk LWIR (2nd and 4th Bns), 1st Pomm LWKR (3 sqns).
Artillery: 1 x FAB.
Cavalry Reserve. GM von Oppen. Brand Drag R, Drag R Koenigin, Westpr Ul R (4 sqns each), 2nd Pomm LWKR, 2nd and 4th Kurmaerk LWKR (4 sqns each).
Artillery: 2 x HAB.
Reserve Artillery: 2 x 12 pdr FABs, 1 x HAB.
Attached were two Don Cossack Pulks (Bychalov II and Illowaisky V) and two Russian 12 pdr FABs.

III Corps total: about 47,000 men. Prussian losses: 1,600 killed and wounded.

Comment. This battle was fought to defend Berlin from Oudinot's thrust at the Prussian capital with the IV, VII and XII Corps. They totalled some 36,000 men with 70 guns. Buelow's corps was part of the Army of the North, but the Swedes and Tauentzien's IV Prussian Corps were scarcely involved at all. Oudinot was soundly beaten, losing 1,700 killed and wounded, 1,500 captured, 14 guns and 58 ammunition wagons. The Saxons of the VII Corps lost 2,100 men.

THE I PRUSSIAN ARMY CORPS AT THE KATZBACH ON 26 AUGUST 1813.

GL von Yorck commanding.
1st Brigade of Colonel von Steinmetz. Leib-Gren-Bn, 1st Ostpr Gren-Bn, Schles Gren-Bn, Ostpr Jaegers (2 coys), 5th and 13th Schles LWIRs (4 and 2 bns resp). 2nd Leib-Hus R (4 sqns).
Artillery: 1 x FAB (6 guns).
2nd Brigade of GM Prinz Karl von Mecklenburg. 1st Ostpr IR and 2nd Ostpr IR (3 bns each), 6th Schles LWIR (3 bns), Mecklenburg-Strelitz Hus R (4 sqns).
Artillery: 2 x FAB (13 guns).
7th Brigade of GM von Horn. Leib-IR (3 bns), Thueringian Bn, Garde-Jaegers (2 coys), 15th Schles LWIR, Brandenburg Hus R (2 sqns), 3rd Schles LWKR (2 sqns).
Pettinkofer, Reichenbach, Sommerfeld.
Artillery: _ x FAB (3 guns).
8th Brigade of GM von Huenerbein. Brandenburg IR, 12th Res IR (3 bns each), 14th Schles LWIR (4 bns), Brandenburg Hus R (2 sqns)3rd Schles LWKR (2 sqns).
Artillery: 1 x FAB (8 guns),
Reserve Cavalry of Colonel Baron von Wahlen-Juergass. Lithuanian Drag R (5 sqns), 1st Westpr Drag R, Brandenburg Ul R, Ostpr NKR, 5th and 10th Schles LWKR, 1st Neumaerk. LWKR, 5th and 10th Scles LWKR (4 sqns each).
Artillery: 2 x HAB (16 guns).
Reserve artillery: 2 x 12 pdr FABs, 1 x HAB, 1 x 3 pdr FAB.
Total: 39,000 men.

Comment. Marshal Macdonald had been given command of the ‘Army of the Bober‘ (V, XI Corps and II Cavalry Corps) and told to observe the Army of Silesia, but to be careful. In this he failed; as his army was crossing the River Neisse in a downpour, Bluecher seized his chance. His army (I (Prussian) Corps, VI, VIII, IX, X, XI and I (Russian) Corps), 86,000 men) inflicted a severe defeat on Macdonald, who lost 12,000 killed and wounded, 18,000 captured, 105 guns, 300 ammunition wagons and two eagles.

THE II PRUSSIAN ARMY CORPS AT DRESDEN ON 26/27 AUGUST 1813.

GL von Kleist commanding.
9th Brigade. GM von Kluex. 1st Westpr LIR (3 bns), 6th Res IR (3bns), 7th Schles LWIR (4 bns), Schles Schuetzen (2 coys), Neumark Drag R (4 sqns).
Artillery: 1 x FAB.
10th Brigade. GM von Pirch I. 2nd Westpr LIR, 7th ResIR, (3 bns each), 9th Schles LWIR (4 bns)
Artillery: 1 x FAB.
11th Brigade. GM von Ziethen. 1st Schles LIR, 10th Res IR (3 bns each), 8th Schles LWIR (4 bns), Schles Schuetzen (2 coys), 1st Schles Hus R (4 sqns).
Artillery: 1 x FAB.
12th Brigade. GL Prinz August von Preussen. 2nd Schles LIR (3 bns), 11th Res IR (3 bns), 10th Schles LWIR (4 bns), 1st Schles LWKR (4 sqns).
Artillery: 1 x FAB.
Reserve Cavalry. GM von Roeder. Ostpr Kuer R, Schles Kuer R, brand Kuer R, Schlkes Ul R (4 sqns each), 2nd Schles Hus R (2 sqns), Schles NKR (2 sqns), 7th and 8th Schles LWKR (4 sqns each).
Artillery: 2 x HABs
Reserve Artillery. Lt-Col von Braun. 2 x 12 pdr FABs, 1 x howitzer battery

Comment. The Prussians were part of the Army of Bohemia under FM Prinz zu Schwarzenberg's command. They missed the opportunity to rush the weak defenders of the city on the first day, allowing Napoleon to arrive and defeat them on the second. The French had 155,000 men, the allies 200,000. French losses were about 10,000, the allies lost about 23,000 men, including 13,000 when Metzko's Austrian division was captured.

THE II PRUSSIAN ARMY CORPS AT KULM ON 29/30 AUGUST 1813.

GL von Kleist commanding.
9th Brigade. GM von Kluex. 1st Westpr LIR (3 bns), 6th Res IR (1st and 3rd Bns).
Artillery: 1 x FAB.
10th Brigade. GM von Pirch I. 2nd Westpr LIR, 7th ResIR, 9th Schles LWIR (2 bns each)
Artillery: 1 x FAB.
12th Brigade. GL Prinz August von Preussen. 2nd Schles LIR (3 bns), 11th Res IR (2 bns), 10th Schles LWIR (4 bns).
Artillery: 1 x FAB.
Reserve Cavalry. GM von Roeder. 7th Schles LWKR (4 sqns).

Comment. The Prussians were part of the Army of Bohemia under Barclay de Tolly's command. They were withdrawing south into the Bohemian mountains after their defeat at Desden on 27/27 August. Vandamme became careless in his wish to catch up with them and was caught in isolation with the I, XIV Corps and the I Cavalry Corps; a total of 37,000 men. He was captured together with two other generals. Total French losses this day were 5,000 killed and wounded, 10,000 captured, 21 guns, 200 ammunition wagons, all his baggage train, two eagles and three colours. The I Corps was very badly damaged. The allies lost 11,319 killed and wounded and 1,000 captured out of 70,000. Of these, the Prussians lost 1,500 men.

THE III AND IV CORPS AT DENNEWITZ ON 6 SEPTEMBER 1813.

III Corps GL von Buelow commanding.

3rd Brigade. GM Prinz Ludwig von Hessen-Homburg. 2nd Ostpr Gren Bn, 3rd Ostpr LIR (2 bns), 4th Res IR (3 bns) 3rd Ostpr LWIR (4 bns), 1st Leib-Hus R (4 sqns).
Artillery: 1 x FAB.
4th Brigade. GM von Thuemen. 4th Ostpr LIR , 5th Res IR (3 bns each), Elb-IR (2nd Bn), Ostpr Jaegers (2 coys), Pomm NKR (4 sqns).
Artillery: 1 x FAB.

5th Brigade. GM von Borstell. 1st Pomm LIR (Fues Bn), Pomm Gren Bn, 2nd Res IR, 2nd Maerk LWIR (3 bns each), Pomm Hus R Nr 5 (4 sqns).
Artillery: 1 x FAB.
6th Brigade. Colonel von Krafft. Colberg IR, 9th Res IR (3 bns each), 1st Macrk LWIR (2nd and 4th Bns), 1st Pomm LWKR (3 sqns).
Artillery: 1 x FAB.
Cavalry Reserve. GM von Oppen. Brand Drag R, Drag R Koenigin, Westpr Ul R (4 sqns each).
III Corps total: 47,000 men.

IV Corps GL von Tauentzein commanding.
1st Brigade. GM von Dobschuetz. 3rd Res IR 94 bns), 1st Kurmark LWIR (3 bns), 2nd and 3rd Neumaerk LWKRs , 2nd and 3rd Ostpr LWKRs (4 sqns each).
Artillery: 1 x FAB.
2nd Brigade. GM Graf von Lindenau. 2nd and 5th Neumaerk LWIRs (4 bns each), 1st Schles LWIR (1 bn), 2nd Schles LWIR (4 bns), Berlin LWKR, 7th Kurmaerk LWKR (2 sqns each), 1st Kurmaerk LWKR and 3rd Pomm LWKR (4 sqns each)
Artillery: 1 x FAB.
Reserve Artillery: 1 x FAB, 1 x HAB.
IV Corps total: 2,200 men.

Comment. This action defeated Marshal Ney's thrust at Berlin with the IV, VII, XII Corps and the III Cavalry Corps, some 45,000 men. The majority of the fighting was done by Buelow's corps alone, Tauentzien came up late in the action and Crown Prince Bernadotte – as usual – was a mere spectator. Ney wrote to the Emperor: 'I have been utterly defeated'. Total Prussian losses were 6,200 men; Ney lost 6,500 killed and wounded, 13,500 captured, 54 guns and several colours and standards.

THE I PRUSSIAN ARMY CORPS AT WARTENBURG ON 3 OCTOBER 1813.

GL von Yorck commanding.
Avantgarde. Colonel von Katzler. Leib-Gren-Bn, Westpr Gren-Bn, 12th Res IR (2nd Bn), Brandenburg IR (1st Bn), 2nd Ostpr IR (Fues Bn), Landwehr IBns Rekowsky, Thiele and Wedel (1 bn each), Jaegers (3 coys).

Brandenburg Hus R Nr 3 (5 sqns), 2nd Leib-Hus R (2 sqns), Brandenburg Ulan R 4 sqns), Ostpr NKR (5 sqns), 5th Schles LWKR (4 sqns),
Artillery: 1 x HAB (8 guns).
Brigade of Colonel von Steinmetz. 1st Ostpr Gren-Bn 2nd Ostpr IR (1st Bn), Landwehr I Bns Cronegk, Fischer, Larisch, Martitz, Mumm, Seidlitz, Walter (1 bn each).
Artillery: 1 x FAB (6 guns).
Brigade of GM Prinz Karl von Mecklenburg. Schles Gren-Bn, 1st Ostpr IR (3 bns), 2nd Ostpr IR (2nd Bn), Landwehr I Bn Kosecky (1 bn), Mecklenburg-Strelitz Hus R (4 sqns), 2nd Leib-Hus-R (3 sqns).
Artillery: 2 x FAB (13 guns).
Brigade of GM von Horn. Leib-IR (3 bns), Thueringian Bn, Landwehr IBns Knorr-Kottulinsky,
Pettinkofer, Reichenbach, Sommerfeld.
Artillery: _ x FAB (3 guns).
Brigade of GM von Huenerbein. Brandenburg IR (2nd and Fues Bns), 12th Res IR (1st and Fues Bns), Landwehr I Bn Kempky.
Artillery: 1 x FAB (8 guns),
Reserve Cavalry of Colonel Baron von Wahlen-Juergass. Lithuanian Drag R (5 sqns), Westpr Drag R, 10th Schles Landwehr K R, 1st Neumaerk. Landwehr K R (4 sqns each).
Artillery: 2 x HAB (16 guns).
Reserve artillery: 1 x HAB, 1 x FAB (10 guns).
Totals: 16,000 men and 64 guns.

Comment. General von Yorck forced the crossing of the River Elbe here against General Count Bertrand's IV Corps of 14,000 men and 24 guns and opened the way for von Bluecher's offensive to join in the fateful Battle of the Nations at Leipzig on 16-18 October. The Prussians lost 2,100 men, Bertrand lost 1,900 casualties, 13 guns and 80 ammunition wagons.

THE I PRUSSIAN ARMY CORPS AT MOECKERN ON 16 OCTOBER 1813.

The order of battle of Yorck's corps is taken from his biography by Droysen and differs greatly from the other sources quoted but seems to reflect the actual tactical groupings as mentioned for 16th October by Seyfert which is highly regarded and very detailed. After the heavy losses suffered at

Moeckern on 16 Oct, Yorck's corps was slightly reorganized from the details shown here:

Commander General-Leutnant von Yorck.
* = wounded on 16 October; + = killed on 16 October.
CoS: Col von Zielinski. Adjutant: Capt Delius.
Oberquartiermeister: Lt Col Freiherr von Valentini.
The General Staff: Maj Wilhelm von Schack, Maj von Klitzling, Capt von Dedenroth, Capt von Lollhoefel, Lt von Wussow. Adjutants: Maj Graf Brandenburg, Maj von Diedrich, Capt von Selasinsky, Rittmeister Ferdinand von Schack, Lt von Below, Lt von Roeder.
Kriegskommissar: von Reiche.
Corps medical officer: Dr Voeltze.
Headquarters medical officer: Dr Hohenhorst.
Padre: Divisionsprediger Schultze.

1st Brigade: Col von Steinmetz*.
Brigade staff. Capt von Kaufberg +. Adjutants: Capt von Luetzow, Lt Henkel von Donnersmark.
1st Grenadier Brigade: Maj Hiller von Gaertringen*.
1st Ostpr Gren Bn (1 bn) Maj von Leslie*, Leibgrenadier - Bn (1 bn) Maj von Carlowitz*, Schl Gren Bn (1 bn) Maj von Burghof, Westpr. Grenadier - Bn (1 bn) Maj von Schon*.
2nd Landwehr Brigade: Col von Losthin*, 5th Schl LWIR: Maj von Maltzahn + (3 bns , the 2nd Bn was in Goerlitz), Maj von Mumm * Maj von Seydlitz *, Maj von Kossecky +. 13th Schl LWIR: Maj von Goedike + (4 bns) Maj von Larisch, Maj Walter von Cronegk, Maj von Rekowsky +, Maj von Martitz. 2nd LeibHusR: Maj von Stoeffel (4 sqns).
2nd 6 pdr FAB (8 guns) Lt Lange.
TOTALS: 11 bns, 4 sqns, 9,270 men, 8 guns.

2nd Brigade: Gen Maj Prinz Karl von Mecklenburg*.
Brigade staff: Maj von Schuetz*, Lt von Riesenburg.
Adjutants: Maj von Folgersberg, Capt von Heinzmann.
1st Line Brigade: Lt Col von Lobenthal*.
1st Ostpr IR (3 bns) Maj von Schleuse+, Maj von Kurnatowsky*, Maj von Pentzig*, 2nd Ostpr IR: Lt Col von Sjoeholm (3 bns) Maj von Dessauniers*, Maj von Krauthof, Maj ? .
2nd Landwehr Brigade: 6th Schl LWIR (1 combined bn) Maj von Fischer*.

Mecklenburg - Strelitz HusR Lt Col von Warburg (4 sqns).
1st 6 pdr FAB (8 guns) Capt Huet.
TOTALS: 7 bns, 4 sqns, 7,673 men, 8 guns.

7th Brigade: Col von Horn.
Brigade Staff: Maj von Rudolphi, Lt von Manstein.
Adjutants: Capt Graf Kanitz, Lt von Barfuss*, Lt von Reibnitz*.
1st Line Brigade: Col von Zepelin. Leibregiment: Col von Zeppelin (3 bns) Maj von Oertzen, Maj von Hagen, Maj von Ledebur. The Thueringian Bn Maj vonLinker; Gardejaeger (_ bn) Capt von Bock.
2nd Landwehr Brigade Col von Welzien .
4th Schl LWIR: Maj Graf Herzberg (2 bns; the 1st Bn was in Wartenburg; the 2nd and 4th Bns had been combined into one in September) Maj Graf Reichenberg, 15th Schl LWIR: Col von Wollzogen *(3 bns; the 3rd Bn was in Wartenburg) Maj von Sommerfeld, Maj von Pettinger, Maj Graf Wedell:
3rd Schl LWCR: Maj von Falkenhausen (2 sqns), Brandenburg Hus R: Maj von Sohr * (3 sqns).
3rd 6pdr FAB (8 guns) Capt Ziegler*.
TOTALS: 8 _ bns, 5 sqns, 8,686 men, 8 guns.

8th Brigade: Gen Maj von Huenerbein.
Brigade Staff: Capt von Arnaud, Lt von Unruh.
Adjutants: Lt von Unruh, Lt von Sellin.
1st Line Brigade: Lt Col von Borcke*. Brandb IR (3 bns) Maj von Buelow, Maj von Othegraven*, Maj von Krosigk+. 12th ResIR: Maj ? (3 bns) Maj von Hermann, Maj von Zeppelin, Maj von Laurens*.
2nd Landwehr Brigade: 14th Schl LWIR Col von Goza (2nd and 3rd Bns; the 1st and 4th Bns were in Wartenburg) Maj von Thile*, Maj von Brixen.
Cavalry: Brandb HusR Maj von Knaublauch (2 sqns), 3rd Schl LWCR Maj von Kanilowsky (2 sqns).
15th 6pdr FAB (8 guns) Lt Anders .
TOTALS: 8 bns, 4 sqns, 7,447 men, 8 guns.

Corps Reserve Cavalry: Col Freiherr von Wahlen - Juergass.
Brigade Staff: Rittmeister Freiherr von Canitz, Lt von Briesen.
Adjutants: Maj von Palusdorf, Graf Reuss XIII, Graf Ingenheim.
1st Brigade: Col Graf Henkel von Donnersmark.

Lith DragR: Lt Col von Below (4 _ sqns including the Freiwillige-Jaegers), 1st Westpr DragR: Lt Col von Wuthenow (4 sqns).
2nd Brigade: Lt Col von Katzeler*.
Brandb UlR: Maj von Stutterheim, Ostpr NCR: Maj Graf Lehndorf (4 sqns each).
3rd Brigade: Maj von Bieberstein.
5th Schl LWCR Maj von Ozorowsky, 10th Schl LWCR Maj von Sohr, 1st Neum LWCR Maj von Sydow (4 sqns each).
1st HAB Capt von Zinken and 3rd HAB Lt von Barowsky (16 guns).
TOTALS: 28 _ sqns, 3,896 men, 16 guns.

Corps Artillery Reserve: Lt Col von Schmidt.
Adjutants: Lt Erhard, Lt von Peucker.
Artillery Staff: Maj von Fiebig , Maj von Rentzell , Maj von Graumann .
Engineers: Maj Markoff , Lt von Huelsen , Lt von Poser .
1st 12 pdr FAB, Lt Witte and 2nd 12 pdr FAB, Lt Simon. 12th 6 pdr FAB, Lt Bully, 24th 6 pdr FAB Lt Varenkampf. 1st 3 pdr FAB Lt von Oppen. 3rd HAB Lt Fischer, 12th HAB Rittmeister von Pfeil. Total 973 men.
Park: Parkkolonnen Nr 1, 3, 5, 11 and 13. Handwerkercolonne Nr 2. Total 126 men.
Pioniers (2 coys) 149 men.
TOTALS: 1,248 men, 56 guns.

TOTALS I (Prussian) Corps: 34 _ bns, 1 coy, 45 _ sqns, 38,220 men (including 1,576 volunteers and 15,236 Landwehr), 13 batteries, 104 guns.

NB ALL TOTALS INCLUDE THE DETACHED TROOPS; THE ACTUAL STRENGTH AT LEIPZIG WAS THUS SOMEWHAT LOWER THAN SHOWN HERE.

Comment. Yorck's aim this day was to fight his way as far as possible into Leipzig from the northeast. He was opposed by Marshal Ney with the IV Corps (Bertrand), part of Souham's III Corps, Marmont's VI Corps, Arrighi's III Cavalry Corps and Dombrowski's independent Polish Division. The house-to-house fighting here this day was among the most dramatic and fierce of the entire 1813 campaign. Yorck's corps lost some 7,000 men this day and was pulled out of the line for some time. Marmont's corps, which did the bulk of the fighting, lost 5,000 killed and wounded, 2,000 captured, 43 guns and an eagle.

Also present at Leipzig, as part of the Army of Bohemia, were the II and III Prussian Corps and the Prussian guards. They were involved in the desperate struggles of 16 – 19 October, on the eastern front of the battlefield.

Prussian Guards. Lt-Col von Alvensleben. 1st and 2nd Garde-Regiment (6 bns), Garde-Jaeger Bn. Garde zu Pferde (Col Laroche von Starkenfeld), Gardes du Corps, Garde-Drag R, Garde-Leichte-Kav-R (4 sqns each).

II Corps GL von Kleist commanding.
9th Brigade. GM von Kluex. 1st Westpr LIR (3 bns), 6th Res IR (1st and 3rd Bns), 7th Schles LWIR (2 bns), Schles Schuetzen (2 coys), 1st Schles LWKR (2 sqns).
Artillery: 1 x FAB.
10th Brigade. GM von Pirch I. 2nd Westpr LIR (3 bns), 7th ResIR, 9th Schles LWIR (2 bns each), 1st Scles LWKR (1 sqn).
Artillery: 1 x FAB.
11th Brigade. GM von Ziethen. Schles Schuetzen (2 coys), 1st Schles LIR (3 bns), 10th Res IR (2 bns), 8th Schles LWIR (2 bns), Neumaerk Drag R (6 sqns), 2nd Schles LWKR (1 sqn)
Artillery: 1 x FAB.
12th Brigade. GL Prinz August von Preussen. 2nd Schles LIR (3 bns), 11th Res IR (3 bns), 10th Schles LWIR (2 bns), Scles Ul R (4 sqns), 2nd Schles LWKR (1 sqn).
Artillery: 1 x FAB.
Reserve Cavalry. GM von Roeder. Ost pr Kuer R, Brandenburg Kuer R, Schles Kuer R (4 sqns each). LWK Brigade. Col von Mutius. 7th and 8th Schles LWKR (5 sqns).
Artillery Reserve. 1 x 12 pdr FAB, 1 x 6 pdr FABs, 1 x HAB.
II Corps totals: 31 _ battalions, 32 sqns, 80 guns.
III Corps GL von Buelow commanding.
3rd Brigade. GM Prinz Ludwig von Hessen-Homburg. 2nd Ostpr Gren Bn, 3rd Ostpr LIR (2 bns), 4th Res IR (3 bns) 4th Ostpr LWIR (4 bns), 1st Leib-Hus R (4 sqns).
Artillery: 1 x FAB.
5th Brigade. GM von Borstell. 1st Pomm LIR (3 bns), Pomm Gren Bn, 2nd Res IR (3 bns), 2nd Maerk LWIR (4 bns), Pomm Hus R Nr 5, Westpr Ul R (4 sqns each).

Artillery: 1 x FAB.

6th Brigade. Colonel von Krafft. Colberg IR, 9th Res IR , 1st Maerk LWIR (3 bns each), Ostpr Schuetzen (2 coys), 1st Pomm LWKR (3 sqns).

Artillery: 1 x FAB.

Cavalry Reserve. GM von Oppen. Brandenburg Drag R, Drag R Koenigin, 2nd Westpr Drag R , 2nd and 4th Kurmeark LWKRs, 2nd Pomm LWKR (4 sqns each).

Artillery: 2 x HAB.

Artillery Reserve: 2 x 12 pdr FABs, 1 x HAB.

III Corps total: 24,619 men, 72 guns.

THE PRUSSIAN ARMY CORPS AT BRIENNE ON 29 JANUARY 1814.

FM von Bluecher commanding an allied army consisting of the IV (Prussian) Corps, the VI, IX and XI (Russian) Corps.

IV (PRUSSIAN) CORPS. Commander GL Graf von Tauentzien.

Brigade of GM Graf Lindenau. 3rd Res IR (3 bns), 3rd Neumark LWIR (2 bns), 11th and 16th Schles LWIRs, 1st Kurmark LWIR (3 bns each). Artillery: 1 x 6 pdr FAB.

Reserve Cavalry Brigade of GM von Bismark. 1st, 2nd, 3rd and 4th Ostpr LWKRs, 3rd Pomm LWKR, 2nd and 3rd Westpr LWKRs, 6th Schles LWKR (4 sqns each). Artillery: _ x HAB.

The other brigade of this corps were blockading various French-held fortresses in the east.

Comment. The action of the day seemed to have ended in favour of the allies as darkness fell, but Napoleon had other ideas. Blucher and Sacken (the Russian commander of the VI and XI Corps) decided to spend the night in the chateau in the town. The *56e de Ligne* overpowered the Prussian guards and the allied commanders were forced to flee for their lives. Despite several violent counter-attacks, the French retained control of the chateau and inflicted heavy losses on the attackers.

Altogether the allies lost 3,000 out of their 28,000 men. Napoleon's losses were 3,500 men and 11 guns.

THE I PRUSSIAN ARMY CORPS AT PARIS ON 30 MARCH 1814.

GdI von Yorck commanding.
1st Brigade. GM von Pirch II. 1st Ostpr Gren-Bn, Leib-Gren-Bn, Westpr Gren-Bn, Schles Gren-Bn, Ostpr Jaeger-Bn (2 coys), 13th Schles LWIR (3 bns). Artillery: 1 x FAB.
2nd Brigade. GL Prinz Karl von Mecklenburg. 1st and 2nd Ostpr IRs, 6th Schles LWIR (3 bns each). Artillery 1 x FAB.
7th Brigade. GM von Horn. Leib-IR, 4th Schles LWIR, 15th Schles LWIR (3 bns each), Brand Hus R (4 sqns). Artillery: 1 x FAB.
8th Brigade. GL Prinz Wilhelm von Preussen. Brand IR, 12th Res IR, 14th LWIR (3 bns each), 1st Leib-Hus R (4 sqns). Artillery: 1 x FAB.

Attached from GL von Kleist's **II Prussian Corps.**
9th Brigade. GM von Kluex. 1st Westpr IR, 6th Res IR (3 bns each), Schles Schuetzen-Bn (2 coys). Artillery: 1 x FAB.
12th Brigade. GL Prinz August von Preussen. 2nd Schles IR, 11th Res IR (3 bns each) Artillery: 1 x FAB.
Totals 30,000 men, 36 guns.

Comment. Most of the heavy fighting this day was done – as usual – by the Russians of the Army of Bohemia. Due to poor allied coordination, they advanced alone and lost 5,050 men out of the 16,000 engaged. The Prussians lost 1,353 men. The city was defended by Napoleon's brother, Joseph; the dedication of the defenders was 'less than 100%' and Joseph surrendered his nation's capital. This was the end (for 1814) of Napoleon's career. He had been caught out by the allies, who swooped on Paris when he was chasing illusory enemy armies in the countryside. His marshals had had enough of war and forced his first abdication on 6 April.

THE PRUSSIAN ARMY CORPS AT LIGNY ON 16 JUNE 1815.

FM Fuerst Bluecher von Wahlstadt commanding.
I Corps. GL von Ziethen.
1st Brigade. GM von Steinmetz. 2nd Brand IR Nr 12, LIR Nr 24, 1st Westphalien LWIR (3 bns each), Schles Schuetzen (2 coys). Artillery: 1 x FAB.

2nd Brigade. GM von Pirch II. 1st Westpr IR Nr 6 (3 bns), 28th LIR (1 bn)*, 2nd Westphalian LWIR (3 bns). Artillery: 1 x FAB.

3rd Brigade. GM von Jagow. 2nd Westpr IR Nr 7, 29th LIR, 3rd Westphalian LWIR (3 bns each), Schles Schuetzen (2 coys). Artillery: 1 x FAB.

4th Brigade. GM von Henkel. 19th LIR, 4th Westphalian LWIR (3 bns each).

Reserve Cavalry. GL von Roeder. Artillery: 1 x FAB.

1st Brigade. GM von Treskow. 1st Westpr Drag R Nr 2, 5th Drag R (Brandenburg), Brandenburg Ulan R Nr 3 (4 sqns each). Artillery: 1 x HAB.

2nd Brigade. Lt Col von Luetzow. 6th Ulan R, 1st and 2nd Kurmark LWKRs, 1st Schles Hus R, 1st Westphalian LWKR (4 sqns each). Artillery: 1 x HAB.

Reserve Artillery. Colonel von Lehmann. 2 x 9 pdr FABs, 1 x howitzer battery, 1 x HAB.

I Corps totals: 26,000 infantry, 1,925 cavalry, 1,019 gunners, 96 guns.

* The fusilier battalions of the 28th LIR and the 2nd Westphalian LWIR ahd lost so many men in the initial clashes of the campaign at Gosselies and Gilly on 15 June, that they were combined into one weak battalion, as were the 1st and 2nd Musketier Bns of the 28th .

II Corps. GL von Pirch I.

5th Brigade. GM von Tippelskirchen. 1st Pomm LIR Nr 2, 25th LIR, 5th Westphalian LWIR (3 bns each). Artillery: 1 x FAB.

6th Brigade. GM von Krafft. Colberg LIR Nr 9, 26th LIR, 7th Elb-LWIR (3 bns each). Artillery: 1 x FAB.

7th Brigade. GM von Brause. 14th and 22nd LIR, 2nd Elb-LWIR (3 bns each). Artillery: 1 x FAB.

8th Brigade. Colonel von Langen. 21st and 23rd LIRs, 3rd Elb-LWIR (3 bns each). Artillery: 1 x FAB.

Reserve Cavalry. LG von Wahlen-Juergass.

1st Brigade. Colonel von Thuemen. Neumark Drag R Nr 6, Hus R Nr 11, Schles Ul R (4 sqns each). Artillery: 1 x HAB.

2nd Brigade. Colonel Graf von Schulenburg. Drag R Koenigin Nr 1, 4th Kurmark LWKR (4 sqns each). Artillery: 1 x HAB.

3rd Brigade. Lt – Col von Sohr. Brandenburg Hus R Nr 3, Pomm Hus R Nr 5, 5th Kurmark LWKR, Elb-LWKR (4 sqns each). Artillery: 1 x HAB.

Reserve Artillery. Colonel von Roehl. 2 x 12 pdr FABs, 1 x 6 pdr FAB.

II Corps totals: 25,836 infantry, 4,468 cavalry, 1,454 gunners, 80 guns.

III Corps. GL Graf von Thielmann.
9th Brigade. GM von Borcke. Leib-IR Nr 8, 30th LIR, 1st Kurmark LWIR (3 bns each). Artillery: 1 x FAB.
10th Brigade. Colonel von Kaempffen. 27th LIR, 2nd Kurmark LWIR (3 bns each). Artillery: 1 x FAB.
11th Brigade. Colonel von Luck. 3rd and 4th Kurmark LWIRs (3 bns each). Artillery: 1 x FAB.
12th Brigade. Colonel von Stuelpnagel. 31st LIR, 5th and 6th Kurmark LWIRs (3 bns each). Artillery: 1 x FAB.
Reserve Cavalry. GM von Hobe.
1st Brigade. Colonel von der Marwitz. 7th Ulan R (3 sqns), 8th Ulan R (4 sqns). Artillery: 1 x HAB.
2nd Brigade. Colonel Graf von Lottum. 7th Drag R (5 sqns), 5th Ulan R (3 sqns), 3rd and 6th Kurmark LWKR (4 sqns each). Artillery: 1 x HAB.
Reserve Artillery. Colonel von Monhaupt. 1 x 12 pdr FAB, 1 x HAB.
III Corps totals: 20,611 infantry, 2,405 cavalry, 964 gunners, 48 guns.
Artillery batteries had 8 guns each.

Comment. The Prussian army occupied a deep position at Ligny and suffered heavily from French artillery fire. Casualty data is contradictory; Bodart gives 16,000 killed and wounded, 8,000 missing and 21 guns lost. Renouard gives 11 – 12,000 casualties and 16 guns lost. The 11th Brigade (III Corps) lost only 98 men. Napoleon brought some 71,000 men and 242 guns to Ligny and lost about 12,000 killed and wounded. Bluecher was thrown and trapped under his horse and unable to exercise command. His Chief of Staff, Gneisenau, assumed command in the crisis as the Prussian army was forced to fall back from Ligny. Although he was deeply suspicious of the Duke of Wellington's aims, he took the courageous decision to fall back north to turn west and join him, as opposed to running off eastwards along his lines of communication. Without that critical Prussian support on 18 June, it is unlikely that Wellington's army would have been able to win the day at Waterloo.

THE PRUSSIAN ARMY CORPS AT WATERLOO ON 18 JUNE 1815.

The I and II Corps also took part in this battle, but it was von Buelow's IV Corps which led the drive into Plancenoit late in the afternoon of this momentous day.

THE IV PRUSSIAN CORPS AT WATERLOO.

General Graf Buelow von Dennewitz commanding.
13th Brigade. General von Hacke. 10th LIR, 2nd and 3rd Neumark LWIRs (3 bns each).
Artillery: 1 x FAB.
14th Brigade. General von Ryssel. 11th LIR, 1st and 2nd Pommeranian LWIRs (3 bns each).
Artillery: 1 x FAB.
15th Brigade. General von Losthin. 18th LIR, 3rd and 4th Schles LWIRs (3 bns each).
Artillery: 1 x FAB.
16th Brigade. Colonel von Hiller. 15th LIR, 1st and 2nd SchlesLWIRs (3 bns each).
Artillery: 1 x FAB.
Reserve Cavalry. General Prinz Wilhelm von Preussen.
1st Brigade. GM von Schwerin. 2nd Schl Hus R Nr 6 (4 sqns), 8th Hus R (3 sqns), Westpr Ul R Nr 1 (4 sqns).
Artillery: 1 x HAB.
2nd Brigade. Lt-Col von Watzdorff. 8th Hus R (4 sqns), 1st and 2nd Neumark LWKRs (4 sqns each) .
Artillery: 1 x HAB.
3rd Brigade. GM von Sydow. 1st Schles LWKR, 1st and 2nd Pommeranian LWKRs (4 sqns each).
Artillery: 1 x HAB.

Reserve Artillery. Lieutenant-Colonel von Bardeleben. 3 x 12 pdr FABs, 1 x 6pdr FAB,

IV Corps totals: 25,381 infantry, 3,081 cavalry, 1,866 gunners, 88 guns.

Comment. The IV Corps was involved in heavy fighting in Plancenoit and suffered 5,461 casualties against 318 in the I Corps and 260 in the II Corps. They began to arrive on the east of the field, through the village of Couture, at about 1530 hours; by 1730 hours they were engaged with Lobau's corps, just east of Plancenoit, deep in the French right flank. At 1900 hours they were still being held, east of Plancenoit, but von Ziethen's corps was now approaching the village of Smohain, to their north. Finally, at about 2000 hours, the assault of the Imperial Guard against Wellington's right flank was repulsed, the guard fell back and the French army crumpled and fled. Von Buelow's IV Corps drove through Plancenoit to harry the fugitives.

Bibliography

Stammliste aller Regimenter und Corps der Koeniglich-Preussischen Armée fuer das Jahr 1806. Reprinted by Bilblio Verlag, Osnabrueck 1975.

Anon. 'Die Schlacht bei Minden 1759'. J C C Bruns Verlag, Minden 1959.

Alt, 'Das Koeniglich Preussische Stehende Heer'. Schrapp, Berlin, 1869.

Bolke, Eberhardt. 'Preussische Fahnen 1740 – 1806.' Dresden, 1944.

Bredow – Wedel. 'Historische Rang- und Stammliste des Deutschen Heeres'. Berlin 1905.

Eckardt, Werner – Morawietz, Otto. 'Die Handwaffen des brandenburgisch-preussisch-deutschen Heeres.' Hamburg, Helmut Gerhard Schulz Verlag, 1973.

Fiebig, H. 'Unsterbliche Treue',

Fiedler, Siegfried. 'Geschichte der Grenadiere Friedrichs des Grossen', Schild Verlag, Munich 1981.

Franke, Ludwig Eberhardt. 'Vorstellung der Koeniglich Preussischen Armée.' Potsdam, 18??

Fraser, David. 'Fredrick the Great', The Penguin Press, London 2000.

Gohlke, W. 'Geschichte der gesamten Feuerwaffen bis 1850' Berlin 1911.

Grossen Generalstab. 'Urkundliche Beitraege und Forschungen zur Geschichte des Preussischen Heeres; Heft 14 / 15. Der Feldzug 1806 / 07 und die Reorganisation der Artillerie'. Berlin 1914

Grossen Generalstab. 'Urkundliche Beitraege und Forschungen zur Geschichte des Preussischen Heeres; Hefte 26 - 30. Die Freikorps und Auslaender-Battailone'. Berlin 1914.

Hoepfner, Edouard von. Oberst. 'Der Krieg von 1806 und 1807'. Berlin, Simon Schropp & Comp. 1850.

Jany, Curt. 'Geschichte der Preussischen Armée vom 15. Jahrhundert bis 1914'. Biblio Verlag, Osnabrueck, 1967.

Kling, C. 'Geshichte der Bekleidung, Bewaffnung und Ausruestung des Koeniglich Preussischen Heeres.' Three volumes. Putzer und Hoeltze, Weimar 1912.

Knoetel – Sieg. 'Handbuch der Uniformkunde'. H. G. Schultz, Hamburg, 1937.

Prussian War Ministry. Fahnen und Standarten der preussischen Armée seit dem Jahre 1806. Berlin 1889.

Ramm, August Leopold. 'Abbildungen von allen Uniformen der Koenigl. Preuss. Armée unter der Regierung Sr. Majestaet Friedrich Wilhelm III' Berlin, J F Unger, 1800.

Voigt, Guenther. 'Deutschlands Heere bis 1918'. Biblio Verlag, Osnabrueck, 1983.

The Author

Digby Smith was born in Aldershot, Hampshire, in 1935. At age sixteen, having been bitten by the wanderlust during two spells in India, following his soldier father around the globe, he left school and joined the army as an apprentice telecommunications technician. In 1961 he was commissioned into the Royal Corps of Signals. Many years of service in the British Army of the Rhine led to him perfecting his German linguistic skills, which helped immensely with his growing interest in the military history of all the old German states. In 1965 he transferred to the Royal Army Ordnance Corps and was employed in computers, work study and logistics. From 1970-1972 he studied at the German Armed Forces Command and Staff College at Blankenese, near Hamburg. By 1978, after a spell in the Ministry of Defence, Whitehall, he had tired of the military and resigned to start a new career in high-tech logistics and customer services in various international computer and telecommunications companies in Germany, Saudi Arabia and Moscow, the latter from 1992-1995. On his return to the mother company from this fascinating stint, he was declared redundant, thus free to concentrate full-time on writing military history books; most clouds have a silver lining if one looks for it hard enough. Among his previously published works are: *Armies of the Napoleonic Era, Navies of the Napoleonic Era, Armies of 1812, The Greenhill Napoleonic Wars Data Book, Napoleon's Regiments, 1813 Leipzig, The Battle of The Nations, Charge!* and *Armies of the Middle East*. Many of his early books, including over a dozen for the Osprey Men at Arms series, were written under the pen name of Otto von Pivka. Digby Smith has three sons, four step-children and seemingly tribes of grandchildren and step-grandchildren, some in America. He now lives in Thetford, Norfolk and is still actively learning much on the Napoleonic era from daily raids onto the internet.